Applied Mathematical Sciences
Volume 91

Editors
S.S. Antman J.E. Marsden L. Sirovich

Advisors
J.K. Hale P. Holmes J. Keener
J. Keller B.J. Matkowsky A. Mielke
C.S. Peskin K.R. Sreenivasan

Springer
New York
Berlin
Heidelberg
Hong Kong
London
Milan
Paris
Tokyo

Applied Mathematical Sciences

1. *John:* Partial Differential Equations, 4th ed.
2. *Sirovich:* Techniques of Asymptotic Analysis.
3. *Hale:* Theory of Functional Differential Equations, 2nd ed.
4. *Percus:* Combinatorial Methods.
5. *von Mises/Friedrichs:* Fluid Dynamics.
6. *Freiberger/Grenander:* A Short Course in Computational Probability and Statistics.
7. *Pipkin:* Lectures on Viscoelasticity Theory.
8. *Giacaglia:* Perturbation Methods in Non-linear Systems.
9. *Friedrichs:* Spectral Theory of Operators in Hilbert Space.
10. *Stroud:* Numerical Quadrature and Solution of Ordinary Differential Equations.
11. *Wolovich:* Linear Multivariable Systems.
12. *Berkovitz:* Optimal Control Theory.
13. *Bluman/Cole:* Similarity Methods for Differential Equations.
14. *Yoshizawa:* Stability Theory and the Existence of Periodic Solution and Almost Periodic Solutions.
15. *Braun:* Differential Equations and Their Applications, 3rd ed.
16. *Lefschetz:* Applications of Algebraic Topology.
17. *Collatz/Wetterling:* Optimization Problems.
18. *Grenander:* Pattern Synthesis: Lectures in Pattern Theory, Vol. I.
19. *Marsden/McCracken:* Hopf Bifurcation and Its Applications.
20. *Driver:* Ordinary and Delay Differential Equations.
21. *Courant/Friedrichs:* Supersonic Flow and Shock Waves.
22. *Rouche/Habets/Laloy:* Stability Theory by Liapunov's Direct Method.
23. *Lamperti:* Stochastic Processes: A Survey of the Mathematical Theory.
24. *Grenander:* Pattern Analysis: Lectures in Pattern Theory, Vol. II.
25. *Davies:* Integral Transforms and Their Applications, 2nd ed.
26. *Kushner/Clark:* Stochastic Approximation Methods for Constrained and Unconstrained Systems.
27. *de Boor:* A Practical Guide to Splines: Revised Edition.
28. *Keilson:* Markov Chain Models—Rarity and Exponentiality.
29. *de Veubeke:* A Course in Elasticity.
30. *Sniatycki:* Geometric Quantization and Quantum Mechanics.
31. *Reid:* Sturmian Theory for Ordinary Differential Equations.
32. *Meis/Markowitz:* Numerical Solution of Partial Differential Equations.
33. *Grenander:* Regular Structures: Lectures in Pattern Theory, Vol. III.
34. *Kevorkian/Cole:* Perturbation Methods in Applied Mathematics.
35. *Carr:* Applications of Centre Manifold Theory.
36. *Bengtsson/Ghil/Källén:* Dynamic Meteorology: Data Assimilation Methods.
37. *Saperstone:* Semidynamical Systems in Infinite Dimensional Spaces.
38. *Lichtenberg/Lieberman:* Regular and Chaotic Dynamics, 2nd ed.
39. *Piccini/Stampacchia/Vidossich:* Ordinary Differential Equations in $\mathbf{R}^n$.
40. *Naylor/Sell:* Linear Operator Theory in Engineering and Science.
41. *Sparrow:* The Lorenz Equations: Bifurcations, Chaos, and Strange Attractors.
42. *Guckenheimer/Holmes:* Nonlinear Oscillations, Dynamical Systems, and Bifurcations of Vector Fields.
43. *Ockendon/Taylor:* Inviscid Fluid Flows.
44. *Pazy:* Semigroups of Linear Operators and Applications to Partial Differential Equations.
45. *Glashoff/Gustafson:* Linear Operations and Approximation: An Introduction to the Theoretical Analysis and Numerical Treatment of Semi-Infinite Programs.
46. *Wilcox:* Scattering Theory for Diffraction Gratings.
47. *Hale/Magalhães/Oliva:* Dynamics in Infinite Dimensions, 2nd ed.
48. *Murray:* Asymptotic Analysis.
49. *Ladyzhenskaya:* The Boundary-Value Problems of Mathematical Physics.
50. *Wilcox:* Sound Propagation in Stratified Fluids.
51. *Golubitsky/Schaeffer:* Bifurcation and Groups in Bifurcation Theory, Vol. I.
52. *Chipot:* Variational Inequalities and Flow in Porous Media.
53. *Majda:* Compressible Fluid Flow and Systems of Conservation Laws in Several Space Variables.
54. *Wasow:* Linear Turning Point Theory.
55. *Yosida:* Operational Calculus: A Theory of Hyperfunctions.
56. *Chang/Howes:* Nonlinear Singular Perturbation Phenomena: Theory and Applications.
57. *Reinhardt:* Analysis of Approximation Methods for Differential and Integral Equations.
58. *Dwoyer/Hussaini/Voigt (eds):* Theoretical Approaches to Turbulence.
59. *Sanders/Verhulst:* Averaging Methods in Nonlinear Dynamical Systems.

(continued following index)

Brian Straughan

The Energy Method, Stability, and Nonlinear Convection

Second Edition

With 30 Illustrations

Springer

Brian Straughan
Department of Mathematical Sciences
Science Laboratories
University of Durham
Durham DH1 3LE
UK
brian.straughan@durham.ac.uk

Editors:

S.S. Antman
Department of Mathematics
and
Institute for Physical Science
and Technology
University of Maryland
College Park, MD 20742-4015
USA
ssa@math.umd.edu

J.E. Marsden
Control and Dynamical
Systems, 107-81
California Institute of
Technology
Pasadena, CA 91125
USA
marsden@cds.caltech.edu

L. Sirovich
Division of Applied
Mathematics
Brown University
Providence, RI 02912
USA
chico@camelot.mssm.edu

Mathematics Subject Classification (2000): 76Exx, 76Rxx, 35K55, 35Q35

Straughan, B. (Brian)
The energy method, stability, and nonlinear convection / Brian Straughan.
p. cm. — (Applied mathematical sciences ; 91)
Includes bibliographical references and index.

1. Fluid dynamics. 2. Differential equations, Nonlinear—Numerical solutions.
3. Heat—Convection—Mathematical models. I. Title. II. Applied mathematical sciences
(Springer-Verlag New York Inc.) ; v. 91.
QA1.A647 vol. 91
[QA911]
510 s—dc21
[532'.05'01515355] 2003053004

ISBN 978-1-4419-1824-6 e-ISBN 978-0-387-21740-6

© 2010 Springer-Verlag New York, Inc.
All rights reserved. This work may not be translated or copied in whole or in part without the written permission of the publisher (Springer-Verlag New York, Inc., 175 Fifth Avenue, New York, NY 10010, USA), except for brief excerpts in connection with reviews or scholarly analysis. Use in connection with any form of information storage and retrieval, electronic adaptation, computer software, or by similar or dissimilar methodology now known or hereafter developed is forbidden.
The use in this publication of trade names, trademarks, service marks, and similar terms, even if they are not identified as such, is not to be taken as an expression of opinion as to whether or not they are subject to proprietary rights.

Printed in the United States of America.

9 8 7 6 5 4 3 2 1

www.springer-ny.com

Springer-Verlag New York Berlin Heidelberg
A member of BertelsmannSpringer Science+Business Media GmbH

Preface

This book is a revised edition of my earlier book of the same title. The current edition adopts the structure of the earlier version but is much changed. The introduction now contains definitions of stability. Chapters 2 to 4 explain stability and the energy method in more depth and new sections dealing with porous media are provided. Chapters 5 to 13 are revisions of those in the earlier edition. However, chapters 6 to 12 are substantially revised, brought completely up to date, and have much new material in.

Throughout the book new results are provided which are not available elsewhere.

Six new chapters, 14 - 19, are provided dealing with topics of current interest. These cover the topics of multi-component convection diffusion, convection in a compressible fluid, convection with temperature dependent viscosity and thermal conductivity, the subject of penetrative convection whereby part of the fluid layer can penetrate into another, nonlinear stability in the oceans, and finally in chapter 19 practical methods for solving numerically the eigenvalue problems which arise are presented.

The book presents convection studies in a variety of fluid and porous media contexts. It should be accessible to a wide audience and begins at an elementary level. Many new references are provided.

It is a pleasure to thank Magda Carr of Durham University for finding several misprints in an early version of this book. It is also a pleasure to thank Achi Dosanjh of Springer for her advice with editorial matters, and also to thank Frank Ganz of Springer for his help in sorting out latex problems for me.

Some of the research work reported in this book was supported in part by the Leverhulme Research Grant number RF & G/9/2000/226. This support is very gratefully acknowledged.

Durham, UK
June 2003

Brian Straughan

Contents

1

Introduction

This book is primarily a presentation of nonlinear energy stability results obtained in convection problems by means of an integral inequality technique we refer to as the energy method. While its use was originally based on the kinetic energy of the fluid motion, subsequent work has, for a variety of reasons, introduced variations of the classical energy. The new functionals have much in common with the Lyapunov method in partial differential equations and standard terminology in the literature would now appear to be *generalized energy methods*. In this book we shall describe many of the new generalizations and explain why such a generalization was deemed necessary. We shall also explain the physical relevance of the problem and indicate the usefulness of an energy technique in this context.

Standard indicial notation is used throughout this book together with the Einstein summation convention for repeated indices. Standard vector or tensor notation is also employed where appropriate. For example, we write

$$u_x \equiv \frac{\partial u}{\partial x} \equiv u_{,x} \qquad u_{i,t} \equiv \frac{\partial u_i}{\partial t} \qquad u_{i,i} \equiv \frac{\partial u_i}{\partial x_i} \equiv \sum_{i=1}^{3} \frac{\partial u_i}{\partial x_i}$$

$$u_j u_{i,j} \equiv u_j \frac{\partial u_i}{\partial x_j} \equiv \sum_{j=1}^{3} u_j \frac{\partial u_i}{\partial x_j}, \qquad i = 1, 2 \text{ or } 3.$$

In the case where a repeated index sums over a range different from 1 to 3 this will be pointed out in the text. Note that

$$u_j u_{i,j} \equiv (\mathbf{u} \cdot \nabla)\mathbf{u} \qquad \text{and} \qquad u_{i,i} \equiv \operatorname{div} \mathbf{u}.$$

As indicated above, a subscript t denotes partial differentiation with respect to time. When a superposed dot is used it means the material derivative, i.e.

$$\dot{u}_i \equiv \frac{\partial u_i}{\partial t} + u_j \frac{\partial u_i}{\partial x_j}\,.$$

The letter Ω will denote a fixed, bounded region of 3-space with boundary, Γ, sufficiently smooth to allow applications of the divergence theorem. A major part of the book concerns motion in a plane layer, say $\{(x, y) \in \mathbb{R}^2\} \times \{z \in (0, d)\}$. In this case, we usually refer to functions that have an (x, y) behaviour which is repetitive in the (x, y) direction, such as regular hexagons. The periodic cell defined by such a shape and its Cartesian product with $(0, d)$ will be denoted by V. The boundary of the period cell V will be denoted by ∂V.

The symbols $\|\cdot\|$ and $< \cdot >$ will denote, respectively, the L^2 norm on Ω and integration over Ω or the L^2 norm on V and integration over V, e.g.,

$$\int_V f^2 dV = \|f\|^2 \qquad \text{and} \int_V fg\, dV =< fg > .$$

(For most of the book the domain is a period cell V.) We also denote by $(\cdot, \cdot)$ the inner product on $L^2(V)$ so that

$$(f, g) \equiv< fg >= \int_V fg\, dV.$$

We sometimes have recourse to use the norm on $L^p(V)$, $1 < p < \infty$, and then we write

$$\|f\|_p = \left(\int_V |f|^p dV\right)^{1/p}.$$

We introduce the ideas of stability and instability in the context of the system of partial differential equations (which would be defined with suitable boundary conditions):

$$\frac{\partial u_i}{\partial t} - a^i \Delta u^i = f_i(\mathbf{u}, \mathbf{x}, t), \quad i = 1, \dots, N, \tag{1.1}$$

where $\mathbf{x} \in \Omega \subset \mathbb{R}^3$, and where $\Delta = \partial^2/\partial x^2 + \partial^2/\partial y^2 + \partial^2/\partial z^2$ is the Laplace operator. As a specific example of such a system which occurs in biology or in a chemical reaction (Segel and Jackson, 1972) we cite

$$\begin{aligned} \frac{\partial C_1}{\partial t} &= k_1 A + k_2 C_1^2 C_2 - k_3 B C_1 - k_4 C_1 + D_1 \Delta C_1, \\ \frac{\partial C_2}{\partial t} &= k_3 B C_1 - k_2 C_1^2 C_2 + D_2 \Delta C_2, \end{aligned} \tag{1.2}$$

where $A, B, k_1, \dots, k_4, D_1, D_2$ are constants, $D_1, D_2 > 0$, and C_1, C_2 are concentrations (of a biological species or of a chemical, for example). System (1.2) is of form (1.1) with $i = 1, 2$, i.e. $N = 2$.

We introduce notation in the context of a steady solution to (1.1), namely a solution $\bar{u}^i$ satisfying

$$a^i \Delta \bar{u}^i + f_i(\bar{\mathbf{u}}, \mathbf{x}) = 0. \tag{1.3}$$

(We could equally deal with the stability of a time-dependent solution, but many of the problems encountered here are for stationary solutions and at this juncture it is as well to keep the ideas as simple as possible.) Let w_i be a perturbation to (1.3), i.e. put $u_i = \bar{u}_i + w_i(\mathbf{x}, t)$. Then, with

$$\mathcal{F}_i(\mathbf{w}, \mathbf{x}, t) = f_i(\mathbf{w}, \mathbf{x}, t) - f_i(\bar{\mathbf{u}}, \mathbf{x})$$

it is seen from (1.1) and (1.3) that w_i satisfies the system

$$\frac{\partial w_i}{\partial t} - a^i \Delta w^i = \mathcal{F}_i(\mathbf{w}, \mathbf{x}, t). \tag{1.4}$$

To discuss linearized instability we linearize (1.4) which means we keep only the terms in $\mathcal{F}_i$ which are linear in $\mathbf{w}$ (the left hand side is already linear). If $\mathcal{F}_i^L$ denotes the linearization of $\mathcal{F}_i$ then the linearized system is

$$\frac{\partial w_i}{\partial t} - a^i \Delta w^i = \mathcal{F}_i^L(\mathbf{w}, \mathbf{x}, t). \tag{1.5}$$

Since (1.5) is a linear equation we may introduce an exponential time depedence in w_i so that $w_i = e^{\sigma t} s_i(\mathbf{x})$. Then (1.5) yields

$$\sigma s_i - a^i \Delta s^i = \mathcal{F}_i^L(\mathbf{s}, \mathbf{x}). \tag{1.6}$$

We say that the steady solution $\bar{u}_i$ to (1.3) is *linearly unstable* if

$$Re(\sigma) > 0, \tag{1.7}$$

where $Re(\sigma)$ denotes the real part of σ. Equation (1.6) (together with appropriate boundary conditions) is an eigenvalue problem for σ. For many of the problems discussed in this book the eigenvalues may be ordered so that

$$Re(\sigma_1) > Re(\sigma_2) > \dots$$

For linear instability we then need only ensure $Re(\sigma_1) > 0$.

The steady solution $\bar{u}_i$ to (1.3) is *nonlinearly stable* if and only if for each $\epsilon > 0$ there is a $\delta = \delta(\epsilon)$ such that

$$\|\mathbf{w}_0\| < \delta \Rightarrow \|\mathbf{w}(t)\| < \epsilon \tag{1.8}$$

and there exists γ with $0 < \gamma \le \infty$ such that

$$\|\mathbf{w}_0\| < \gamma \Rightarrow \lim_{t\to\infty} \|\mathbf{w}(t)\| = 0. \tag{1.9}$$

If $\gamma = \infty$, we say the solution is *unconditionally* nonlinearly stable (or simply refer to it as being asymptotically stable), otherwise for $\gamma < \infty$ the solution is *conditionally* (nonlinearly) stable. In this book we discuss at length examples of conditional and unconditional nonlinear stability in convection.

We have only defined stability with respect to the $L^2(\Omega)$ norm in (1.8) and (1.9). However, sometimes it is convenient to use an analogous definition with respect to some other norm or positive-definite solution measure. It will be clear in the text when this is the case.

It is important to realise that the linearization as in (1.5) and (1.6) can only yield linear *instability*. It tells us nothing whatsoever about stability. There are many equations for which nonlinear solutions will become unstable well before the linear instability analysis predicts this. Also, when an analysis is performed with $\gamma < \infty$ in (1.9) this yields conditional nonlinear stability, i.e. nonlinear stability for only a resricted class of initial data. This condition is frequently encountered in nonlinear energy stability analyses. However, for very small γ it may represent a severe restriction and we spend time discussing this. In particular, therefore, in this edition of the book we emphasize recent work in which unconditional nonlinear stability has been found. This means $\gamma = \infty$ and the stability so achieved is for all initial data.

A brief outline of the contents is now given. In chapter 2 we develop the ideas of the energy method on some simple one-space dimensional equations. Conditional stability is explained and studied in depth in two examples from mathematical biology which illustrate why different energies may have to be considered. Chapter 3 then extends the energy method to the equations for a linear viscous, incompressible, heat-conducting fluid. Additionally models for thermal convection in a porous medium are introduced and an energy analysis is developed for these. In chapters 4 to 18 we systematically develop the energy method in a variety of contexts: half-space problems, geophysical problems, convection driven by surface tension, convection in classes of fluid other than simply the linear viscous one, time-dependent convection problems, electrohydrodynamics, magnetohydrodynamics, and ferrohydrodynamic convection, convection with chemical reactions, multi-component convection, compressible convection, convection with temperature dependent viscosity and thermal diffusivity, penetrative convection, water circulation in the oceans, and we study the connection with the Lyapunov method in partial differential equations. Chapter 19 includes an account of two highly accurate but very efficient techniques for solving eigenvalue problems which occur in nonlinear energy stability problems. Full details are given for a specific problem in thermal convection and for some porous convection problems.

Before proceeding to the body of the book and nonlinear energy stability we point out that there is another technique in continuum mechanics also sometimes referred to as the energy method. This method seeks to find the configuration of minimum potential energy and then attempts to establish it as the stable configuration. These ideas go back at least as far as the work of (Kelvin, 1887) on a perfect fluid and have been extended by (Arnold, 1965a) -(Arnold, 1966b). A comprehensive and authoritative review of the Arnold method up to 1985 is given by (Holm et al., 1985): this paper covers many interesting topics in fluid mechanics. A later connected work developing and extending this technique to an idea known as the energy - momentum method is that of (Simo et al., 1990). The energy - Casimir (convexity) method is an interesting one, which is appropriate to the investigation of the stability of inviscid flows and the stability of water waves (among other things). It would appear that this method is attributed to V.I. Arnold and (Holm et al., 1985) provides an extensive account. This is a stimulating paper and draws distinct attention to differences in the ideas of formal stability and true nonlinear stability. Due to the numerous and diverse applications of these ideas to magnetohydrodynamics, rotating stars, rotating self-gravitating systems, and many other practical problems (see the references in (Simo et al., 1991a)), the energy - Casimir method is undoubtedly an important one. The review of (Simo et al., 1990) develops and applies the energy-momentum method to the problem of nonlinear stability of relative equilbria. Although the report concentrates on detailed applications to the stability of uniformly rotating states of models for geometrically exact rods and to a rigid body with an attached flexible appendage, it is relevant to mention it here. The energy-momentum technique of (Simo et al., 1990) introduces a block diagonalization procedure, in which the second variation of the energy augmented with the linear and angular momentum block diagonalizes, separating the rotational from the internal vibration modes. Further references to the extensive applications of these ideas may be found in (Simo et al., 1990). (Simo et al., 1991a) presents a general approach to deriving rigorous nonlinear stability for relative equilibria in Hamiltonian systems. They refer to their technique as the reduced energy-momentum method and show how it substantially extends previous approaches to the subject. This is a very readable paper which gives a concise historical account of the subject. In (Simo et al., 1991b) their technique is further developed in nonlinear elasticity. We do not describe these ideas and concentrate on the dynamic theory of Energy Stability whose modern revival has been inspired by the work of (Serrin, 1959a; Serrin, 1959b; Serrin, 1959c; Serrin, 1959d), and by (Joseph, 1965; Joseph, 1966; Joseph, 1970).

We should point out that while we do concentrate on thermal convection problems, energy stability theory is being successfully applied in other areas of mathematics. For example, in chapter 18 we look at work on energy stability theory in ocean circulation models. This is a relatively new area where energy theory can lead to more progress. Due to questions such as what effect does thermohaline circulation in the oceans have on climate change, see e.g. (Clark et al., 2002), bounds derived by energy theory are likely to prove very useful. Mathematical biology is another field where nonlinear energy stability like arguments are proving effective and some of these are mentioned/reviewed in section 14.4. Another area where energy theory is proving successful is in control theory, see (Alvarez-Ramirez et al., 2001), (Ydstie, 2002). These and other exciting new areas of application will undoubtedly form part of future work in nonlinear energy stability theory.

2

Illustration of the energy method on simple examples and discussion of linear theory

2.1 The diffusion equation

Before considering the nonlinear partial differential equations encountered in convection we approach the problem of stability or instability for the diffusion equation.

Let u be a solution to the diffusion equation

$$u_t = u_{xx}, \tag{2.1}$$

where for now, $-\infty < x < \infty$, $t > 0$, and suppose we wish to investigate the behaviour of u subject to initial data

$$u(x, 0) = u_0(x), \qquad -\infty < x < \infty. \tag{2.2}$$

The zero solution $u \equiv 0$ is a solution to (2.1), and it is the stability of this solution we wish to investigate. The key to the procedure is that for a solution (here $u \equiv 0$) to be stable it must be stable against *any* disturbance to which it may be subjected. In fact, we shall understand stability to be in the sense of asymptotic stability so that for the zero solution to (2.1) to be stable we mean all perturbations decay to zero as time progresses. From a practical viewpoint it is necessary that disturbances are damped out sufficiently rapidly. In this respect the energy method is very useful as it usually guarantees exponential decay.

To demonstrate instability of the steady solution $u \equiv 0$, on the other hand, it is sufficient to find *one* disturbance that either progressively grows in amplitude or at least remains bounded away from zero in the case of instability of the zero solution.

With the above in mind, since (2.1) is linear we consider a perturbation to the zero solution to (2.1) of the form

$$u(x,t) = e^{\sigma t} e^{ikx}, \tag{2.3}$$

where k is any real number. Despite the special appearance of this perturbation, we are able to study the instability of the zero solution to (2.1) against any periodic (in x) disturbance. The idea behind this is that we suppose u is periodic in x, but allow *any* periodic behaviour, then $u(x,t)$ may be written as

$$u(x,t) = \sum_{n=0}^{\infty} A_n e^{\sigma_n t} e^{ik_n x}, \tag{2.4}$$

i.e., a Fourier series. The term in (2.3) is referred to as a Fourier mode. As it is sufficient for only one destabilizing disturbance to cause instability we need consider only (2.3), because by varying over all real numbers k we hence would pick up the most destabilizing term in (2.4).

For a representation of the form (2.3), equation (2.1) reduces to

$$\sigma = -k^2.$$

Since $k \in \mathbb{R}$, $\sigma < 0$ and so there is no unstable mode and all solutions of form (2.3) decay. Therefore, for the periodic disturbance situation considered above the zero solution to (2.1) is always stable.

Of course, the above analysis is for an infinite spatial region whereas many practical convection studies usually require a finite region of disturbance. Hence, without loss of generality, suppose now (2.1) holds on the spatial region $(0,1)$ with boundary conditions

$$u(0,t) = u(1,t) = 0. \tag{2.5}$$

The only difference with the infinite region case for (2.1) is that k can no longer take all values in $\mathbb{R}$, for the class of periodic functions in x must all vanish at $x = 0, 1$. Hence, here we have a subclass of (2.3). As *all* solutions of form (2.3) are stable, so are all solutions of this subclass.

Even though we know $u \equiv 0$ is a stable solution to (2.1), (2.5) we shall obtain this result directly using an *energy method*, choosing an extremely simple example to illustrate the technique.

Let now u be a solution to (2.1), (2.5) that satisfies arbitrary initial data $u_0(x)$. We define an *energy* $E(t)$ by

$$E(t) = \frac{1}{2}\|u(t)\|^2, \tag{2.6}$$

where $\|\cdot\|$ denotes the norm on $L^2(0,1)$, i.e.,

$$\|f\|^2 = \int_0^1 f^2\, dx.$$

Differentiate $E(t)$ and use (2.1),

$$\begin{aligned}\frac{dE}{dt} &= \frac{1}{2}\int_0^1 \frac{\partial}{\partial t}u^2\,dx\\ &= \int_0^1 uu_t\,dx\\ &= \int_0^1 uu_{xx}\,dx.\end{aligned}$$

Next, integrate by parts,

$$\begin{aligned}\int_0^1 uu_{xx}\,dx &= -\int_0^1 u_x^2\,dx + uu_x\Big|_0^1\\ &= -\|u_x\|^2.\end{aligned}$$

Therefore, the energy equation becomes

$$\frac{dE}{dt} + \|u_x\|^2 = 0. \tag{2.7}$$

Next, use the Poincaré inequality

$$\|u_x\|^2 \geq \pi^2\|u\|^2,$$

for u with $u = 0$ at $x = 0, 1$. Using this in (2.7) gives

$$\frac{d}{dt}\|u\|^2 + 2\pi^2\|u\|^2 \leq 0$$

or, equivalently,

$$\frac{d}{dt}\left(e^{2\pi^2 t}\|u\|^2\right) \leq 0,$$

which leads to

$$\|u(t)\|^2 \leq e^{-2\pi^2 t}\|u_0\|^2. \tag{2.8}$$

Hence, $\|u(t)\| \to 0$ at least exponentially and the zero solution to (2.1), (2.5) is stable.

2.2 The diffusion equation with a linear source term

The problem is to study the stability of the zero solution to the equation

$$u_t = u_{xx} + au, \tag{2.9}$$

where a is a positive constant, with given initial data

$$u(x, 0) = u_0(x). \tag{2.10}$$

When $a = 0$ it has been shown in section 2.1 that the zero solution is stable always, regardless of whether the spatial domain is finite or infinite. Here things are different.

2.2.1 Spatial Region $x \in \mathbb{R}$.

We use a *normal mode* analysis, and as before substitute

$$u(x,t) = e^{\sigma t + ikx}.$$

Then, from (2.9) we obtain

$$\sigma = -k^2 + a. \tag{2.11}$$

Now, $x \in \mathbb{R}$, therefore we have to analyse stability among *all* periodic disturbances, i.e., $k \in \mathbb{R}$. From (2.11) since $\sigma > 0 \quad \Rightarrow$ instability, we conclude

$$k^2 < a \quad \Rightarrow \quad \text{instability},$$

i.e., any mode for which $k^2 < a$ is sufficient to yield an unstable solution. Therefore, if $a > 0$, there is always some $k^2 < a$ and so $a > 0 \Rightarrow$ instability always.

So, with the infinite spatial region, $x \in \mathbb{R}$, we have $u \equiv 0$ always stable, for u satisfying

$$u_t = u_{xx},$$

but if

$$u_t = u_{xx} + au,$$

then $u \equiv 0$ is always unstable no matter how small a $(a > 0)$.

2.2.2 Finite Spatial Region.

Suppose now $x \in (0,1)$, u satisfies (2.9), and $u(0,t) = u(1,t) = 0, \quad \forall t > 0$. Here (2.11) is still valid, but now $u = 0$ whenever $x = 0, 1$; i.e., the solutions may be thought of as being spatially periodic over $\mathbb{R}$ but *must* vanish at $x = 0, 1$, and at every multiple of 1.

For example,

$$u(x,t) = e^{\sigma t} \sin kx,$$

and then

$$u = 0 \quad \text{at} \quad x = 0, 1 \qquad \Rightarrow \qquad k = n\pi, \qquad n = \pm 1, \pm 2, \dots .$$

The Fourier-series is here a half-range series because cosine terms do not satisfy the boundary conditions. Therefore,

$$\sigma = -k^2 + a, \qquad \text{where} \qquad k^2 = n^2\pi^2, \qquad n = \pm 1, \pm 2, \dots .$$

For $\sigma > 0$, $k^2 < a$, but $k^2_{\min} = \pi^2$. Hence, provided $a > \pi^2$, there is *always* instability, i.e., the mode $e^{\sigma t} \sin \pi x$ will grow. For $a < \pi^2$, all Fourier modes decay, and so there is linear stability. So $a = \pi^2$ is the stability-instability boundary. (It should be observed that the size of the region is crucial; for $x \in (0, L)$ we find a different stability boundary.)

2.2.3 Energy Stability for a Solution to (2.9).

We apply the energy method to study the stability of the solution to the boundary-initial value problem,

$$\begin{aligned} &\frac{\partial u}{\partial t} = \frac{\partial^2 u}{\partial x^2} + au, && x \in (0,1),\ t > 0, \\ &u(0,t) = u(1,t) = 0, && \forall t \geq 0, \\ &u(x,0) = u_0(x). \end{aligned} \tag{2.12}$$

Multiply the differential equation $(2.12)_1$ by u and integrate over (0,1) to find

$$\int_0^1 u \frac{\partial u}{\partial t}\, dx = \int_0^1 u \frac{\partial^2 u}{\partial x^2}\, dx + a \int_0^1 u^2\, dx.$$

Hence, with

$$E(t) = \frac{1}{2}\|u(t)\|^2 \quad \left(= \frac{1}{2} \int_0^1 u^2(x,t)\, dx \right),$$

this equation becomes (after integration by parts as before)

$$\begin{aligned} \frac{dE}{dt} &= -\|u_x\|^2 + a\|u\|^2, \\ &= -a\|u_x\|^2 \Big(\frac{1}{a} - \frac{\|u\|^2}{\|u_x\|^2} \Big), \qquad (\|u_x\|^2 \neq 0), \\ &\leq -a\|u_x\|^2 \Big(\frac{1}{a} - \max_{\mathcal{H}} \frac{\|u\|^2}{\|u_x\|^2} \Big), \end{aligned} \tag{2.13}$$

where $\mathcal{H}$ is the space of admissible functions over which we seek a maximum. Set

$$\mathcal{H} = \big\{ u \in C^2(0,1) \,\big|\, u = 0 \text{ when } x = 0, 1 \big\}.$$

Now define R_E by

$$\frac{1}{R_E} = \max_{\mathcal{H}} \frac{\|u\|^2}{\|u_x\|^2};$$

then the energy *inequality* (2.13) may be rewritten

$$\frac{dE}{dt} \leq -a\|u_x\|^2 \Big(\frac{1}{a} - \frac{1}{R_E} \Big).$$

If $a < R_E$, then $1/a - 1/R_E > 0$, say $1/a - 1/R_E = c\,(> 0)$, and so

$$\frac{dE}{dt} \leq -ac\|u_x\|^2.$$

Again using the Poincaré inequality, $\|u_x\|^2 \geq \pi^2\|u\|^2$, from (2.13) we deduce

$$\frac{dE}{dt} \leq -ac\pi^2\|u\|^2 = -2\pi^2 acE,$$

from which it follows that

$$\frac{d}{dt}(e^{2\pi^2 act}E) \leq 0.$$

This may be integrated to see that

$$E(t) \leq e^{-2\pi^2 act}E(0). \tag{2.14}$$

We have thus shown that if $a < R_E$,

$$E(t) = \frac{1}{2}\|u(t)\|^2 \to 0 \quad \text{as} \quad t \to \infty,$$

with the decay at least exponential in time.

The problem remains to find R_E. Recall that

$$R_E^{-1} = \max_{\mathcal{H}} \frac{\|u\|^2}{\|u_x\|^2}.$$

Let $I_1 = \|u\|^2, \quad I_2 = \|u_x\|^2$. The Euler-Lagrange equations are found from

$$\begin{aligned}\frac{d}{d\epsilon}\frac{I_1(u+\epsilon\eta)}{I_2(u_x+\epsilon\eta_x)}\bigg|_{\epsilon=0} = \delta\Big(\frac{I_1}{I_2}\Big) &= \frac{I_2\delta I_1 - I_1\delta I_2}{I_2^2} \\ &= \frac{1}{I_2}\Big(\delta I_1 - \frac{I_1}{I_2}\Big|_{\max}\delta I_2\Big) \\ &= \frac{1}{I_2}\{\delta I_1 - \frac{1}{R_E}\delta I_2\}.\end{aligned}$$

(Since δ refers to the "derivative" evaluated at $\epsilon = 0$, I_1/I_2 is here understood to be at the stationary value.) Therefore,

$$\delta I_1 - \frac{1}{R_E}\delta I_2 = 0. \tag{2.15}$$

Here

$$\delta I_1 = \frac{d}{d\epsilon}\int_0^1 (u+\epsilon\eta)^2 dx\bigg|_{\epsilon=0},$$

where η is an arbitrary $C^2(0,1)$ function with $\eta(0) = \eta(1) = 0$, and

$$\delta I_2 = \frac{d}{d\epsilon}\int_0^1 (u_x+\epsilon\eta_x)^2 dx\bigg|_{\epsilon=0}.$$

So (2.15) leads to

$$\int_0^1 (u\eta - R_E^{-1}\eta_x u_x)dx = 0.$$

Integration by parts shows that

$$\int_0^1 \eta(u_{xx} + R_E u)dx = 0.$$

Since η is arbitrary apart from the continuity and boundary condition requirements, we must have

$$\frac{d^2u}{dx^2} + R_E u = 0, \qquad u(0) = u(1) = 0. \tag{2.16}$$

Equation (2.16) is the Euler equation that gives an eigenvalue problem for R_E.

The general solution to (2.16) is

$$u = A \sin R_E^{1/2} x + B \cos R_E^{1/2} x,$$

where A, B are constants to be determined. The boundary condition $u(0) = 0$ shows $B = 0$, and so

$$u = \sin R_E^{1/2} x,$$

where we have taken $A = 1$, since we are primarily interested in R_E not u. The condition $u(1) = 0$ then shows that

$$\sqrt{R_E} = n\pi, \qquad n = \pm 1, \pm 2, \dots .$$

This gives an infinite sequence of values for R_E (corresponding to stationary values of the quotient $\|u\|^2/\|u_x\|^2$),

$$R_E = \pi^2, 4\pi^2, 9\pi^2, \dots .$$

For stability, we need $a < R_E(\min)$, and so $R_E = \pi^2$. In particular, therefore,

$$a < \pi^2$$

yields stability of the zero solution to (2.9), (2.10). This criterion has been derived by using energy inequalities. It is the same as the one found by the "normal mode" method. This is partly because the basic differential equation is linear. We are primarily interested in nonlinear differential equations. In this situation the energy method is very useful since it does yield precise practical information regarding the *nonlinear* stability of a basic solution (the solution $u \equiv 0$ in the examples above).

2.3 Conditional energy stability

In order to investigate the effect of a nonlinear term on the stability of a solution to a partial differential equation (pde), we shall consider the boundary-initial value problem for the diffusion equation (2.1) but now with a quadratic nonlinear term forcing the right-hand side and also with a convective nonlinear term uu_x added to the left (such a convective nonlinearity is typical in hydrodynamic stability questions).

We examine the stability of the zero solution to the boundary-initial value problem,

$$\begin{aligned} &\frac{\partial u}{\partial t} + u\frac{\partial u}{\partial x} = \frac{\partial^2 u}{\partial x^2} + \beta u^2, \qquad x \in (0,1),\ t > 0,\\ &u(0,t) = u(1,t) = 0, \qquad \forall\, t \geq 0,\\ &u(x,0) = u_0(x), \end{aligned} \tag{2.17}$$

where β is a positive constant.

If we attempt a linear analysis, that is linearize about the solution $u \equiv 0$, then since any perturbation u is assumed such that $|u|, |u_x| << 1$, u^2 and uu_x may be neglected. We are then left with the linear stability analysis of (2.1). Therefore, the zero solution is always *linearly stable*. The nonlinear terms, however, cannot be neglected. The effect of the u^2 term is to destabilize as we shall now show. The uu_x term in certain respects acts to stabilize and this is shown in section 2.4.

To see the effect of the nonlinear term u^2 we multiply the differential equation $(2.17)_1$ by u and integrate over (0,1) to obtain

$$\frac{1}{2}\frac{d}{dt}\,\|u\|^2 = \int_0^1 u\frac{\partial^2 u}{\partial x^2}\,dx + \beta\int_0^1 u^3\,dx.$$

Note that the convective term integrates to zero,

$$\int_0^1 u^2\frac{\partial u}{\partial x}\,dx = \frac{1}{3}\big[u^3(1) - u^3(0)\big] = 0.$$

This is a key point as this feature carries over to fluid mechanics problems and allows us to remove otherwise troublesome terms.

Again, as before, integrating by parts,

$$\int_0^1 u\frac{\partial^2 u}{\partial x^2}\,dx = -\,\|u_x\|^2.$$

So, the energy equation in this case becomes

$$\frac{1}{2}\frac{d}{dt}\,\|u\|^2 = -\,\|u_x\|^2 + \beta\int_0^1 u^3\,dx. \tag{2.18}$$

Since $u(x,t)$ may be positive or negative we do not know a priori if $\int u^3\,dx$ is one signed. In general, this term will lead to an instability. Therefore, may we recover any stability? Motivated by linear theory, which shows that if $|u| << 1$ the solution is always stable, we might guess that we shall have stability provided the initial data $u_0(x)$ is small enough.

We write

$$\int_0^1 u^3\,dx = \int_0^1 u^2 u\,dx \le \Big(\int_0^1 u^4\,dx\Big)^{1/2}\Big(\int_0^1 u^2\,dx\Big)^{1/2},$$

by use of the Cauchy-Schwarz inequality. From the Sobolev embedding inequality we know that (see the Appendix)

$$\int_0^1 u^4\,dx \le \frac{1}{4}\Big(\int_0^1 u_x^2\,dx\Big)^2.$$

Using this leads to

$$\begin{aligned}\int_0^1 u^3\,dx &\le \frac{1}{2}\Big(\int_0^1 u_x^2\,dx\Big)\Big(\int_0^1 u^2\,dx\Big)^{1/2}\\ &= \frac{1}{2}\|u\|\,\|u_x\|^2.\end{aligned} \tag{2.19}$$

Put (2.19) into (2.18) to find

$$\frac{1}{2}\frac{d}{dt}\|u\|^2 \le -\|u_x\|^2\Big(1-\frac{1}{2}\beta\|u(t)\|\Big). \tag{2.20}$$

Next, assume that

$$\|u_0\| < 2\beta^{-1} \qquad \Big(\text{i.e., } \int_0^1 u_0^2(x)\,dx < 4/\beta^2\Big).$$

Then either

(i) $$\|u(t)\| < 2\beta^{-1}, \qquad \forall\, t>0$$

or

(ii) there exists an $\eta < \infty$ such that

$$\begin{aligned}\|u(\eta)\| &= 2\beta^{-1}, \quad \text{with}\\ \|u(\eta)\| &< 2\beta^{-1}, \quad \text{on} \quad [0,\eta).\end{aligned}$$

Suppose (ii) holds. Then on $[0,\eta)$, $1-\frac{1}{2}\beta\|u(t)\|>0$ so (2.20) shows

$$\frac{d}{dt}\|u\|^2 < 0, \quad \text{for} \quad 0\le t<\eta. \tag{2.21}$$

Hence,

$$\|u(t)\|^2 \le \|u(0)\|^2 = \|u_0\|^2 < 4\beta^{-2}, \qquad t\in[0,\eta).$$

Since $\|u(t)\|$ is assumed continuous in t, this means $\|u(\eta)\| \ne 2/\beta$, a contradiction. Hence, (ii) is false and (i) holds. (We are assuming the solutions we are dealing with are "classical", and so $u\in C^2$ in x, $u\in C^1$ in t.) Therefore, provided

$$\|u_0\| < 2/\beta,$$

it follows that

$$\|u(t)\| < 2/\beta, \qquad \forall\, t \geq 0.$$

Further, (2.21) now holds $\forall\, t \geq 0$, and hence

$$\|u(t)\|^2 \leq \|u_0\|^2, \qquad \forall\, t \geq 0.$$

We have shown that

$$1 - \frac{1}{2}\beta\|u(t)\| \geq 1 - \frac{1}{2}\beta\|u_0\| \ (> 0).$$

Now, use this in (2.20),

$$\begin{aligned}\frac{1}{2}\frac{d}{dt}\|u\|^2 &\leq -\|u_x\|^2\Big(1 - \frac{1}{2}\beta\|u(t)\|\Big),\\ &\leq -\|u_x\|^2\Big(1 - \frac{1}{2}\beta\|u_0\|\Big).\end{aligned}$$

Next, from Poincaré's inequality, $\|u_x\|^2 \geq \pi^2\|u\|^2$, and since $(1 - \frac{1}{2}\beta\|u_0\|) > 0$, we find

$$\begin{aligned}\frac{1}{2}\frac{d}{dt}\|u\|^2 &\leq -\pi^2\Big(1 - \frac{1}{2}\beta\|u_0\|\Big)\|u\|^2\\ &= -A\|u\|^2,\end{aligned}$$

where we have set $A = \pi^2(1 - \frac{1}{2}\beta\|u_0\|)$.

Using an integrating factor we may obtain from this inequality

$$\frac{d}{dt}(\|u(t)\|^2 e^{2At}) \leq 0,$$

which in turn integrates to

$$\|u(t)\|^2 \leq e^{-2At}\|u_0\|^2.$$

What we have shown is that if $\|u_0\| < 2/\beta$, then $\|u(t)\| \to 0$ at least exponentially fast. We refer to this as *nonlinear conditional stability.* Nonlinear since the nonlinear terms are handled in the analysis and conditional since the initial data has to be small enough; although it is important to note that we find a threshold for u_0 even though it is not necessarily the best (i.e., it is possible $u \to 0$ even if $\|u_0\| \geq 2\beta^{-1}$).

2.4 Weighted energy and boundedness

In the last section we saw that if $\|u_0\| < 2\beta^{-1}$, then a solution to (2.17) will decay to zero in L^2 norm, at least exponentially fast. It is natural to wonder what will happen if this restriction on the initial data is exceeded, especially as *when the convective uu_x term is not present,* a u^2 term can lead to global nonexistence of the solution in finite time by some norm of the solution blowing up, see e.g., (Straughan, 1998) and the references therein.

In this section we shall show by using an L^2 "energy" with a weight that finite time blow-up for a solution to (2.17) cannot occur, at least not in L^2 norm. The technique is simple but appealing as it uses the convective term to stabilize the problem.

Let u be a solution to the boundary-initial value problem (2.17), which for clarity we rewrite here,

$$\begin{aligned} &\frac{\partial u}{\partial t} + u\frac{\partial u}{\partial x} = \frac{\partial^2 u}{\partial x^2} + \beta u^2, \qquad x \in (0,1),\ t > 0,\\ &u(0,t) = u(1,t) = 0, \qquad \forall\, t \geq 0,\\ &u(x,0) = u_0(x). \end{aligned} \tag{2.22}$$

We shall restrict attention to positive solutions, $u(x,t) \geq 0$, which we might expect to be most destabilizing. Furthermore, although we consider any positive β, we are primarily interested in β large, when the conditional result becomes restrictive.

Let us introduce the notation $< \cdot >$ to mean the integral over (0,1), and let $\mu(x) = e^{-kx}$ for $k(> 0)$ to be defined, and then define

$$F(t) = < \mu u^2 > = \int_0^1 \mu(x) u^2(x,t)\, dx. \tag{2.23}$$

Differentiate this with respect to t, then use (2.22), integrate by parts, and use the boundary conditions to find

$$\begin{aligned} F_t &= 2 < \mu u(-uu_x + u_{xx} + \beta u^2) >\\ &= 2\beta < \mu u^3 > + \frac{2}{3} < \mu_x u^3 > -2 < \mu u_x^2 > + < \mu_{xx} u^2 >\\ &= -2\Big(\frac{k}{3} - \beta\Big) < \mu u^3 > -2 < \mu u_x^2 > + k^2 < \mu u^2 > . \end{aligned} \tag{2.24}$$

Now, by Hölder's inequality,

$$< \mu u^2 > \leq < \mu u^3 >^{2/3} < \mu >^{1/3},$$

or evaluating $< \mu >$ and rearranging,

$$- < \mu u^3 > \leq - \frac{k^{1/2} < \mu u^2 >^{3/2}}{(1 - e^{-k})^{1/2}} . \tag{2.25}$$

We also need the value for

$$\tilde{\lambda}_1 = \min_{H_0^1(0,1)} \frac{< \mu(x) u_x^2 >}{< \mu u^2 >} . \tag{2.26}$$

The Euler equation for this problem is

$$u_{xx} - ku_x + \tilde{\lambda}_1 u = 0,$$

and substituting $u = e^{mx}$ we find

$$m = \frac{1}{2}k \pm i\sqrt{\tilde{\lambda}_1 - \frac{k^2}{4}}.$$

(The possibility that $m \in \mathbf{R}$ is quickly seen to lead to a zero solution.) Hence,

$$u = A\exp\Big(\frac{1}{2}kx\Big)\cos\theta x \;+\; B\exp\Big(\frac{1}{2}kx\Big)\sin\theta x\,,$$

where $\theta = [\tilde{\lambda}_1 - k^2/4]^{1/2}$ give the extremals. The boundary conditions require $A = 0$ and

$$\theta^2 = \pi^2, 4\pi^2, 9\pi^2, \dots .$$

In consequence we find

$$\tilde{\lambda}_1 = \frac{k^2}{4} + \pi^2. \tag{2.27}$$

We next choose $k = 3\beta + 1$, and use (2.25), (2.27) in (2.24) to obtain

$$F_t \le -\frac{2k^{1/2}}{3(1 - e^{-k})^{1/2}}\,F^{3/2} + \Big(\frac{1}{2}k^2 - 2\pi^2\Big)F.$$

(If $k^2 < 4\pi^2$, then from this we can deduce unconditional decay of F, i.e., for all u_0. However, we can do better than this, and this is shown in section 2.5.) To solve the above inequality divide by $F^{3/2}$, define $v = F^{-1/2}$ to obtain

$$\frac{dv}{dt} + Av \ge B,$$

where

$$A = \frac{k^2}{4} - \pi^2, \qquad B = \frac{k^{1/2}}{3(1 - e^{-k})^{1/2}}\,.$$

After using an integrating factor on the last inequality and then integrating, we solve for $F^{1/2}$ to see that

$$F^{1/2}(t) \le \frac{1}{F^{-1/2}(0)e^{-At} + (B/A)(1 - e^{-At})}\,. \tag{2.28}$$

Without explicitly putting in A, B, it is easy to see that $F(t)$ is bounded for all t. As $t \to \infty$, the right hand side approaches

$$\frac{A}{B} = \Big(\frac{k^2}{4} - \pi^2\Big)\frac{3(1 - e^{-k})^{1/2}}{k^{1/2}} \sim O\,(\beta^{3/2}).$$

It follows that for large time (and large β),

$$F(t) \le O\,(\beta^3).$$

Furthermore,

$$F(t) = <\mu u^2> \ge e^{-k} < u^2 > .$$

Since $k = 3\beta + 1$, this yields the asymptotic behaviour

$$\|u(t)\|^2 \le O\,(\beta^3 e^{3\beta}).$$

Note that $\|u\|$ can become very large, but is always bounded. In fact, for β above a threshold u tends to a positive steady state that has a steep boundary layer near $x = 1$. Numerical results showing the steady states for systems with equations like (2.22) are shown in section 2.9.

2.5 Weighted energy and unconditional decay

The object of this section is to show that by modifying the argument of the last section we may obtain a value for β, such that for β less than this, all solutions to (2.22) decay (in a suitable norm) regardless of the size of the initial data.

Consider for now a positive, $u \ge 0$, solution to (2.22) and for a constant $\delta (> 0)$ to be specified choose

$$F(t) = < \mu u^{1+\delta} >, \tag{2.29}$$

where again $\mu = e^{-kx}$. By differentiating F, using the differential equation of (2.22), integrating by parts, and using the boundary conditions we find

$$\begin{aligned} F_t =& (1+\delta) < \mu u^\delta(-uu_x + u_{xx} + \beta u^2) > \\ =& (1+\delta)\beta < \mu u^{2+\delta} > + \Big(\frac{1+\delta}{2+\delta}\Big) < \mu_x u^{2+\delta} > \\ & - \frac{4\delta}{(1+\delta)} < \mu w_x^2 > + < \mu_{xx} u^{1+\delta} >, \end{aligned}$$

where we have set $w = u^{(1+\delta)/2}$. Recalling the definition of μ, this leads to

$$\begin{aligned} F_t \le& (1+\delta)\Big(\beta - \frac{k}{2+\delta}\Big) < \mu u^{2+\delta} > \\ & - \frac{4\delta}{(1+\delta)}\Big(1 - \frac{k^2(1+\delta)}{4\delta}\tilde{\lambda}_1^{-1}\Big) < \mu w_x^2 >, \end{aligned} \tag{2.30}$$

where $\tilde{\lambda}_1^{-1} = \max_{H_0^1(0,1)} < \mu w^2 > / < \mu w_x^2 >$ and is given by (2.27), and we have used the fact that

$$< \mu_{xx} u^{1+\delta} > \le k^2 \Big(\max_{H_0^1(0,1)} \frac{< \mu w^2 >}{< \mu w_x^2 >} \Big) < \mu w_x^2 > .$$

If δ and k are now chosen such that

$$\frac{k^2(1+\delta)}{4\delta\tilde{\lambda}_1} < 1 \tag{2.31}$$

and $k = \beta(2+\delta)$, then from (2.30) we use Poincaré's inequality to assert that $F(t) \to 0$, (at least) exponentially, as $t \to \infty$.

Now, recall from (2.27) that $\tilde{\lambda}_1 = \pi^2 + k^2/4$, then (2.31) and the choice for k require

$$\beta^2 < \frac{4\delta\pi^2}{(2+\delta)^2} = f(\delta), \quad \text{say.}$$

$f(\delta)$ achieves a maximum where $\delta = 2$ and so this is our optimum choice for δ in (2.29). With this choice, we deduce that provided

$$\beta^2 < \frac{1}{2}\pi^2, \tag{2.32}$$

solutions to (2.22) for arbitrarily large initial data decay to zero in the F measure of (2.29). Of course, from this we may also deduce $\|u\|_{L^2}$ decay with the aid of Hölder's inequality.

Finally in this section we observe that the argument with

$$F(t) =< \mu u^3 >$$

requires $u \geq 0$. To circumvent this we take

$$F(t) =< \mu u^{3+1/m} >,$$

m an odd integer; i.e., take a real root $u^{1/m}$ and then consider the non-negative quantity $u^{(3m+1)/m}$. Apply the argument of this section to this F and let $m \to \infty$ to see that (2.32) is sufficient for the decay of all solutions to (2.22), not just positive ones.

2.6 A stronger force in the diffusion equation

In section 2.4 we saw that if the force in (2.22) is βu^2, then $\|u\|_{L^2}$ remains bounded for all t regardless of the size of β. It is natural to wonder if this result continues to hold for a larger power of u. The answer is no, as we now show.

Consider now the following boundary-initial value problem for the forced convection-diffusion equation,

$$\begin{aligned} &\frac{\partial u}{\partial t} + u\frac{\partial u}{\partial x} = \frac{\partial^2 u}{\partial x^2} + \beta u^{2+\delta}, \qquad x \in (0,1),\ t > 0, \\ &u(0,t) = u(1,t) = 0, \qquad \forall\, t \geq 0, \\ &u(x,0) = u_0(x), \end{aligned} \tag{2.33}$$

where δ is *any positive* constant.

We shall show that it is possible to prescribe initial data for which no global solution exists. Let ϕ be the first eigenfunction in the membrane problem for (0,1), so here

$$\phi = \sin \pi x,$$

and consider for n to be determined

$$F(t) = <\phi^n u> .$$

By direct calculation and use of (2.33),

$$\begin{aligned} F_t &= <\phi^n u_t> \\ &= <\phi^n(-uu_x + u_{xx} + \beta u^{2+\delta})> \\ &= -\pi^2 nF + n(n-1) <\phi^{n-2}\phi_x^2 u> + \beta <\phi^n u^{2+\delta}> + \frac{n}{2} <\phi^n \phi_x u^2>, \end{aligned}$$

where the definition of ϕ has also been employed. From the last equation we may deduce that

$$F_t \geq -n\pi^2 F + \beta <\phi^n u^{2+\delta}> - \frac{1}{2} n\pi <\phi^{n-1} u^2> . \tag{2.34}$$

Since from Hölder's inequality

$$<\phi^n u^{2+\delta}>^{2/(2+\delta)} \cdot <1>^{\delta/(2+\delta)} \;\geq\; <\phi^{2n/(2+\delta)} u^2>,$$

the choice $n = 1 + 2/\delta$ gives

$$<\phi^{1+2/\delta} u^{2+\delta}> \;\geq\; <\phi^{2/\delta} u^2>^{(2+\delta)/2},$$

and this in (2.34) allows us to see that

$$\begin{aligned} F_t \geq &- \Big(\frac{\delta+2}{\delta}\Big)\pi^2 F \\ &+ \beta <\phi^{2/\delta} u^2>^{(2+\delta)/2} - \frac{1}{2}\pi\Big(\frac{\delta+2}{\delta}\Big) <\phi^{2/\delta} u^2> . \end{aligned} \tag{2.35}$$

Further, by the Cauchy-Schwarz inequality

$$<\phi^{2/\delta} u^2> \;\geq\; \frac{<\phi^{(\delta+2)/\delta} u>^2}{<\phi^{2(\delta+1)/\delta}>} \;\geq\; <\phi^{(\delta+2)/\delta} u>^2 .$$

Use of this on the first term on the right in (2.35) gives

$$F_t \geq -\Big(\frac{\delta+2}{\delta}\Big)\pi^2 G^{\frac{1}{2}} + \beta G^{(2+\delta)/2} - \frac{1}{2}\pi\Big(\frac{\delta+2}{\delta}\Big)G, \tag{2.36}$$

where

$$G(t) = <\phi^{2/\delta} u^2> .$$

Let us further assume that $u_0(x)$ is chosen such that

$$\beta F(0)^{2+\delta} - \frac{1}{2}\pi\Big(\frac{\delta+2}{\delta}\Big)F^2(0) - \Big(\frac{\delta+2}{\delta}\Big)\pi^2 F(0) \;>\; 0. \tag{2.37}$$

Then by continuity there exists a t_1 such that for $0 < t < t_1$,

$$R(F) = \beta F^{2+\delta}(t) - \frac{1}{2}\pi\Big(\frac{\delta+2}{\delta}\Big)F^2(t) - \Big(\frac{\delta+2}{\delta}\Big)\pi^2 F(t) > 0.$$

For t in this range we check that $R'(F) > 0$, and so since $G^{1/2} \geq F$, $R(G^{1/2}) \geq R(F)$, for $0 \leq t < t_1$. Hence, in this t-interval we may replace G on the right of (2.36) by F to find

$$F_t \geq R(F).$$

Since F' is increasing on $(0, t_1)$ and $R(F)$ is also increasing on $(0, t_1)$, $R(F(t_1)) \neq 0$. Hence, t_1 must either be infinity or the limit of existence of the solution. Therefore, separating variables,

$$\infty > \int_{F(0)}^{\infty} \frac{dF}{R(F)} \geq \int_{F(0)}^{F(t_1)} \frac{dF}{R(F)} \geq t_1 .$$

This leads to a contradiction if t_1 were infinite, and we deduce the solution ceases to exist globally in a finite time.

The behaviour expected is blow-up of the solution in finite time, as indeed it is, see (Levine et al., 1989). While the material of this section is not strictly in keeping with the rest of the book, we include it since it is important to realize that global existence need not follow. Convection problems with non-Boussinesq or phase change effects can lead to quadratic and higher nonlinearities, and this simple example indicates care has to be exercised.

2.7 A polynomial heat source in three dimensions

So far we have concentrated on examples in one spatial dimension. We here introduce a different *energy* that allows us to produce a decay result for an arbitrary polynomial type source term, provided we suitably restrict the initial data. It also allows us to see treatment of a problem in three dimensions, for which the "embedding" inequalities are different from those in one dimension.

In this section we study the asymptotic behaviour of a solution u to the boundary-initial value problem,

$$\begin{aligned} \frac{\partial u}{\partial t} &= \Delta u + f(u), && \text{in} \quad \Omega \times (0, \infty), \\ u &= 0, && \text{on} \quad \Gamma \times (0, \infty), \\ u &= u_0(\mathbf{x}), && t = 0, \quad \mathbf{x} \in \bar{\Omega}, \end{aligned} \tag{2.38}$$

where Ω is a bounded domain in $\mathbb{R}^3$ with (sufficiently smooth) boundary Γ.

The nonlinear function f satisfies the conditions $f(0) = 0$ and

$$f(u) \leq \gamma u^k, \tag{2.39}$$

for $u \geq 0$, constant γ and k with $\gamma > 0,\ k > 1$. (For $k > 1$ it is known that unless the initial data is small enough the solution u can blow up in finite time, see e.g., (Levine, 1973), (Straughan, 1998).)

Again we employ an energy argument and the bound on the initial data is analogous to the bound required on the initial energy in section 2.3, where only conditional asymptotic stability is found. The *energy* we employ is no longer the L^2-integral of u. To dominate the arbitrarily high power of u in $f(u)$, we find it necessary to use a higher power of u as energy: this was suggested by (Bailey et al., 1984).

Henceforth, we suppose $u \geq 0$ and $k > 5/3$. As we are primarily interested in the decay behaviour for large k this causes no loss in generality.

Theorem 2.7.1 *Suppose $u(\geq 0)$ is a solution to* (2.38) *with f satisfying* (2.39). *If*

$$\|u_0^p\| < \left(\frac{(3k-5)\pi^{2/3}2^{7/3}}{3^{3/2}\gamma(k-1)^2}\right)^{3/4}, \tag{2.40}$$

where

$$4p = 3(k-1), \tag{2.41}$$

then $\|u^p\| \to 0$ at least exponentially, as $t \to \infty$, where $\|\cdot\|$ denotes the norm on $L^2(\Omega)$.

Proof.
Define E_p by

$$E_p(t) = \|u^p(t)\|^2 = \int_\Omega u^{2p}\,dx\,,$$

for p a positive constant. Then

$$\begin{aligned}\frac{dE_p}{dt} &= 2p\int_\Omega u^{2p-1}u_t\,dx\\ &= 2p\int_\Omega u^{2p-1}\left[\Delta u + f(u)\right]dx\,. \end{aligned}\tag{2.42}$$

Define I_1 and I_2 by

$$I_1 = -2p(2p-1)\int_\Omega u^{2p-2}|\nabla u|^2\,dx, \qquad I_2 = 2\gamma p\int_\Omega u^{2p+k-1}\,dx.$$

Then note that since $u = 0$ on Γ,

$$2p\int_\Omega u^{2p-1}\Delta u\,dx = I_1, \tag{2.43}$$

and from (2.39),

$$2p\int_\Omega u^{2p-1}f(u)\,dx \leq I_2. \tag{2.44}$$

Use of (2.43) and (2.44) in (2.42) shows that

$$\frac{dE_p}{dt} \leq I_1 + I_2. \tag{2.45}$$

Next, I_1 may be rearranged as

$$I_1 = -\frac{2}{p}(2p-1)\|\nabla u^p\|^2. \tag{2.46}$$

Furthermore, with the aid of Hölder's inequality,

$$I_2 \le 2p\gamma\|u^{3p}\|^{2/3}\|u^{3(k-1)/4}\|^{4/3}. \tag{2.47}$$

Since $u = 0$ on Γ,

$$\|u^{3p}\|^{2/3} \le \frac{2^{2/3}}{\pi^{2/3}3^{1/2}}\|\nabla u^p\|^2,$$

where the Sobolev embedding of $W_0^{1,2}(\Omega) \subset L^6(\Omega)$ has been employed (see e.g., (Gilbarg and Trudinger, 1977) or the Appendix).

The last inequality is substituted into (2.47) and the resulting inequality is used together with (2.46) in (2.45); then if p is chosen as in (2.41) we find

$$\frac{dE_p}{dt} \le -2\|\nabla u^p\|^2\left(\frac{2p-1}{p} - \frac{2^{2/3}p\gamma}{\pi^{2/3}3^{1/2}}E_p^{2/3}\right). \tag{2.48}$$

Since (2.40) and (2.41) hold, the coefficient of $-2\|\nabla u^p\|^2$ is positive, at least in an open neighbourhood of 0, say $[0,\eta)$. If $\eta \ne \infty$, and since (2.48) shows E_p decreases on $[0,\eta)$, a contradiction is obtained. Hence, (2.40) ensures

$$\frac{dE_p}{dt} \le -2\left(\frac{2p-1}{p} - \frac{2^{2/3}p\gamma}{\pi^{2/3}3^{1/2}}E_p^{2/3}(0)\right)\|\nabla u^p\|^2.$$

Denote the coefficient of $-2\|\nabla u^p\|^2$ in the above by A, and since $\|\nabla u^p\|^2 \ge \xi_1 E_p$, (Poincaré inequality for Ω) for $\xi_1 > 0$, we find

$$\frac{dE_p}{dt} \le -2A\xi_1 E_p,$$

and the theorem is proved.

The key to the theorem is the bound (2.40), which is a testable criterion since $u_0(x)$ is a measurable quantity. It should be observed that it depends on both $u_0(x)$ and Ω, and so it is a restriction on the size of both the initial data and the domain.

2.8 Sharp conditional stability

Even though throughout the book we are stressing the use of the energy method to derive unconditional nonlinear stability results, there are many instances where significant progress appears possible only with a conditional analysis. In sections 2.3 and 2.7 we have seen that it is possible, for solutions to the nonlinear parabolic equations considered there, to obtain conditional decay (stability) results. In their nonlinear stability analysis of the problem of Couette flow between concentric cylinders (Joseph and Hung, 1971) derived very sharp estimates on the Reynolds number, although they too found it necessary to restrict the size of the initial kinetic energy, thereby introducing the concept of conditional stability into the theory of nonlinear energy stability in fluid mechanics as a means of obtaining very sharp *quantitative* results. Since then, it has been found that very sharp nonlinear stability results may also be obtained in various problems of practical interest. However, some of this work, like (Joseph and Hung, 1971) obtains sharp conditions on the Rayleigh number at the expense of restricting the size of the amplitude of the initial perturbation. One of the open problems in these cases of energy stability theory is to find out what happens when the Rayleigh number is less than the *energy limit*, but in excess of the initial energy restriction. We should point out that the concept of conditional stability in the context of the Navier-Stokes equations was developed extensively by (Prodi, 1962) and by (Sattinger, 1970). These are fundamental papers in the mathematical theory of energy stability.

We now present a simple but relevant example where the conditions on the Rayleigh number and initial energy *must* be used together to determine the subsequent solution behaviour. This example is taken from (Galdi and Straughan, 1987).

On page 458 of (Drazin and Reid, 1981) they use a nonlinear Burgers' equation to illustrate a bifurcation example. Here we study a modification of their problem, namely,

$$\frac{\partial u}{\partial t}+\alpha u\frac{\partial u}{\partial x}=R^{-1}\frac{\partial^2 u}{\partial x^2}+u+R\|u\|^\epsilon u, \qquad x\in(0,1),\ t>0, \tag{2.49}$$

where R is a positive parameter, ϵ is a positive constant, α is 1 or 0, and $\|\cdot\|$ again denotes the norm on $L^2(0,1)$. The solution u satisfies homogeneous boundary conditions

$$u(0,t)=u(1,t)=0, \qquad \forall\, t\geq 0, \tag{2.50}$$

and initial data

$$u(x,0)=u_0(x), \qquad x\in(0,1). \tag{2.51}$$

We first establish

Theorem 2.8.1 *For $\alpha=0$ or 1, if*

$$\|u_0\|^\epsilon \quad < \quad C(R)\equiv\frac{(\pi^2-R)}{R^2}, \tag{2.52}$$

and $R<\pi^2$, then $\|u\|\to 0, \quad t\to\infty$.

Proof.

Using (2.49), (2.50), the energy equation is

$$\begin{aligned}\frac{1}{2}\frac{d}{dt}\|u\|^2 &= -R^{-1}\|u_x\|^2 + \|u\|^2 + R\|u\|^{2+\epsilon},\\ &\le -\|u\|^2\Big(\frac{\pi^2}{R} - 1 - R\|u\|^\epsilon\Big),\end{aligned}$$

where in the last step Poincaré's inequality has been employed. If now $R < \pi^2$ and (2.52) holds, the theorem follows directly from this inequality by an argument similar to that of section 2.3.

We now show that (2.52) is sharp in the sense that if it is violated there is non-uniqueness of the steady solution, while a catastrophic instability may result in the time-dependent problem; of course, it must be observed that this is the situation for which $u \equiv 0$ is judged to be *linearly* stable. In fact we study the case $\alpha = 0,\ R < \pi^2$.

Theorem 2.8.2 *Suppose $\alpha = 0$ and $R < \pi^2$.*

(A) If $\|u_0\|^\epsilon = C(R)$, then there are at least two steady solutions to (2.49), (2.50).

(B) If $\|u_0\|^\epsilon > C(R)$, then there are an infinite number of steady solutions to (2.49), (2.50).

(C) If $\|u_0\|^\epsilon > \big[(2+\epsilon)/2\big]C(R)$ and $u_0 = \sin \pi x$, then u ceases to exist in a finite time.

Proof.
Let $\bar{u} = A \sin n\pi x$. The steady form of equation (2.49) is

$$0 = R^{-1}\bar{u}_{xx} + B\bar{u},$$

where $B = 1 + R\|\bar{u}\|^\epsilon$, and so from these equations we must have

$$B = R^{-1}n^2\pi^2 \qquad \text{and} \qquad B = 1 + RA^\epsilon 2^{-\epsilon/2}.$$

Hence,

$$\frac{A^\epsilon}{2^{\epsilon/2}} \quad = \quad \|\bar{u}\|^\epsilon \quad = \quad \frac{n^2\pi^2 - R}{R^2}\,. \tag{2.53}$$

Part (A) follows immediately from (2.53) with $n = 1$, since $\bar{u} = 0$ is a second solution.

Part (B) follows directly from (2.53), which defines $A(n)$.

To establish (C) we may use a concavity argument of (Levine, 1973): his abstract equation $Pu_t = -Au + \mathcal{F}(u)$, covers (2.49) with $\alpha = 0$, provided

$$Au = -R^{-1}u_{xx} - u \qquad \text{and} \qquad \mathcal{F}(u) = Ru\|u\|^\epsilon.$$

The potential $\mathcal{G}(u)$ defined by (Levine, 1973) is here $\mathcal{G}(u) = (R/[2+\epsilon])\|u\|^{2+\epsilon}$. The conditions of theorem 1 of (Levine, 1973), which establishes nonexistence, are satisfied provided

$$(u, Au) \ge 0 \tag{2.54}$$

and

$$\mathcal{G}(u_0) > \frac{1}{2}(u_0, Au_0), \tag{2.55}$$

the brackets denoting a suitable inner product; the inner product on $L^2(0,1)$ for the present example. Inequality (2.54) is equivalent to $R^{-1}\|u_x\|^2 \geq \|u\|^2$; this is certainly true if $R < \pi^2$. Recalling that $u_0 = \sin \pi x$, inequality (2.55) is equivalent to

$$\|u_0\|^\epsilon \quad > \quad \Big(\frac{2+\epsilon}{2}\Big)C(R).$$

The theorem is thus proved.

2.9 Interaction diffusion systems

2.9.1 A general Segel-Jackson system

Up to this point we have considered examples for a single dependent variable satisfying one equation. As a prelude to fluid dynamical stability where the partial differential equations form a coupled system we here investigate how the techniques outlined in sections 2.2 and 2.3 may be adapted to a system of interaction-diffusion equations of interest in many chemical and biological processes.

Consider two species of concentration C_1 and C_2 whose interaction and diffusion in a one-dimensional spatial region are governed by the equations,

$$\begin{aligned}\frac{\partial C_1}{\partial t} &= D_1\frac{\partial^2 C_1}{\partial x^2} + R_1(C_1, C_2),\\ \frac{\partial C_2}{\partial t} &= D_2\frac{\partial^2 C_2}{\partial x^2} + R_2(C_1, C_2).\end{aligned} \tag{2.56}$$

In (2.56), D_i, $\quad i = 1, 2$, are positive, constant diffusion coefficients, R_i are interaction terms, in general, nonlinear functions of C_1 and C_2.

Such a system may, for example, govern the evolutionary behaviour of an organic pollutant in water, in which case C_1 and C_2 might represent the pollutant and oxygen concentrations, respectively.

We suppose (2.56) are defined on $x \in \mathbb{R}$, $t > 0$. Differently from the equations studied earlier, (2.56) is a *system* linking C_1 and C_2 together. Our aim is to investigate using linear theory the stability of a *constant*, but in general *non-zero* solution to the time independent version of (2.56). This presupposes that the functions R_i are such that there exists a *constant* equilibrium solution $\bar{C}_1$, $\bar{C}_2$ to

$$R_1(\bar{C}_1, \bar{C}_2) = 0, \qquad R_2(\bar{C}_1, \bar{C}_2) = 0. \tag{2.57}$$

To study the instability of the solution $(\bar{C}_1, \bar{C}_2)$ we let $c_1(x,t)$, $c_2(x,t)$ be perturbations so that in (2.56) we set

$$C_1 = \bar{C}_1 + c_1, \qquad C_2 = \bar{C}_2 + c_2. \tag{2.58}$$

Next expand R_i in a Taylor series about $(\bar{C}_1, \bar{C}_2)$

$$\begin{aligned} R_i(C_1, C_2) =& R_i(\bar{C}_1, \bar{C}_2) + \frac{\partial R_i}{\partial C_m}\Big|_{\bar{C}_j} c_m \\ &+ \frac{1}{2}\frac{\partial^2 R_i}{\partial C_m \partial C_n}\Big|_{\bar{C}_j} c_m c_n + O(c^3), \end{aligned} \tag{2.59}$$

where a repeated index denotes summation over 1 and 2. For linear stability, $|c_i| << 1$, so we neglect terms like $c_m c_n$ and higher order.

Define the (constant) matrix (a_{ij}) by $a_{im} = \partial R_i/\partial C_m\big|_{\bar{C}_j}$, then, since $R_i(\bar{C}_j) = 0$, (2.59) is, in the linear approximation

$$R_i(C_1, C_2) = a_{im} c_m. \tag{2.60}$$

Equations (2.56) for c_1, c_2 become

$$\begin{aligned} \frac{\partial c_1}{\partial t} &= D_1 \frac{\partial^2 c_1}{\partial x^2} + a_{11} c_1 + a_{12} c_2, \\ \frac{\partial c_2}{\partial t} &= D_2 \frac{\partial^2 c_2}{\partial x^2} + a_{21} c_1 + a_{22} c_2. \end{aligned} \tag{2.61}$$

(We have already studied in section 2.2 the case for $a_{12} = a_{21} = 0$, which would give linear instability always for a_{11}, $a_{22} > 0$.)

Now use a normal mode analysis, i.e., write

$$c_1 = \gamma_1 e^{(\sigma t + ikx)}, \qquad c_2 = \gamma_2 e^{(\sigma t + ikx)}, \tag{2.62}$$

with γ_i, $k \in \mathbb{R}$, and σ possibly complex. (Again we are looking at one term in the Fourier series for c_1, c_2, just as in the analysis for the diffusion equation.)

Putting (2.62) into (2.61) yields

$$\begin{aligned} \sigma\gamma_1 &= -D_1 k^2 \gamma_1 + a_{11}\gamma_1 + a_{12}\gamma_2, \\ \sigma\gamma_2 &= -D_2 k^2 \gamma_2 + a_{21}\gamma_1 + a_{22}\gamma_2, \end{aligned} \tag{2.63}$$

which is a pair of simultaneous equations in γ_1, γ_2 that may be rewritten as

$$\begin{pmatrix} a_{11} - D_1 k^2 - \sigma & a_{12} \\ a_{21} & a_{22} - D_2 k^2 - \sigma \end{pmatrix} \begin{pmatrix} \gamma_1 \\ \gamma_2 \end{pmatrix} = \mathbf{0}.$$

For a non-zero solution (γ_1, γ_2) to this system, we require

$$\begin{vmatrix} a_{11} - D_1 k^2 - \sigma & a_{12} \\ a_{21} & a_{22} - D_2 k^2 - \sigma \end{vmatrix} = 0, \tag{2.64}$$

and, therefore, letting

$$\hat{a}_{11} = a_{11} - D_1 k^2, \qquad \hat{a}_{22} = a_{22} - D_2 k^2,$$

(2.64) becomes

$$\sigma^2 - \sigma(\hat{a}_{11} + \hat{a}_{22}) + \hat{a}_{11}\hat{a}_{22} - a_{12}a_{21} = 0.$$

Hence,

$$\sigma = \frac{1}{2}(\hat{a}_{11} + \hat{a}_{22}) \pm \frac{1}{2}\left[(\hat{a}_{11} + \hat{a}_{22})^2 + 4(a_{12}a_{21} - \hat{a}_{11}\hat{a}_{22})\right]^{1/2}. \qquad (2.65)$$

2.9.2 *Exchange of Stabilities.*

In general

$$\sigma = \sigma_1 + i\sigma_2, \qquad \sigma_1, \sigma_2 \in \mathbb{R}.$$

If $\sigma_2 \neq 0 \Rightarrow \sigma_1 < 0$ it then is said that the Principle of *Exchange of Stabilities* holds.

Of course, if $\sigma_2 = 0$ always, i.e., $\sigma \in \mathbb{R}$, then exchange of stabilities always holds. It is useful to know when $\sigma \in \mathbb{R}$ (if at all).

If $\sigma_2 \neq 0$ and exchange of stabilities does not hold, then any instability in that case is called an *overstable oscillation.*

If $\sigma \in \mathbb{R}$, overstability can be neglected and the analysis is usually much simpler.

Rewriting (2.65) as

$$\sigma = \frac{1}{2}(\hat{a}_{11} + \hat{a}_{22}) \pm \frac{1}{2}\left[(\hat{a}_{11} - \hat{a}_{22})^2 + 4a_{12}a_{21}\right]^{1/2}, \qquad (2.66)$$

an important case of exchange of stabilities is easily seen to be when $a_{12} = a_{21}$, i.e., when the system is symmetric. (More generally,

$$\sigma \in \mathbb{R} \qquad \text{if} \qquad \operatorname{sgn} a_{12} = \operatorname{sgn} a_{21} .$$

This is equivalent, however, to a_{ij} symmetric since if $a_{12} = \alpha a_{21}$ ($\alpha > 0$) then we multiply the second of (2.61) by α and the resulting system has a_{ij} symmetric.)

If $a_{12}a_{21} < 0$, then there is a possibility that $\sigma_2 \neq 0$, i.e., there is a chance of overstability.

Suppose now a_{ij} symmetric. Then $\sigma \in \mathbb{R}$ and as $\sigma > 0$ implies instability we see, from (2.66), with the positive sign (most destabilizing),

$$\sigma = \frac{1}{2}(a_{11} - k^2D_1 + a_{22} - k^2D_2) + \frac{1}{2}\left[(\hat{a}_{11} - \hat{a}_{22})^2 + 4a_{12}a_{21}\right]^{1/2}. \quad (2.67)$$

If $a_{11}, a_{22} > 0$, there is always instability since as $k^2 \to 0$, σ becomes positive (c.f. section 2.2, the diffusion equation for $x \in \mathbb{R}$).

It follows that a disturbance of infinite (spatial) wavelength is always destabilizing (i.e., in the limit as $k^2 \to 0$).

When there was only one equation, e.g.,

$$\frac{\partial c_1}{\partial t} = D_1\frac{\partial^2 c_1}{\partial x^2} + a_{11}c_1, \qquad x \in \mathbb{R}, t > 0,$$

then $a_{11} < 0$ would always lead to stability of the solution $c_1 \equiv 0$. Is an analogous result true for the system (2.61)? No. It is sufficient to give one example where the $a_{12}a_{21}$ term can overcome the negativity of a_{11}, a_{22} and lead to an instability. Suppose next, therefore, that

$$D_1 = D_2 = D\,(> 0), \qquad a_{11}, a_{22} < 0,$$

where the conditions on the a's are both stabilizing in the single equation case, and

$$a_{12}a_{21} > a_{11}a_{22}\,(> 0). \tag{2.68}$$

Then, (2.67) reduces to

$$\sigma = \frac{1}{2}(a_{11} + a_{22}) - k^2 D + \frac{1}{2}\left[a_{11}^2 + a_{22}^2 + 4a_{12}a_{21} - 2a_{11}a_{22}\right]^{1/2}. \tag{2.69}$$

Using (2.68) we may deduce from (2.69),

$$\begin{aligned} \sigma &> \frac{1}{2}(a_{11} + a_{22}) - k^2 D + \frac{1}{2}|a_{11} + a_{22}| + \epsilon_1, \\ &\geq \epsilon_1 - k^2 D, \end{aligned} \tag{2.70}$$

for a positive ϵ_1. As $k^2 \to 0$, σ becomes positive, so infinitely long wavelength disturbances are again unstable.

It should be observed that this is in complete contrast to the single equation case for which $a_{11} < 0$ or $a_{22} < 0$, i.e., $a_{12} = a_{21} = 0$, is always stable.

A deeper analysis of (2.66) is possible, but we do not pursue this here. Details may be found in (Segel and Jackson, 1972).

2.9.3 Conditional Nonlinear Stability in a Two-Constituent System.

Consider now two species of concentrations C_1 and C_2 whose interaction and diffusion in a three-dimensional region are governed by the equations (c.f. (2.56)),

$$\begin{aligned} \frac{\partial C_1}{\partial t} &= D_1 \Delta C_1 + R_1(C_1, C_2), \\ \frac{\partial C_2}{\partial t} &= D_2 \Delta C_2 + R_2(C_1, C_2). \end{aligned} \tag{2.71}$$

In the above, D_i, $i = 1, 2$, are again positive diffusion coefficients and R_i are interaction terms.

We now take (2.71) to be defined on a bounded domain $\Omega \subset \mathbb{R}^3$ for each instant of time and on the boundary assume

$$C_i = g_i, \quad \text{on} \quad \Gamma_1 \times [0, \infty), \qquad \frac{\partial C_i}{\partial n} = h_i, \quad \text{on} \quad \Gamma_2 \times [0, \infty), \tag{2.72}$$

where $\Gamma_1 \cup \Gamma_2 = \Gamma$, $\Gamma_1 \neq \emptyset$, and g_i and h_i are prescribed. In addition, the initial values of C_1 and C_2 are given.

We wish to generalize the method of section 2.3 to study the nonlinear stability of the steady solution to (2.71), i.e., the stability of $\bar{C}_i$ where

$$D_1 \Delta \bar{C}_1 + R_1(\bar{C}_j) = 0, \qquad D_2 \Delta \bar{C}_2 + R_2(\bar{C}_j) = 0, \tag{2.73}$$

with $\bar{C}_j$ satisfying the boundary conditions (2.72).

Our analysis is appropriate only to quadratic R_i. We write $C_i = \bar{C}_i + c_i$ where c_i is a perturbation to $\bar{C}_i$ and then

$$R_i(C_j) = R_i(\bar{C}_j) + a_{im} c_m + b_{imn} c_m c_n \,, \tag{2.74}$$

where $a_{im} = \partial R_i / \partial C_m \big|_{\bar{C}_j}$, $2b_{imn} = \partial^2 R_i / \partial C_m \partial C_n \big|_{\bar{C}_j}$, b_{imn} being constants and a_{ij} functions of $\bar{C}_k$.

From (2.71), (2.73), (2.74), c_i satisfy the equations

$$\frac{\partial c_\alpha}{\partial t} = a_{\alpha m} c_m + b_{\alpha mn} c_m c_n + D_\alpha \Delta c_\alpha \,, \tag{2.75}$$

where α signifies no summation although the summation convention applies to lower case Roman subscripts. Furthermore, c_i satisfy (2.72) with $g_i \equiv h_i \equiv 0$.

Let L be a length scale and introduce the non-dimensional variables

$$t^* = tM/L^2; \qquad M = D_1 + D_2; \qquad \mathbf{x} = L\mathbf{x}^*;$$
$$\lambda_i = M/2D_i\,; \qquad A_{\alpha j} = a_{\alpha j} L^2 / D_\alpha; \qquad B_{\alpha mn} = b_{\alpha mn} L^2 / D_\alpha \,.$$

The non-dimensional version of (2.75) is (omitting all stars)

$$2\lambda_\alpha \frac{\partial c_\alpha}{\partial t} = A_{\alpha m} c_m + B_{\alpha mn} c_m c_n + \Delta c_\alpha \,. \tag{2.76}$$

We multiply (2.76) by c_α, integrate over Ω, and add the two equations to find

$$\frac{dE}{dt} = I - D + F(c), \tag{2.77}$$

where

$$E(t) = \lambda_1 \|c_1\|^2 + \lambda_2 \|c_2\|^2, \tag{2.78}$$

$$I = \int_\Omega A_{im} c_i c_m \, dx, \qquad D = \int_\Omega \frac{\partial c_i}{\partial x_q} \frac{\partial c_i}{\partial x_q} \, dx, \tag{2.79}$$

in which the summation on i and m is over 1 and 2 whereas q sums from 1 to 3. Moreover,

$$F(c) = \int_\Omega B_{imn} c_i c_m c_n \, dx. \tag{2.80}$$

We consider only the case A_{im} symmetric. For this case the idea of the proof leading to (2.67) shows the time factor σ is real and so the equations governing the *linear stability boundary* are

$$0 = A_{im}c_m + \Delta c_i. \tag{2.81}$$

These equations determine a set of relations between the a_{ij} and D_i.

To examine energy stability we return to (2.77) and derive

$$\frac{dE}{dt} \leq -D\Big(1 - \max\frac{I}{D}\Big) + F(c), \tag{2.82}$$

where the maximum is over the space of admissible solutions. We leave for the moment the determination of this maximum and suppose

$$k = -\max\frac{I}{D} + 1 > 0.$$

Denote by B the maximum of the constants B_{imn}. The nonlinear term is estimated using the Cauchy-Schwarz inequality

$$F(c) \leq 4\lambda B E^{1/2}\Big[\Big(\int_\Omega c_1^4\,dx\Big)^{1/2} + \Big(\int_\Omega c_2^4\,dx\Big)^{1/2}\Big], \tag{2.83}$$

where $\lambda = \max\{\lambda_1^{-1}, \lambda_2^{-1}\}$. Again, the Sobolev inequality $\|u^2\| \leq \mu\|\nabla u\|^2$ holds, although μ now depends on Ω (see the Appendix). Using this in (2.83) leads to

$$F(c) \leq 4\lambda\mu B E^{1/2} D. \tag{2.84}$$

Therefore,

$$\frac{dE}{dt} \leq -kD + 4\lambda\mu B E^{1/2} D. \tag{2.85}$$

From this point the argument follows that after inequality (2.20), to show that if

$$E(0) < \frac{k^2}{16(\lambda\mu B)^2}, \tag{2.86}$$

then $E \to 0$, $t \to \infty$.

Thus, if $k > 0$ we have established a conditional stability result. It remains to investigate the condition $k > 0$.

We note that if the entries of A_{ij} are bounded, with $\Lambda = \max(I/D)$, the Euler-Lagrange equations for the maximum are

$$A_{im}c_m + \Lambda\Delta c_i = 0. \tag{2.87}$$

The limit of energy stability is $\Lambda = 1$; i.e., the boundary for $k > 0$, and for this case the Euler-Lagrange equations (2.87) of energy theory are exactly the same as those of the linear theory (2.81). Hence, the criteria for stability obtained by both methods are the same and so we conclude that when A_{ij} are symmetric and (2.86) holds the energy stability and linear instability boundaries coalesce and sub-critical instabilities do not occur. This point is taken up in further detail in section 4.3.

2.9.4 Specific Segel-Jackson systems

In the present section we use two examples of (Segel and Jackson, 1972) to illustrate some of the methods which generalize to the fluid dynamical situations later.

The first is a predator-prey model of (Segel and Jackson, 1972),

$$\begin{aligned}\frac{\partial v}{\partial t} &= v + \kappa v^2 - aev + \delta^2 \Delta v,\\ \frac{\partial e}{\partial t} &= ev - e^2 + \Delta e,\end{aligned} \tag{2.88}$$

where κ, a, δ^2 are positive constants and where v denotes the population of a victim, while e denotes the population of its exploiter.

The second model of (Segel and Jackson, 1972) is:

$$\begin{aligned}\frac{\partial C_1}{\partial t} &= k_1 A + k_2 C_1^2 C_2 - k_3 B C_1 - k_4 C_1 + D_1 \Delta C_1,\\ \frac{\partial C_2}{\partial t} &= k_3 B C_1 - k_2 C_1^2 C_2 + D_2 \Delta C_2,\end{aligned} \tag{2.89}$$

where $A, B, k_1, ..., k_4, D_1, D_2$ are constants and C_1, C_2 are concentrations (of a species or of a chemical, for example).

(Segel and Jackson, 1972) study by linear theory a diffusive instability for (2.88), (2.89). They are interested in the stability of *constant equilibrium solutions.*

System (2.88) has a constant equilibrium solution $e = v = L$ with $L = 1/(a-\kappa)$, provided $a > \kappa$. A constant equilibrium solution to system (2.89) is $\bar{C}_1 = (k_1/k_4)A$, $\bar{C}_2 = (Bk_3k_4/Ak_1k_2)$. To study the stability of these solutions we introduce perturbations and so for (2.88) seek a solution of the form $L + v(\mathbf{x}, t)$, $L + e(\mathbf{x}, t)$ while for (2.89) $\bar{C}_1 + \phi(\mathbf{x}, t)$, $\bar{C}_2 + \psi(\mathbf{x}, t)$. The equations governing the perturbations (e, v) are:

$$\begin{aligned}\frac{\partial v}{\partial t} &= (1 + 2\kappa L - aL)v - aLe + \delta^2 \Delta v + \kappa v^2 - aev,\\ \frac{\partial e}{\partial t} &= Lv - Le + \Delta e + ev - e^2.\end{aligned} \tag{2.90}$$

We choose to study (2.90) on a bounded domain, Ω, of $\mathbb{R}^2$ or $\mathbb{R}^3$, with on the boundary of Ω, Γ,

$$v(\mathbf{x}, t) = e(\mathbf{x}, t) = 0. \tag{2.91}$$

The perturbations (ϕ, ψ) to (2.89) satisfy the system:

$$\begin{aligned}\frac{\partial \phi}{\partial t} &= a_{11}\phi + a_{22}\psi + D_1 \Delta \phi + \alpha \phi^2 + 2\beta \phi \psi + \gamma \phi^2 \psi,\\ \frac{\partial \psi}{\partial t} &= -a_{21}\phi - a_{22}\psi + D_2 \Delta \psi - \alpha \phi^2 - 2\beta \phi \psi - \gamma \phi^2 \psi,\end{aligned} \tag{2.92}$$

where the coefficients are given by: $a_{11} = Bk_3 - k_4$, $a_{22} = k_2 k_1^2 A^2/k_4^2$, $a_{21} = Bk_3$, $\alpha = Bk_3k_4/k_1 A$, $\beta = k_1 k_2 A/k_4$, $\gamma = k_2$. Since we are using

(2.90), (2.92) solely to illustrate approaches to conditional stability results via an energy method, we choose *purely for simplicity* to simplify (2.92) and study instead

$$\frac{\partial \phi}{\partial t} = a_{11}\phi + a_{22}\psi + D_1\Delta\phi,$$
$$\frac{\partial \psi}{\partial t} = -a_{21}\phi - a_{22}\psi + D_2\Delta\psi - \gamma\phi^2\psi, \qquad (2.93)$$

where we shall take the coefficients $a_{11}, a_{21}, a_{22}, D_1, D_2, \gamma$ to be positive.

Equations (2.93) are studied on the bounded domain $\Omega \subset \mathbb{R}^3$, together with the boundary conditions

$$\phi(\mathbf{x}, t) = \psi(\mathbf{x}, t) = 0, \qquad \mathbf{x} \in \Gamma.$$

The reason for here examining (2.90), (2.93), is that in convection problems, if a nonlinear density-temperature relationship is employed in the buoyancy term, then (2.90), (2.93), represent mathematically similar, albeit simpler, prototypes of the nonlinear systems arising in fluid dynamics.

It is interesting to observe that the linear operator in (2.90), is not symmetric. However, the skew-symmetric part is comprised of "cross" interaction terms not containing derivatives, i.e. the $-aLe$ term in $(2.90)_1$, and the Lv term in $(2.90)_2$. These terms will be stabilizing. We do not dwell on this point here and proceed to show how a nonlinear, conditional energy stability result may be found for a solution to (2.90).

Form individual energy identities by multiplying $(2.90)_1$, by v, $(2.90)_2$, by e, integrate over Ω and use the boundary conditions to find:

$$\begin{aligned}\frac{1}{2}\frac{d}{dt}\|v\|^2 =& \big[1 + (2\kappa - a)L\big]\|v\|^2 - aL < ev > \\ &- \delta^2\|\nabla v\|^2 + \kappa < v^3 > - a < ev^2 >, \\ \frac{1}{2}\frac{d}{dt}\|e\|^2 =& L < ev > - L\|e\|^2 - \|\nabla e\|^2 + < e^2 v > - < e^3 >,\end{aligned}$$

where $\|\cdot\|$ and $< \cdot >$ denote the norm on $L^2(\Omega)$ and integration over Ω, respectively. For a positive number λ (a coupling parameter) which we are free to select in such a way as to obtain as sharp a stability result as possible, we define an "energy", $E(t)$, by

$$E(t) = \frac{1}{2}\|v\|^2 + \frac{1}{2}\lambda\|e\|^2.$$

Then, the energy equation for E is:

$$\frac{dE}{dt} = I - D + \mathcal{N}, \qquad (2.94)$$

where the quadratic terms I and D are given by

$$\begin{aligned} I =& \frac{\kappa}{a-\kappa}\|v\|^2 + L(\lambda - a) < ev >, \\ D =& \lambda L\|e\|^2 + \delta^2\|\nabla v\|^2 + \lambda\|\nabla e\|^2, \end{aligned} \tag{2.95}$$

and where the cubic (nonlinear) term $\mathcal{N}$ is:

$$\mathcal{N} = \kappa < v^3 > -a < ev^2 > +\lambda < e^2 v > -\lambda < e^3 > . \tag{2.96}$$

To establish a nonlinear stability result we must investigate the maximum problem $\Lambda = \max I/D$ where the maximum is over a suitable space of functions. We require $\Lambda < 1$. In fact $\Lambda = 1$ is the energy stability threshold. Suppose now $\Lambda < 1$ and put $b = 1 - \Lambda\, (> 0)$. Then, from (2.94)

$$\frac{dE}{dt} \le -bD + \mathcal{N}. \tag{2.97}$$

To proceed beyond (2.97) we need the Sobolev inequality

$$< s^4 >^{1/2} \le c_1\|\nabla s\|^2, \tag{2.98}$$

for functions s which vanish on Γ. We then estimate the terms in $\mathcal{N}$ by employing the Cauchy-Schwarz inequality and (2.98) as:

$$\begin{aligned} \mathcal{N} &= \kappa < v^3 > -a < ev^2 > +\lambda < e^2 v > -\lambda < e^3 > \\ &\le \kappa\|v\| < v^4 >^{1/2} + a\|e\| < v^4 >^{1/2} + \lambda\|v\| < e^4 >^{1/2} + \lambda\|e\| < e^4 >^{1/2} \\ &\le c_1\big(\kappa\|v\|\|\nabla v\|^2 + a\|e\|\|\nabla v\|^2 + \lambda\|v\|\|\nabla e\|^2 + \lambda\|e\|\|\nabla e\|^2\big). \end{aligned}$$

Then recalling the definition of E and D, we see that

$$\mathcal{N} \le \sqrt{2}\Big[\frac{1}{\delta^2}\Big(\kappa + \frac{a}{\sqrt{\lambda}}\Big) + 1 + \frac{1}{\sqrt{\lambda}}\Big]E^{1/2}D. \tag{2.99}$$

Hence, inserting (2.99) in (2.97) we obtain

$$\frac{dE}{dt} \le -D(b - AE^{1/2}), \tag{2.100}$$

where the constant A is defined by $A = \sqrt{2}[\delta^{-2}(\kappa + a\lambda^{-1/2}) + 1 + \lambda^{-1/2}]$. Provided $E^{1/2}(0) < b/A$, it is now straightforward to show $E(t) \to 0$ as $t \to \infty$. Thus, provided $E^{1/2}(0) < b/A$ and $\Lambda < 1$ the constant equilibrium solution to (2.88) is nonlinearly stable. The stability is *conditional* since it is for a restricted class of initial perturbations.

The equivalent analysis for (2.93) commences by multiplying $(2.93)_1$ by ϕ and integrating over Ω, and by multiplying $(2.93)_2$ by ψ and again integrating over Ω. Using the boundary conditions we find

$$\begin{aligned} &\frac{d}{dt}\frac{1}{2}\|\phi\|^2 = a_{11}\|\phi\|^2 + a_{22} < \psi\phi > -D_1\|\nabla\phi\|^2, \\ &\frac{d}{dt}\frac{1}{2}\|\psi\|^2 = -a_{21} < \psi\phi > -a_{22}\|\psi\|^2 - D_2\|\nabla\psi\|^2 - \gamma < \phi^2\psi^2 > . \end{aligned}$$

We may again define

$$E(t) = \frac{1}{2}\|\phi\|^2 + \frac{1}{2}\lambda\|\psi\|^2$$

and derive

$$\frac{dE}{dt} = I - D - \gamma\lambda < \phi^2\psi^2 >, \tag{2.101}$$

where now

$$\begin{aligned} I =& a_{11}\|\phi\|^2 + (a_{22} - \lambda a_{21}) < \phi\psi >, \\ D =& \lambda a_{22}\|\psi\|^2 + D_1\|\nabla\phi\|^2 + \lambda D_2\|\nabla\psi\|^2. \end{aligned}$$

The difficulty is now to dominate the nonlinear term in (2.101) by a term of the form $E^\alpha D$. I don't know how to do this. Instead, let us form extra energy identities for $\zeta = \phi^2$ and $\eta = \psi^2$. We may show that

$$\begin{aligned} \frac{d}{dt}\frac{1}{4}\|\zeta\|^2 =& a_{11}\|\zeta\|^2 + a_{22} < \psi\phi^3 > - \frac{3}{4}D_1\|\nabla\zeta\|^2, \\ \frac{d}{dt}\frac{1}{4}\|\eta\|^2 =& - a_{21} < \psi^3\phi > - a_{22}\|\eta\|^2 \\ & - \frac{3}{4}D_2\|\nabla\eta\|^2 - \gamma < \phi^2\psi^4 > . \end{aligned} \tag{2.102}$$

Define now, for $\lambda_1, \lambda_2\,(> 0)$ at our disposal,

$$E_1(t) = \frac{1}{4}\lambda_1\|\zeta\|^2 + \frac{1}{4}\lambda_2\|\eta\|^2.$$

Then define a generalized energy, $\mathcal{E}(t)$, by

$$\mathcal{E}(t) = E(t) + E_1(t). \tag{2.103}$$

The energy maximum problem arises purely from the quadratic terms in (2.101), namely $\Lambda = \max I/D$. We again require $\Lambda < 1$. Suppose $a \stackrel{\text{def}}{=} 1 - \Lambda\,(> 0)$. Then we derive:

$$\begin{aligned} \frac{d}{dt}\mathcal{E} \leq & - aD - a_{22}\lambda_2\|\eta\|^2 - \frac{3}{4}\lambda_2 D_2\|\nabla\eta\|^2 \\ & - \frac{3}{4}\lambda_1 D_1\|\nabla\zeta\|^2 + a_{11}\lambda_1\|\zeta\|^2 - \gamma\lambda < \zeta\eta > \\ & - \lambda_2 a_{21} < \phi\psi\eta > + \lambda_1 a_{22} < \psi\phi\zeta > - \gamma\lambda_2 < \zeta\eta^2 > . \end{aligned} \tag{2.104}$$

Assume now $a_{11} < 3D_1\lambda^*/4$, where λ^* is the constant in Poincaré's inequality $\lambda^*\|s\|^2 \leq \|\nabla s\|^2$, for functions s which vanish on Γ.

The parameter λ is used to obtain the best stability threshold. In contrast, λ_1 and λ_2 are used to dominate terms in (2.104) we wish to remove.

The last four terms on the right of (2.104) are estimated as follows:

$$-\lambda\gamma < \zeta\eta > \leq \frac{\lambda\gamma}{2\alpha}\|\zeta\|^2 + \frac{1}{2}\lambda\gamma\alpha\|\eta\|^2, \tag{2.105}$$

for $\alpha(>0)$ to be chosen.

$$\begin{aligned}-\gamma\lambda_2 < \zeta\eta^2 > &\le \gamma\lambda_2\|\zeta\| < \eta^4 >^{1/2}\\ &\le c_1\gamma\lambda_2\|\zeta\|\|\nabla\eta\|^2\\ &\le c_1\gamma\lambda_2\frac{2}{\sqrt{\lambda_1}}E_1^{1/2}\|\nabla\eta\|^2. \end{aligned}\tag{2.106}$$

$$\begin{aligned}-\lambda_2 a_{21} < \phi\psi\eta > &\le \lambda_2 a_{21}\|\eta\| < \phi^2\psi^2 >^{1/2}\\ &\le \lambda_2 a_{21}\|\eta\| < \phi^4 >^{1/4} < \psi^4 >^{1/4}\\ &\le c_1\lambda_2 a_{21}\|\eta\|\|\nabla\phi\|\|\nabla\psi\|. \end{aligned}\tag{2.107}$$

$$\begin{aligned}\lambda_1 a_{22} < \phi\zeta\psi > &\le \lambda_1 a_{22}\|\zeta\| < \phi^2\psi^2 >^{1/2}\\ &\le c_1\lambda_1 a_{22}\|\zeta\|\|\nabla\phi\|\|\nabla\psi\|. \end{aligned}\tag{2.108}$$

We now pick $\alpha = \lambda\gamma/\lambda_1 D_1^*\lambda^*$, and choose λ_1, λ_2 such that $(1/2\lambda_1 D_1^*)(\lambda\gamma/\lambda^*)^2 = (3\lambda_2/8)D_2$, where $D_1^* = (3/4)D_1 - (a_{11}/\lambda^*)$.

Next, define $\mathcal{D}$ by

$$\mathcal{D} = aD + a_{22}\lambda_2\|\eta\|^2 + \frac{3\lambda_2 D_2}{8}\|\nabla\eta\|^2 + \frac{\lambda_1 D_1^*}{2}\|\nabla\zeta\|^2,$$

and then use of (2.105) - (2.108) in (2.104) leads to:

$$\begin{aligned}\frac{d}{dt}\mathcal{E} \le &-\mathcal{D} + \frac{2c_1\gamma\lambda_2}{\sqrt{\lambda_1}}\mathcal{E}^{1/2}\|\nabla\eta\|^2\\ &+ c_1(\lambda_2 a_{21}\|\eta\| + \lambda_1 a_{22}\|\zeta\|)\|\nabla\phi\|\|\nabla\psi\|. \end{aligned}\tag{2.109}$$

We finally set $c = \min\{\lambda_1 D_1^*/2, 3\lambda_2 D_2/8\}$ and then the last term in (2.109) may be bounded by $(c_1/c)(a_{21}\sqrt{\lambda_2} + a_{22}\sqrt{\lambda_1})\mathcal{E}^{1/2}\mathcal{D}$. Hence, from (2.109) we deduce

$$\frac{d}{dt}\mathcal{E} \le -\mathcal{D}(1 - A\mathcal{E}^{1/2}),\tag{2.110}$$

where $A = (c_1/c)(a_{21}\sqrt{\lambda_2}+a_{22}\sqrt{\lambda_1})+(8c_1\gamma/3D_2\sqrt{\lambda_1})$. Provided $\mathcal{E}^{1/2}(0) < 1/A$ we may now deduce nonlinear, conditional energy stability from (2.110).

Again, the point of the foregoing examples is to show that (2.90) and (2.93) apparently need different generalized energies. We should point out that the full system of (Segel and Jackson, 1972), (2.92), may also be handled by employing an energy of form (2.103).

2.9.5 Numerical results

To complete this chapter we primarily present some numerically computed steady states to systems of two interaction diffusion equations with convective terms. These are generalisations of steady states obtained for single equations like (2.22) and clearly demonstrate the boundary layers which may form. In fact, the systems we describe are all covered in (Straughan, 1998), chapter 2, where further details of the analysis and numerical techniques may be found. (Straughan, 1998) studies eight systems starting with the form

System 2.9.1

$$\begin{aligned} &\frac{\partial u}{\partial t}+u\frac{\partial u}{\partial x}-\frac{1}{R}\frac{\partial^2 u}{\partial x^2}=\alpha u+\gamma v, && x\in(0,1),t>0,\\ &\frac{\partial v}{\partial t}+v\frac{\partial v}{\partial x}-\frac{1}{R}\frac{\partial^2 v}{\partial x^2}=\delta u+\beta v, && x\in(0,1),t>0,\\ &u(0,t)=u(1,t)=0, && v(0,t)=v(1,t)=0,\\ &u(x,0)=u_0(x)=0, && v(x,0)=v_0(x). \end{aligned} \tag{2.111}$$

The other seven systems studied in (Straughan, 1998) arise by replacing the u and v terms on the right of (2.111) by combinations of u^2 or v^2. We include steady states for

System 2.9.2

$$\begin{aligned} &\frac{\partial u}{\partial t}+u\frac{\partial u}{\partial x}-\frac{1}{R}\frac{\partial^2 u}{\partial x^2}=\alpha u^2+\gamma v, && x\in(0,1),t>0,\\ &\frac{\partial v}{\partial t}+v\frac{\partial v}{\partial x}-\frac{1}{R}\frac{\partial^2 v}{\partial x^2}=\delta u+\beta v, && x\in(0,1),t>0,\\ &u(0,t)=u(1,t)=0, && v(0,t)=v(1,t)=0,\\ &u(x,0)=u_0(x)=0, && v(x,0)=v_0(x). \end{aligned} \tag{2.112}$$

System 2.9.3

$$\begin{aligned} &\frac{\partial u}{\partial t}+u\frac{\partial u}{\partial x}-\frac{1}{R}\frac{\partial^2 u}{\partial x^2}=\alpha u^2+\gamma v^2, && x\in(0,1),t>0,\\ &\frac{\partial v}{\partial t}+v\frac{\partial v}{\partial x}-\frac{1}{R}\frac{\partial^2 v}{\partial x^2}=\delta u+\beta v, && x\in(0,1),t>0,\\ &u(0,t)=u(1,t)=0, && v(0,t)=v(1,t)=0,\\ &u(x,0)=u_0(x)=0, && v(x,0)=v_0(x). \end{aligned} \tag{2.113}$$

System 2.9.4

$$\begin{aligned} &\frac{\partial u}{\partial t}+u\frac{\partial u}{\partial x}-\frac{1}{R}\frac{\partial^2 u}{\partial x^2}=\alpha u^2+\gamma v^2, && x\in(0,1),t>0,\\ &\frac{\partial v}{\partial t}+v\frac{\partial v}{\partial x}-\frac{1}{R}\frac{\partial^2 v}{\partial x^2}=\delta u^2+\beta v^2, && x\in(0,1),t>0,\\ &u(0,t)=u(1,t)=0, && v(0,t)=v(1,t)=0,\\ &u(x,0)=u_0(x)=0, && v(x,0)=v_0(x). \end{aligned} \tag{2.114}$$

In the figures shown the initial data are all large enough that the solution approaches a non-zero steady state. The steady states chosen show

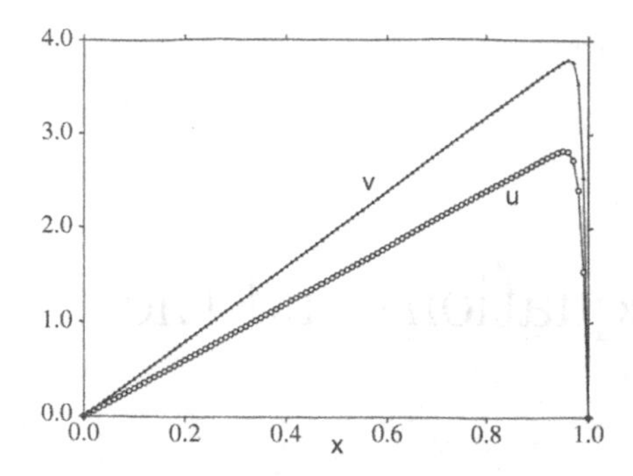

Figure 2.1. Steady solution to (2.111). $R = 40$, $\alpha = 1$, $\gamma = 1.5$, $\delta = 5$, $\beta = 0.2$.

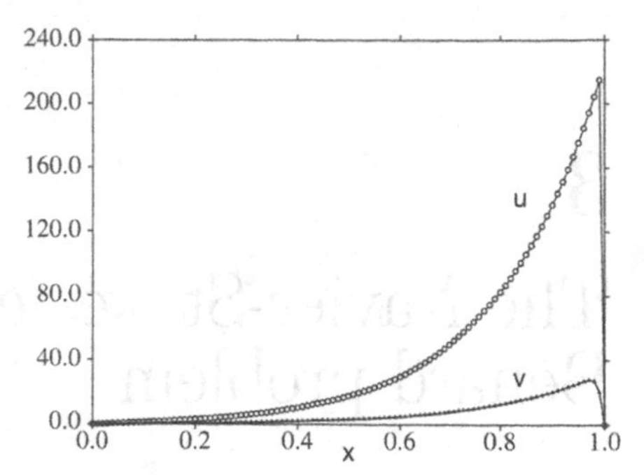

Figure 2.2. Steady solution to (2.114). $R = 5$, $\alpha = 5$, $\gamma = 0.1$, $\delta = 0.1$, $\beta = 0.1$.

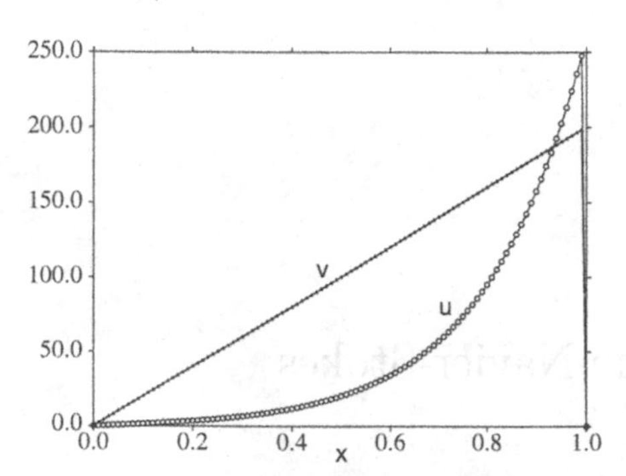

Figure 2.3. Steady solution to (2.112). $R = 5$, $\alpha = 5$, $\gamma = 0.1$, $\delta = 1$, $\beta = 200$.

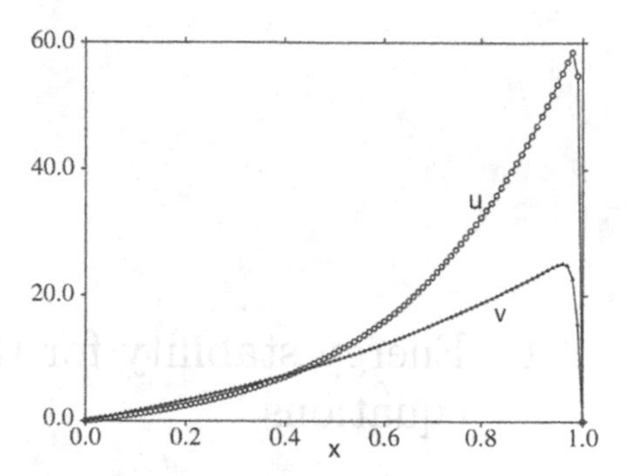

Figure 2.4. Steady solution to (2.113). $R = 5$, $\alpha = 3$, $\gamma = 1$, $\delta = 20$, $\beta = 2$.

how the solution behaves across the layer depending on the nonlinearity (or linear term) on the right hand side of the system, and then diffusion dominates near $x = 1$ to create a boundary layer. Much more detailed analysis and many more computed steady states may be found in chapter 2 of (Straughan, 1998).

3

The Navier-Stokes equations and the Bénard problem

3.1 Energy stability for the Navier-Stokes equations

For many flows it is sufficient to regard the fluid as incompressible. In this situation, with a constant temperature throughout, the velocity field, $\mathbf{v}$, and pressure, p, are determined from the *Navier-Stokes* equations,

$$\begin{aligned} v_{i,t} + v_j v_{i,j} &= -\frac{1}{\rho} p_{,i} + \nu \Delta v_i + f_i, \\ v_{i,i} &= 0. \end{aligned} \tag{3.1}$$

Here ρ is the (constant) density, ν (positive constant) is the kinematic viscosity, $\mathbf{f}$ is the external body force, and Δ is the Laplacian operator.

We shall study equations (3.1) on a bounded spatial region $\Omega\,(\subset \mathbb{R}^3)$ with $\mathbf{v}$ given on the boundary Γ, i.e.,

$$v_i(\mathbf{x},t) = \hat{v}_i(\mathbf{x},t), \qquad \mathbf{x} \in \Gamma. \tag{3.2}$$

3.1.1 Stability of the zero solution.

The stability of the zero solution to (3.1) when $\mathbf{f}$ is given by a potential and $\hat{\mathbf{v}} \equiv \mathbf{0}$ is now investigated.

The physical situation here is one of a liquid in a container that is arbitrarily moved around and then at a preassigned time, say $t = 0$, the container is held fixed so that the only force acting is gravity. Then $\mathbf{f} = -\nabla\phi$, and **f** is given by a potential. Physically, we expect the existing disturbance to decay, no matter what its initial value.

Mathematically the problem is to investigate the behaviour of $\mathbf{v}$ that solves (3.1) with $\mathbf{f} = -\nabla\phi$, $\hat{\mathbf{v}} \equiv \mathbf{0}$ in (3.2) subject to the initial condition

$$v_i(\mathbf{x},0) = v_i^0(\mathbf{x}), \qquad \mathbf{x} \in \Omega. \tag{3.3}$$

Consider the (kinetic) energy $E(t)$ defined by

$$E(t) = \frac{1}{2}\|\mathbf{v}\|^2, \qquad \left(= \frac{1}{2}\int_\Omega v_i v_i\, dx\right). \tag{3.4}$$

We differentiate (3.4),

$$\frac{dE}{dt}(t) = < v_i v_{i,t} > .$$

Then from (3.1),

$$\frac{dE}{dt} = - < v_i v_j v_{i,j} > - < P_{,i} v_i > + \nu < v_i \Delta v_i > ,$$

where we have put $P = p/\rho + \phi$. An integration by parts and use of the divergence theorem then yields

$$\begin{aligned}\frac{dE}{dt} = &- \oint_\Gamma n_i P v_i\, dA + \int_\Omega P v_{i,i}\, dx + \nu \oint_\Gamma n_j v_i v_{i,j}\, dA \\ &- \nu\|\nabla\mathbf{v}\|^2 - \frac{1}{2}\oint_\Gamma n_j v_j v_i v_i\, dA + \frac{1}{2}\int_\Omega v_{j,j} v_i v_i\, dx .\end{aligned}$$

Since $v_i \equiv 0$ on Γ, the boundary terms vanish. Further, from $(3.1)_2$, $\operatorname{div}\mathbf{v} \equiv 0$ and so we find

$$\frac{dE}{dt} = -\nu\|\nabla\mathbf{v}\|^2 . \tag{3.5}$$

From the Poincaré inequality (see page 387),

$$\lambda_1\|\mathbf{v}\|^2 \le \|\nabla\mathbf{v}\|^2 , \tag{3.6}$$

since here $\mathbf{v} \equiv \mathbf{0}$ on Γ, where $\lambda_1 > 0$. So use of (3.6) in (3.5) gives

$$\frac{dE}{dt} \le -\nu\lambda_1\|\mathbf{v}\|^2 = -2\nu\lambda_1 E(t). \tag{3.7}$$

Therefore,

$$E(t) \le e^{-2\nu\lambda_1 t} E(0). \tag{3.8}$$

Hence, $\int_\Omega v_i v_i\, dx \to 0$ as $t \to \infty$.

We have shown that all disturbances to the zero solution, therefore, decay, as is physically expected. This is a classical result usually associated with Kampe de Feriet. Energy decay results of L^2 solutions to the Navier-Stokes equations on exterior domains pose a difficult question. For example, Poincaré's inequality no longer holds (in the form we have used it) and one can certainly not expect exponential decay in time. This mathematically technical problem is addressed, for example, in the papers by (Borchers and Miyakawa, 1995), (Kato, 1984), (Kozono and Ogawa, 1994), (Maremonti, 1985; Maremonti, 1988), (Okamoto, 1997), (Schonbek, 1985), (Schonbek and Schonbek, 2000), (Tsai, 1998).

We are usually interested in the stability of some particular *non-zero* flow. We indicate how one may develop an energy procedure for investigating the stability of a *non-zero* but *steady* solution to (3.1), (3.2).

3.1.2 Steady Solutions.

Such a solution satisfies, with $P = p/\rho + \phi$,

$$\begin{aligned} V_j V_{i,j} &= -P_{,i} + \nu \Delta V_i, \\ V_{i,i} &= 0, \end{aligned} \tag{3.9}$$

in Ω, with

$$V_i(\mathbf{x}) = \hat{V}_i(\mathbf{x}), \tag{3.10}$$

on Γ, $\hat{V}_i$ being prescribed.

The idea is now to investigate the stability of $\mathbf{V}$. To do this we suppose $V_i^*(\mathbf{x},t)$, $P^*(\mathbf{x},t)$ is a solution to (3.1) that also satisfies the boundary data (3.10), and define

$$u_i(\mathbf{x},t) = V_i^* - V_i\,, \tag{3.11}$$

where u_i is a perturbation to the steady velocity field $\mathbf{V}$. We write the pressure perturbation as p so that

$$p = P^* - P. \tag{3.12}$$

From (3.1) and (3.9) we then find u_i satisfies

$$\begin{aligned} u_{i,t} + u_j V_{i,j} + V_j^* u_{i,j} &= -p_{,i} + \nu \Delta u_i, \\ u_{i,i} &= 0, \end{aligned} \tag{3.13}$$

in $\Omega \times (0,\infty)$, together with the homogeneous boundary data,

$$u_i = 0, \tag{3.14}$$

on $\Gamma \times [0,\infty)$.

The study of the stability of $\mathbf{V}$ is equivalent now to studying the stability of the zero solution ($\mathbf{u} \equiv \mathbf{0}$) to (3.13) subject to arbitrary initial disturbances,

$$u_i(\mathbf{x},0) = u_i^0(\mathbf{x}). \tag{3.15}$$

3.1.3 Nonlinear (Energy) Stability of V.

The idea is to determine a *sufficient* condition on $\mathbf{V}$ such that all disturbances $\mathbf{u}$ tend to zero as $t \to \infty$. No attempt is here made to optimize the analysis, since such refinements follow later.

Again consider the kinetic energy, but now of the perturbation $\mathbf{u}$,

$$E(t) = \frac{1}{2}\|\mathbf{u}\|^2. \tag{3.16}$$

Differentiate E and then substitute for $u_{i,t}$ from (3.13) to find

$$\begin{aligned}\frac{dE}{dt} &= < u_i u_{i,t} >,\\ &= - < u_i p_{,i} > + \nu < u_i \Delta u_i >\\ &\quad - < V_j^* u_i u_{i,j} > - < u_j u_i V_{i,j} >,\end{aligned} \tag{3.17}$$

and denote the terms on the right of (3.17) by $I_1 - I_4$, respectively.

We consider $I_1 - I_4$ in turn. After integration by parts, use of the divergence theorem, and use of the solenoidal condition $(3.13)_2$ and the boundary condition (3.14), we derive the following equalities:

$$I_1 = - < u_i p_{,i} > = - \oint_\Gamma n_i u_i p \, dA + < p u_{i,i} > = 0.$$

$$\begin{aligned}I_2 &= \nu < u_i \Delta u_i >\\ &= -\nu < u_{i,j} u_{i,j} > + \oint_\Gamma \nu n_j u_i u_{i,j} \, dA\\ &= -\nu \|\nabla \mathbf{u}\|^2.\end{aligned}$$

$$\begin{aligned}I_3 &= - < V_j^* u_i u_{i,j} > = -\frac{1}{2} < V_j^* (u_i u_i)_{,j} >\\ &= -\frac{1}{2} \oint_\Gamma n_j V_j^* |\mathbf{u}|^2 \, dA + \frac{1}{2} < V_{j,j}^* u_i u_i >\\ &= 0.\end{aligned}$$

$$I_4 = - < V_{i,j} u_i u_j > = - < D_{ij} u_i u_j >,$$

where $D_{ij} = \frac{1}{2}(V_{i,j} + V_{j,i})$ is the symmetric part of the velocity gradient of the base solution. These relations in (3.17) yield

$$\frac{dE}{dt} = -\nu \|\nabla \mathbf{u}\|^2 - < D_{ij} u_j u_i > . \tag{3.18}$$

It is convenient now to non-dimensionalize this equation, i.e., to work with dimensionless variables. Let U be a "typical velocity" for the problem, and let L be a "typical lengthscale", e.g., L might be the maximum dimension of Ω, U may be $\max_\Omega |\mathbf{V}|$. Then since

$$\dim \nu \equiv [\nu] = \frac{L^2}{T},$$

we introduce,

$$\mathbf{x} = \hat{\mathbf{x}}L, \qquad t = \frac{L^2}{\nu}\hat{t}, \qquad \mathbf{u} = \hat{\mathbf{u}}U, \qquad \mathbf{V} = \hat{\mathbf{V}}U. \tag{3.19}$$

Note that the "hat" variables have no dimensions.

With these transformations the terms in (3.18) become

$$\begin{aligned}
\frac{dE}{dt} &= \frac{\nu U^2}{2L^2}L^3\frac{d}{d\hat{t}}\int_{\hat{\Omega}} \hat{u}_i\hat{u}_i\,d\hat{x}\,,\\
\nu\int_\Omega u_{i,j}u_{i,j}\,dx &= \frac{\nu U^2}{L^2}L^3\int_{\hat{\Omega}} \hat{u}_{i,j}\hat{u}_{i,j}\,d\hat{x}\,,\\
\int_\Omega D_{ij}u_iu_j\,dx &= \frac{U^3}{L}L^3\int_{\hat{\Omega}} \hat{D}_{ij}\hat{u}_i\hat{u}_j\,d\hat{x}.
\end{aligned}$$

Hence, the *dimensionless* form of (3.18) becomes (although for ease in writing we now *omit* all hats)

$$\frac{dE}{dt} = -\mathcal{D} - \frac{UL}{\nu}\int_\Omega D_{ij}u_iu_j\,dx\,, \tag{3.20}$$

where

$$\mathcal{D} = \int_\Omega u_{i,j}u_{i,j}\,dx. \tag{3.21}$$

The dimensionless quantity UL/ν is denoted by Re, the *Reynolds number.* (After Osborne Reynolds who saw its relevance in stability theory — indeed, (3.20) is often called the Reynolds-Orr energy equation.) So, let

$$I = -< D_{ij}u_iu_j >, \tag{3.22}$$

then (3.20) is

$$\begin{aligned}
\frac{dE}{dt} &= -\mathcal{D} + RI,\\
&\le -\mathcal{D}R\Big(\frac{1}{R} - \max_{\mathcal{H}}\frac{I}{\mathcal{D}}\Big)\,,
\end{aligned} \tag{3.23}$$

$\mathcal{H}$ being the space of admissible functions $\mathbf{u}$ over which the maximum is sought. Here, $\mathcal{H}$ is the space of functions

$$\mathcal{H} = \left\{\mathbf{u}\middle|\mathbf{u} \in \left(H_0^1(\Omega)\right)^3, \nabla\cdot\mathbf{u} = 0\right\}.$$

Denote

$$\frac{1}{R_E} = \max_{\mathcal{H}}\frac{I}{\mathcal{D}}\,; \tag{3.24}$$

then if $R < R_E$, $1/R - 1/R_E > 0$. Therefore, from (3.23)

$$\frac{dE}{dt} \le -\left(\frac{R_E - R}{R_E}\right)\mathcal{D} \le -2a\lambda_1 E, \tag{3.25}$$

where we have also used Poincaré's inequality (3.6) and have set

$$a = \frac{R_E - R}{R_E} (> 0).$$

This may be integrated to yield

$$E(t) \leq e^{-2a\lambda_1 t} E(0). \tag{3.26}$$

If $R < R_E$, then $E \to 0$ as $t \to \infty$ at least exponentially fast, and there is nonlinear stability of the steady solution $\mathbf{V}$.

The criterion of importance is (3.24), since if $R < R_E$, the base solution $\mathbf{V}$ is nonlinearly stable *for all initial disturbances* $\mathbf{u}_0$, regardless of how large they may be.

To calculate the maximum in (3.24), we use the calculus of variations. From a similar analysis to that leading to (2.15), we know that the maximizing solution satisfies

$$\delta I_2 - R_E \delta I_1 = 0, \tag{3.27}$$

where now $I_2 = I_2(\nabla \mathbf{u}),\ I_1 = I_1(\mathbf{u})$. So,

$$\delta I_2 = \frac{d}{d\epsilon} I_2(\nabla \mathbf{u} + \epsilon \nabla \mathbf{h})\Big|_{\epsilon=0}$$

for all admissible $\mathbf{h}$ and

$$\delta I_1 = \frac{d}{d\epsilon} I_1(\mathbf{u} + \epsilon \mathbf{h})\Big|_{\epsilon=0}$$

for all admissible $\mathbf{h}$. Hence,

$$\begin{aligned} \delta I_2 &= 2 < (u_{i,j} + \epsilon h_{i,j}) h_{i,j} > \big|_{\epsilon=0} \\ &= 2 < u_{i,j} h_{i,j} > \\ &= -2 < h_i \Delta u_i >, \end{aligned}$$

since $\mathbf{h} = \mathbf{0}$ on Γ. Also,

$$\begin{aligned} \delta I_1 &= 2 < D_{ij}(u_i + \epsilon h_i) h_j > \big|_{\epsilon=0} \\ &= 2 < D_{ij} h_i u_j > . \end{aligned}$$

However, since $\mathcal{H}$ is restricted to those functions that are divergence free, we must add into the maximum problem the constraint $u_{i,i} = 0$ by a Lagrange multiplier. This is done by adding a term

$$\int_\Omega \pi(\mathbf{x}) u_{i,i}\, dx = 0$$

in the maximization: π depends on $\mathbf{x}$ since otherwise no contribution arises in the Euler-Lagrange equations from π. By adding in the above contribution, equation (3.27) reduces to

$$-2 < h_i(\Delta u_i + R_E D_{ij} u_j + \pi_{,i}) >= 0.$$

Since $\mathbf{h}$ is arbitrary, we need

$$\begin{aligned} \Delta u_i + R_E D_{ij} u_j &= -\pi_{,i} \,, \\ u_{i,i} &= 0 \,, \end{aligned} \tag{3.28}$$

in the region Ω together with the boundary conditions

$$u_i = 0, \qquad \text{on} \quad \Gamma. \tag{3.29}$$

As we want the *smallest* value of R_E, R_E is then determined as the lowest eigenvalue $R_E^{(1)}$ of (3.28), (3.29). (System (3.28), (3.29) is essentially a three-dimensional Sturm-Liouville problem. Since $\mathbf{V}(\mathbf{x})$ is known, (3.28), (3.29) is a *linear* problem for $\mathbf{u}^{(1)}$ and $R_E^{(1)}$.)

System (3.28), (3.29) for the energy eigenvalue R_E should be contrasted with the corresponding problem for the critical Reynolds number of linear theory. From (3.13), (3.14) the linearized equations for *linear instability* of the steady solution, $\mathbf{V}$, to (3.9), (3.10) are

$$\begin{aligned} \sigma u_i + u_j V_{i,j} + V_j u_{i,j} &= -p_{,i} + \nu \Delta u_i, \\ u_{i,i} &= 0, \end{aligned}$$

in Ω, with

$$u_i = 0$$

on Γ. Here, σ is the (possibly complex) growth rate in the representation $u_i(\mathbf{x}, t) = e^{\sigma t} u_i(\mathbf{x})$. If we use the non-dimensionalization (3.19), then this linearized system becomes

$$\begin{aligned} \sigma u_i + Re(u_j V_{i,j} + V_j u_{i,j}) &= -p_{,i} + \Delta u_i, \\ u_{i,i} &= 0, \end{aligned} \tag{3.30}$$

in Ω, with

$$u_i = 0 \tag{3.31}$$

on Γ, where Re is again the Reynolds number, $Re = UL/\nu$.

The logic is to determine the lowest eigenvalue R_E of (3.28), (3.29) and the corresponding lowest eigenvalue (value of Re), R_L, to (3.30), (3.31). If $R < R_E$ we are assured certain nonlinear stability, whereas if $R > R_L$ we know there is definitely instability. One of the main objectives of energy stability theory is to try to arrange that R_E is as close to R_L as possible. Of course, system (3.30), (3.31) is not symmetric (in a sense made precise in chapter 4), and is very different from the one of nonlinear energy stability theory, (3.28), (3.29). There are many problems in fluid mechanics, for example shear flow, where R_L and R_E are currently very different, especially if one is looking for an unconditional stability result, i.e. one which holds for all initial data. In this book we describe several convection examples where by a suitable choice of an *energy*, it has been possible to arrange that R_E is very close to R_L and thus deliver results that are practically useful.

The determination of R_E for specific problems is considered later in the context of convection. A crude, but often useful, estimate may be obtained by directly deriving an upper bound for $I/\mathcal{D}$ from (3.23), instead of proceeding to (3.24), c.f. (Serrin, 1959a). From the spectral theorem of linear algebra we know that since $\mathbf{D}$ is a symmetric tensor,

$$I = - < D_{ij}u_iu_j > \leq \lambda_m \|\mathbf{u}\|^2 ,$$

where λ_m is the maximum of the three eigenvalues of $\mathbf{D}$, maximized over Ω. (In general, $D_{ij} = D_{ij}(\mathbf{x})$ so its eigenvalues too depend on $\mathbf{x}$, and λ_m denotes the maximum over Ω.)

Thus, from (3.23)

$$\begin{aligned} \frac{dE}{dt} &\leq -\mathcal{D} + R\lambda_m \|\mathbf{u}\|^2 \\ &\leq -(\lambda_1 - R\lambda_m)\|\mathbf{u}\|^2, \end{aligned} \tag{3.32}$$

using also Poincaré's inequality (3.6).

Inequality (3.32) is

$$\frac{dE}{dt} \leq -2(\lambda_1 - R\lambda_m)E,$$

and this integrates to

$$E(t) \leq \exp\{-2(\lambda_1 - R\lambda_m)t\}E(0).$$

Thus,

$$R < \frac{\lambda_1}{\lambda_m}, \tag{3.33}$$

represents a *sufficient* condition for nonlinear stability. Estimates for λ_1/λ_m are usually easy to determine, and may constitute a useful nonlinear stability estimate. In general, however, a criterion like (3.33) is much weaker than the variational result obtained from (3.28), (3.29).

3.2 The Balance of Energy and the Boussinesq Approximation

When we deal with motion of a fluid driven by buoyancy forces, such as those caused by heating the fluid, equations (3.1) are not sufficient and it is necessary to also add an equation for the temperature field, T. This is the equation for the *balance of energy*,

$$\rho T\dot{S} = \sigma_{ij}d_{ij} - q_{i,i}, \tag{3.34}$$

where S, σ_{ij}, d_{ij}, and q_i are the entropy, stress tensor, symmetric part of the velocity gradient, and heat flux, and

$$\dot{S} = \frac{\partial S}{\partial t} + \mathbf{v}.\nabla S\,.$$

For a linear viscous fluid we take

$$\sigma_{ij} = -p\delta_{ij} + 2\mu d_{ij},$$

where μ $(= \rho\nu)$ is the dynamic viscosity. In general, the fluid is compressible, but for many convective motions the system may be considerably simplified by assuming the motion is isochoric, i.e., essentially incompressible flow, except in the body force term $\mathbf{f}$ in $(3.1)_1$. This approximation has caused a lot of concern in the literature and is known as the *Boussinesq approximation.*

In general, ρ is not constant and pressure variations have to be taken into account. Thus $S = S(\rho, p)$, and ρ, p, T are connected by an equation of state

$$f(\rho, p, T) = 0.$$

One may then show (cf. (Batchelor, 1967) pp. 164–171)

$$\rho T\dot{S} = \rho c_p \dot{T} - \alpha T\dot{p}, \tag{3.35}$$

where c_p is the specific heat at constant pressure and

$$\alpha = -\frac{1}{\rho}\Big(\frac{\partial\rho}{\partial T}\Big)_p \tag{3.36}$$

is the thermal expansion coefficient of the fluid. An order of magnitude argument like that of (Batchelor, 1967) pp. 164–171, may then be employed to show that provided

$$\frac{U^2}{c^2} << 1, \tag{3.37}$$

where U is a typical velocity in the problem and c is the local sound speed, the $\dot{p}$ term in (3.35) may be neglected as can the $\sigma_{ij}d_{ij}$ term in (3.34) by comparison with the other terms. Since for air at 15°C and one atmosphere pressure, $c = 340.6$ m sec^{-1} and, for water at the same temperature and pressure, $c = 1470$ m sec^{-1}, (3.37) is certainly satisfied in the convection problems we envisage here.

Thus, with a Fourier heat flux law,

$$\mathbf{q} = -k\nabla T\,,$$

(3.34) reduces to

$$\rho c_p \dot{T} = \nabla(k\nabla T). \tag{3.38}$$

We shall for now regard ρ, c_p, and k as constant. For many fluids this is very realistic.

The pressure in (3.1) is regarded as an unknown to be solved for in the problem, and the density is assumed constant except in the body force term $\mathbf{f}$, where

$$\rho \mathbf{f} = -g\mathbf{k}\rho(T), \tag{3.39}$$

assuming the only force acting is gravity, and where $\mathbf{k} = (0,0,1)$. The density in (3.39) is expanded in a Taylor series, and at this stage we consider only the first term, so

$$\rho = \rho_0\Big[1 - \alpha(T - T_R)\Big], \tag{3.40}$$

where ρ_0 is the density at temperature T_R, and α is the coefficient of thermal expansion given by (3.36).

Thus, taking account of (3.38)-(3.40), and (3.1), the equations for a linearly viscous, heat conducting, incompressible fluid (utilizing the Boussinesq approximation) are

$$\begin{aligned} v_{i,t} + v_j v_{i,j} &= -\frac{1}{\rho_0} p_{,i} + \nu\Delta v_i - k_i g\big(1 - \alpha[T - T_R]\big), \\ v_{i,i} &= 0, \\ T_{,t} + v_i T_{,i} &= \kappa\Delta T, \end{aligned} \tag{3.41}$$

where $\kappa = k/\rho_0 c_p$ is the thermal diffusivity.

For water between 0°C and 100°C (see e.g., (Batchelor, 1967) pp. 596, 597) c_p varies from 4.27 joule gm^{-1} deg C^{-1} to 4.216 joule gm^{-1} deg C^{-1}, κ varies from $1.33 \times 10^{-3}\,\text{cm}^2\,\text{sec}^{-1}$ to $1.66 \times 10^{-3}\,\text{cm}^2\,\text{sec}^{-1}$, so it is certainly reasonable to treat them as constant in many convection studies. The kinematic viscosity ν varies from $1.787 \times 10^{-2}\,\text{cm}^2\,\text{sec}^{-1}$ to $0.295 \times 10^{-2}\,\text{cm}^2\,\text{sec}^{-1}$ so its variation is greater. We shall primarily look at thermal convection with ν constant, but there are practical situations where a strongly varying viscosity with temperature is necessary. Energy stability arguments applied to such problems are discussed in chapter 16.

More elaborate treatments of the Boussinesq approximation are available and some of these are discussed in chapter 15. However, they all basically arrive at a system like (3.41). Without a Boussinesq approximation to yield a simplified system like (3.41) with a solenoidal velocity field, one is left with treating convection in a compressible fluid, and this is a far more complicated issue, as we see in chapter 15.

3.3 Energy Stability and the Bénard Problem

In the majority of situations if a layer of fluid is heated from below the fluid in the lower part of the layer expands as it becomes hotter, and when the temperature gradient or layer depth is sufficiently large to overcome the effect of gravity the fluid rises and a pattern of cellular motion may be seen.

This is called Bénard convection, after (Bénard, 1900); his experiments are now thought to have been driven by surface tension, see chapter 8. It is of historical interest to point out that the tesselated structure was previously observed by (Thomson, 1882). To describe this phenomenon mathematically we begin with equations (3.41).

We suppose the fluid is contained in the infinite layer $z \in (0, d)$ and the temperatures of the planes $z = 0, d$ are kept fixed,

$$T = T_0, \quad z = 0; \qquad T = T_1, \quad z = d; \tag{3.42}$$

with $T_0 > T_1$. Then, the conduction solution (which is motionless) to (3.41), (3.42) is

$$\bar{\mathbf{v}} \equiv \mathbf{0}, \qquad \bar{T} = -\beta z + T_0 \,, \tag{3.43}$$

where

$$\beta = \frac{T_0 - T_1}{d} \,,$$

and the pressure is determined from

$$\frac{d\bar{p}}{dz} = -\rho_0 g \big(1 - \alpha\{-\beta z + T_0 - T_R\}\big),$$

i.e.,

$$\bar{p} = -\rho_0 g \big[1 + \alpha(T_R - T_0)\big] z - \frac{\rho_0 g \alpha \beta}{2} \, z^2 \,, \tag{3.44}$$

selecting the pressure scale to vanish at $z = 0$.

We wish to study the stability of the conduction solution (3.43), (3.44). To this end we introduce perturbations $\mathbf{u}, \theta, \pi$ to $\bar{\mathbf{v}}, \bar{T}$, and $\bar{p}$, respectively, i.e.,

$$\mathbf{v} = \bar{\mathbf{v}} + \mathbf{u}, \qquad T = \bar{T} + \theta, \qquad p = \bar{p} + \pi.$$

The perturbations are *not* assumed to be small. From (3.41) the perturbations are found to satisfy

$$\begin{aligned} u_{i,t} + u_j u_{i,j} &= -\frac{1}{\rho_0} \pi_{,i} + \nu \Delta u_i + k_i g \alpha \theta, \\ u_{i,i} &= 0, \\ \theta_{,t} + u_i \theta_{,i} &= \beta w + \kappa \Delta \theta, \end{aligned} \tag{3.45}$$

where $w = u_3$; in general, we write $\mathbf{u} = (u, v, w)$.

It is convenient to non-dimensionalize (3.45) according to the scales:

$$t = t^* \frac{d^2}{\nu}, \qquad \mathbf{x} = \mathbf{x}^* d, \qquad \theta = T^\sharp \theta^*, \qquad \mathbf{u} = \mathbf{u}^* U,$$

$$U = \frac{\nu}{d}, \qquad Pr = \frac{\nu}{\kappa}, \qquad T^\sharp = U\sqrt{\frac{\beta\nu}{\kappa\alpha g}},$$

$$R = \sqrt{\frac{\alpha g \beta d^4}{\kappa\nu}}, \qquad p^* P = \pi, \qquad P = \frac{\rho_0 \nu^2}{d^2}.$$

Here Pr is the Prandtl number and $Ra = R^2$ is the Rayleigh number. With this scaling the *non-dimensional* form of (3.45) becomes (we, as usual, omit all stars even though the *dimensionless* form is understood)

$$\begin{aligned} u_{i,t} + u_j u_{i,j} &= -p_{,i} + \Delta u_i + k_i R\theta, \\ u_{i,i} &= 0, \\ Pr(\theta_{,t} + u_i\theta_{,i}) &= Rw + \Delta\theta. \end{aligned} \tag{3.46}$$

The boundary conditions on the planes $z = 0, 1$ are that there is no slip in velocity, and from (3.42) the temperatures remain constant, so

$$u_i = 0, \quad \theta = 0; \qquad z = 0, 1. \tag{3.47}$$

In addition, we assume that $\mathbf{u}, \theta, p$ have an (x, y)-dependence consistent with one that has a repetitive shape that tiles the plane, such as two-dimensional rolls or hexagons. The hexagon solution was originally given by (Christopherson, 1940) namely,

$$u(x,y) = \cos\frac{1}{2}a(\sqrt{3}\,x + y) + \cos\frac{1}{2}a(\sqrt{3}\,x - y) + \cos ay\,. \tag{3.48}$$

In particular, the (x, y)-dependence is consistent with a wavenumber, a, for which with

$$\Delta^* = \frac{\partial^2}{\partial x^2} + \frac{\partial^2}{\partial y^2},$$

u satisfies the relation

$$\Delta^* u = -a^2 u.$$

Whatever shape the cell has in the (x, y)-plane, its Cartesian product with $(0, 1)$ is the period cell V. (In this book we do not discuss the problem of which cell shape is actually taken up when convection commences in the fluid. This requires an analysis of the possible patterns which may occur in the nonlinear theory once convective motion ensues. We refer the reader to the concise mathematical analysis of (Mielke, 1997) for a very clear account of this.)

Before discussing nonlinear energy stability of a solution to (3.46) we briefly digress into linearized instability theory. The governing equations are obtained from (3.46) by omitting the nonlinear terms $u_j u_{i,j}$ and $Pr u_i \theta_{,i}$. The resulting linearized equations possess solutions of type

$$\mathbf{u}(\mathbf{x},t) = \mathbf{u}(\mathbf{x})e^{\sigma t}, \qquad \theta(\mathbf{x},t) = \theta(\mathbf{x})e^{\sigma t}, \qquad p(\mathbf{x},t) = p(\mathbf{x})e^{\sigma t},$$

so that $\mathbf{u}(\mathbf{x}), \theta(\mathbf{x}), p(\mathbf{x})$ satisfy

$$\begin{aligned} \sigma u_i &= -p_{,i} + \Delta u_i + k_i R\theta, \\ u_{i,i} &= 0, \\ \sigma Pr\theta &= Rw + \Delta\theta. \end{aligned} \tag{3.49}$$

The boundary conditions are still (3.47) and "periodicity" in the (x, y)-directions.

We now multiply $(3.49)_1$ by u_i^* (the complex conjugate of u_i), $(3.49)_3$ by θ^* and integrate over V, to find, integrating by parts and using the boundary conditions,

$$\begin{aligned} \sigma(< u_i u_i^* > + Pr < \theta\theta^* >) =& R(< w\theta^* > + < w^*\theta >) \\ & - (< \theta_{,i}\theta_{,i}^* > + < u_{i,j}u_{i,j}^* >). \end{aligned} \tag{3.50}$$

The right hand side of (3.50) is real and so if we let $\sigma = \sigma_r + i\sigma_1$, then taking the imaginary part of (3.50) we find

$$\sigma_1\big(\|\mathbf{u}\|^2 + Pr\|\theta\|^2\big) = 0,$$

the norm temporarily denoting the norm on the complex Hilbert space $L^2(V)$. Hence,

$$\sigma_1 = 0.$$

Therefore, in Bénard convection the growth rate σ is real. Thus, to find the instability boundary, the lowest value of R^2 in (3.49) for which $\sigma > 0$, we solve (3.49) for the smallest eigenvalue $R^2_{(1)}$ *with* $\sigma = 0$.

The linearized equations of Bénard convection satisfy the so-called *Principle of exchange of stabilities.* This is said to hold if in a system the growth rate σ is such that $\sigma \in \mathbf{R}$ or $\sigma_1 \neq 0 \Rightarrow \sigma_r < 0$. When this principle holds, convection sets in as *stationary convection.* If on the other hand, at the onset of instability, $\sigma = i\sigma_1$ with $\sigma_1 \neq 0$, the convection mechanism is referred to as *oscillatory convection.* It must be emphasized that the linearized theory only yields a boundary for *instability,* i.e., whenever $R > R_L$ the solution grows in time and is unstable (for (3.49) R_L is the lowest eigenvalue). In particular, the linearized equations do not yield any information on *nonlinear* stability: it is, in general, possible for the solution to become unstable at a value of R lower than R_L, and in this case subcritical instability (bifurcation) is said to occur. Details of how R_L^2 is found are given in chapter 19, where numerical solutions of eigenvalue problems relevant to convection are described.

For the *standard Bénard problem,* (3.46), we shall now show by energy stability theory that sub-critical instability is not possible.

At this point we consider the simplest, natural "energy", formed by adding the kinetic and thermal energies of the perturbations, and so define

$$E(t) = \frac{1}{2}\|\mathbf{u}\|^2 + \frac{1}{2}Pr\|\theta\|^2. \tag{3.51}$$

We differentiate E, substitute for $u_{i,t}, \theta_{,t}$ from (3.46), and use the boundary conditions to find

$$\frac{dE}{dt} = 2R < w\theta > -\big[D(\mathbf{u}) + D(\theta)\big], \tag{3.52}$$

where $D(\cdot)$ denotes the Dirichlet integral, e.g.,

$$D(f) = \|\nabla f\|^2 = \int_V |\nabla f|^2\, dV.$$

It is convenient to put

$$I = 2 < w\theta >, \qquad \mathcal{D} = D(\mathbf{u}) + D(\theta).$$

Then, from (3.52) we see that

$$\frac{dE}{dt} = RI - \mathcal{D} \le -\mathcal{D}R\Big(\frac{1}{R} - \frac{1}{R_E}\Big), \tag{3.53}$$

where R_E is defined by

$$\frac{1}{R_E} = \max_{\mathcal{H}} \frac{I}{\mathcal{D}}, \tag{3.54}$$

with $\mathcal{H}$ being the space of admissible solutions.

If now

$$R < R_E, \tag{3.55}$$

then

$$R\Big(\frac{1}{R} - \frac{1}{R_E}\Big) = \frac{R_E - R}{R_E} > 0;$$

and if we let

$$a = \frac{R_E - R}{R_E},$$

then from (3.53),

$$\frac{dE}{dt} \le -a\mathcal{D} \le -2a\lambda_1 E, \tag{3.56}$$

where we have also used Poincaré's inequality. Inequality (3.56) easily integrates to obtain

$$E(t) \le e^{-2a\lambda_1 t}E(0),$$

from which we see that $E \to 0$ at least exponentially fast as $t \to \infty$. This demonstrates that provided (3.55) is satisfied, the conduction solution (3.43) is nonlinearly stable for all initial disturbances.

The quantity of importance, therefore, is R_E as defined by (3.54). The Euler-Lagrange equations for this maximum are found by a similar procedure to that leading to (3.28). From (3.27) we know the Euler-Lagrange equations are found from

$$\delta D - R_E \delta I = 0. \tag{3.57}$$

Here $I = 2 < w\theta >$ and $D = \|\nabla \mathbf{u}\|^2 + \|\nabla \theta\|^2$ and so

$$\begin{aligned} \delta D &= \frac{d}{d\epsilon} D(\nabla \mathbf{u} + \epsilon \nabla \mathbf{h}, \nabla \theta + \epsilon \nabla \eta)\Big|_{\epsilon=0} \\ \delta I &= \frac{d}{d\epsilon} I(\mathbf{u} + \epsilon \mathbf{h}, \theta + \epsilon \eta)\Big|_{\epsilon=0}, \end{aligned}$$

for all admissible $\mathbf{h}$ and η. Thus,

$$\begin{aligned} \delta D =& 2 < (u_{i,j} + \epsilon h_{i,j})h_{i,j} > \big|_{\epsilon=0} + 2 < (\theta_{,i} + \epsilon \eta_{,i})\eta_{,i} > \big|_{\epsilon=0} \\ =& 2 < u_{i,j}h_{i,j} > +2 < \theta_{,i}\eta_{,i} > \\ =& -2\big[< h_i \Delta u_i > + < \eta \Delta \theta >\big], \end{aligned}$$

where in the last line we have integrated by parts. Furthermore, if we recall the constraint $u_{i,i} = 0$ and incorporate this into I we find

$$\begin{aligned} \delta I =& 2 < (w + \epsilon h_3)\eta > \big|_{\epsilon=0} + 2 < h_3(\theta + \epsilon \eta) > \big|_{\epsilon=0} \\ & - \frac{1}{R_E}\frac{d}{d\epsilon} 2 < p(u_{i,i} + \epsilon h_{i,i}) > \big|_{\epsilon=0} \\ =& 2 < w\eta > +2 < h_3\theta > - \frac{2}{R_E} < p h_{i,i} > \\ =& 2 < w\eta > +2 < h_3\theta > + \frac{2}{R_E} < p_{,i} h_{,i} > \end{aligned}$$

where p is a Lagrange multiplier and we have integrated by parts. Hence, (3.57) is equivalent to

$$(\eta, R_E w + \Delta\theta) + (h_i, \Delta u_i + R_E \delta_{i3}\theta + p_{,i}) = 0.$$

In this manner we find by using the arbitrariness of h_i and η, the Euler-Lagrange equations which arise from the variational problem (3.54) are

$$\begin{aligned} \Delta u_i + k_i R_E \theta &= p_{,i}, \\ u_{i,i} &= 0, \\ \Delta\theta + R_E w &= 0, \end{aligned} \tag{3.58}$$

together with the same boundary conditions as (3.47) and the "periodicity" conditions. It is immediately evident that R_E satisfies the same eigenvalue problem as (3.49) with $\sigma = 0$. Thus, for the standard Bénard problem $R_E \equiv R_L$, i.e., the linear instability boundary $\equiv$ the nonlinear stability boundary, and so no sub-critical instabilities are possible. This result was first established by (Joseph, 1965; Joseph, 1966).

It is tempting to generalize the natural energy in (3.51) by adding a parameter, $\lambda(>0)$ say, and considering a functional

$$E(t)=\frac{1}{2}\|\mathbf{u}\|^2+\frac{1}{2}\lambda Pr\|\theta\|^2\,,$$

with the idea of employing λ in an optimal way. This is, in fact, precisely what (Joseph, 1965; Joseph, 1966) did, and his technique is described in the next chapter.

It is pertinent to mention very interesting and not unrelated work using energy - integral - like estimates in fluid and porous convection contexts. One area is in bounding turbulence flows using variational turbulence methods or background bounding techniques. Another is in length scale arguments and connected regularity studies. A very readable account of both areas is given in (Doering and Gibbon, 1995). Recent relevant references to turbulence techniques include (Doering and Constantin, 1996; Doering and Constantin, 1998), (Doering and Foias, 2002), (Doering and Hyman, 1997), (Ierley and Worthing, 2001), (Kerswell, 2000; Kerswell, 2001), (Nicodemus et al., 1998; Nicodemus et al., 1999), and (Vitanov and Busse, 2001). Details of the length scale, existence and regularity results may be found in (Bartuccelli et al., 1993), (Bartuccelli, 2002), (Bartuccelli and Woolcock, 2001), (Gibbon, 1995), (Gourley and Bartuccelli, 1995), (Ly and Titi, 1999), (Oliver and Titi, 2000), (Rodrigues, 1986; Rodrigues, 1994), (Rodrigues and Urbano, 1999). (Doering and Gibbon, 2002) focus on flows driven by a time-independent body force. They derive very interesting long time average estimates of various norms of the solution relating them rigorously to the dimension, Reynolds number, and length scale of the problem.

Other energy like techniques for partial differential equations are described in (Antontsev et al., 2001) and (Flavin and Rionero, 1995). (Antontsev et al., 2001), in particular, describe many finite time extinction results, and (Flavin and Rionero, 1995) cover many topics across a range of problems modelled by partial differential equations. (Payne and Song, 1997; Payne and Song, 2000; Payne and Song, 2002), (Philippin and Vernier-Piro, 2001), and (Song, 2002) derive spatial decay results in porous media theories via energy-like (integral) estimates, and interesting spatial decay bounds for the problem when a fluid interfaces a saturated porous medium are produced in (Ames et al., 2001).

3.4 The equations for convection in a porous medium

3.4.1 The Darcy equations

(Darcy, 1856) is usually credited with deriving the equations for fluid motion in a porous body. He concluded in the one-dimensional situation that the fluid velocity is proportional to the pressure gradient in the direction of flow, thus $u \propto dp/dx$, where $\mathbf{v} = (u, v, w)$. He deduced the relation

$$\mu u = -k \frac{dp}{dx}, \tag{3.59}$$

where μ is the dynamic viscosity of the fluid and k is the permeability of the porous medium. (The permeability is a function of the porosity = fluid volume / total volume, and effectively measures the ease with which fluid flows through the porous medium.) When (3.59) is extended to three dimensions we find

$$\frac{\mu}{k} v_i = -\frac{\partial p}{\partial x_i}. \tag{3.60}$$

This equation is now known as Darcy's law. In this equation, v_i is the pore averaged velocity (i.e. the velocity averaged over the pores of the porous medium), so if $\hat{v}_i$ is the actual fluid velocity in a pore, then $v_i = \phi \hat{v}_i$ in the continuum approximation, ϕ being the porosity. For thermal convection problems the driving mechanism for instability is provided by the buoyancy force and (3.60) must be extended to incorporate this. Thus, we write

$$\frac{\mu}{k} v_i = -\frac{\partial p}{\partial x_i} - k_i g \rho(T).$$

Here g is gravity, $\mathbf{k} = (0, 0, 1)$, and $\rho(T)$ is for now taken to be linear in T as given by (3.40). To complete the derivation of the Darcy equations for non-isothermal motion in a porous medium we recall $v_{i,i} = 0$ and the equation for the temperature field,

$$\frac{\partial T}{\partial t} + v_i T_{,i} = \kappa \Delta T.$$

Thus, the complete system of equations for convection in a porous medium according to Darcy's law is

$$\begin{aligned} 0 &= -p_{,i} - \frac{\mu}{k} v_i - k_i g \rho(T) \\ v_{i,i} &= 0 \\ T_{,t} + v_i T_{,i} &= \kappa \Delta T. \end{aligned} \tag{3.61}$$

3.4.2 The Brinkman equations

It is often argued that (3.61) are insufficient to describe porous flow situations near a solid wall, or when the flow rate becomes large. For such a scenario we may replace (3.61) by the Brinkman equations (3.62)

$$\begin{aligned} 0 &= -p_{,i} - \frac{\mu}{k} v_i + \lambda \Delta v_i - k_i g \rho(T) \\ v_{i,i} &= 0 \\ T_{,t} + v_i T_{,i} &= \kappa \Delta T. \end{aligned} \tag{3.62}$$

Frequently the argument is advanced that the Brinkman equations, due to the inclusion of the viscosity term λ, may be more relevant for flows involving a solid boundary, see e.g. (Nield and Bejan, 1999).

3.4.3 The Forchheimer equations

When the flow rate is large the Forchheimer equations may be more appropriate than the Darcy ones, and these are

$$\begin{aligned} 0 &= -p_{,i} - \frac{\mu}{k} v_i - b|\mathbf{v}|v_i - k_i g \rho(T) \\ v_{i,i} &= 0 \\ T_{,t} + v_i T_{,i} &= \kappa \Delta T. \end{aligned} \tag{3.63}$$

The Forchheimer equations are believed more appropriate when the velocity is not small, the idea being that the pressure gradient is no longer proportional to the velocity itself.

The Darcy equations are usually attributed to (Darcy, 1856), the Forchheimer equations to (Forchheimer, 1901), although see also (Dupuit, 1863), while Brinkman's equations are due to (Brinkman, 1957). Modern derivations and verifications of these equations may be found in e.g. (Firdaouss et al., 1997), (Givler and Altobelli, 1994), (Giorgi, 1997), (Kladias and Prasad, 1991), (Nield and Bejan, 1999), and (Whitaker, 1996). The nonlinear drag term $|\mathbf{v}|v_i$ in (3.63) and its relevance to viscous dissipation is discussed in depth by (Nield, 2000). We are here only interested in describing their use in porous convection problems and the associated energy stability calculations which arise thereupon.

A model of flow in a porous elastic solid which allows deformation of the elastic solid is derived in (Ambrosi, 2002). A model of flow in a porous elastic solid which allows swelling of the elastic matrix is analysed in (Quintanilla, 2002c; Quintanilla, 2002a; Quintanilla, 2002b).

3.4.4 The Darcy equations with anisotropic permeability

For many practical situations the permeability k is not isotropic. (Just imagine rock strata. Many times this has a preferred direction and so anisotropic permeability is a realistic thing to consider.) When the permeability is not isotropic we can generalise the Darcy equations in a straightforward manner. This then yields a system of equations with rich mathematical properties which is very amenable to linear and energy stability techniques but which also applies to many real life mechanisms. For an anisotropic permeability we replace the velocity equation in (3.61) by

$$\mu v_i = -K_{ij}p_{,j} - K_{ij}k_j g\rho(T),$$

where K_{ij} is the permeability tensor. We shall require K_{ij} to be invertible so that the inverse tensor M_{ij} satisfies $MK = I$. Then the system of equations for convective motion in an anisotropic porous medium of Darcy type is given by

$$\begin{aligned} &\mu M_{ij}v_j = -p_{,i} - k_i g\rho(T) \\ &v_{i,i} = 0 \\ &T_{,t} + v_i T_{,i} = \kappa\Delta T. \end{aligned} \tag{3.64}$$

In section 4.2.3 we concentrate on the case of a transversely isotropic material where the axis of isotropy is the vertical (z) axis so that

$$K_{ij} = \begin{pmatrix} k_1 & 0 & 0 \\ 0 & k_1 & 0 \\ 0 & 0 & k \end{pmatrix}$$

and then

$$M_{ij} = \frac{1}{k}\begin{pmatrix} \zeta & 0 & 0 \\ 0 & \zeta & 0 \\ 0 & 0 & 1 \end{pmatrix}$$

with $\zeta = k/k_1$, where k, k_1 are positive constants.

There are situations in which one also needs the thermal diffusivity to be anisotropic. We do not study this explicitly, but we draw attention to work of (Storesletten, 1993) where this effect is investigated at length.

4

Symmetry, Competing Effects, and Coupling Parameters

4.1 Coupling parameters and the classical Bénard problem

In this section we explain an important idea in the theory of energy stability. This is the idea of coupling parameters which was briefly introduced in the last chapter and was originally developed by Professor D.D. Joseph.

Our exposition of the method of coupling parameters begins in the context of the standard Bénard problem introduced in section 3.3. Recall that in section 3.3 we derived the linearised equations for instability in the Bénard problem, equations (3.49), and we showed that the growth rate σ is real. Then, the instability boundary is found from the eigenvalue problem for the linear critical Rayleigh number, R_L^2, governed by (3.49) with $\sigma = 0$, namely

$$\begin{aligned} 0 &= -p_{,i} + \Delta u_i + k_i R_L \theta, \\ u_{i,i} &= 0, \\ 0 &= R_L w + \Delta\theta. \end{aligned} \tag{4.1}$$

To describe the coupling parameter method in the theory of nonlinear energy stability we begin with the *nonlinear* equations (3.46). Multiply $(3.46)_1$ by u_i and integrate over the period cell V to obtain (after integration by parts)

$$\frac{d}{dt}\frac{1}{2}\|\mathbf{u}\|^2 = R(\theta, w) - \|\nabla\mathbf{u}\|^2. \tag{4.2}$$

Next, multiply $(3.46)_3$ by θ and integrate over V to now derive

$$\frac{d}{dt}\frac{1}{2}Pr\|\theta\|^2 = R(\theta, w) - \|\nabla\theta\|^2. \tag{4.3}$$

In deriving (4.2) and (4.3) we have employed boundary conditions (3.47) and periodicity in the x, y directions. The idea is to add λ times (4.3) to (4.2) and use λ in an optimal way. The constant λ is positive and is called a *coupling parameter*. We define an energy functional which depends on λ by

$$E(t) = \frac{1}{2}\|\mathbf{u}(t)\|^2 + \frac{\lambda}{2}Pr\|\theta(t)\|^2. \tag{4.4}$$

From (4.2) - (4.4) we now deduce

$$\frac{dE}{dt} = RI - D, \tag{4.5}$$

where $I(t)$ is a non-definite production term

$$I(t) = (1+\lambda)(\theta, w) \tag{4.6}$$

and $D(t)$ is the positive-definite dissipation term

$$D(t) = \|\nabla\mathbf{u}\|^2 + \lambda\|\nabla\theta\|^2. \tag{4.7}$$

Next, define R_E by

$$\frac{1}{R_E} = \max_H \frac{I}{D} \tag{4.8}$$

where H is the space of admissible solutions, in this case

$$H = \{u_i, \theta | u_i, \theta \in L^2(V), u_{i,i} = 0 \text{ in } V\},$$

and in addition u_i, θ satisfy the boundary conditions (3.47), together with periodicity in the x, y directions.

Then, from (4.5)

$$\begin{aligned}\frac{dE}{dt} &= -D\Big(1 - R\frac{I}{D}\Big)\\ &\leq -D\Big(1 - \frac{R}{R_E}\Big)\\ &= -\frac{D}{R_E}(R_E - R).\end{aligned} \tag{4.9}$$

From the Poincaré inequality (see A.1) we know

$$D \geq \pi^2(\|\mathbf{u}\|^2 + \lambda\|\theta\|^2) \geq aE \tag{4.10}$$

where $a = 2\pi^2\min\{1, Pr\}$. We now require

$$R < R_E \tag{4.11}$$

and then put $a(R_E - R)/R_E = \omega > 0$. Thus, from (4.9) and (4.10) we have

$$\frac{dE}{dt} \leq -\omega E$$

and so by integrating this from 0 to t and rearranging we find

$$E(t) \leq \exp(-\omega t)\, E(0).$$

Thus, E decays at least exponentially fast and nonlinear stability is assured for all values of $E(0)$. It is important to note that this result holds for all initial data. In this book we are stressing uses of the energy method for global (unconditional) nonlinear stability, whenever possible.

The condition for unconditional nonlinear stability is, therefore, (4.11) and R_E is found from (4.8). The determination of R_E is a variational problem and the Euler-Lagrange equations arising from (4.8) are found by proceeding as in section 3.3. We first redefine θ to remove λ from the denominator in (4.8). Thus, put $\hat{\theta} = \sqrt{\lambda}\,\theta$ and

$$\max_H \frac{I}{D} = \max_H \frac{f(\lambda)(\hat{\theta}, w)}{\|\nabla \mathbf{u}\|^2 + \|\nabla \hat{\theta}\|^2}$$

where $f(\lambda) = (1+\lambda)/\sqrt{\lambda}$. Since this expression seeks the maximum for $u_i, \theta \in H$ we simply now replace (4.8) by

$$\frac{1}{R_E} = \max_H \frac{f(\lambda)(\theta, w)}{\|\nabla \mathbf{u}\|^2 + \|\nabla \theta\|^2} = \max_H \frac{I}{D}\,. \tag{4.12}$$

Following the details of section 3.3 the Euler-Lagrange equations from (4.12) are given by

$$R_E \delta I - \delta D = 0,$$

where we add the constraint in I, and so

$$\delta I = \frac{d}{d\epsilon} f(\lambda)\,(\theta + \epsilon\eta, w + \epsilon h_3)\Big|_{\epsilon=0} - \frac{2}{R_E}\frac{d}{d\epsilon}(p, u_{i,i} + \epsilon h_{i,i})\Big|_{\epsilon=0}$$

and

$$\delta D = \frac{d}{d\epsilon}\Big(\|\nabla(\mathbf{u} + \epsilon\mathbf{h})\|^2 + \|\nabla(\theta + \epsilon\eta)\|^2\Big)\Big|_{\epsilon=0}$$

Performing the calculations as in section 3.3 we find

$$\begin{aligned} \delta I &= f(\lambda)\,(w, \eta) + f(\lambda)\,(h_3, \theta) + \frac{2}{R_E}(p_{,i}, h_i) \\ \delta D &= -2\big[(\Delta u_i, h_i) + (\Delta\theta, \eta)\big]. \end{aligned}$$

This leads to the Euler-Lagrange equations

$$\begin{aligned} \Delta u_i + \frac{1}{2}R_E f\theta &= p_{,i} \\ u_{i,i} &= 0 \\ \Delta\theta + \frac{1}{2}R_E f w &= 0. \end{aligned} \qquad (4.13)$$

In general, we now wish to use λ to maximise R_E and so find the critical Rayleigh number Ra_E below which instability is not possible, as

$$Ra_E = \max_{\lambda>0} R_E^2. \qquad (4.14)$$

It is necessary to find the value of λ which maximises R_E^2 in (4.14).

The best value of λ, $\bar{\lambda}$, satisfies

$$\bar{\lambda} = 1. \qquad (4.15)$$

This result is due to (Joseph, 1965; Joseph, 1966). To prove this we construct a relation from (4.13) for $\partial R_E/\partial\lambda$ and put this equal to zero (as it is for the maximum value in (4.14)). We let (u_i^2, θ^2, p^2) be a solution to (4.13) on V for $\lambda = \lambda^2$ and likewise let (u_i^1, θ^1, p^1) be a solution for $\lambda = \lambda^1$, $\lambda^2 \neq \lambda^1$. Multiply $(4.13)_1$ holding for λ^2 by u_i^1, and $(4.13)_1$ holding for λ^1 by u_i^2. Integrate the results over V to find, with $f^\alpha = f(\lambda^\alpha) = (1+\lambda^\alpha)/\sqrt{\lambda^\alpha}$, $\alpha = 1, 2$,

$$\frac{1}{2}R_E^2 f^2(\theta^2, w^1) - (\nabla u_i^2, \nabla u_i^1) = 0 \qquad (4.16)$$

$$\frac{1}{2}R_E^1 f^1(\theta^1, w^2) - (\nabla u_i^1, \nabla u_i^2) = 0 \qquad (4.17)$$

$$\frac{1}{2}R_E^2 f^2(w^2, \theta^1) - (\nabla\theta^2, \nabla\theta^1) = 0 \qquad (4.18)$$

$$\frac{1}{2}R_E^1 f^1(w^1, \theta^2) - (\nabla\theta^1, \nabla\theta^2) = 0. \qquad (4.19)$$

Now, form the combination (4.16) - (4.19) + (4.18) -(4.17) to derive

$$(R_E^2 f^2 - R_E^1 f^1)\big[(w^1, \theta^2) + (w^2, \theta^1)\big] = 0. \qquad (4.20)$$

We write $R_E^2 f^2 - R_E^1 f^1 = R_E^2(f^2 - f^1) + f^1(R_E^2 - R_E^1)$ and divide by $\lambda^2 - \lambda^1$ to obtain

$$\left[R_E^2\left(\frac{f^2 - f^1}{\lambda^2 - \lambda^1}\right) + f^1\left(\frac{R_E^2 - R_E^1}{\lambda^2 - \lambda^1}\right)\right]\big[(w^1, \theta^2) + (w^2, \theta^1)\big] = 0.$$

Let now $\lambda^2 \to \lambda^1$ and we find

$$\left(R_E\frac{df}{d\lambda} + f\frac{\partial R_E}{\partial\lambda}\right)(w, \theta) = 0. \qquad (4.21)$$

By multiplying $(4.13)_1$ by u_i, $(4.13)_3$ by θ and integrating over V one sees that

$$\|\nabla\mathbf{u}\|^2 + \|\nabla\theta\|^2 = R_E f(w, \theta). \qquad (4.22)$$

Hence, (4.21) and (4.22) yield

$$\Big(R_E\,\frac{df}{d\lambda}+f\,\frac{\partial R_E}{\partial\lambda}\Big)\,(\|\nabla\mathbf{u}\|^2+\|\nabla\theta\|^2)=0, \tag{4.23}$$

where we have noted $R_E f > 0$. For the value of λ which maximises in (4.14), $\partial R_E/\partial\lambda = 0$ and thus (4.23) yields

$$\frac{df}{d\lambda}=0. \tag{4.24}$$

Solving (4.24) one finds $\bar{\lambda} = 1$ as claimed.

With $\lambda = 1$ the Euler-Lagrange equations (4.13) become

$$\begin{aligned}&\Delta u_i + R_E\theta = p_{,i}\\ &u_{i,i}=0\\ &\Delta\theta + R_E w = 0.\end{aligned} \tag{4.25}$$

These are exactly the same as the equations of linearised instability, equations (4.1). Thus, we have again established that for the standard Bénard problem $R_E = R_L$ and so no subcritical instabilities are possible. (This was shown in section 3.3, but we have here allowed for a more general "energy" $E(t)$ involving λ. This idea works with more complicated problems for which λ is not necessarily equal to 1.)

Note that (4.13) is contained in the more general eigenvalue problem of form

$$L(\mathbf{x})u + R_E f(\mathbf{x};\lambda)u = G, \tag{4.26}$$

where L is a symmetric, linear differential operator on a suitable Hilbert space and $< G, u >= 0$, $<,>$ being the inner product on the associated Hilbert space. Let now u_1, u_2 be the first eigenfunctions to (4.26) corresponding to parameters λ_1, λ_2. Take the inner product of the equation for u_1 with u_2 and the inner product of the equation for u_2 with u_1. Using the symmetry of L leads to

$$R_E(\lambda_2) < f(\lambda_2)u_2, u_1 > - R_E(\lambda_1) < f(\lambda_1)u_1, u_2 >= 0.$$

Next, add and subtract $R_E(\lambda_1) < f(\lambda_2)u_2, u_1 >$, divide by $\lambda_2 - \lambda_1$, and let $\lambda_2 \to \lambda_1$ to obtain

$$\frac{\partial R_E}{\partial\lambda} < f(\lambda)u, u > + R_E\Big\langle\frac{\partial f}{\partial\lambda}u, u\Big\rangle = 0.$$

The maximum value of $R_E(\lambda)$ satisfies $\partial R_E/\partial\lambda = 0$ and so

$$\Big\langle\frac{\partial f}{\partial\lambda}u, u\Big\rangle = 0.$$

The optimal value of λ may sometimes be deduced from this equation and the above strategy is what we essentially apply in more complicated problems than the standard Bénard one.

4.2 Energy stability in porous convection

In this section we investigate the Bénard problem in a porous medium. In fact, we handle each of the porous systems introduced in section 3.4, namely the equations of Darcy, Brinkman, and Forchheimer, and the equations for Darcy convection with anisotropic permeability. We shall show that one can establish the same optimal results as in section 4.1, that the linear instability and nonlinear stability Rayleigh numbers are the same, i.e. $R_L \equiv R_E$.

The physical picture is one of a saturated porous medium of infinite extent in the x and y directions bounded by the planes $z = 0$ and $z = d(> 0)$ with gravity, g, in the negative $z-$direction. The upper boundary is held at fixed temperature $T = T_U$ while the lower is held at constant temperature $T = T_L$, with $T_L > T_U$. The problem is to determine under what conditions heating from below will lead to convective (cellular) fluid motion in the porous medium. We begin with the equations of Darcy.

4.2.1 *The Bénard problem for the Darcy equations*

We commence with the steady solution to equations (3.61) when the density is given by the relation $\rho(T) = \rho_0(1 - \alpha[T - T_0])$. The steady solution is again

$$\bar{v}_i = 0, \qquad \bar{T} = -\beta z + T_L, \tag{4.27}$$

with $\bar{p}$ determined from $d\bar{p}/dz = g\rho_0(1 - \alpha[\bar{T}(z) - T_0])$. The nonlinear perturbation equations which arise from (3.61) using (4.27) as steady solution, with $v_i = \bar{v}_i + u_i$, $T = \bar{T} + \theta$, $p = \bar{p} + \pi$, are

$$\begin{aligned}
0 &= -\pi_{,i} - \frac{\mu}{k} u_i + g\rho_0\alpha k_i\theta, \\
u_{i,i} &= 0, \\
\theta_{,t} + u_i\theta_{,i} &= \beta w + \kappa\Delta\theta.
\end{aligned}$$

We non-dimensionalize these equations with the scalings $t = t^*\mathcal{T}$, $u_i = u_i^*U$, $x_i = x_i^*d$, $\theta = T^\sharp\theta^*$, $\mathcal{T} = d^2/\kappa$, $P = \mu U d/k$, $U = \kappa/d$, $T^\sharp = dU\sqrt{\mu\beta/\kappa g\rho_0\alpha k}$, and the Rayleigh number R^2 defined by

$$R^2 = \frac{d^2 g\rho_0\alpha k\beta}{\mu\kappa}. \tag{4.28}$$

Next, the stars are omitted and the non-dimensional (fully nonlinear) perturbation equations for the thermal convection problem arising from Darcy's equations are

$$\begin{aligned}
0 &= -\pi_{,i} - u_i + Rk_i\theta, \\
u_{i,i} &= 0, \\
\theta_{,t} + u_i\theta_{,i} &= Rw + \Delta\theta.
\end{aligned} \tag{4.29}$$

The boundary conditions are now that

$$\theta = w = 0, \qquad z = 0, 1, \tag{4.30}$$

with (u_i, θ, π) satisfying a plane tiling periodicity as in section 3.3. Note that since we are using Darcy's law, which is lower order than Navier-Stokes, we can only prescribe boundary conditions on the normal component of u_i.

To determine linear instability from (4.29) we drop the $u_i\theta_{,i}$ term and write $u_i = u_i(\mathbf{x})e^{\sigma t}$, $\theta = \theta(\mathbf{x})e^{\sigma t}$, $\pi = \pi(\mathbf{x})e^{\sigma t}$, to find

$$\begin{aligned} 0 &= -\pi_{,i} - u_i + Rk_i\theta, \\ u_{i,i} &= 0, \\ \sigma\theta &= Rw + \Delta\theta. \end{aligned} \tag{4.31}$$

We now show $\sigma \in \mathbb{R}$. To do this let V be a period cell for the solution (u_i, θ, π) and then multiply $(4.31)_1$ by the complex conjugate of u_i, u_i^*, and integrate over V. The result is, after using the boundary conditions

$$\|\mathbf{u}\|^2 = R(\theta, w^*) \tag{4.32}$$

where $w^* = u_3^*$. Now multiply $(4.31)_3$ by the complex conjugate of θ, θ^*, and integrate over V. One may now show that

$$\sigma\|\theta\|^2 = R(w, \theta^*) - \|\nabla\theta\|^2. \tag{4.33}$$

In (4.32) and (4.33) it is understood that u_i, θ are complex and so $\|\cdot\|$ is to be interpreted accordingly, e.g.

$$\|\theta\|^2 = \int_V \theta\theta^* dx.$$

Next, add (4.32) to (4.33) to find

$$\sigma\|\theta\|^2 = R\big[(\theta, w^*) + (w, \theta^*)\big] - \|\mathbf{u}\|^2 - \|\nabla\theta\|^2.$$

If now $\sigma = \sigma_r + i\sigma_1$ then the imaginary part of this equation shows that

$$\sigma_1\|\theta\|^2 = 0.$$

Thus, $\sigma_1 = 0$ and $\sigma \in \mathbb{R}$. Exchange of stabilities holds and it is sufficient to take $\sigma = 0$ in (4.31) to have the equations which govern the boundary for linearised instability, i.e.

$$\begin{aligned} 0 &= -\pi_{,i} - u_i + Rk_i\theta, \\ u_{i,i} &= 0, \\ 0 &= Rw + \Delta\theta. \end{aligned} \tag{4.34}$$

To investigate nonlinear energy stability we multiply $(4.29)_1$ by u_i, $(4.29)_3$ by θ and integrate each over V. This gives

$$0 = R(\theta, w) - \|\mathbf{u}\|^2, \tag{4.35}$$

$$\frac{d}{dt}\frac{1}{2}\|\theta\|^2 = R(w, \theta) - \|\nabla\theta\|^2. \tag{4.36}$$

To apply Joseph's coupling parameter method we again multiply one of (4.35) or (4.36) by a positive parameter, add the equations and then select this parameter optimally. So, add λ times (4.36) to (4.35). The result is

$$\frac{dE}{dt} = RI - D \tag{4.37}$$

where now

$$E(t) = \frac{\lambda}{2}\,\|\theta(t)\|^2, \tag{4.38}$$

$$I(t) = (1+\lambda)\,(w,\theta), \tag{4.39}$$

$$D(t) = \|\mathbf{u}\|^2 + \lambda\|\nabla\theta\|^2. \tag{4.40}$$

The idea is now to proceed as in section 4.1 and define

$$\frac{1}{R_E} = \max_H \frac{I}{D} \tag{4.41}$$

where now H is such that $u_i \in L^2(V)$, $\theta \in H^1(V)$, and $w = 0, \theta = 0$, on $z = 0, 1$. In this way we find from (4.37)

$$\frac{dE}{dt} \le -D\left(\frac{R_E - R}{R_E}\right). \tag{4.42}$$

Since $D \ge \lambda\pi^2\|\theta\|^2 + \|\mathbf{u}\|^2$ by using the Poincaré inequality, we see that $D \ge 2\pi^2 E$. Thus our nonlinear stability criterion is now

$$R < R_E \tag{4.43}$$

for then $\omega = 2\pi^2(R_E - R)/R_E > 0$ and (4.42) leads to

$$\frac{dE}{dt} \le -\omega E. \tag{4.44}$$

Then by integration and rearranging

$$E(t) \le \exp(-\omega t)E(0). \tag{4.45}$$

Thus, (4.41), (4.43) yield *unconditional* nonlinear stabilty in the porous Bénard problem when the equations for the porous medium are those of Darcy. Of course, for the Darcy problem, all (4.45) shows is that $\|\theta(t)\|$ decays exponentially. However, from (4.35) we may use the arithmetic-geometric mean inequality to deduce

$$\|\mathbf{u}\|^2 = R(\theta, w) \le \frac{R^2}{2}\|\theta\|^2 + \frac{1}{2}\|w\|^2 \le \frac{R^2}{2}\|\theta\|^2 + \frac{1}{2}\|\mathbf{u}\|^2,$$

and this leads to

$$\|\mathbf{u}\|^2 \le R^2\|\theta\|^2.$$

Hence, (4.43) leads also to exponential decay of $\|\mathbf{u}(t)\|$.

The nonlinear stability threshold is now given by the variational problem (4.41). It is typical of nonlinear energy stability calculations, as in section

4.1 and here, that we find a variational problem like (4.41), determine the Euler-Lagrange equations and maximize in the coupling parameter λ to obtain the best value of R_E. The maximum problem (4.41) is

$$\frac{1}{R_E} = \max_H \frac{(1+\lambda)(\theta, w)}{\|\mathbf{u}\|^2 + \lambda\|\nabla\theta\|^2}. \tag{4.46}$$

It is again convenient to rescale θ by putting $\hat{\theta} = \sqrt{\lambda}\,\theta$. Then, since we are seeking the maximum over a space H we find, in a similar manner to the calculation leading to (4.12), that (4.46) is equivalent to

$$\frac{1}{R_E} = \max_H \frac{f(\lambda)\,(\theta, w)}{\|\mathbf{u}\|^2 + \|\nabla\theta\|^2}$$

where $f(\lambda) = (1+\lambda)/\sqrt{\lambda}$. Hence, the Euler-Lagrange equations arising from (4.41) are determined from

$$R_E\delta I - \delta D = 0$$

where now (we incorporate the constraint $u_{i,i} = 0$ in I)

$$\begin{aligned}\delta I =& \frac{d}{d\epsilon} f\,(\theta + \epsilon\eta, w + \epsilon h_3)\Big|_{\epsilon=0} - 2\frac{d}{d\epsilon}\,(\pi, u_{i,i} + \epsilon h_{i,i})\Big|_{\epsilon=0}\\ =& f(\lambda)(w,\eta) + f(\lambda)(h_3,\theta) + 2(\pi_{,i}, h_i)\end{aligned}$$

and

$$\begin{aligned}\delta D =& \frac{d}{d\epsilon}\left(\|\mathbf{u} + \epsilon\mathbf{h}\|^2 + \|\nabla(\theta + \epsilon\eta)\|^2\right)\Big|_{\epsilon=0}\\ =& 2\big[(u_i, h_i) - (\Delta\theta, \eta)\big].\end{aligned}$$

Thus, the Euler-Lagrange equations from (4.41) are

$$\begin{aligned}0 &= -\pi_{,i} - u_i + \frac{1}{2}R_E f\theta k_i,\\ u_{i,i} &= 0,\\ 0 &= \frac{1}{2}fR_E w + \Delta\theta.\end{aligned} \tag{4.47}$$

One now uses the variation of parameters method as in 4.1 to show $\lambda = 1$ is the optimal value, i.e. that value which maximises R_E. The details are very similar to those following from (4.16) and so we only outline the changes. In fact (4.16), (4.17) must be replaced by

$$\begin{aligned}R_E^2 f^2(\theta^2, w^1) + (u_i^2, u_i^1) = 0,\\ R_E^1 f^1(\theta^1, w^2) + (u_i^1, u_i^2) = 0,\end{aligned}$$

while (4.18), (4.19) remain unchanged. One again arrives at (4.21). Instead of (4.23) we find

$$\left(R_E\frac{df}{d\lambda} + f\frac{\partial R_E}{\partial\lambda}\right)(\|\mathbf{u}\|^2 + \|\nabla\theta\|^2) = 0.$$

However, for the best value of R_E, $\partial R_E/\partial\lambda = 0$ and so again we need $df/d\lambda = 0$ and so we find $\lambda = 1$ yields the sharpest nonlinear stability boundary.

With $\lambda = 1$ equations (4.47) become

$$\begin{aligned} 0 &= -\pi_{,i} - u_i + R_E\theta k_i, \\ u_{i,i} &= 0, \\ 0 &= R_E w + \Delta\theta. \end{aligned} \tag{4.48}$$

These are the same as the linear instability equations (4.34). Thus, again we have the optimal result that $R_L = R_E$. This means that the linear instability critical Rayleigh number is the same as the nonlinear stability Rayleigh number. This result holds for all initial data. It is to be stressed that this is for the Darcy equations of porous convection.

We now calculate the critical Rayleigh number $R_L = R_E$ arising from (4.48). To do this we first remove the pressure by taking curlcurl of $(4.48)_1$. This gives

$$0 = \Delta u_i + R_E(k_j\theta_{,ij} - k_i\Delta\theta), \tag{4.49}$$

since $u_{i,i} = 0$. Now, pick $i = 3$ in this equation. Thus, we find a coupled system in w and θ of form

$$\begin{aligned} 0 &= \Delta w - R_E\Delta^*\theta, \\ 0 &= R_E w + \Delta\theta, \end{aligned} \tag{4.50}$$

where $\Delta^* = \partial^2/\partial x^2 + \partial^2/\partial y^2$, and $w = \theta = 0$ at $z = 0, 1$. Since $R_E = R_L$ we omit the E and simply solve (4.50) for R. We seek a solution of form $w = W(z)f(x,y)$, $\theta = \Theta(z)f(x,y)$ where $f(x,y)$ is a planform which tiles the plane. Typically f is the hexagonal solution given by (3.48). So,

$$\Delta^* f = -a^2 f, \tag{4.51}$$

where a is the horizontal wavenumber.

Let $D = d/dz$ then from (4.50) we eliminate θ to find

$$\Delta^2 w = -R^2\Delta^* w.$$

Thus,

$$(D^2 - a^2)^2 W = R^2 a^2 W. \tag{4.52}$$

Since $w = \theta = 0$ on $z = 0, 1$, we may show from (4.52) that $D^{(2n)}W = 0$ on $z = 0, 1$. Hence, in (4.52) we may select $W = \sin n\pi z$. Thus (4.52) becomes

$$\frac{(n^2\pi^2 + a^2)^2}{a^2} = R^2.$$

We want the smallest value of R^2 as a function of n and so take $n = 1$. Then we must minimise $R^2(a^2)$ in a^2, i.e. minimise $R^2 = (\pi^2 + a^2)^2/a^2$.

This leads to $a_c^2 = \pi^2$, where c denotes the critical value. Then, $R_c^2 = 4\pi^2$. Thus,

$$R_E^2 = R_L^2 = 4\pi^2.$$

For a Rayleigh number Ra less than $4\pi^2$ we cannot have any instability, i.e. all perturbations decay rapidly to zero. This is a nonlinear result which holds for all initial data. On the other hand if the Rayleigh number Ra is greater than $4\pi^2$ instability occurs and cellular convective motion is witnessed.

4.2.2 The Bénard problem for the Forchheimer equations

Suppose now that the porous medium is governed by the Forchheimer equations (3.63). We assume the porous material fills the three-dimensional region $\{(x,y) \in \mathbb{R}^2\} \times \{z \in (0,d)\}$ as in section 4.2.1. The boundary conditions are the same, the steady state solution is (4.27), and the same non-dimensionalisation leading to (4.29) is employed except we additionally need to account for the Forchheimer term in $(3.63)_1$, $b|\mathbf{v}|v_i$. Thus, the new non-dimensional variable $F = kb\kappa/\mu d$ arises. Instead of the dimensionless perturbation equations (4.29), when we employ the Forchheimer equations (3.63) we arrive at the dimensionless perturbation equations

$$\begin{aligned} 0 &= -\pi_{,i} - u_i + R\theta k_i - F|\mathbf{u}|u_i, \\ u_{i,i} &= 0, \\ \theta_{,t} + u_i\theta_{,i} &= Rw + \Delta\theta, \end{aligned} \tag{4.53}$$

where $w = u_3$. These equations hold on $\{(x,y) \in \mathbb{R}^2\} \times \{z \in (0,1)\} \times \{t > 0\}$. The boundary conditions are again (4.30), i.e.

$$\theta = w = 0, \qquad z = 0, 1,$$

with (u_i, θ, π) satisfying a plane tiling periodicity in the (x, y) directions.

To determine the linear instability boundary from (3.63) we now discard the $u_i\theta_{,i}$ term and also the Forchheimer term $F|\mathbf{u}|u_i$. After writing $u_i = u_i(\mathbf{x})e^{\sigma t}$ with a similar representation for θ and π the linearized equations are derived. These are exactly the same as those of Darcy theory, i.e. (4.31). Thus, the linear instability boundary is exactly the same as that found by employing Darcy's law.

To develop an energy theory from (4.53) we proceed exactly as in the Darcy case, i.e. equations (4.37) - (4.41), the only difference is that we must now include a term of form $F < |\mathbf{u}|^3 >$. Thus, with E, I and D defined by (4.38) - (4.40) we find instead of (4.37) the equation

$$\frac{dE}{dt} = RI - D - F < |\mathbf{u}|^3 > . \tag{4.54}$$

Next, we discard the non-positive Forchheimer term from the right of (4.54) and find E satisfies the differential inequality

$$\frac{dE}{dt} \le RI - D.$$

The development from this point is exactly the same as for the Darcy equations. Thus, we again find the optimal result that the nonlinear critical Rayleigh number $Ra_E = R_E^2$ is the same as the linear critical Rayleigh number $Ra_L = R_L^2$ and indeed, we have $Ra_E = Ra_L = 4\pi^2$.

Hence, in the current situation the Forchheimer term plays no role in the instability or stability process since the optimum result is achieved which is the same as that found in the Darcy case. However, the Forchheimer term can make a difference when the fluid properties vary with temperature. For example, in section 16.6, it is shown how the Forchheimer theory leads to different conclusions from the Darcy theory when the viscosity is a function of temperature.

4.2.3 The Bénard problem for the Darcy equations with anisotropic permeability

In this subsection we treat the problem of stability of a layer of saturated porous medium heated from below when the permeability is anisotropic. The anisotropy we allow is the one discussed in section 3.4.4 where the permeability in the vertical direction may be different from that in the horizontal directions. The relevant equations governing convection are then (3.64).

Again, the steady solution whose stability is under investigation is

$$\bar{v}_i \equiv 0, \quad \bar{T} = -\beta z + T_L\,,$$

where the porous medium occupies the infinite plane layer $\mathbb{R}^2 \times \{z \in (0,d)\}$ with upper and lower boundary conditions as given in (4.30). Let (u_i, θ, π) be perturbations to $(\bar{v}_i, \bar{T}, \bar{p})$ and then the perturbation equations are

$$\begin{aligned} \mu M_{ij} u_j &= -\pi_{,i} + \rho_0 g \alpha k_i \theta, \qquad u_{i,i} = 0, \\ \theta_{,t} + u_i \theta_{,i} &= \beta w + \kappa \Delta \theta. \end{aligned} \tag{4.55}$$

Let now $m_{ij} = \mathrm{diag}\,(\zeta, \zeta, 1)$ where $\zeta = k/k_1$. Then (4.55) are non-dimensionalized with the scalings $\mathbf{x} = \mathbf{x}^* d$, $t = t^* \mathcal{T}$, $\mathcal{T} = d^2/\kappa$, $u_i = u_i^* U$, $U = \kappa/d$, $\pi = \pi^* P$, $P = \mu U d/k$, $\theta = \theta^* T^\sharp$, $T^\sharp = Ud\sqrt{\mu\beta/\kappa g \rho_0 \alpha k}$, with the Rayleigh number $Ra = R^2$ being defined by

$$R^2 = \frac{d^2 g \rho_0 \alpha k \beta}{\mu \kappa}\,.$$

The perturbation equations in non-dimensional form are

$$\begin{aligned} m_{ij}u_j &= -\pi_{,i} + Rk_i\theta, \\ u_{i,i} &= 0, \\ \theta_{,t} + u_i\theta_{,i} &= Rw + \Delta\theta. \end{aligned} \tag{4.56}$$

We note that the linearised instability equations become

$$\begin{aligned} m_{ij}u_j &= -\pi_{,i} + Rk_i\theta, \\ u_{i,i} &= 0, \\ \sigma\theta &= Rw + \Delta\theta, \end{aligned} \tag{4.57}$$

where u_i has been written as $u_i(\mathbf{x})e^{\sigma t}$, with similar forms for θ and π.

To develop a nonlinear energy stability analysis we multiply $(4.56)_1$ by u_i and integrate over a period cell V to obtain

$$< m_{ij}u_ju_i >= R(\theta, w). \tag{4.58}$$

Likewise, we multiply $(4.56)_3$ by θ and integrate over V to find

$$\frac{d}{dt}\frac{1}{2}\|\theta\|^2 = R(\theta, w) - \|\nabla\theta\|^2. \tag{4.59}$$

By adding λ(4.58) to (4.59) we derive an energy equation of form

$$\frac{dE}{dt} = RI - D \tag{4.60}$$

where now

$$\begin{aligned} E &= \frac{1}{2}\|\theta\|^2, \qquad I = (1+\lambda)(\theta, w), \\ D &= \|\nabla\theta\|^2 + \lambda < m_{ij}u_iu_j > . \end{aligned}$$

Since $m_{ij} = \mathrm{diag}\,(\zeta, \zeta, 1)$, $< m_{ij}u_iu_j > \geq k_0\|\mathbf{u}\|^2$ where $k_0 = \min\{k/k_1, 1\}$, and from Poincaré's inequality $\|\nabla\theta\|^2 \geq \pi^2\|\theta\|^2$. Thus I/D is bounded and one can show a maximising solution exists to the problem

$$\frac{1}{R_E} = \max_H \frac{I}{D} \tag{4.61}$$

where H is the same space of admissible solutions as in the Darcy Bénard problem. Then from (4.60) we derive

$$\frac{dE}{dt} \leq -D\Big(\frac{R_E - R}{R_E}\Big) \leq -2\pi^2\Big(\frac{R_E - R}{R_E}\Big)E$$

provided $R < R_E$. From this inequality we find

$$E(t) \leq E(0)\,\exp\Big(-\frac{2\pi^2}{R_E}\,(R_E - R)t\Big)$$

and nonlinear stability follows, for all initial data. It remains to find R_E to solve the nonlinear stability problem. The Euler-Lagrange equations from

(4.61) are (replacing u_i by $\hat{u}_i = \sqrt{\lambda}\, u_i$ and then dropping the hat),

$$\begin{aligned} &R_E \frac{f}{2}\theta k_i - m_{ij}u_j = \pi_{,i}, \qquad u_{i,i} = 0, \\ &R_E \frac{f}{2} w + \Delta\theta = 0 \end{aligned} \tag{4.62}$$

where $f(\lambda) = (1+\lambda)/\sqrt{\lambda}$.

One now uses variation of parameters. Thus, let $f/2 = h$, and multiply $(4.62)_1$ evaluated at λ^2 by u_i^1, $(4.62)_1$ evaluated at λ^1 by u_i^2, $(4.62)_3$ evaluated at λ^2 by θ^1, and $(4.62)_3$ evaluated at λ^1 by θ^2. After integration over V the results are

$$R_E^2 h^2(\theta^2, w^1) - < m_{ij}u_j^2 u_i^1 > = 0, \tag{4.63}$$

$$R_E^1 h^1(\theta^1, w^2) - < m_{ij}u_j^1 u_i^2 > = 0, \tag{4.64}$$

$$R_E^2 h^2(w^2, \theta^1) - (\nabla\theta^1, \nabla\theta^2) = 0, \tag{4.65}$$

$$R_E^1 h^1(w^1, \theta^2) - (\nabla\theta^1, \nabla\theta^2) = 0. \tag{4.66}$$

We now form the combination (4.63)-(4.64)+(4.65)-(4.66) to find

$$(R_E^2 h^2 - R_E^1 h^1)\big[(\theta^2, w^1) + (w^2, \theta^1)\big] = 0.$$

Next, recall $\lambda_1 \neq \lambda_2$, divide by $\lambda_2 - \lambda_1$ to obtain

$$\left[\frac{R_E^2(h^2 - h^1) + (R_E^2 - R_E^1)h^1}{\lambda_2 - \lambda_1}\right]\big[(\theta^2, w^1) + (w^2, \theta^1)\big] = 0.$$

Take the limit $\lambda_2 \to \lambda_1$ to find

$$\left[R_E \frac{\partial h}{\partial\lambda} + h\frac{\partial R_E}{\partial\lambda}\right](\theta, w) = 0. \tag{4.67}$$

Directly from (4.62) one shows

$$2R_E h(\theta, w) = \|\nabla\theta\|^2 + < m_{ij}u_i u_j > . \tag{4.68}$$

Thus, rearranging between (4.67) and (4.68) we see that

$$\left[\frac{\|\nabla\theta\|^2 + < m_{ij}u_i u_j >}{R_E f}\right]\Big[f\frac{\partial R_E}{\partial\lambda} + R_E\frac{\partial f}{\partial\lambda}\Big] = 0$$

where we note $f = 2h$. At the maximum value of R_E, $\partial R_E/\partial\lambda = 0$, and so $\partial f/\partial\lambda = 0$ gives the best value of λ. It is easily seen that $\lambda = 1$. Hence, for the maximum value of R_E as a function of λ we set $\lambda = 1$ in (4.62). Thus the Euler-Lagrange equations become

$$\begin{aligned} &R_E\theta k_i - m_{ij}u_j = \pi_{,i}, \qquad u_{i,i} = 0, \\ &R_E w + \Delta\theta = 0. \end{aligned} \tag{4.69}$$

Observe that equations (4.69) are the same as the linear instability equations (4.57) if $\sigma = 0$. In fact, by multiplying $(4.57)_1$ by u_i^*, $(4.57)_3$ by θ^* and integrating, one finds as in the Darcy case that $\sigma \in \mathbb{R}$. Hence, the equations for linear instability are exactly the same as those for nonlinear energy stability. Hence, the linear critical Rayleigh number Ra_L is the same as the nonlinear critical Rayleigh number Ra_E, even when the permeability is transversely isotropic in the $z-$direction. Since this is an unconditional nonlinear stability result this means no subcritical instabilities can arise even in this anisotropic case.

Recall that equations (4.69) hold on $\mathbb{R}^2 \times (0,1)$ and are to be solved subject to the boundary conditions

$$w = \theta = 0, \qquad z = 0,1.$$

The pressure π is removed by taking curlcurl of $(4.57)_1$ and then the third component of the result is

$$-R_E \Delta^* \theta - m_{jr} u_{r,3j} + m_{3r} u_{r,jj} = 0.$$

Recollecting $m_{ij} = \text{diag}\,(\zeta,\zeta,1)$ and using $u_{i,i} = 0$, this reduces to

$$-R_E \Delta^* \theta + \Delta^* w + \zeta w_{,33} = 0. \tag{4.70}$$

Thus we must solve $(4.69)_3$ and (4.70) for θ and w. The plane tiling form $w = Wf, \theta = \Theta f$ is assumed as in the Darcy problem, where f satisfies (4.51). Upon removal of θ we find

$$\Delta\left(\Delta^* + \zeta \frac{\partial^2}{\partial z^2}\right) w = -R_E^2 \Delta^* w.$$

Hence the form $w = Wf$ followed by putting $W(z) = \sin n\pi z$ (due to the boundary conditions) yields

$$R_E^2 = \frac{(n^2\pi^2 + a^2)(\zeta n^2\pi^2 + a^2)}{a^2}.$$

For R_E^2 smallest in n we need $n = 1$. Then taking dR_E^2/da^2 yields $a_c^2 = \sqrt{\zeta}\,\pi^2$. This, in turn, leads to

$$R_{E\,crit}^2 = R_{L\,crit}^2 = (1 + \sqrt{\zeta})^2 \pi^2. \tag{4.71}$$

From (4.71) several deductions follow. Firstly, note $\zeta = k/k_1$. Thus, when $k = k_1$, $\zeta = 1$ and then $a_c^2 = \pi^2$, $Ra_{crit} = 4\pi^2$, which is the Darcy result, as it should be. Also, as $\zeta \to \infty$, $Ra_{crit} \to \infty$, $a_c^2 \to \infty$ (this corresponds to the permeability all being in the vertical direction). Finally, we observe that as $\zeta \to 0$, $Ra_{crit} \to \pi^2$, $a_c^2 \to 0$ (which corresponds to the permeability all being in the horizontal direction).

4.2.4 The Bénard problem for the Brinkman equations

The equations for thermal convection according to the Brinkman model are given in (3.62). These are rewritten for clarity

$$\begin{aligned} 0 &= -p_{,i} - \frac{\mu}{k}v_i + \tilde{\lambda}\Delta v_i - k_i g\rho(T), \\ v_{i,i} &= 0, \\ T_{,t} + v_i T_{,i} &= \kappa\Delta T, \end{aligned} \tag{4.72}$$

where $\rho(T) = \rho_0(1 - \alpha(T - T_0))$. These equations are fundamentally different from the three other porous systems we have analysed in that the order of (4.72) is two higher due to the $\tilde{\lambda}\Delta v_i$ (Brinkman) term. Thus, in prescribing boundary conditions on the planar boundaries $z = 0, d$, we must specify all components of $\mathbf{v}$, not just v_3.

For the thermal convection problem in hand (4.72) hold on $\mathbb{R}^2 \times (0, d) \times \{t > 0\}$ and the boundary conditions are $v_i = 0$, $z = 0, d$, $T = T_U$, $z = d$, $T = T_L$, $z = 0$, with $T_L > T_U$. The steady solution whose stability we investigate is

$$\bar{T} = -\beta z + T_L, \qquad \bar{v}_i = 0,$$

where $\beta = (T_L - T_U)/d$. Letting $v_i = \bar{v}_i + u_i$, $T = \bar{T} + \theta$, $p = \bar{p} + \pi$, the perturbation equations arising from (4.72) are

$$\begin{aligned} 0 &= -\pi_{,i} - \frac{\mu}{k}u_i + \tilde{\lambda}\Delta u_i + k_i g\rho_0\alpha\theta, \\ u_{i,i} &= 0, \\ \theta_{,t} + u_i\theta_{,i} &= \beta w + \kappa\Delta\theta. \end{aligned} \tag{4.73}$$

These equations are non-dimensionalized with the scalings $\mathbf{x} = \mathbf{x}^* d$, $t = t^*\mathcal{T}$, $\pi = \pi^* P$, $u_i = u_i^* U$, $P = Ud\mu/k$, $\mathcal{T} = d^2/\kappa$, $U = \kappa/d$, $\lambda = \tilde{\lambda}k/d^2\mu$, and the Rayleigh number $Ra = R^2$ is defined as

$$Ra = \frac{d^2 g\rho_0\alpha k\beta}{\mu\kappa}.$$

The non-dimensional perturbation equations arising from (4.73) are (dropping $*$'s)

$$\pi_{,i} = -u_i + \lambda\Delta u_i + R\theta k_i, \tag{4.74}$$

$$u_{i,i} = 0, \tag{4.75}$$

$$\theta_{,t} + u_i\theta_{,i} = Rw + \Delta\theta, \tag{4.76}$$

which hold on $\mathbb{R}^2 \times \{z \in (0,1)\} \times \{t > 0\}$. The boundary conditions are

$$u_i = 0, \theta = 0, \qquad \text{on } z = 0, 1,$$

and u_i, θ, π satisfy a plane tiling periodicity.

Firstly we note that the linearized equations which follow from (4.74)-(4.76) are

$$\pi_{,i} = -u_i + \lambda\Delta u_i + R\theta k_i, \tag{4.77}$$

$$u_{i,i} = 0, \tag{4.78}$$

$$\sigma\theta = Rw + \Delta\theta. \tag{4.79}$$

One may show $\sigma \in \mathbb{R}$. To do this one multiplies (4.77) by u_i^*, (4.79) by θ^*, integrates each over a period cell V and adds. Taking the imaginary part of the result leads to the stated conclusion.

A nonlinear energy analysis may be developed by multiplying (4.74) by u_i and integrating over V to find

$$0 = -\|\mathbf{u}\|^2 - \lambda\|\nabla\mathbf{u}\|^2 + R(\theta, w). \tag{4.80}$$

In a similar manner we multiply (4.76) by θ and integrate over V to obtain

$$\frac{d}{dt}\frac{1}{2}\|\theta\|^2 = R(\theta, w) - \|\nabla\theta\|^2. \tag{4.81}$$

For a positive coupling parameter ξ, form ξ(4.81) + (4.80) to find

$$\frac{dE}{dt} = RI - D, \tag{4.82}$$

where now

$$E = \frac{1}{2}\xi\|\theta\|^2, \qquad I = (1+\xi)(w, \theta),$$
$$D = \|\mathbf{u}\|^2 + \lambda\|\nabla\mathbf{u}\|^2 + \xi\|\nabla\theta\|^2.$$

One sets

$$\frac{1}{R_E} = \max_H \frac{I}{D} \tag{4.83}$$

where $H = \{u_i, \theta \in H^1(V) | u_{i,i} = 0\}$ and the solutions satisfy a horizontal plane tiling periodicity. Then from (4.82)

$$\frac{dE}{dt} \leq -D\Big(1 - \frac{R}{R_E}\Big).$$

If $R < R_E$ then put $a = (R_E - R)/R_E (> 0)$ and note that $D \geq \xi\pi^2\|\theta\|^2 = 2\pi^2 E$. Thus, one derives

$$\frac{dE}{dt} \leq -2\pi^2 aE.$$

This yields

$$E(t) \leq \exp(-2\pi^2 at)\, E(0)$$

from which global nonlinear energy stability follows (i.e. for all initial data). The only condition imposed is $R < R_E$.

We next put $f(\xi) = (1+\xi)/2\sqrt{\xi}$ and let $\sqrt{\xi}\,\theta \to \theta$ in (4.83) to scale out the ξ from the denominator. The Euler-Lagrange equations which then arise from (4.83) are

$$R_E f(\xi)\theta k_i - u_i + \lambda\Delta u_i = \pi_{,i}, \tag{4.84}$$

$$u_{i,i} = 0, \tag{4.85}$$

$$R_E f(\xi) w + \Delta\theta = 0, \tag{4.86}$$

where in (4.84), π is a Lagrange multiplier.

The steps in the variation of parameters proof of the previous subsections may be followed to find

$$\begin{aligned}
&R_E^2 f^2(\theta^2, w^1) - (u_i^2, u_i^1) - \lambda(\nabla u_i^2, \nabla u_i^1) = 0,\\
&R_E^1 f^1(\theta^1, w^2) - (u_i^1, u_i^2) - \lambda(\nabla u_i^1, \nabla u_i^2) = 0,\\
&R_E^2 f^2(w^2, \theta^1) - (\nabla\theta^1, \nabla\theta^2) = 0,\\
&R_E^1 f^1(w^1, \theta^2) - (\nabla\theta^2, \nabla\theta^1) = 0,
\end{aligned}$$

where now λ is constant and f^i denotes $f(\xi^i)$, $i = 1, 2$. From the above equations one arrives at, for $\partial R_E/\partial\xi = 0$,

$$f^{-1}\frac{df}{d\xi}\left(\|\mathbf{u}\|^2 + \lambda\|\nabla\mathbf{u}\|^2 + \|\nabla\theta\|^2\right) = 0.$$

Thus, the optimal value of ξ is $\xi = 1$.

With $\xi = 1$ equations (4.84) - (4.86) reduce to

$$\begin{aligned}
&R_E\theta k_i - u_i + \lambda\Delta u_i = \pi_{,i},\\
&u_{i,i} = 0,\\
&R_E w + \Delta\theta = 0.
\end{aligned} \tag{4.87}$$

Notice that (4.87) are the same as (4.77) - (4.79) with $\sigma = 0$, which we adopt since $\sigma \in \mathbb{R}$. Thus we may again conclude that $Ra_L = Ra_E$, i.e. the linear instability boundary coincides with the nonlinear stability one. Since this result is unconditional, i.e. it holds for all initial data, this precludes any subcritical instabilities. Note that this optimal result holds for the Darcy theory, Forchheimer theory, the transversely isotropic theory covered earlier, and also for the Brinkman theory.

To find $R_E(= R_L)$ we set $R_E = R$ and take curlcurl $(4.87)_1$ and then (4.87) reduce to the system

$$\begin{aligned}
-R\Delta^*\theta + \Delta w - \lambda\Delta^2 w &= 0,\\
Rw + \Delta\theta &= 0,
\end{aligned} \tag{4.88}$$

with

$$w = \theta = 0 \qquad \text{at} \quad z = 0, 1. \tag{4.89}$$

Two more boundary conditions are needed on w at $z = 0, 1$ and these depend on whether the surfaces are fixed or free of tangential stress. If either one is fixed then numerical solution of (4.88) is recommended.

For purposes of illustration we here consider two stress free surfaces and then in addition to (4.89) we impose

$$w_{,zz} = 0 \qquad \text{at} \quad z = 0, 1.$$

The plane tiling (x, y) form of (4.51) is assumed and then due to evenness in (4.88) and the boundary conditions, $W(z)$ is taken as $\sin n\pi z$. If we eliminate θ from (4.88), w satisfies

$$(\lambda\Delta^3 - \Delta^2)w = R^2\Delta^* w$$

and then $w = f(x, y) \sin n\pi z$ leads to

$$R^2 = \frac{\lambda(n^2\pi^2 + a^2)^3 + (n^2\pi^2 + a^2)^2}{a^2}.$$

It is worth observing that as $\lambda \to 0$ we obtain the equivalent expression for a Darcy porous material, whereas if we let $\lambda \to \infty$ we approach that for a fluid. To minimize R^2 in n we take $n = 1$ and then $dR^2/da^2 = 0$ yields the critical value of a^2 as

$$a_c^2 = \frac{-(\lambda\pi^2 + 1) + (\lambda\pi^2 + 1)\sqrt{1 + 8\pi^2\lambda/(\lambda\pi^2 + 1)}}{4\lambda}.$$

When $\lambda \to \infty$, $a_c^2 \to \pi^2/2$ as in the fluid case, whereas when $\lambda \to 0$, $a_c^2 \to \pi^2$ which is the Darcy case.

4.3 Symmetry and competing effects

In this section we examine an important idea in the theory of energy stability: the concept of symmetry, which is a major reason why the energy method works so well for convection problems.

It is convenient to investigate the above topic in the setting of Bénard convection with an internal heat source, while we also allow the gravity g to depend on the vertical coordinate z. To this end we introduce the relevant equations for a fluid *and* for fluid flow in a porous solid.

We shall adopt the Boussinesq approximation and so in the body force term we write the density as

$$\rho = \rho_0\Big[1 - \alpha(T - T_0)\Big], \tag{4.90}$$

where ρ_0 is a constant, T is temperature, T_0 is a reference temperature and α $(= -\rho_0^{-1}(\partial\rho/\partial T)(T_0))$ is the thermal expansion coefficient. For convective motion in an incompressible Newtonian fluid, with spatially varying gravity field and an imposed internal heat source, the relevant equations are then

$$\begin{aligned} v_{i,t} + v_j v_{i,j} &= -\frac{1}{\rho_0} p_{,i} + \nu \Delta v_i - k_i g(z)\big(1 - \alpha[T - T_0]\big), \\ v_{i,i} &= 0, \\ T_{,t} + v_i T_{,i} &= \kappa \Delta T + Q(z), \end{aligned} \tag{4.91}$$

where $\mathbf{v}, p, \nu, \Delta, \kappa$, and Q are, respectively, velocity, pressure, viscosity, the Laplacian operator, thermal diffusivity, and internal heat source, and $\mathbf{k} = (0, 0, 1)$. The analogous equations for convective fluid motion in a porous solid, according to Darcy's law, cf. 3.4.1, may be written

$$\begin{aligned} 0 &= -p_{,i} - \frac{\mu}{k} v_i - k_i \rho_0 g(z)\big(1 - \alpha[T - T_0]\big), \\ v_{i,i} &= 0, \\ T_{,t} + v_i T_{,i} &= \kappa \Delta T + Q(z). \end{aligned} \tag{4.92}$$

In these equations $\mathbf{v}$ is the seepage velocity, μ is dynamic viscosity, k is the permeability of the porous medium, and Q and the time t are adjusted for porosity and specific heat variations between the fluid and solid. Both sets of equations (4.91) and (4.92) are defined on the spatial region $\{z \in (0, d)\} \times \mathbb{R}^2$.

The conduction solution to either set of equations that satisfies the boundary conditions

$$T = T_\ell, \quad z = 0; \qquad T = T_u, \quad z = d; \tag{4.93}$$

where $T_\ell > T_u$, is

$$\bar{\mathbf{v}} \equiv \mathbf{0}, \qquad \bar{T} = -\frac{1}{\kappa} \int_0^z \int_0^\xi Q(\eta) d\eta \, d\xi - cz + T_\ell, \tag{4.94}$$

where the constant c is given by

$$c = \frac{T_\ell - T_u}{d} - \frac{1}{\kappa d} \int_0^d \int_0^\xi Q(\eta) d\eta \, d\xi,$$

and where the hydrostatic pressure, $\bar{p}(z)$, is determined from either (4.91) or (4.92), whichever is relevant.

To investigate the stability/instability of (4.94) we introduce perturbations (not necessarily small) such that $\mathbf{v} = \bar{\mathbf{v}} + \mathbf{u}$, $T = \bar{T} + \theta$, $p = \bar{p} + p$, and then calculate the appropriate equations for the perturbations $\mathbf{u}(\mathbf{x}, t)$, $\theta(\mathbf{x}, t)$, $p(\mathbf{x}, t)$. The equations for the fluid case (4.91) are non-dimensionalized according to the scalings:

$$\begin{aligned} &\mathbf{x} = \mathbf{x}^* d, \qquad \mathbf{u} = \mathbf{u}^* U, \qquad U = \frac{\nu}{d}, \qquad \theta = \theta^* T^\sharp, \\ &T^\sharp = \left(\frac{Pr\, c}{g\alpha}\right)^{1/2} U, \qquad P = \frac{U \rho_0 \nu}{d}, \qquad p = p^* P, \qquad Pr = \frac{\nu}{\kappa}, \\ &t^* = \frac{t\nu}{d^2}, \qquad \delta q(z) = \frac{F(z)}{c}, \qquad R = \left(\frac{cg\alpha d^4}{\kappa\nu}\right)^{1/2}, \end{aligned} \tag{4.95}$$

where the starred quantities are dimensionless, Pr is the Prandtl number, R^2 is the Rayleigh number, $\delta q(z)$ is defined as indicated with

$$F(z) = \frac{1}{\kappa}\int_0^z Q(\xi)d\xi, \tag{4.96}$$

and δ is a constant being a scaling for $q(z)$.

The non-dimensional perturbation equations for the fluid are then (omitting all stars)

$$u_{i,t} + u_j u_{i,j} = -p_{,i} + H(z)R\theta k_i + \Delta u_i\,, \tag{4.97}$$

$$u_{i,i} = 0, \tag{4.98}$$

$$Pr(\theta_{,t} + u_i\theta_{,i}) = RN(z)w + \Delta\theta, \tag{4.99}$$

where $\mathbf{u} = (u, v, w)$, $N(z) = 1 + \delta q(z)$ and $H(z) = 1 + \epsilon h(z)$, $g(z)$ having been defined by $g(z) = g[1 + \epsilon h(z)]$, g constant, and ϵ being a scale for h. These equations hold on the region $\{(x, y) \in \mathbb{R}^2\} \times \{z \in (0, 1)\} \times \{t > 0\}$ and

$$\theta = 0, \quad z = 0, 1; \qquad \mathbf{u} = \mathbf{0}, \quad z = 0, 1. \tag{4.100}$$

A word of caution is in order concerning the Rayleigh number. We have tacitly assumed $c > 0$; if this is not the case we employ $|c|$ with an appropriate insertion of the signum function for R. Also, since c involves Q, which defines $q(z)$, care must be taken in interpreting results involving R^2; however, our definition is consistent with that of (Joseph and Shir, 1966) who studied the equivalent problem for g constant.

An equivalent analysis for the porous medium equations (4.92) commences again with the steady solution (4.94). The scalings change in so far as ν is replaced by μ/ρ_0 in P, U, and Pr. The temperature scale and Rayleigh number are replaced by $T^\sharp = U(c\mu d^2/k\rho_0 g\alpha\kappa)^{1/2}$, $R = (d^2 k\rho_0 g\alpha/\mu\kappa)^{1/2}$, the other scalings remaining the same. With this replacement, the perturbation equations arising from (4.92) are (again all variables are dimensionless),

$$p_{,i} = H(z)R\theta k_i - u_i\,, \tag{4.101}$$

$$u_{i,i} = 0, \tag{4.102}$$

$$Pr(\theta_{,t} + u_i\theta_{,i}) = RN(z)w + \Delta\theta, \tag{4.103}$$

with θ satisfying (4.100) and,

$$u_i n_i = \pm w = 0, \quad \text{on} \quad z = 0, 1, \tag{4.104}$$

with $\mathbf{n}$ being the unit outward normal to the planes $z = 0, 1$, and where the plus sign is taken if $z = 1$, the minus sign when $z = 0$.

In section 4.1 we saw that the critical Rayleigh numbers for linear instability and for nonlinear energy stability are identical for the classical Bénard problem. In fact, the key to the coincidence of the critical Rayleigh numbers for energy and linear theory is the fact that the operator attached to the linear theory is symmetric. This evidently was recognised by (Joseph and Carmi, 1966), (Davis, 1971), and independently and in a different context by (Galdi and Straughan, 1985a). We outline the ideas of the last contribution.

Let $\mathcal{H}$ be a Hilbert space endowed with a scalar product $(\cdot,\cdot)$ and associated norm $\|\cdot\|$. We consider in $\mathcal{H}$ the following initial-value problem,

$$u_t + Lu + N(u) = 0, \qquad u(0) = u_0. \tag{4.105}$$

Here L represents a linear operator (possibly unbounded), and N is a nonlinear operator with $N(0) = 0$ in order that (4.105) admits the null solution. We assume:

(i) L is a densely defined, closed, sectorial operator such that $(L - \lambda I)^{-1}$ is compact for some complex number λ (I is the identity operator in $\mathcal{H}$), that is, L is an operator with compact resolvent;

(ii) The bilinear form associated with L is defined (and bounded) on a space $\mathcal{H}^*$, which is compactly embedded in $\mathcal{H}$.

(iii) The nonlinear operator N verifies the condition

$$(N(u), u) \geq 0, \qquad \forall u \in D(N), \tag{4.106}$$

where $D(\cdot)$ denotes the domain of the associated operator.

Actually, for the Bénard problem the u_t term in (4.105) should be premultiplied by a bounded linear operator A. For the Bénard problem $A = \text{diag}\{1, 1, 1, Pr\}$. The structure of A means we can effectively work with (4.105). Also, the sectorial condition on L in (i) is satisfied for virtually all of the cases one meets in fluid mechanics. However, if one considers operators such as those with traction-like boundary conditions in elasticity, then great care must be taken to demonstrate the operator is sectorial. Details of this may be found in (Simpson and Spector, 1989), and I am indebted to Henry Simpson for pointing this out to me.

Thanks to (i) the following result is true ((Kato, 1976), pp. 185–187).

Lemma 4.3.1 *The spectrum of the operator L consists entirely of an at most denumerable number of eigenvalues $\{\sigma_n\}_{n\in\mathbb{N}}$ with finite (both algebraic and geometric) multiplicities and, moreover, such eigenvalues can cluster only at infinity.*

Since the operator L is in general non-symmetric the eigenvalues, which satisfy the equation

$$L\phi = \sigma\phi,$$

are not necessarily real; they may, however, be ordered in the following manner:

$$\text{Re}\,(\sigma_1) \leq \text{Re}\,(\sigma_2) \leq \cdots \leq \text{Re}\,(\sigma_n) \leq \cdots. \tag{4.107}$$

We have defined ideas of stability in chapter 1, see (1.8), (1.9). However, in the interests of clarity we include the following definitions in the context of (4.105).

Definition 4.3.1 *The zero solution to* (4.105) *is said to be linearly stable if and only if*

$$\mathrm{Re}\,(\sigma_1) > 0. \tag{4.108}$$

Definition 4.3.2 *The zero solution to* (4.105) *is said to be nonlinearly stable if and only if for each* $\epsilon > 0$ *there is a* $\delta = \delta(\epsilon)$ *such that*

$$\|u_0\| < \delta \quad \Rightarrow \quad \|u(t)\| < \epsilon, \tag{4.109}$$

and there exists γ *with* $0 < \gamma \leq \infty$ *such that*

$$\|u_0\| < \gamma \quad \Rightarrow \quad \lim_{t\to\infty} \|u(t)\| = 0. \tag{4.110}$$

If $\gamma = \infty$, we say the zero solution is unconditionally nonlinearly stable (or simply refer to it as being nonlinearly asymptotically stable), otherwise for $\gamma < \infty$ the solution is conditionally stable. The value of γ is called the size of the *attracting radius.*

The relationship between linear and nonlinear stability is now investigated.

The operator L is in general non-symmetric, although it allows a decomposition into two parts L_1 and L_2 such that

(i) $L = L_1 + L_2$, $\qquad D(L_2) \supset D(L_1) = D(L)$;

(ii) L_1 is symmetric, with compact resolvent;

(iii) L_2 is skew-symmetric and bounded in $\mathcal{H}^*$.

From (ii) it follows that the spectrum of L_1 satisfies the same type of result as that given above for the spectrum of L. Moreover, because of the symmetry, the eigenvalues $\{\lambda_n\}_{n\in\mathbb{N}}$ associated with L_1 are *all* real and may be ordered

$$\lambda_1 \leq \lambda_2 \leq \cdots \leq \lambda_n \leq \cdots.$$

Let $L_1[\phi,\phi]$, $\phi \in \mathcal{H}^*$, be the bilinear form associated with the operator L_1, i.e.,

$$(L_1\phi,\phi) = L_1[\phi,\phi], \qquad \forall\phi \in D(L_1).$$

Under the above conditions, the following lemma holds.

Lemma 4.3.2 *Let* $\bar{\phi}$ *be a (normalized) eigenfunction associated with the eigenvalue* λ_1. *Then*

$$\lambda_1 \;=\; L_1[\bar{\phi},\bar{\phi}] \;=\; \min_{\phi\in\mathcal{H}^*} \frac{L_1[\phi,\phi]}{\|\phi\|^2}.$$

The following result establishes unconditional nonlinear stability.

Theorem 4.3.1 *Suppose*

$$\lambda_1 > 0. \tag{4.111}$$

Then the zero solution to (4.105) *is unconditionally nonlinearly stable.*

Proof. Form the scalar product of (4.105) with u to obtain

$$\frac{1}{2}\frac{d}{dt}\|u\|^2 + (Lu, u) + (N(u), u) = 0. \tag{4.112}$$

Since L_2 is skew-symmetric and since by (4.106) N is non-negative, there follows

$$\frac{1}{2}\frac{d}{dt}\|u\|^2 + \frac{L_1[u, u]}{\|u\|^2}\|u\|^2 \le 0.$$

With the aid of Lemma 4.3.2 we thus derive

$$\frac{1}{2}\frac{d}{dt}\|u\|^2 + \lambda_1\|u\|^2 \le 0, \tag{4.113}$$

and so

$$\|u(t)\|^2 \le \|u_0\|^2 e^{-2\lambda_1 t}.$$

In the light of (4.111), the result follows at once.

From the above arguments it follows that while the linear stability problem is reduced to studying the eigenvalue problem associated with all of L, nonlinear stability according to the standard energy method described in the above result involves the eigenvalues of the *symmetric* part L_1 only. Moreover, whenever $L_2 = 0$, the two eigenvalue problems coincide and *linear stability always implies nonlinear stability.* (The converse is also true.)

(Galdi and Straughan, 1985a) show that convection between spherical shells with the gravitational potential proportional to the steady temperature distribution results in a symmetric linearized problem. Also, in section 11.1 we see that magnetohydrodynamic convection in a "quasi-static" electric field approximation with zero limit of magnetic Prandtl number also enjoys such a property. Nevertheless, symmetry is too strong a condition to expect in general. Indeed, (Busse, 1967) uses weakly nonlinear analysis to show that sub-critical bifurcation will occur if either the fluid viscosity or expansion coefficient depends on temperature. In reality fluid properties are not constant and so the linearized system will not, in general, be symmetric. Fortunately, however, for what would appear to be the majority of convection problems, the non-symmetric part acts like a bounded operator when compared to the symmetric (unbounded) operator. This fact is closely connected to the reason why the energy method often yields nonlinear stability results very close to those of linear theory, thereby yielding a small band of Rayleigh numbers where possible sub-critical instability may arise. In such a case energy theory is also very useful in that it shows the linear theory has essentially captured the physics of the onset of convection.

In precisely the situation outlined above (Davis, 1969b) established an interesting result. It roughly says that if one adds a bounded (non-symmetric) operator to (4.105) then the eigenvalue of linear theory remains real provided the non-symmetric operator is small enough. More precisely (Davis, 1969b) considers the reduced eigenvalue problem

$$Au_0 = \sigma_0 Bu_0 , \qquad (4.114)$$

with associated homogeneous boundary conditions where

(i) u_0 is an N-component vector of functions;

(ii) A is a selfadjoint linear operator mapping real vectors into real vectors;

(iii) B is a real $N \times N$ constant, diagonal matrix $\mathrm{diag}(b_i)$, with $0 < b_i < \infty$ $(i = 1, ..., N)$;

(iv) (4.114) has a discrete spectrum $\{\sigma_0^{(n)}\}$ each $\sigma_0^{(n)}$ having at most finite multiplicity and ordered so that

$$|\sigma_0^{(1)}| \le |\sigma_0^{(2)}| \le \cdots \le |\sigma_0^{(k)}| \le \cdots$$

and corresponding eigenfunctions $\{u_{0n}\}$ that form a complete set under the inner product (u_{0n}, Bu_{0n}) where $(u_{0k}, u_{0k})^{1/2}$ is the L^2 norm of u_{0k}.

He compares this to the perturbed eigenvalue problem

$$Au - \epsilon Mu = \sigma Bu \qquad (\epsilon > 0), \qquad (4.115)$$

with associated homogeneous boundary conditions where

(v) M is an operator of L^2 norm one,

$$\|M\| = \sup\{\|Mx\| \,\Big|\, \|x\| \le 1\} = 1,$$

which maps real vectors into real vectors and Mu belongs to the complete space of (iv).

By writing the solution to (4.115) as a power series in ϵ and finding a radius of convergence for this power series, (Davis, 1969b), p. 343 shows that

Theorem 4.3.2 *The kth eigenvalue $\sigma^{(k)}$ of the perturbed problem* (4.115) *for which*

$$\lim_{\epsilon \to 0} \sigma^{(k)} = \sigma_0^{(k)}$$

is real, provided that $\sigma_0^{(k)}$ is simple and

$$\epsilon < \epsilon_c^{(k)} = \mathcal{Q}\mathcal{L}_k^{-1} \qquad (k = 1, 2, ...), \qquad (4.116)$$

where

$$\mathcal{Q} = B_m\big[1 + 2B_M B_m^{-1} - 2(B_M^2 B_m^{-2} + B_M B_m^{-1})^{1/2}\big], \qquad (4.117)$$

$$B_M = \max\{b_1, ..., b_N\}, \qquad B_m = \min\{b_1, ..., b_N\}, \qquad (4.118)$$

and $\mathcal{L}_k^{-1}$ *is the minimum non-zero value of* $|\sigma_0^{(k)} - \sigma_0^{(n)}|$, *minimized over* n, $n \neq k$.

This is certainly a very interesting result. However, its practical use is limited largely due to the need to calculate $\mathcal{L}_k^{-1}$ in (4.116). Nevertheless, it does show there is a range of ϵ for which the first eigenvalue of linear theory remains real.

We now return to (4.97)-(4.99), the linearized form of which satisfy Davis' theorem for ϵ, δ small enough. With $\epsilon = 0$, nonlinear energy stability of (4.97)-(4.99) has been studied in detail by (Joseph and Shir, 1966) while a nonlinear energy analysis for (4.101)-(4.103) with $\epsilon = 0$ is contained in (Joseph, 1976a). On the other hand, a nonlinear energy stability analysis for (4.97)-(4.99) with $\delta = 0$ is given by (Straughan, 1989). These problems, i.e., when $\epsilon = 0$ or $\delta = 0$ but not both zero are very special in that exchange of stabilities holds, at least for two *stress free* boundaries. This follows from E.A. Spiegel's method, given in (Veronis, 1963). (If we write $u(\mathbf{x}, t) = e^{\sigma t}\phi(\mathbf{x})$ in the linearized version of (4.105) then ϕ satisfies

$$\sigma\phi + L\phi = 0,$$

where, in general, $\sigma = \sigma_r + i\sigma_1$, $\sigma_r, \sigma_1 \in \mathbb{R}$: in the present context exchange of stabilities is said to hold if $\sigma_1 \neq 0 \Rightarrow \sigma_r < 0$.) We outline the idea of Spiegel's method here.

Suppose $\delta = 0$ in (4.99), linearize by removing the convective terms from (4.97) and (4.99) and assume a solution like $\mathbf{u} = e^{\sigma t}\mathbf{u}(\mathbf{x})$ with a similar representation for θ and p. Next, take curlcurl of the equation arising from (4.97) and take the third component of the resulting equation to obtain the system

$$\sigma\Delta w = \Delta^2 w + RH(z)\Delta^*\theta, \tag{4.119}$$

$$\sigma Pr\theta = \Delta\theta + Rw, \tag{4.120}$$

where $\Delta^* = \partial^2/\partial x^2 + \partial^2/\partial y^2$. For two free boundaries the relevant boundary conditions are

$$w = \Delta w = \theta = 0, \qquad z = 0, 1. \tag{4.121}$$

The variable w is eliminated from (4.119), (4.120) to yield

$$\Delta^3\theta - \sigma(1 + Pr)\Delta^2\theta + \sigma^2 Pr\Delta\theta - R^2 H(z)\Delta^*\theta = 0. \tag{4.122}$$

Multiply (4.122) by θ^* (complex conjugate of θ) and integrate over a cell, V, of solution periodicity. With $\|\cdot\|$ and $<\cdot>$ again denoting the norm on $L^2(V)$ and integration over V, respectively, we find

$$-\|\nabla\Delta\theta\|^2 - \sigma(1 + Pr)\|\Delta\theta\|^2 - \sigma^2 Pr\|\nabla\theta\|^2 + R^2 < H|\nabla^*\theta|^2 > = 0,$$

where $\nabla^* = (\partial/\partial x, \partial/\partial y)$. The imaginary part of this equation leads to

$$\sigma_1\big[(1 + Pr)\|\Delta\theta\|^2 + 2\sigma_r Pr\|\nabla\theta\|^2\big] = 0,$$

where it has been assumed that $\sigma = \sigma_r + i\sigma_1$. Hence, if $\sigma_1 \neq 0$ then $\sigma_r < 0$. This shows it is sufficient to consider the stationary convection boundary $\sigma \equiv 0$.

The same conclusion is easily arrived at for (4.97)-(4.99) when $\epsilon = 0$, $\delta \neq 0$. Here, instead of eliminating w, θ is eliminated to obtain instead of (4.122), a sixth order equation in w. The rest of the proof is similar.

It would appear that the importance of this result for energy stability is that the linear and nonlinear boundaries are very close, even when δ or ϵ is large (with one of them zero), thus rendering the nonlinear results very useful, even for situations where Davis' theorem does not apply. If $H \propto N$ in either (4.97)-(4.99) or (4.101)-(4.103), then it is straightforward to symmetrize these equations, i.e. arrange that the linear operator L is symmetric. Thus, in this situation $\sigma \in \mathbb{R}$. The work of (Herron, 2000; Herron, 2001) investigates the exchange of stabilities question further. He argues that when $H(z)N(z) \geq 0$ the growth rate $\sigma \in \mathbb{R}$. For arbitrary choices of H and N (which do not satisfy $HN \geq 0$) the conclusion is false, as we see in section 17.5. In the last situation there are cases where overstable convection is the dominant mode of instability.

This section is completed with an energy analysis of (4.101)-(4.103). The separate energy identities are derived by multiplying (4.101) by u_i, (4.103) by θ and integrating over V to obtain

$$\frac{1}{2}Pr\frac{d}{dt}\|\theta\|^2 = R < N\theta w > -D(\theta), \tag{4.123}$$

$$R < Hw\theta > = \|\mathbf{u}\|^2. \tag{4.124}$$

Hence, for $\lambda(>0)$ to be chosen we may form the *energy* identity,

$$\frac{dE}{dt} = R < M(z)w\theta > -D(\theta) - \lambda\|\mathbf{u}\|^2, \tag{4.125}$$

where

$$E(t) = \frac{1}{2}Pr\|\theta\|^2, \qquad M(z) = 1 + \delta q + \lambda(1 + \epsilon h). \tag{4.126}$$

If we define R_E by

$$R_E^{-1} = \max_{\mathcal{H}} \frac{I}{\mathcal{D}} \tag{4.127}$$

with

$$I =< Mw\theta >, \qquad \mathcal{D} = D(\theta) + \lambda\|\mathbf{u}\|^2,$$

then as before it is straightforward to show $E(t) \to 0$ at least exponentially provided $R < R_E$.

It is expedient to put $\mathbf{v} = \lambda^{1/2}\mathbf{u}$ in (4.127), then with $F(z) = M(z)/\lambda^{1/2}$ and setting $v_3 = w$, the Euler-Lagrange equations arising from (4.127), for the determination of R_E, are

$$\Delta\theta + \frac{1}{2}FR_E w = 0, \qquad v_i - \frac{1}{2}FR_E\theta k_i = \pi_{,i}\,, \tag{4.128}$$

v_i again being solenoidal.

In this case by employing parametric differentiation we may show that with ξ standing for δ, ϵ, or λ,

$$R_E^2\Big\langle \frac{\partial F}{\partial \xi} w\theta\Big\rangle + \mathcal{D}\frac{\partial R_E}{\partial \xi} = 0, \tag{4.129}$$

where $\mathcal{D} = \|\mathbf{v}\|^2 + D(\theta)$. At the best value of λ, $\partial R_E/\partial\lambda = 0$ and then (4.129) gives rise to $\langle w\theta\{\bar{\lambda}(1+\epsilon h) - (1+\delta q)\}\rangle = 0$. This suggests that a good "guess" (useful when searching numerically) is $\tilde{\lambda} = (1+\delta q_a)/(1+\epsilon h_a)$, where q_a , h_a denote average values. If $h = -z$, $q = z$, then

$$\tilde{\lambda} = \frac{1+\frac{1}{2}\delta}{1-\frac{1}{2}\epsilon}. \tag{4.130}$$

Taking ξ to be ϵ or δ in (4.129) yields

$$\begin{aligned} \mathcal{D}\frac{\partial R_E}{\partial \epsilon} &= -\,\lambda^{1/2}R_E^2 < hw\theta >, \\ \mathcal{D}\frac{\partial R_E}{\partial \delta} &= -\,\lambda^{-1/2}R_E^2 < qw\theta > . \end{aligned} \tag{4.131}$$

Experience with numerical eigenvalue problems of this type suggests terms like $< hw\theta >$ are frequently one signed and thus (4.131) are useful guides to the qualitative behaviour of R_E with ϵ or δ.

Since F depends on z, (4.128) is solved numerically. The periodicity is exploited to look for an x and y dependence with

$$\Delta^*\theta = -a^2\theta, \tag{4.132}$$

where a is a wavenumber. After removing π by operating by curlcurl equations (4.128) may then be reduced to

$$D^2W = a^2W - \frac{1}{2}R_EFa^2\Theta, \qquad D^2\Theta = a^2\Theta - \frac{1}{2}R_EFW, \tag{4.133}$$

where $w = W(z)w(x,y)$, $\theta = \Theta(z)\theta(x,y)$, the x,y parts satisfying a relation like (4.132), and $D = d/dz$. The boundary conditions are

$$W = \Theta = 0 \quad \text{on} \quad z = 0, 1. \tag{4.134}$$

System (4.133), (4.134) constitutes a fourth order eigenvalue problem with non-constant coefficients. It was solved numerically by the compound matrix method (see section 19.2). Since $x, y \in \mathbb{R}^2$, we seek the most unstable wavenumber. On the other hand, the "energy" parameter λ may be chosen to make R_E as large as possible. Hence we must solve the numerical problem

$$\max_{\lambda}\ \min_{a^2} R_E(a^2;\lambda). \tag{4.135}$$

The derivatives in the optimization problem are not known and so a quasi-Newton (or similar) technique that does not require derivatives would have to be used. I have found the golden section search method works well on both the maximum and minimum problems. It may be a little more expensive in computer time than other techniques, but for problems like (4.135) it has been very reliable.

Numerical results for (4.133) and its linear equivalent are presented in (Rionero and Straughan, 1990). The agreements between the linear critical Rayleigh numbers R_L^2 and the nonlinear critical Rayleigh numbers R_E^2 are generally very good.

4.4 Convection with internal heat generation

In this section we now investigate some problems where the linear system is not symmetric, but where an energy analysis yields sharp results. We shall concentrate on finding *unconditional* nonlinear stability thresholds. In particular, in this section we obtain *quantitative nonlinear* stability estimates which guarantee nonlinear stability for the problem of convection in a plane layer with a non-uniform heat source, a constant temperature upper surface, and thermally insulated from below. (Joseph and Shir, 1966) and (Joseph and Carmi, 1966) have developed a very sharp nonlinear stability theory for the Bénard problem with a constant heat source. These are inspiring articles which illustrate the use of coupling parameters beautifully. The boundary conditions employed by these writers take the temperature to be prescribed on the upper and lower boundaries and so we here employ a lower boundary condition of thermal insulation. There is a one to one correspondence between the steady temperature $\bar{T}(z)$ for the Joseph boundary conditions where T is prescribed on $z = 0$ and $z = d$ and the corresponding steady temperature which holds for the boundary conditions where T is only prescribed at $z = d$ while $\partial T/\partial z = 0$ at $z = 0$. However, the perturbation boundary conditions are different in each case and so the stability problems are not equivalent. This is explained in more depth in section 17.4. In fact, (Roberts, 1967a) dealt with the case of a constant heat source when the temperature is prescribed on the upper plane while the lower is thermally insulated. The motivation of (Roberts, 1967a) was to explain the experimental findings of (Tritton and Zarraga, 1967) and so in addition to the technique of linear instability theory he also employed truncated expansions and mean field theory to examine the structure of the finite amplitude solution. For the internally heated convection problem sub-critical instabilities may occur see e.g. (Joseph and Shir, 1966), (Joseph and Carmi, 1966). Nonlinear energy stability results establish a critical Rayleigh number below which convection cannot occur provided the stability analysis is unconditional as it is here. In this way the energy thresholds are useful, especially when they are close to the linear instability values (albeit below them). The parameter region of possible sub-critical instability is the region below the linear instability curve and above the nonlinear energy stability one. It is found in this section that the parameter region of possible sub-critical instability is small and so the energy results are of practical value.

We commence with a layer of heat-conducting linear viscous fluid occupying the horizontal layer $\mathbb{R}^2 \times \{z \in (0,d)\}$. The lower boundary $z=0$ is thermally insulated and the temperature scale is chosen in order that the temperature at $z=d$ remains zero. This is (Roberts, 1967a) problem. For a non-uniform heat source, $Q(z)$, the partial differential equations governing this situation are those of (4.91),

$$\begin{aligned} v_{i,t} + v_j v_{i,j} &= -\frac{1}{\rho_0} p_{,i} + \nu \Delta v_i + g k_i \alpha T, \qquad v_{i,i} = 0, \\ T_{,t} + v_i T_{,i} &= \kappa \Delta T + Q, \end{aligned} \tag{4.136}$$

where g is now constant and the 1 and T_0 terms in the buoyancy have been absorbed into the pressure gradient. The other notation is as in (4.91). Equations (4.136) are defined on the domain $\mathbb{R}^2 \times \{z \in (0,d)\} \times \{t > 0\}$.

The boundary conditions appropriate to the physical situation studied by (Roberts, 1967a) require no-slip on the velocity field, the temperature zero at $z=d$, and no heat flux at $z=0$. Thus,

$$\mathbf{v} = \mathbf{0}, \quad z = 0, d; \qquad T = 0, \quad z = d; \qquad \frac{\partial T}{\partial z} = 0, \quad z = 0. \tag{4.137}$$

(Roberts, 1967a), pp. 34–35, points out that the electrical resistivity of the fluid employed in the experiments of (Tritton and Zarraga, 1967) is strongly temperature dependent. This means that it is difficult experimentally to achieve a constant heat source in practice. In order to accommodate the possibility just outlined, we also investigate the non-uniform heat source problem and obtain the nonlinear energy stability – linear instability critical Rayleigh numbers. In particular, we consider four heat supply functions, namely:

$$\begin{aligned} &\text{A.} \quad Q = Q_0 \, (\text{constant}), \\ &\text{B.} \quad Q = Q_0 \left(\frac{1}{2} + \frac{z}{d} \right), \\ &\text{C.} \quad Q = Q_0 \left(2 + \frac{3z^2}{2d^2} - 3\frac{z}{d} \right), \\ &\text{D.} \quad Q = Q_0 \left(1 + \sin \frac{2\pi z}{d} + \sin \frac{4\pi z}{d} \right), \end{aligned} \tag{4.138}$$

where in B-D, Q_0 is a constant.

Other choices of Q are considered in (Straughan, 1990). While this writer does study the case of Q being constant he also analyses $Q = Q_0[\exp(z/d) - 1]$ and $Q = Q_0 \sin(2\pi z/d)$. The current forms for $Q(z)$ are chosen so that the non-dimensional form of $Q(z)$, $Q^*(z^*)$, is such that

$$Q^*_{average} = \int_0^1 Q^*(z^*)dz^* = Q^* \,(\text{constant}).$$

However, the heat sources display a wide range of behaviour. Case A is constant, for $Q_0 > 0$ case B increases across the layer while case C decreases. The final case, D, heats and cools, in a non-uniform way.

The steady solutions $(\bar{v}_i, \bar{T}, \bar{p})$, in whose stability we are interested, which correspond to boundary conditions (4.137) are:

$$\text{Case A} \qquad \bar{v}_i \equiv 0; \qquad \bar{T} = \frac{Q_0}{2\kappa}(d^2 - z^2);$$

$$\text{Case B} \qquad \bar{v}_i \equiv 0; \qquad \bar{T} = \frac{Q_0}{2\kappa}\Big(\frac{5d^2}{6} - \frac{1}{2}z^2 - \frac{z^3}{3d}\Big);$$

$$\text{Case C} \qquad \bar{v}_i \equiv 0; \qquad \bar{T} = \frac{Q_0 d^2}{\kappa}\Big(\frac{5}{8} - \frac{z^2}{d^2} - \frac{z^4}{8d^4} + \frac{z^3}{2d^3}\Big);$$

and

Case D

$$\bar{v}_i \equiv 0; \quad \bar{T} = \frac{Q_0 d^2}{2\pi\kappa}\Big(\pi + \frac{3}{2} - \frac{3z}{2d} - \pi\frac{z^2}{d^2} + \frac{1}{2\pi}\sin\frac{2\pi z}{d} + \frac{1}{8\pi}\sin\frac{4\pi z}{d}\Big);$$

with $\bar{p} = \rho g \alpha \int_0^z \bar{T}(s)\,ds$, where the appropriate expression for $\bar{T}$ must be substituted in each of cases A–D.

To investigate the stability of these solutions perturbations (u_i, θ, π) are introduced according to the definitions $v_i = \bar{v}_i + u_i, T = \bar{T} + \theta, p = \bar{p} + \pi$. The perturbation equations are non-dimensionalized with the scalings (stars denote dimensionless quantities), $u_i = u_i^* U$, $t = t^* d^2/\nu$, $U = \nu/d$, $x_i = x_i^* d$, $\theta = \theta^* T^\sharp$ and $T^\sharp = (U/\kappa)\sqrt{Q_0 d\nu/\alpha g}$. The Prandtl number, Pr, and the Rayleigh number, Ra, are here selected as

$$Pr = \frac{\nu}{\kappa}, \qquad Ra = R^2 = \frac{Q_0 d^5 g \alpha}{\kappa^2 \nu}.$$

The above non-dimensionalization leads to the following system of nonlinear perturbation equations

$$\begin{aligned} &u_{i,t} + u_j u_{i,j} = -\pi_{,i} + \Delta u_i + R\theta k_i; \qquad u_{i,i} = 0; \\ &Pr(\theta_{,t} + u_i \theta_{,i}) = Rm(z)w + \Delta\theta; \end{aligned} \tag{4.139}$$

where all stars have been omitted, $k_i = \delta_{i3}$ and $\mathbf{u} = (u, v, w)$. The function $m(z)$ corresponds to a non-dimensional temperature gradient and for cases A-D takes the form,

$$m(z) = \begin{cases} z, & \text{caseA;} \\ \frac{1}{2}(z + z^2), & \text{caseB;} \\ 2z + \frac{1}{2}z^3 - \frac{3}{2}z^2, & \text{caseC;} \\ z + \frac{1}{4\pi}(3 - 2\cos 2\pi z - \cos 4\pi z), & \text{caseD,} \end{cases} \tag{4.140}$$

where it is understood that in (4.140), $0 < z < 1$. The spatial domain for equations (4.139) is $\mathbb{R}^2 \times (0,1)$ and the non-dimensional boundary conditions for the perturbations, which arise from (4.137), become:

$$u_i = 0, \quad z = 0, 1; \quad \theta = 0, \quad z = 1; \quad \frac{\partial \theta}{\partial z} = 0, \quad z = 0. \tag{4.141}$$

In addition, u_i, θ, π have a periodic structure in x, y so that they satisfy a plane tiling planform such as that described in section 3.3.

The penetrative convection effect (if any, cf. chapter 17) is here entirely due to the heat source Q. Since the only nonlinearities in (4.139) are those associated with the convective terms an energy functional comprised of L^2 integrals is adequate. Hence, to develop a nonlinear energy stability analysis we multiply $(4.139)_1$ by u_i, $(4.139)_3$ by θ, and integrate over a period cell V, to find the identities

$$\frac{d}{dt}\frac{1}{2}\|\mathbf{u}\|^2 = R(\theta, w) - \|\nabla \mathbf{u}\|^2 \tag{4.142}$$

and

$$\frac{d}{dt}\frac{1}{2}Pr\|\theta\|^2 = R(m\theta, w) - \|\nabla \theta\|^2. \tag{4.143}$$

We multiply (4.143) by a coupling parameter $\lambda(> 0)$ and add the result to (4.142). This leads to an energy equation of form

$$\frac{dE}{dt} = RI - \mathcal{D}, \tag{4.144}$$

where the energy functional E, the production term I, and the dissipation term $\mathcal{D}$ are defined by:

$$E = \frac{1}{2}\|\mathbf{u}\|^2 + \frac{1}{2}Pr\lambda\|\theta\|^2, \quad I = \big([1 + \lambda m]\theta, w\big), \quad \mathcal{D} = \|\nabla \mathbf{u}\|^2 + \lambda\|\nabla \theta\|^2.$$

From (4.144) we follow the procedure leading to (4.9) to deduce

$$\frac{dE}{dt} \le -\mathcal{D}\left(\frac{R_E - R}{R_E}\right), \tag{4.145}$$

where the number R_E is defined by

$$R_E^{-1} = \max_{\mathcal{H}} \frac{I}{\mathcal{D}}, \tag{4.146}$$

with $\mathcal{H}$ the space of admissible functions. Suppose now that $R < R_E$, and then put $a = (R_E - R)/R_E$, so $a > 0$. Upon use of Poincaré's inequality we may show that $\mathcal{D} \ge \pi^2(\|\mathbf{u}\|^2 + \lambda\|\theta\|^2) \ge kE$, where $k = 2\pi^2/Pr$ when $Pr \ge 1$, while $k = 2\pi^2$ for $Pr < 1$. If we now employ this inequality in (4.145) we see that $dE/dt \le -akE$ which integrates to find

$$E(t) \le E(0)\exp\{-akt\}. \tag{4.147}$$

This leads to exponential decay of $E(t)$ for all initial data and so to unconditional nonlinear stability.

The nonlinear stability threshold is the value of R_E which is obtained from (4.146). Thus, our goal is now to calculate R_E. To do this we remove the coefficient λ from $\mathcal{D}$ by putting $\psi = \theta\sqrt{\lambda}$ and then the Euler-Lagrange equations for R_E are found as:

$$\Delta u_i + GR_E k_i \psi = \varpi_{,i}, \qquad u_{i,i} = 0, \qquad \Delta\psi + GR_E w = 0, \tag{4.148}$$

where ϖ is a Lagrange multiplier and G is a function which arises due to the non-uniformity of the temperature gradient and is given by $G = (1+\lambda m)/2\sqrt{\lambda}$. The determination of R_E from (4.148) mimics the procedure in the linear case. Remove ϖ from $(4.148)_1$ and reduce to a system in w and ψ, namely,

$$\Delta^2 w + R_E G \Delta^* \psi = 0, \qquad \Delta\psi + GR_E w = 0,$$

where Δ^* is the horizontal Laplacian. Now put w and ψ in normal mode form, i.e. set $w = W(z)f(x,y)$, $\psi = \Psi(z)f(x,y)$, where $f(x,y)$ specifies the planform of the cell, which is typically a two-dimensional roll or a three-dimensional hexagon. The function f satisfies $\Delta^* f = -a^2 f$, where a is a horizontal wavenumber, cf. the discussion following (3.48). The determination of R_E then requires that we solve the system

$$(D^2 - a^2)^2 W = R_E G a^2 \Psi, \qquad (D^2 - a^2)\Psi = -GR_E W, \tag{4.149}$$

for $z \in (0,1)$, where $D = d/dz$, subject to the boundary conditions:

$$W = DW = 0,\ z = 0,1; \quad \Psi = 0,\ z = 1; \quad D\Psi = 0,\ z = 0. \tag{4.150}$$

The critical Rayleigh number of energy stability theory requires determination of the lowest eigenvalue $R_E(a^2;\lambda)$. This is achieved by performing the optimization

$$Ra_E = \max_{\lambda} \min_{a^2} R_E^2(a^2;\lambda). \tag{4.151}$$

The number Ra_E so found is the critical Rayleigh number of energy theory.

While the nonlinear energy stability critical Rayleigh number is found from (4.149), the eigenvalue problem of linear theory for stationary convection arises from (4.139). The nonlinear terms are discarded in (4.139) as are the $\partial/\partial t$ terms (stationary convection takes the growth rate σ to be zero). Then the pressure perturbation is removed. After seeking a normal mode form the equations governing the eigenvalue problem of linear theory are

$$(D^2 - a^2)^2 W = Ra^2\Theta, \qquad (D^2 - a^2)\Theta = -Rm(z)W. \tag{4.152}$$

The appropriate boundary conditions for which (4.152) are to be solved are

$$W = DW = 0,\ z = 0,1; \quad \Theta = 0,\ z = 1; \quad D\Theta = 0,\ z = 0. \tag{4.153}$$

To calculate the critical Rayleigh number of linear instability theory we must perform the optimization

$$Ra_L = \min_{a^2} R^2(a^2). \tag{4.154}$$

Case	Ra_L	Ra_E	a_L^2	a_E^2	λ^*
A	2772.28	2737.16	6.913	6.946	2.068
B	3611.57	3538.31	7.037	7.091	2.638
C	2045.21	2032.24	6.786	6.803	1.553
D	1647.63	1638.98	6.760	6.772	1.254

Table 4.1. Critical Rayleigh numbers of linear theory, Ra_L, and nonlinear energy theory, Ra_E, with their respective critical wavenumbers a_L, a_E. λ^* denotes the optimum value of λ in (4.151).

The quantity Ra_L so found is the critical Rayleigh number of linear instability theory (assuming stationary convection, $\sigma = 0$). In fact, we ought not to treat σ as real. One should really solve the appropriate linearized equations from (4.139) treating σ as complex from the outset. In this problem we have not done this.

The values of Ra_E in (4.151) and Ra_L in (4.154) were obtained numerically with the compound matrix method being used to determine the eigenvalue at each stage.

Table 4.1 shows that the linear and energy results are certainly close enough to be useful from a practical viewpoint despite the wide variation in $Q(z)$. There is definitely instability if $Ra > Ra_L$ whereas there is definitely global nonlinear stability when $Ra < Ra_E$. It is also worth observing that a variation in Q may influence the critical Rayleigh number enormously. This would suggest that the experiment needs to be performed very carefully to determine the correct form for Q. Clearly care must be taken with the interpretation of the data since the definition of Ra involves Q. However, the average of $Q(z)$ is here always the *same* constant.

4.5 Convection in a variable gravity field

When the gravity field acting in the z-direction perpendicular to a horizontal fluid layer is itself varying with the spatial coordinate z, then different points of the fluid experience different buoyancy forces. This can result in a situation where part of a fluid layer has a tendency to become unstable while the rest wishes to remain stable. In this chapter we investigate the stability of a horizontal layer of fluid heated from below when the gravity field is varying across the layer.

This problem was studied by (Pradhan and Samal, 1987) on the basis of inviscid, linearized stability theory. Their work was concerned with estimating the growth rate σ in a time like dependence $e^{\sigma t}$ of the velocity and temperature fields. They point out that variable gravity effects have been investigated on a laboratory scale and are also likely to be important in the large scale convection of planetary atmospheres. (King-Hele, 1977) uses experimental measurements of the Earth's upper atmosphere to conclude that the density there is an approximately exponentially decreasing function of the vertical height. If we use this in the body force in the Boussinesq approximation, then the models studied here do have relevance. The writer, in (Straughan, 1989) took up the question of finding the linear instability and nonlinear energy stability critical Rayleigh numbers for the problem of (Pradhan and Samal, 1987) but when viscosity was also present. (Straughan, 1989) concentrated on a gravity field which decreases linearly across the layer.

In this section we concentrate on nonlinear stability results and allow several gravity fields, thereby assessing the effect of a nonlinear gravity variation on the convection threshold.

Consider then a layer of heat conducting viscous fluid contained between the planes $z = 0$ and $z = d$. The equations governing the convective fluid motion are (4.136) *with* $Q = 0$, and $g(z)$ variable, namely:

$$\begin{aligned} v_{i,t} + v_j v_{i,j} &= -\frac{1}{\rho} p_{,i} + \nu \Delta v_i + g\big[1 + \hat{\epsilon}\hat{h}(z)\big] k_i \alpha T, \\ v_{i,i} &= 0, \\ T_{,t} + v_i T_{,i} &= \kappa \Delta T, \end{aligned} \tag{4.155}$$

where $\nu, \rho, \alpha, \kappa$ are viscosity, density, thermal expansion coefficient and thermal diffusivity, p is pressure, v_i velocity, T temperature, Δ the three-dimensional Laplacian, g is a constant, and $\hat{\epsilon}\hat{h}(z)$ represents the gravity variation. Equations (4.155) are defined on the spatial domain $\mathbb{R}^2 \times \{z \in (0, d)\}$ with $t > 0$.

The boundary conditions appropriate to fixed surfaces and specified constant temperatures, T_U on $z = d$, T_L on $z = 0$, are,

$$\mathbf{v} = \mathbf{0}, \quad z = 0, d; \qquad T = T_L, \quad z = 0; \qquad T = T_U, \quad z = d. \tag{4.156}$$

We here assume $T_L > T_U$.

The stationary solution of (4.155), subject to (4.156), in whose stability we are interested is

$$\bar{v}_i \equiv 0, \qquad \bar{T} = -\beta z + T_0, \tag{4.157}$$

where β is the temperature gradient given by $\beta = (T_L - T_U)/d$. The equations for a perturbation to the solution (4.157) are non - dimensionalized with the scalings

$$t = t^* \frac{d^2}{\nu}, \qquad \pi = \pi^* P, \qquad P = \frac{U \nu \rho}{d}, \qquad Pr = \frac{\nu}{\kappa} \tag{4.158}$$

$$u_i = u_i^* U, \qquad \theta = \theta^* T^\sharp, \qquad T^\sharp = U \left(\frac{\beta \nu}{\kappa g \alpha} \right)^{1/2}, \qquad R^2 = \frac{\alpha g \beta d^4}{\kappa \nu}, \tag{4.159}$$

where u_i, θ, π are perturbations to the velocity, temperature and pressure, and the non-dimensional perturbation equations are (stars omitted):

$$\begin{aligned} &u_{i,t} + u_j u_{i,j} = -\pi_{,i} + \Delta u_i + R\theta k_i\big[1 + \epsilon h(z)\big]; \qquad u_{i,i} = 0; \\ &Pr(\theta_{,t} + u_i\theta_{,i}) = Rw + \Delta\theta. \end{aligned} \tag{4.160}$$

In these equations $\mathbf{k} = (0,0,1)$, $w = u_3$, $Ra = R^2$ is the Rayleigh number, Pr is the Prandtl number, and $\epsilon h(z)$ is the non-dimensional form of $\hat{\epsilon}\hat{h}(z)$.

The spatial domain for equations (4.160) is $\mathbb{R}^2 \times (0,1)$ and the non-dimensional boundary conditions on the perturbations are taken to be:

$$\begin{aligned} &u_i = 0, \;\; \theta = 0, \quad z = 0, 1; \\ &\mathbf{u}, \theta, \pi \text{ have a periodic structure in } x, y. \end{aligned} \tag{4.161}$$

Before dealing with a nonlinear energy stability analysis a few remarks are in order concerning the linear instability problem.

To investigate linear instability the nonlinear terms are removed from (4.160), i.e. the terms $u_j u_{i,j}$ and $u_i\theta_{,i}$. Then, a solution like $u_i = e^{\sigma t}u_i(\mathbf{x}), \theta = e^{\sigma t}\theta(\mathbf{x}), \pi = e^{\sigma t}\pi(\mathbf{x})$, is assumed. (Straughan, 1989) notes that with this representation we may use the technique of E.A.Spiegel (see section 4.3) to prove that it is sufficient to restrict attention to stationary convection, when the boundaries are *stress free.* Hence it is sufficient to consider the stationary convection boundary $\sigma \equiv 0$ when the surfaces are free of tangential stress. The proof leading to $\sigma \in \mathbb{R}$ breaks down if we instead consider boundary conditions for fixed surfaces. The reason is that the condition $\Delta w = 0$ on $z = 0, 1$ which holds in the free surface case must be replaced by $\partial w/\partial z = 0$ and we can no longer use the differential equations to deduce appropriate boundary conditions on θ to allow the derivation of (4.122). Despite this, numerical evidence would suggest a similar conclusion may be true for fixed surfaces.

Since numerical calculations show the stationary convection linear boundary is very close to the nonlinear energy one we consider only the case $\sigma = 0$ in deriving the critical Rayleigh numbers of linear theory. Therefore, under this premise, to calculate the critical Rayleigh number of linear theory from (4.160) requires solving for the lowest eigenvalue of the system

$$\begin{aligned} &\Delta u_i + R\theta k_i\big\{1 + \epsilon h(z)\big\} = \pi_{,i}, \qquad u_{i,i} = 0, \\ &Rw + \Delta\theta = 0; \end{aligned} \tag{4.162}$$

subject to the boundary conditions (4.161).

Owing to the presence of the variable coefficient $h(z)$ in (4.162) this system is solved numerically. However, we note that (Straughan, 1989) argues by an asymptotic analysis that $h < 0$ results in a larger critical Rayleigh number than the classical situation and so is stabilizing whereas $h > 0$ yields the opposite conclusion.

The eigenvalue problem (4.162) is reduced by employing normal modes to:

$$(D^2 - a^2)^2 W = R\big[1 + \epsilon h(z)\big] a^2 \Theta, \qquad (D^2 - a^2)\Theta = -RW, \tag{4.163}$$

where $z \in (0,1)$, $D = d/dz$, and numerical results are given for the boundary conditions

$$W = DW = \Theta = 0, \qquad \text{on} \quad z = 0, 1,$$

in tables 4.2–4.6.

4.5.1 Nonlinear energy stability

Linear instability theory produces a critical Rayleigh number such that if the Rayleigh number exceeds this critical value the system (4.160) is unstable and the fluid commences a cellular motion. When $\epsilon \neq 0$ the presence of the $\epsilon h(z)$ term in (4.160) means that the linear operator associated with this system is not symmetric. This leaves open the possibility that convection could commence for Rayleigh numbers below the critical value of linear theory. Since the nonlinearities in (4.160) are simply the convective ones we can work with an energy $E(t)$ defined by

$$E(t) = \frac{1}{2}\|\mathbf{u}\|^2 + \frac{1}{2} Pr\lambda \|\theta\|^2, \tag{4.164}$$

where $\lambda (> 0)$ is a coupling parameter to be optimally selected.

The energy equation is obtained by differentiating E and using (4.160) and the boundary conditions to find:

$$\frac{dE}{dt} = RI - D, \tag{4.165}$$

where the production term I, and the dissipation term D are defined by:

$$I = \big([\lambda + 1 + \epsilon h]\theta, w\big), \qquad D = \|\nabla \mathbf{u}\|^2 + \lambda \|\nabla \theta\|^2. \tag{4.166}$$

Define, as usual,

$$R_E^{-1} = \max_{\mathcal{H}} \frac{I}{D}, \tag{4.167}$$

with $\mathcal{H}$ being the space of admissible solutions. Then, we may obtain from (4.165)

$$\frac{dE}{dt} \leq -D\left(\frac{R_E - R}{R_E}\right). \tag{4.168}$$

We may now use Poincaré's inequality as in the proof leading to (4.147) to show that provided $R < R_E$, $E(t) \to 0$ as $t \to \infty$, at least exponentially. Thus, for the present problem, $R < R_E$ guarantees unconditional nonlinear stability. In particular, we have at least exponential decay of $E(t)$ for *all initial amplitudes*.

To find R_E we transform θ in (4.167) to $\psi = \sqrt{\lambda}\,\theta$ since this removes λ explicitly from the denominator. We then transform the variational problem of (4.167) to

$$\frac{1}{R_E} = \max_{\mathcal{H}} \frac{(M(z)w, \psi)}{\mathcal{D}}, \tag{4.169}$$

where $M(z) = [\lambda + 1 + \epsilon h(z)]/\sqrt{\lambda}$ and the new denominator $\mathcal{D}$ has form $\mathcal{D} = \|\nabla \mathbf{u}\|^2 + \|\nabla \psi\|^2$. The Euler-Lagrange equations arising from (4.169) are

$$2\Delta u_i + MR_E k_i \psi = \varpi_{,i}, \qquad 2\Delta\psi + MR_E w = 0, \qquad u_{i,i} = 0, \tag{4.170}$$

where ϖ is a Lagrange multiplier. This system is solved numerically subject to boundary conditions (4.161). (Straughan, 1989) argues on the basis of parametric differentiation that the value $\lambda = 1 + \epsilon h_{\text{average}}$ will be a good initial guess in the numerical optimization programme. This is found to be true in computations.

Employing normal modes the energy eigenvalue problem (4.170) becomes

$$(D^2 - a^2)^2 W = \frac{1}{2} R_E M a^2 \Psi, \qquad (D^2 - a^2)\Psi = -\frac{1}{2} MR_E W, \tag{4.171}$$

together with the boundary conditions:

$$W = DW = \Psi = 0, \quad \text{at} \quad z = 0, 1. \tag{4.172}$$

System (4.171), (4.172) is solved for the lowest eigenvalue R_E and then the optimization is carried out of

$$Ra_E = \max_{\lambda} \min_{a^2} R_E^2(a^2; \lambda). \tag{4.173}$$

We now present five tables of numerical results. Tables 4.3–4.6 present findings of the optimization (4.173) for various functions $h(z)$, and compare the nonlinear stability Rayleigh number Ra_E against the corresponding one of linear instability theory Ra_L. It is found in tables 4.3–4.5 that in all cases where $\epsilon \leq 1$, Ra_E is close to Ra_L and so we may conclude that the linear theory has captured the essential content of the onset of convection. The other cases allow for changing sign of gravity and hence we might expect a strong penetrative effect and hence the possibility of subcritical bifurcation. (Penetrative convection is discussed in much greater detail in chapter 17.) Table 4.2 concentrates on linear results, but is presented to show the effect of various functions h on the convection boundary.

4.5.2 Discussion of numerical results.

The numerical results investigate a variety of gravity fields including those which change sign. Such fields are of interest in laboratory experiments in areas of crystal growth and other applications.

Ra_L	a_L^2	$h(z)$
1996.70	9.713	z^2-2z
1897.34	9.712	$-z$
1807.39	9.712	$-z^2$
1765.59	9.711	$-z^3$
1743.70	9.711	$-z^4$
1731.28	9.711	$-z^5$

Table 4.2. Critical Rayleigh number of linear theory, Ra_L, with the respective wavenumber a_L. The gravity variation $h(z)$ is as indicated, with $g(z)=1+\epsilon h(z)$, and $\epsilon=0.2$.

ϵ	Ra_L	Ra_E	a_L^2	a_E^2	λ^*
0.0	1707.76	1707.76	9.712	9.712	1.000
0.2	1807.39	1807.15	9.712	9.712	0.945
0.4	1918.98	1917.86	9.713	9.716	0.891
0.6	2044.68	2041.70	9.718	9.723	0.838
0.8	2187.35	2180.86	9.726	9.736	0.787
1.0	2350.35	2337.92	9.739	9.756	0.736
1.2	2538.13	2515.98	9.758	9.786	0.689

Table 4.3. Critical Rayleigh numbers of linear theory, Ra_L, and nonlinear energy theory, Ra_E, with their respective critical wavenumbers a_L, a_E. λ^* denotes the optimum value of λ in (4.173): $h(z)=-z^2$.

ϵ	Ra_L	Ra_E	a_L^2	a_E^2	λ^*
0.0	1707.76	1707.76	9.712	9.712	1.000
0.2	1996.70	1996.37	9.713	9.713	0.856
0.4	2402.46	2400.24	9.718	9.721	0.712
0.6	3012.82	3002.99	9.733	9.744	0.571
0.8	4030.33	3988.63	9.775	9.808	0.444
1.0	6036.75	5832.57	9.930	10.031	0.309
1.2	11346.07	9921.15	10.845	10.906	0.212

Table 4.4. Critical Rayleigh numbers of linear theory, Ra_L, and nonlinear energy theory, Ra_E, with their respective critical wavenumbers a_L, a_E. λ^* denotes the optimum value of λ in (4.173): $h(z)=z^2-2z$.

Only linear results are included in table 4.2, since for $\epsilon=0.2$, tables 4.3-4.6 show that Ra_L and Ra_E are for practical purposes indistinguishable. However, the point of table 4.2 is to show just how strong an effect the change of $h(z)$ has on Ra_L. It is also of interest to note that the wavenumber remains unchanged. The profiles are all for decreasing gravity across the layer and vary from a convex one, $h(z)=z^2-2z$, through linear and concave ones up to $h(z)=-z^5$.

ϵ	Ra_L	Ra_E	a_L^2	a_E^2	λ^*
0.0	1707.76	1707.76	9.712	9.712	1.000
0.2	1552.41	1552.26	9.712	9.712	1.101
0.4	1422.83	1422.38	9.713	9.714	1.201
0.6	1313.13	1312.32	9.715	9.717	1.302
0.8	1219.07	1217.95	9.717	9.720	1.404
1.0	1137.55	1136.12	9.719	9.723	1.507

Table 4.5. Critical Rayleigh numbers of linear theory, Ra_L, and nonlinear energy theory, Ra_E, with their respective critical wavenumbers a_L, a_E. λ^* denotes the optimum value of λ in (4.173): $h(z) = z$.

ϵ	Ra_L	Ra_E	a_L^2	a_E^2	λ^*
0.0	1707.76	1707.76	9.712	9.712	1.000
0.2	1705.56	1702.21	9.722	9.729	1.007
0.4	1699.01	1686.05	9.755	9.780	1.027
0.6	1688.28	1660.58	9.808	9.858	1.057
0.8	1673.67	1627.58	9.881	9.956	1.098
1.0	1655.53	1589.04	9.970	10.067	1.146
1.2	1634.29	1546.74	10.075	10.184	1.204

Table 4.6. Critical Rayleigh numbers of linear theory, Ra_L, and nonlinear energy theory, Ra_E, with their respective critical wavenumbers a_L, a_E. λ^* denotes the optimum value of λ in (4.173): $h(z) = \sin 2\pi z$.

Tables 4.3, 4.4 present values for decreasing gravity fields, but with entirely different structure. The wavenumbers again show little change. The concave profile $h(z) = -z^2$ clearly shows an increase in Ra for increasing ϵ, but the convex profile $h(z) = z^2 - 2z$ shows a very marked increase in Ra. The Ra_E values in table 4.3 are all close to the corresponding Ra_L ones. In table 4.4 this is also true except when $\epsilon = 1.2$.

Table 4.5 is included to show the effect of an increasing gravity profile. There is hardly any variation in the wavenumbers and the Ra_E, Ra_L results are almost identical. The quantitative decrease in Ra is clearly seen.

Table 4.6 is for a situation where the gravity is stabilizing for $0 \leq z \leq 1/2$, whereas for $1/2 \leq z \leq 1$ the effect is the opposite. The wavenumbers again show little change. The Ra_L values show a relatively small change for increasing ϵ. Here, however, a penetrative effect is definitely possible, and the nonlinear stability boundary does begin to diverge from the linear instability one as ϵ increases. Nevertheless, the results are practically useful as they do delimit a region where possible subcritical instabilities may be found.

We point out that (Alex et al., 2001) have studied the onset of thermal convection in a fluid saturated porous layer with a varying gravity field. They also include the effects of an internal heat source and an inclined temperature gradient induced by thermal boundary conditions which vary spatially. This is an interesting piece of work which shows that the variation of gravity has a very large effect on the convection threshold.

(Saravanan and Kandaswamy, 2002) analyse the effect of horizontal mass flow on convection induced by an inclined temperature gradient in a horizontal layer of saturated porous material. They find that positive values of gravity gradient in the presence of horizontal mass flux and high horizontal temperature gradient promote instability of the system. Negative values delay instability. Transverse rolls are discovered to be the favourable mode of instability.

(Siddheshwar and Pranesh, 1999) also study the effect of gravity modulation on the onset of convection in a fluid. They additionally include magnetic effects and specifically include angular momentum (particle spin). They derive conditions which allow the possibility of subcritical instabilities.

4.6 Multiparameter eigenvalue problems

If there are more than two fields such as velocity and temperature in a convection problem, say a salt field as a third, then it is natural to use an energy that involves more than one coupling parameter. Thermal convection which incorporates a salt field is known as thermohaline convection, or more generally, as double diffusive convection. From the mathematical and physical viewpoint, problems such as thermohaline convection are interesting; they definitely exhibit sub-critical bifurcation, see (Veronis, 1965), (Proctor, 1981) and, therefore, a global nonlinear stability threshold is always desirable.

It is appropriate at this point to discuss the application of the energy method when there are three (or more) fields present, a situation frequently encountered in practice.

We let u, v, w be solutions to the system

$$\begin{aligned}\frac{\partial u}{\partial t} &= \Delta u + \alpha w + \beta v,\\ \frac{\partial v}{\partial t} &= \Delta v + \beta u + \gamma w,\\ \frac{\partial w}{\partial t} &= \Delta w + \alpha u.\end{aligned} \tag{4.174}$$

These equations are defined on the domain $\Omega \times \{t > 0\}$, where Ω is a bounded domain in $\mathbb{R}^3$, and the coefficents α, β, γ are constants. Let Γ be the boundary of Ω. On Γ, we assume that

$$u = v = w = 0. \tag{4.175}$$

To investigate the stability of the zero solution to (4.174) by the energy method, we choose as an energy functional

$$E(t) = \frac{1}{2}\|u\|^2 + \frac{1}{2}\lambda_1\|v\|^2 + \frac{1}{2}\lambda_2\|w\|^2, \tag{4.176}$$

where λ_1, λ_2 are positive coupling parameters to be selected.

To view (4.174) as an operator equation we write it as

$$\frac{\partial}{\partial t}\begin{pmatrix} u \\ v \\ w \end{pmatrix} = \begin{pmatrix} \Delta & 0 & 0 \\ 0 & \Delta & 0 \\ 0 & 0 & \Delta \end{pmatrix}\begin{pmatrix} u \\ v \\ w \end{pmatrix} + \begin{pmatrix} 0 & \beta & \alpha \\ \beta & 0 & \gamma \\ \alpha & 0 & 0 \end{pmatrix}\begin{pmatrix} u \\ v \\ w \end{pmatrix}.$$

This shows that the operator on the right hand side is symmetric if $\gamma = 0$. Note that the skew-symmetry only arises in the v-equation in (4.174). Thus, one might think that it is sufficient to take $\lambda_2 = 1$ in (4.176). We now employ a parametric differentiation argument to show that both λ_1 and λ_2 are not equal to 1 at the optimal value for stability. This demonstrates that for problems involving three or more fields, the energy stability analysis will involve multi-dimensional optimization in coupling parameters and the wavenumbers.

We form energy equations from (4.174), (4.175), as follows,

$$\begin{aligned}
\frac{1}{2}\frac{d}{dt}\|u\|^2 &= -\|\nabla u\|^2 + \alpha < uw > + \beta < uv >,\\
\frac{1}{2}\frac{d}{dt}\|v\|^2 &= -\|\nabla v\|^2 + \beta < uv > + \gamma < vw >,\\
\frac{1}{2}\frac{d}{dt}\|w\|^2 &= -\|\nabla w\|^2 + \alpha < uw > .
\end{aligned}$$

Adding these equations with suitable coupling parameters we derive an energy equation as

$$\begin{aligned}
\frac{dE}{dt} = &-(\|\nabla u\|^2 + \lambda_1\|\nabla v\|^2 + \lambda_2\|\nabla w\|^2) + \alpha(1+\lambda_2) < uw > \\
&+ \beta(1+\lambda_1) < uv > + \gamma\lambda_1 < vw > .
\end{aligned} \tag{4.177}$$

Now define the indefinite production term I by

$$I = \alpha(1+\lambda_2) < uw > + \beta(1+\lambda_1) < uv > + \gamma\lambda_1 < vw >,$$

and the dissipation D by

$$D = \|\nabla u\|^2 + \lambda_1\|\nabla v\|^2 + \lambda_2\|\nabla w\|^2.$$

Further, define Λ by

$$\Lambda = \max \frac{I}{D}. \tag{4.178}$$

Then, from (4.177) we may show that

$$\frac{dE}{dt} \leq -D(1-\Lambda), \tag{4.179}$$

and with Poincaré's inequality, one deduces nonlinear stability provided $\Lambda < 1$. From (4.179) we see that the optimal stability threshold is achieved when $\Lambda = 1$. We now put $\phi = \sqrt{\lambda_1}v$, $\psi = \sqrt{\lambda_2}w$, in (4.178), and then with $\Lambda = 1$ the Euler-Lagrange equations for the maximizing solution are,

$$\begin{aligned} A\psi + B\phi + 2\Delta u &= 0, \\ Bu + C\psi + 2\Delta\phi &= 0, \\ Au + C\phi + 2\Delta\psi &= 0. \end{aligned} \tag{4.180}$$

The coefficients A, B, C are given by $A = \alpha(1+\lambda_2)/\sqrt{\lambda_2}$, $B = \beta(1+\lambda_1)/\sqrt{\lambda_1}$, $C = \gamma\sqrt{\lambda_1/\lambda_2}$.

We derive useful information using parametric differentiation. Let (u^1, ϕ^1, ψ^1), (u^2, ϕ^2, ψ^2), be solutions to (4.180) for $(\lambda_1^{(1)}, \lambda_2^{(1)})$ and $(\lambda_1^{(2)}, \lambda_2^{(2)})$, respectively. Then, multiply each of (4.180) with superscript 1, respectively by u^2, ϕ^2, ψ^2 and integrate over Ω. Repeat the procedure with 1 and 2 reversed. After dividing by $\lambda_1^{(2)} - \lambda_1^{(1)}$, then $\lambda_2^{(2)} - \lambda_2^{(1)}$ and taking the limit $\lambda_i^{(2)} \to \lambda_i^{(1)}$ one may show

$$\frac{\partial A}{\partial \lambda_2} < u\psi > + \frac{\partial C}{\partial \lambda_2} < \phi\psi > = 0, \qquad \frac{\partial B}{\partial \lambda_1} < u\phi > + \frac{\partial C}{\partial \lambda_1} < \phi\psi > = 0.$$

From these equations we may conclude

$$\begin{aligned} &\alpha\left(\frac{\lambda_2 - 1}{2\lambda_2^{3/2}}\right) < u\psi > - \frac{\gamma\sqrt{\lambda_1}}{2\lambda_2^{3/2}} < \phi\psi > = 0, \\ &\beta\left(\frac{\lambda_1 - 1}{2\lambda_1^{3/2}}\right) < u\phi > + \frac{\gamma}{2\sqrt{\lambda_1\lambda_2}} < \phi\psi > = 0. \end{aligned} \tag{4.181}$$

We do not expect the eigenfunctions u, ϕ, ψ to be orthogonal to each other, and so from (4.181) we find that $\lambda_1 = \lambda_2 = 1$ only when $\gamma = 0$, i.e. when system (4.174) is symmetric. If $\gamma \neq 0$, then we will inevitably find $\lambda_1, \lambda_2 \neq 1$ and the nonlinear stability problem will involve an optimization procedure in λ_1, λ_2, to derive the sharpest result.

The above is a simple model of what is found in fluid dynamical contexts involving three or more competing fields.

To illustrate the point in a physical example we might, consider convection with a temperature and a concentration field in the presence of a Soret effect, see e.g. (Straughan and Hutter, 1999). (The Soret effect is that whereby mass diffusion is influenced by a temperature gradient, i.e., manifested by the $\kappa_S\Delta T$ term in (4.184) below.) If the Soret effect is only *linear* the equations are

$$\dot{v}_i = -\frac{1}{\rho_0} p_{,i} + \nu \Delta v_i - g k_i \big(1 - \alpha[T - T_0] + \gamma[C - C_0]\big), \tag{4.182}$$

$$\dot{T} = \kappa \Delta T, \tag{4.183}$$

$$\dot{C} = \kappa_C \Delta C + \kappa_S \Delta T, \tag{4.184}$$

$$v_{i,i} = 0. \tag{4.185}$$

It is assumed these equations occupy the plane layer $\mathbb{R}^2 \times z \in (0, d)$, C is the concentration field, $C_0, \gamma, \kappa_C, \kappa_S$ are constants, and a superposed dot denotes the convective derivative $\partial/\partial t + v_i \partial/\partial x_i$.

Under prescribed boundary conditions on T, C a steady solution linear in z is possible, and the perturbation equations from this solution will be

$$\begin{aligned} \dot{u}_i &= -p_{,i} + \Delta u_i + R\theta k_i + H R_S \phi k_i, \\ u_{i,i} &= 0, \\ Pr\dot{\theta} &= \hat{H} R w + \Delta\theta, \\ Pc\dot{\phi} &= \tilde{H} R_S w + \Delta\phi + \xi \Delta\theta, \end{aligned} \tag{4.186}$$

where a non-dimensionalization has been employed, R_S is a Rayleigh number associated with the concentration field, $H, \hat{H}, \tilde{H}$ are ± 1 according to the steady solution (which depends on the boundary conditions), Pc is a concentration Prandtl number, ξ is a constant, and ϕ is the perturbation to C. We assume $\kappa_S \le 2(\kappa\kappa_C)^{1/2}$ to ensure $\xi \le 2$.

To study the nonlinear stability of this system we may follow (Shir and Joseph, 1968) who studied this problem with $\xi = 0$ and introduce an "energy" $E(t)$ of the form

$$E(t) = \frac{1}{2}\|\mathbf{u}\|^2 + \frac{1}{2} Pr \lambda_1 \|\theta\|^2 + \frac{1}{2} Pc \lambda_2 \|\phi\|^2, \tag{4.187}$$

where λ_1 and λ_2 are positive coupling parameters to be selected optimally to give as sharp a stability boundary as possible. The energy equation that arises has the form

$$\frac{dE}{dt} = RI - \mathcal{D}, \tag{4.188}$$

where we have put $R_S = \alpha R$ and

$$\begin{aligned} I &= (1 + \hat{H}\lambda_1) < \theta w > + \alpha(H + \tilde{H}\lambda_2) < \phi w >, \\ \mathcal{D} &= D(\mathbf{u}) + \lambda_1 D(\theta) + \lambda_2 \big[D(\phi) + \xi D(\theta, \phi)\big], \end{aligned}$$

where $D(\theta, \phi) = < \nabla\theta . \nabla\phi >$. Again, for nonlinear stability it may be demonstrated from (4.188) that $R < R_E$ is a sufficient condition, where R_E is defined by $R_E^{-1} = \max_{\mathcal{H}} (I/\mathcal{D})$. The five Euler-Lagrange equations that arise from this maximum problem together with the solenoidal constraint on $\mathbf{u}$ define a three-parameter eigenvalue problem for the lowest eigenvalue $R_E(a^2; \lambda_1, \lambda_2)$, a being again a wavenumber. The critical value of the nonlinear stability Rayleigh number in this case is $Ra_E = R_E^2$, where

$$R_E = \max_{\lambda_1, \lambda_2} \min_{a^2} R_E(a^2; \lambda_1, \lambda_2). \tag{4.189}$$

An efficient way to solve (4.189) is to use a quasi-Newton technique on the maximization problem, such as Broyden's method described e.g., in (Dennis and Schnabel, 1983). This is certainly a promising area for tackling energy stability problems when there are many equations and as it also seems that the maximization/minimization routines will be well suited to be spread over the processors in a parallel array of computers, further work should prove rewarding.

The paper of (Rionero and Mulone, 1987) contains some interesting results concerning a linearization principle for the coincidence of the linear and nonlinear stability boundaries in certain situations, for a system like (4.186), but which contains both linear Soret and Dufour effects. (A Dufour effect essentially adds a term like $\zeta\Delta\phi$ to the right-hand side of $(4.186)_3$.)

Other areas where successful nonlinear energy stability results have been obtained from max-min problems like (4.189) are in double-diffusive convection where the layer is heated and salted from below, (Lombardo and Mulone, 2002a; Lombardo et al., 2000; Lombardo et al., 2001a; Lombardo et al., 2001b), in penetrative convection in a micropolar fluid, (Lindsay and Straughan, 1992), and in the interesting penetrative convection model of (Krishnamurti, 1997), see (Straughan, 2002b). The last model is considered in section 17.8. Another numerical max-min problem which yielded sharp nonlinear energy results was found in anisotropic porous penetrative convection by (Straughan and Walker, 1996a). This is described in section 17.6, and is interesting in that unlike (4.189) it involves maximising in one coupling parameter λ, but minimising in two wave numbers k, m corresponding to the x and y directions.

4.7 Finite geometries and numerical basic solutions

Up to this point we have described convection problems in which the fluid occupies an infinite plane layer. For many real situations this is inadequate. When, however, we abandon the infinite layer and examine finite regions, the minimization is no longer over a continuous set of wavenumbers. One has to be careful and minimise over the right set of discrete wavenumbers. Also, it is quite likely that even the basic solution will have to be determined numerically, although this depends on the boundary conditions. Having to numerically determine the basic solution in the region and then use this in the energy eigenvalue analysis clearly gives a more complicated numerical task.

A fundamental step in the lines of numerically oriented energy analyses was made by (Munson and Joseph, 1971) in their work on the stability problem of flow between two spheres: this is well documented by (Joseph, 1976b).

In the field of convection the work of (Neitzel et al., 1991a; Neitzel et al., 1991b) on a crystal growth problem derives interesting linear instability and nonlinear stability thresholds where the basic state is numerically determined.

Another interesting contribution is the paper of (Reddy and Voyé, 1988). They write out the general equations for convection in an arbitrary shaped finite region and develop an energy analysis for a general base flow, which could be numerically determined. They first show, by using a Galerkin (finite element) penalty technique, that the energy stability problem is well posed. In particular, they show the existence of a maximizing solution to the problem for $R_E(\lambda)$ for a fixed coupling parameter λ. The first existence proof of a maximizing solution was given by (Rionero, 1967a; Rionero, 1968a). (Reddy and Voyé, 1988) point out that it is not yet known whether the value of λ that yields a maximum in $R_E(\lambda)$ is positive. From a practical point of view one usually solves for R_E with λ fixed and then treats the maximum problem as a constrained one. Nevertheless, Reddy & Voyé's point is correct and care must be taken. Indeed, in the next chapter we draw attention to the fact that when the region is infinite in all directions the maximizing solution may not exist for all λ.

(Reddy and Voyé, 1988) quote a convergence result for their penalty approximation to the energy eigenvalue problem. They also illustrate their results by examining the two-dimensional Bénard convection problem with an internal heat source. Their results are certainly sharp and very detailed. They include several graphs of critical Rayleigh number against the aspect ratio ($\ell/d =$ length/depth of the two-dimensional rectangular layer), and these are revealing. They indicate that when $\ell/d \approx 8$ the results are nearly identical to those obtained by normal modes for an infinite layer. Also interesting are the results on the number of convection cells at the boundary of stability. For their energy results they include graphs of R_E against λ for $\ell/d = 10$, and these clearly indicate a maximum.

5

Convection problems in a half space

5.1 Existence in the energy maximum problem

It is often useful to define a convection problem on a half space. For example, (Hurle et al., 1982) use the velocity in a phase change problem to transform their stability analysis to one on a half-space; also (Hurle et al., 1967) have heat conducting half-spaces bounding a fluid layer to investigate the effects of finite conductivity at the boundary. While it may offer some simplicity to deal with a half-space configuration, from the mathematical point of view it does introduce new complications. In particular, (Galdi and Rionero, 1985) derive a very sharp result on the asymptotic behaviour of the base solution for which the energy maximum problem for R_E admits a maximizing solution. Roughly speaking, either the base solution must decay at least linearly at infinity, or the gradient of the base solution must decay at least like $1/z^2$ (if $z > 0$ is the half-space.) To describe this result and related ones in geophysics it is convenient to return to the general equations for a heat conducting linearly viscous fluid.

Let now $V_i(\mathbf{x}), P(\mathbf{x}), T(\mathbf{x})$ be a steady solution to (3.41), with g constant, on the unbounded domain $\Omega = \{z > 0\} \times \mathbb{R}^2$. We suppose V_i, T are prescribed on $z = 0$ and asymptotic behaviour is prescribed as $x, y, z \to \infty$. If we introduce perturbations $u_i(\mathbf{x}, t), p(\mathbf{x}, t), \theta(\mathbf{x}, t)$ to this solution and nondimensionalize with length and time scales $d, d^2/\nu$, then select base velocity and temperature scales V, $dT'(0)$, where d is some length; select perturbation velocity and temperature scales $U = \nu/d$ and $U(T'(0)\nu/\kappa g\alpha)^{1/2}$, where $T'(0) = dT/dz(0)$; and define the Rayleigh number, $Ra = R^2$, Reynolds number, Re, and Prandtl number, Pr, as

$$R^2 = \frac{T'(0)\alpha g d^4}{\kappa\nu}, \qquad Re = \frac{Vd}{\nu}, \qquad Pr = \frac{\nu}{\kappa}; \tag{5.1}$$

then the non-dimensional perturbation equations are

$$\begin{aligned} u_{i,t} + Re(V_j u_{i,j} + V_{i,j}u_j) + u_j u_{i,j} &= -p_{,i} + \Delta u_i + R\theta k_i, \\ u_{i,i} &= 0, \\ Pr(\theta_{,t} + u_i\theta_{,i}) + PrReV_i\theta_{,i} &= -RT_{,i}u_i + \Delta\theta, \end{aligned} \tag{5.2}$$

holding on the region $\Omega = \{z > 0\} \times \mathbb{R}^2$. We now briefly describe theorem 2.3 of (Galdi and Rionero, 1985); however, they only establish a result for the Navier-Stokes equations and so we here interpret their result for Bénard convection. Furthermore, we stress that (Galdi and Rionero, 1985) establish existence of a suitable steady solution and rigorously establish existence of the energy maximizing function and a suitable energy theory. Here we only include sufficient technical details to understand the difficulty imposed by the half-space aspect of the problem. Extensions of the results of (Galdi and Rionero, 1985) are given by (De Angelis, 1990) and by (Capone and De Angelis, 1993).

We can assume periodicity in the (x, y) directions, and let V denote $\{z > 0\}\times$ {a period cell}. We then multiply $(5.2)_1$ by u_i, $(5.2)_3$ by θ and integrate over V. With u_i, θ vanishing on $z = 0$ and decaying sufficiently rapidly at infinity in the sense that

$$\Big\langle \frac{u_i u_i + \theta^2}{(1+z)^2} \Big\rangle + D(\mathbf{u}) + D(\theta) < \infty, \tag{5.3}$$

where $< \cdot >$ denotes integration over V, $D(\cdot)$ the Dirichlet integral over V, we formally obtain the equations

$$\frac{1}{2}\frac{d}{dt}\|\mathbf{u}\|^2 = -D(\mathbf{u}) + R < \theta w > - Re < D_{ij}u_i u_j >, \tag{5.4}$$

$$\frac{1}{2}Pr\frac{d}{dt}\|\theta\|^2 = -D(\theta) - R < T_{,i}u_i\theta >, \tag{5.5}$$

where $D_{ij} = \frac{1}{2}(V_{i,j} + V_{j,i})$.

Of course, it is *not* clear that these equations make sense since the

$$< \theta w >, \quad < D_{ij}u_i u_j >, \quad < T_{,i}u_i\theta >$$

terms do not in general satisfy (5.3). It is precisely this point that needs great care. The $< \theta w >$ term requires $\theta, w \in L^2(V)$. Even then, we cannot dominate this term by the $D(\theta)$ term since the form of Poincaré's inequality that holds for the half-space has the form ((Galdi and Rionero, 1985), lemma 2.1)

$$\Big\langle \frac{u_i u_i}{(1+z)^2} \Big\rangle \le 4D(\mathbf{u}) \quad \text{and} \quad \Big\langle \frac{\theta^2}{(1+z)^2} \Big\rangle \le 4D(\theta). \tag{5.6}$$

Even a decay behaviour at infinity in dT/dz like z^{-2} is not sufficient, at least from (5.4), (5.5). One solution to what is needed is after a finite z–length $T_{,i} \sim \beta$ (positive constant), i.e., the *temperature gradient is destabilizing only for a finite z–length* and after that it stabilizes linearly. In this way, the

$$R < \theta w > -\lambda R < T_{,i} u_i \theta > \tag{5.7}$$

terms, which will arise in a coupling parameter energy formulation, may be reduced to integrals over a finite region. *Assuming this to be so* we form

$$\begin{aligned} \frac{d}{dt}\Big(\frac{1}{2}\|\mathbf{u}\|^2 + \frac{1}{2}\lambda Pr\|\theta\|^2\Big) = &- \big[D(\mathbf{u}) + \lambda D(\theta)\big] + R < \theta w > \\ &- R\lambda < T_{,i} u_i \theta > - Re < D_{ij} u_i u_j > . \end{aligned} \tag{5.8}$$

Note that λ has already been selected to remove the troublesome terms in (5.7). The maximum problem is then to determine

$$\max_{\mathcal{H}} \left[\frac{-\alpha < D_{ij} u_i u_j > + < \theta w > -\lambda < T_{,i} u_i \theta >}{D(\mathbf{u}) + \lambda D(\theta)} \right], \tag{5.9}$$

where α is a parameter such that $Re = \alpha R$. The θ terms in the numerator are bounded and to ensure the $D_{ij} u_i u_j$ term is likewise well behaved, (Galdi and Rionero, 1985) require that either

$$|D_{ij}| = O\Big(\frac{1}{(1+z)^2}\Big), \qquad \text{as } z \to \infty \tag{5.10}$$

or

$$|\mathbf{V} - \mathbf{V}_\infty| = O\Big(\frac{1}{(1+z)}\Big), \qquad \text{as } z \to \infty, \tag{5.11}$$

where $\mathbf{V}_\infty$ is the asymptotic value of $\mathbf{V}$ as $z \to \infty$. Essentially the idea is that (5.10) guarantees the $D(\cdot)$ terms can control the numerator via (5.6), while (5.11) does likewise with the rearrangement

$$< D_{ij} u_i u_j > = < (V_i - V_{i\infty}) u_j u_{i,j} >,$$

and use of the arithmetic-geometric mean inequality. What is important is that (Galdi and Rionero, 1985) show (5.6) *cannot be weakened* by producing a counterexample when $(1+z)^2$ is replaced by $(1+z)^{2-\epsilon}$. This, of course, means their existence result for the maximum problem does not hold under weaker behaviour either.

Whitehead & Chen (1970) report an experimental study together with a linear analysis for a penetrative convection model on a half-space where they achieve a nonlinear basic temperature profile by use of a heat source $Q(z)$. It is interesting to note that all of their basic temperature profiles have the property required above, namely, that the temperature is destabilizing on only a finite layer of the half-space.

This section is concluded by reviewing a piece of work by (Dudis and Davis, 1971) where again the maximizing solution does not in general exist; it does so for only one value of the coupling parameter, the value they selected. The basic solution whose stability they examine, the buoyancy boundary layer solution of (Gill, 1966), is an exact solution to the Boussinesq equations of motion. As they point out, Prandtl's "mountain and valley winds in stratified air" contains this solution as a special case when the surface under consideration is vertical rather than slanted.

The fluid occupies the half-space $x > 0$ bounded by an infinite vertical wall at $x = 0$, the acceleration due to gravity acts in the negative $z-$direction (vertically downward) and a stratification is established by a temperature difference ΔT across the half-space, the temperature field varying linearly with z at each z height, thus the boundary conditions are

$$\begin{aligned} &V_i = 0, \qquad T = \Delta T + Gz, \quad \text{at } x = 0,\\ &V_i \to 0, \qquad T \to Gz, \quad \text{as } x \to \infty, \end{aligned} \tag{5.12}$$

where G is a positive constant.

Under the non-dimensionalization of (Dudis and Davis, 1971) the basic state is

$$U = V = 0, \qquad W = e^{-x}\sin x, \qquad T = e^{-x}\cos x + \frac{2z}{PrRe}\,. \tag{5.13}$$

The energy equation derived by (Dudis and Davis, 1971) is:

$$\begin{aligned} \frac{d}{dt}&\Big(\frac{1}{2}\|\mathbf{u}\|^2 + \frac{1}{2}\lambda Pr\|\theta\|^2\Big)\\ &= -\big[D(\mathbf{u}) + \lambda D(\theta)\big]\\ &\quad + \frac{2}{Re} < \theta w > -Pr\lambda < T_{,i}u_i\theta > -Re < D_{ij}u_iu_j > . \end{aligned}$$

They choose $\lambda = 1$ for apparently no good reason, although we point out it does precisely remove the troublesome $< w\theta >$ term, since

$$-\lambda PrT_{,i}u_i\theta = -\frac{2}{Re}\lambda w\theta - \lambda Pr\frac{\partial T}{\partial x}\,u\theta.$$

According to theory of (Galdi and Rionero, 1985) the choice $\lambda = 1$ is necessary and for the very reason that the $< w\theta >$ terms are removed and the energy maximum problem is then well set. (In this case essentially the periodicity is in the y, z directions and T, W decay sufficiently rapidly in x.)

The paper of (Dudis and Davis, 1971) finds the solution to the Euler-Lagrange equations for the energy maximum problem with $\lambda = 1$, and their results are sharp when compared to linear theory. The paper describes a non-trivial numerical procedure.

5.2 The salinity gradient heated vertical sidewall problem

(Kerr, 1990) correctly applies energy theory to an interesting half-space problem. He studies the theory of an infinite body of fluid with a vertical salinity gradient heated from a vertical sidewall. In his linear instability analysis, (Kerr, 1989), the predictions of onset agree well with experimental results. However, it takes a weakly nonlinear analysis, (Kerr, 1990), to correctly predict that the bifurcation into instability is sub-critical and that co-rotating cells are found, in agreement with experiments. He also includes an energy analysis, and it is this aspect that fits into the material of this chapter; and hence we include a description of the relevant work.

The object is to determine nonlinearly a stability criterion for the problem of fluid occupying the half-space $x > 0$, with a vertical linear salinity gradient superimposed and heated at the sidewall $x = 0$. The basic solution of (Kerr, 1989; Kerr, 1990) is

$$\bar{\mathbf{v}} = (0, 0, \bar{w}(x)), \quad \bar{T} = f(x), \quad \bar{S} = f(x) - z,$$

where $\bar{S}$ is the steady salt concentration, z is in the vertical direction, and $\bar{w}$ and f are determined by (Kerr, 1989), p. 329,

$$\bar{w}(x) = \frac{x}{2\sqrt{\pi}} \exp\Big(-\frac{1}{4}x^2\Big), \quad f'(x) = \frac{1}{\sqrt{\pi}} \exp\Big(-\frac{1}{4}x^2\Big). \tag{5.14}$$

($\bar{w}$ and f do actually have a time dependence, but this is effectively disregarded in the energy analysis. Also, in the scaling of (Kerr, 1990) a factor δ^*, which measures the ratio of a vertical to a horizontal length scale, has a presence in (5.14).)

With Pr Prandtl number, τ the salt to thermal diffusivity ratio, and defining a Rayleigh number $\mathcal{R} = g\alpha\Delta T h^3/\kappa^2$, where κ is the thermal diffusivity and h is a vertical length scale, the perturbation equations, in terms of perturbations u_i, p, θ and salt perturbation s, are

$$\begin{aligned}
\frac{\partial u_i}{\partial t} + u_j u_{i,j} + \bar{v}_j u_{i,j} + u_j \bar{v}_{i,j} &= -p_{,i} + \mathcal{R}(\theta - s)k_i + Pr\Delta u_i\,,\\
u_{i,i} &= 0,\\
\frac{\partial \theta}{\partial t} + u_j\theta_{,j} + \bar{v}_j\theta_{,j} + u_j\bar{T}_{,j} &= \Delta\theta\,,\\
\frac{\partial s}{\partial t} + u_j s_{,j} + \bar{v}_j s_{,j} + u_j\bar{S}_{,j} &= \tau\Delta s\,.
\end{aligned} \tag{5.15}$$

These equations are defined on the half-space $\{x > 0\}$ for all positive time. The boundary conditions employed are

$$u_i = \theta = \frac{\partial s}{\partial x} = 0, \quad \text{at } x = 0, \qquad u_i, \theta, s \to 0, \quad \text{as } x \to \infty. \tag{5.16}$$

To derive the energy identity, (Kerr, 1990) introduces the notation

$$\bar{A}(x,t) = \lim_{K,L\to\infty} \frac{1}{4KL}\int_{-L}^{L}\int_{-K}^{K} A(x,y,z,t)\,dy\,dz,$$
$$< A > = \int_0^\infty \bar{A}(x,t)\,dx.$$

Then, assuming u_i, p, θ, s, and their derivatives as required, belong to the appropriate L^2 class of functions such that the energy equations make sense, he derives four energy identities; namely, those formed by multiplying $(5.15)_1$ by u_i, $(5.15)_3$ by θ, $(5.15)_4$ by s and integrating, and a mixed one formed by multiplying $(5.15)_3$ by s, $(5.15)_4$ by θ and adding. The identities are

$$\frac{1}{2}\frac{d}{dt} < u_iu_i > = - < wu\bar{w}_x > + \mathcal{R}(< \theta w > - < sw >) - Pr < u_{i,j}u_{i,j} >,$$
$$\frac{1}{2}\frac{d}{dt} < \theta^2 > = - < \theta u f_x > - < \theta_{,j}\theta_{,j} >,$$
$$\frac{1}{2}\frac{d}{dt} < s^2 > = - < suf_x > + < sw > - \tau < s_{,j}s_{,j} >,$$
$$\frac{d}{dt} < s\theta > = - < (\theta + s)uf_x > + < \theta w > - (1+\tau) < s_{,j}\theta_{,j} > .$$

An energy $E(t)$ is defined as

$$E(t) = \frac{1}{2} < u_iu_i > + \frac{1}{2}\lambda_1 < \theta^2 > + \frac{1}{2}\lambda_2 < s^2 > + \lambda_3 < s\theta >,$$

for coupling parameters $\lambda_1, \lambda_2, \lambda_3\, (> 0)$ to be chosen, and this is observed to satisfy the equation

$$\frac{dE}{dt} = I - \mathcal{D}, \tag{5.17}$$

where

$$\begin{aligned}\mathcal{D} =& Pr < u_{i,j}u_{i,j} > + \lambda_1 < \theta_{,i}\theta_{,i} >\\ &+ \lambda_2\tau < s_{,i}s_{,i} > + \lambda_3(1+\tau) < s_{,i}\theta_{,i} >,\end{aligned}$$

$$\begin{aligned}I =& - < wu\bar{w}_x > - \lambda_1 < \theta u f_x > - \lambda_2 < suf_x >\\ &- \lambda_3 < (\theta+s)uf_x > + (\mathcal{R} + \lambda_3) < \theta w >\\ &+ (\lambda_2 - \mathcal{R}) < sw > .\end{aligned} \tag{5.18}$$

The stability criterion is then found to be

$$\sup \frac{I}{\mathcal{D}} < 1. \tag{5.19}$$

(Kerr, 1990) observes that since the usual Poincaré inequality (with no weight) does not hold for a half-space then the $< \theta w >$ and $< sw >$ terms in (5.18) must not be present, and he selects

$$\lambda_3 = -\mathcal{R}, \qquad \lambda_2 = \mathcal{R} \tag{5.20}$$

to remove them. Of course, the theory of (Galdi and Rionero, 1985) requires precisely the choice (5.20) before the maximum problem arising from (5.19) admits a solution; the decay behaviour of $\bar{w}$ and f_x as $x \to \infty$ is very rapid and certainly satisfies the criteria of (Galdi and Rionero, 1985). Kerr derives a general result from (5.19), but to achieve a stronger result he then assumes the perturbations are periodic in y and z, a restriction that is meaningful since such behaviour is observed experimentally. (Kerr, 1990) p. 543 notes that,... *in experiments the instabilities are observed to be thin, almost horizontal convection cells with a vertical length scale of order* h, ..., *in none of the experiments was there a layer thickness greater than* h. His further analysis proceeds from (5.17) - (5.19) by splitting u_i, θ, s into two parts, one averaged over the y, z directions, the other being the remainder. By use of a variety of inequalities and estimates, and with the choice $\lambda_1 = (1+\tau^2)/2\tau$, he then arrives at the nonlinear energy stability criterion (Kerr, 1990), inequality (3.44),

$$\frac{1}{2}(1-\tau)\delta^{*3} + \frac{(1-\tau)}{\sqrt{Pr\tau}}\mathcal{R}^{1/2}\delta^* < \frac{8\pi^{5/2}}{(\Delta z)^2}, \tag{5.21}$$

where Δz is the vertical periodicity of the disturbance and δ^* is the ratio of the vertical length scale h to the horizontal length scale $\sqrt{\kappa t}$ (the $\sqrt{t}$ relates to the heating time). Inequality (5.21) is a useful threshold against which the experimental occurrence of subcritical instability may be compared.

6
Generalized energies and the Lyapunov method

6.1 The stabilizing effect of rotation in the Bénard problem

Thus far in convection studies we have explored uses of the energy method that have concentrated on employing some form of kinetic-like energy, involving combinations of L^2 integrals of perturbation quantities. While this is fine and yields strong results for a large class of problem, there are many situations where such an approach leads only to weak results if it works at all, and an alternative device must be sought. In fact, recent work has often employed a variety of integrals rather than just the squares of velocity or temperature perturbations. (Drazin and Reid, 1981), p. 431, point out that this natural extension of the energy method is essentially the method advocated by Lyapunov for the stability of systems of ordinary differential equations some 60 years or so ago. In this chapter we indicate where a variety of different *generalized energies* (or *Lyapunov functionals*) have been employed to achieve several different effects.

A good example of a situation where the standard L^2 energy theory yields very conservative results is the rotating Bénard problem, especially for large rotation rates. In the beginning of this chapter we shall briefly review the work of (Galdi and Straughan, 1985b) who constructed a generalized energy for the Bénard problem in a rotating fluid layer that brings out the stabilizing effect of rotation.

The physical aspect of convection in a rotating fluid layer is the driving force for analysis. (Rossby, 1969) reports detailed results for his experiments on the rotating Bénard problem for a wide range of rotation rates (Taylor numbers), for Prandtl numbers, $Pr\ (= \nu/\kappa)$, which correspond to water and to mercury. He observes p. 329: *Measurements with water revealed two striking features. The first is the presence of a subcritical instability at Taylor numbers greater than 5×10^4 which becomes more pronounced at larger Taylor numbers...The second feature is that water exhibits a maximum heat flux not without rotation but at a Taylor number which is an increasing function of the Rayleigh number...It was conjectured that this increase is due to an 'Ekman-layer-like' modification of the viscous boundary layer.* Clearly, therefore, in the light of Rossby's work, any useful *quantitative* nonlinear analysis is desirable.

The equations for a perturbation (u_i, θ, p) to the stationary solution (referred to a constant rotation reference state) $u_i = 0$, $T = -\beta z + T_0$, are given in (Chandrasekhar, 1981), p. 87. The fluid is contained in the layer $z \in (0, d)$ and the temperature, T, is kept constant on the bounding planes $z = 0, d$, with $T = T_0$ when $z = 0$, and $T = T_1$ when $z = d$, where $T_0 > T_1$. The quantities u_i, θ, p are the perturbations in velocity, temperature, and pressure (incorporating centrifugal forces and absorbing the constant density), and $\beta = (T_0 - T_1)/d$ is the temperature gradient across the layer.

The energy analysis of (Galdi and Straughan, 1985b) selects the non-dimensionalization (stars indicating non-dimensional variables):

$$
\begin{aligned}
&x_i = x_i^* d, \qquad && u_i = u_i^* U, \qquad && U = \frac{\nu}{d}, \qquad && \theta = \theta^* T^\sharp, \\
&T^\sharp = \Big(\frac{Pr\beta}{g\alpha}\Big)^{1/2} U, \qquad && p = p^* P, \qquad && P = \frac{U\rho_0 \nu}{d}, \qquad && t^* = t\frac{\nu}{d^2}, \\
&R^2 = \alpha\beta g d^4 \kappa\nu\,, \qquad && T = \frac{2d^2\Omega}{\nu}, \qquad && Pr = \frac{\nu}{\kappa};
\end{aligned}
$$

where the non-dimensional parameters $Ra = R^2$, $\mathcal{T} = T^2$, Pr are the Rayleigh, Taylor, and Prandtl numbers, and Ω is the magnitude of the angular velocity of the layer, which is in the direction $\mathbf{k} = (0, 0, 1)$.

Omitting the stars, the non-dimensional equations for the perturbations are

$$
\begin{aligned}
u_{i,t} + u_j u_{i,j} &= -p_{,i} + R\theta k_i + \Delta u_i + \epsilon_{ijk} T u_j \delta_{k3}\,, \\
u_{i,i} &= 0, \\
Pr(\theta_{,t} + u_i \theta_{,i}) &= Rw + \Delta\theta,
\end{aligned}
\tag{6.1}
$$

where the spatial region is the layer $z \in (0, 1)$.

The boundary conditions adopted are that

$$
u_i, \theta, p \text{ are periodic in } x, y, \text{ with periods } 2a_1, 2a_2, \text{ respectively.} \tag{6.2}
$$

The planes $z = 0, 1$ are *free surfaces* on which no tangential stresses act (cf. (Chandrasekhar, 1981), pp. 21–22), so that

$$\frac{\partial u}{\partial z} = \frac{\partial v}{\partial z} = w = \theta = 0, \qquad \text{on} \quad z = 0, 1, \tag{6.3}$$

where $\mathbf{u} = (u, v, w)$. To exclude rigid motions we assume the mean values of u, v are zero, i.e., $< u >=< v >= 0$, where $< \cdot >$ again denotes integration over a periodicity cell V, and here we take $V = [0, 2a_1) \times [0, 2a_2) \times (0, 1)$. The assumption of free surfaces is not unrealistic. For high rotation rates Ekman layers form and the convecting layer is bounded by the boundary layers that may act like free surfaces.

In (Galdi and Straughan, 1985b) a nonlinear energy analysis is developed, which determines a stability boundary $Ra_E = Ra_E(T)$, such that for $R^2 < Ra_E$ we are able to obtain sufficient conditions to guarantee nonlinear stability of the stationary solution given above.

A standard energy method for the rotating Bénard problem employing Joseph's coupling parameter idea, described in chapter 4, would commence with an energy of form

$$E = \frac{1}{2}\|\mathbf{u}\|^2 + \frac{1}{2}\gamma Pr\|\theta\|^2,$$

where $\|\cdot\|$ again denotes the L^2 norm on V and $\gamma\,(> 0)$ is a coupling parameter to be chosen in such a way as to obtain the sharpest stability boundary. (Joseph, 1966), pp. 174–175, points out that this method cannot give any information regarding the stabilizing effect of rotation as found by the linear analyses of (Chandrasekhar, 1953; Chandrasekhar, 1981). The reason for this is that the Coriolis term associated with the rotation, the last term in $(6.1)_3$, when multiplied by u_i and integrated over V contributes to the energy equation only via the term $T < \epsilon_{ijk}u_iu_j\delta_{3k} > (= 0)$. In fact, (Galdi and Straughan, 1985b) show that the optimum result is achieved with $\gamma = 1$, and this is the same as when there is no rotation present. Therefore, the standard energy analysis yields only $Ra_E = 27\pi^4/4$ and no stabilization is observed. When $\mathcal{T}$ is of order 10^5 (Chandrasekhar, 1953; Chandrasekhar, 1981) shows that the critical Rayleigh number of linear theory, Ra_L, is of order 21000, and so there is a huge variation between the linear and energy results.

Since the inhibiting effect of rotation on the instability of a fluid layer heated from below has long been recognized as a phenomenon of major importance in Bénard convection and as convection in a rotating system is relevant to many geophysical applications and to such industrial applications as semiconductor crystal growing, it is not surprising that there have been many articles dealing with theoretical or experimental analyses of this problem; see, for example, (Chandrasekhar, 1953; Chandrasekhar, 1981), (Chandrasekhar and Elbert, 1955), (Kloeden and Wells, 1983), (Langlois,

1985), (Roberts, 1967b), (Roberts and Stewartson, 1974), (Rossby, 1969), and (Veronis, 1959; Veronis, 1966a; Veronis, 1968b).

Therefore, a nonlinear analysis using an *energy* theory is very desirable. It is clear though that a different energy functional must be chosen to encompass the stabilizing effect of rotation. Vorticity is inherently a quantity associated with rotation and it plays an important role in the linear analysis of (Chandrasekhar, 1953; Chandrasekhar, 1981), and we might expect its inclusion also to be important in any worthwhile energy functional, and this point strongly influenced the choice of *generalized energy* in (Galdi and Straughan, 1985b).

Depending on what theory is studied, and what aim is to be achieved, a very different Lyapunov functional from a combination of L^2 integrals may well be needed. For instance in thermoelasticity, (Quintanilla and Straughan, 2001) find it necessary to use the L^2 norm of the displacement field coupled to a weighted H_0^1 norm of the second time integral of the temperature perturbation field. To include stabilizing effects in a nonlinear analysis it would appear necessary to introduce a generalized energy that involves higher derivatives of u_i, θ and so we need the relevant equations for the higher derivatives. We start with a vorticity component equation and to this end take the curl of $(6.1)_1$ and with $\boldsymbol{\omega} = (\text{curl}\,\mathbf{u})$, $\omega = \omega_3$ find

$$\frac{\partial \omega}{\partial t} + u_i\omega_{,i} = \Delta\omega + Tw_{,z} + u_{r,y}u_{,r} - u_{r,x}v_{,r}\,. \tag{6.4}$$

An evolution equation for Δw is obtained by applying the operator curl curl to $(6.1)_1$ and then taking the third component of the resulting equation to find

$$\frac{\partial}{\partial t}\Delta w = R\Delta^*\theta + (u_a u_{i,a})_{,zi} - \Delta(u_i w_{,i}) + \Delta^2 w - T\omega_{,z}, \tag{6.5}$$

where Δ^* denotes $\partial^2/\partial x^2 + \partial^2/\partial y^2$. We also need an equation for $\theta_{,z}$ and this follows by differentiating $(6.1)_3$,

$$Pr\frac{\partial}{\partial t}\theta_{,z} = -Pr(u_{i,z}\theta_{,i} + u_i\theta_{,iz}) + Rw_{,z} + \Delta\theta_{,z}. \tag{6.6}$$

In fact, the vorticity only enters the analysis of (Galdi and Straughan, 1985b) through the combination $F = \omega - (T/R)\,\xi Pr\theta_{,z}$, for $0 < \xi < 1$ a coupling parameter to be selected. The evolution equation for F is calculated from (6.4) and (6.6) as

$$\begin{aligned}\frac{\partial F}{\partial t} =& \Delta F + T(1-\xi)w_{,z} + \frac{\xi T}{R}(Pr-1)\Delta\theta_{,z}\\ &+ u_{i,y}u_{,i} - u_{i,x}v_{,i} - u_iF_{,i} + Pr\frac{\xi T}{R}u_{i,z}\theta_{,i}\,.\end{aligned} \tag{6.7}$$

The generalized energy of (Galdi and Straughan, 1985b) commences with

$$E(t) = \frac{1}{2}\{\lambda(\|\mathbf{u}\|^2 + Pr\|\theta\|^2) + c_1 Pr\|\theta_{,z}\|^2 + c_2\|F\|^2 + \|\nabla w\|^2\},$$

where there are four coupling parameters, $\lambda, c_1, c_2 (> 0)$, and $0 < \xi < 1$ to be chosen.

The equation for dE/dt is obtained as

$$\begin{aligned}\frac{dE}{dt} =& \lambda\big(2R < \theta w > - < u_{i,j}u_{i,j} > -\|\nabla\theta\|^2\big) + c_1 R < w_{,z}\theta_{,z} > \\ &+ c_2 T(1-\xi) < w_{,z}F > + T < F_{,z}w > \\ &+ \frac{T^2}{R}\xi Pr < w\theta_{,zz} > + c_1 < \theta_{,z}\Delta\theta_{,z} > \\ &+ c_2 < F\Delta F > + c_2\frac{T}{R}\xi(Pr-1) < F\Delta\theta_{,z} > \\ &- R < w\Delta^*\theta > - < w\Delta^2 w > + N,\end{aligned} \tag{6.8}$$

where N contains cubic perturbation quantities and may be arranged to have the form

$$\begin{aligned}N =& c_1 Pr < u_i(\theta_{,i}\theta_{,zz} + \theta_{,z}\theta_{,zi}) > - c_2 Pr\xi\frac{T}{R} < F_{,i}u_{i,z}\theta > \\ &+ c_2 < F_{,i}(vu_{i,x} - uu_{i,y}) > + < u_i w_{,i}\Delta w - u_i u_{j,i}w_{,jz} > .\end{aligned} \tag{6.9}$$

The coupling parameters c_1, c_2 are selected to be $c_1 = \xi PrT^2/R^2$, $c_2 = 1/(1-\xi)$ (> 0). Equation (6.8) may then be reduced to

$$\frac{dE}{dt} = RI - D + N, \tag{6.10}$$

where I and D are the terms

$$\begin{aligned}I =& 2\lambda < \theta w > - c_2\xi\frac{T}{R^2}(Pr-1) < F_{,i}\theta_{,iz} > + < \theta_{,\alpha}w_{,\alpha} >, \\ D =& \lambda\{< u_{i,j}u_{i,j} > + < \theta_{,i}\theta_{,i} >\} \\ &+ c_1 < \theta_{,iz}\theta_{,iz} > + c_2 < F_{,i}F_{,i} > + \|\Delta w\|^2.\end{aligned} \tag{6.11}$$

We wish to develop an energy analysis from equation (6.10). In the case of (6.10) the cubic nature of the nonlinearities necessitates the introduction of a stronger dissipation term D to control the nonlinear terms in N, i.e. D as given by (6.11) is not sufficient to handle N. This, therefore, involves the introduction of another part to the "energy", and this in turn leads only to a conditional nonlinear analysis. Nevertheless, to the best of our knowledge the work of (Galdi and Straughan, 1985b) was the first to incorporate the stabilizing effect of rotation into a nonlinear energy stability theory.

6.2 Construction of a generalised energy

The cubic nonlinearities in N contain terms like $\theta_{,zz}$ and $w_{,iz}$, and we control these by introducing a dissipative term that contains higher order derivatives, such as $\|\Delta\mathbf{u}\|^2$ and $\|\Delta\theta\|^2$. We, therefore, define another part,

$E_1(t)$, of the energy by

$$E_1(t) = \frac{1}{2} < u_{i,j}u_{i,j} > + \frac{1}{2}Pr < \theta_{,i}\theta_{,i} > . \tag{6.12}$$

This part of the energy plays no role in the variational problem for determining the critical Rayleigh number, Ra_E, of energy theory, it serves simply to dominate the nonlinear terms.

An evolution equation for E_1 is found as

$$\frac{dE_1}{dt} = F_1 - D_1 + N_1 ,$$

where F_1, D_1, N_1 are given by

$$F_1 = 2R < w_{,i}\theta_{,i} >, \qquad D_1 = \|\Delta\theta\|^2 + < \Delta u_i \Delta u_i >,$$
$$N_1 = < u_j u_{i,j} \Delta u_i > + Pr < u_i \theta_{,i} \Delta\theta > .$$

The final energy functional employed by (Galdi and Straughan, 1985b) has the form

$$\mathcal{E} = E(t) + bE_1(t),$$

where $b\,(> 0)$ is yet another coupling parameter to be selected judiciously. It is important to reiterate that the two parts to $\mathcal{E}(t)$ play different roles. The part $E(t)$ is the important one that yields the stability boundary and its choice is crucial, whereas the part $E_1(t)$ plays the role of a piece to dominate nonlinearities. In applying a generalized energy technique such as this to other problems the difficult part is always to select the $E(t)$ piece in such a way that a sharp stability boundary is achieved. There *may* be several ways to choose the E_1 component to dominate the nonlinear terms.

The energy equation for $\mathcal{E}(t)$ may be written

$$\frac{d\mathcal{E}}{dt} = RI - D + bF_1 - bD_1 + N + bN_1. \tag{6.13}$$

The energy maximum problem involves only those quadratic terms arising from E, i.e., I and D, and so we define

$$\frac{1}{R_E} = \max_{\mathcal{H}} \frac{I}{D}, \tag{6.14}$$

where the space of admissible functions is

$$\mathcal{H} = \big\{F, u_i, \theta \big| u_{i,i} = 0,\ u_i, \theta, F \text{ periodic in } x, y$$
$$\text{and satisfy (6.3), and } F_{,z} = \theta_{,zz} = w_{,zz} = 0 \text{ on } z = 0, 1\big\}.$$

The stability threshold is R_E and so we require $R < R_E$, and we set $m = R/R_E$, where $0 < m < 1$. One now defines $\mathcal{D}$ by $\mathcal{D} = \frac{1}{2}bD_1 + \frac{1}{2}(1 - m)D$, where since $R < R_E$, $m < 1$, and so the coefficient of D is positive. One may then derive an inequality of form (Galdi and Straughan, 1985b),

$$\frac{d\mathcal{E}}{dt} \leq -\mathcal{D}(1 - A\mathcal{E}^{1/2}), \tag{6.15}$$

where A is a constant that behaves like $b^{-1/2}$.

Inequality (6.15) yields sufficient conditions to ensure that the generalized energy *decays monotonically to zero.* To see this, suppose that $\mathcal{E}(0) < A^{-2}$. Then, one may deduce

$$\mathcal{E}(t) \leq \mathcal{E}(0) \exp\left\{-\pi^2[1 - A\mathcal{E}^{1/2}(0)](1 - R/R_E)t/Pr\right\}. \qquad (6.16)$$

From this inequality it follows that $\mathcal{E}(t) \to 0$ as $t \to \infty$ in a rapid, monotonic manner.

6.2.1 *Major open problems for unconditional stability*

We have just seen some of the progress made with the nonlinear rotating Bénard problem. However, there are two fundamental weaknesses. The first is that the energy stability found is conditional. The second is that the bounding surfaces have been assumed stress free. In fact, many articles have appeared dealing with energy stability in the rotating Bénard problem, see e.g. (Kaiser and Xu, 1998), (Mulone, 1988), (Mulone and Rionero, 1989; Mulone and Rionero, 1993; Mulone and Rionero, 1994; Mulone and Rionero, 1997), (Qin and Kaloni, 1992b), (Rionero and Mulone, 1988b; Rionero and Mulone, 1989). The paper by (Kaiser and Xu, 1998) is an interesting one dealing with the situation where the Prandtl number takes the value of 1. The analysis has been achieved by construction of novel Lyapunov functions. However, to my knowledge there is always a restriction on $E(0)$, and this restriction is severe. Full details may be found in (Mulone and Rionero, 1997). To the best of my knowledge the nonlinear problem for unconditional stability is still wide open.

The magnetic Bénard problem is in some ways analogous to the rotating Bénard problem in the nonlinear energy stability sense. In particular, the magnetic field stabilizes and one must incorporate this effect into a suitable generalized energy. This is another problem which has attracted a lot of attention, see e.g. (Mulone and Rionero, 1993; Mulone and Rionero, 2003), (Qin and Kaloni, 1994b), (Rionero, 1967a; Rionero, 1968a; Rionero, 1968b; Rionero, 1971), (Rionero and Maiellaro, 1995), (Rionero and Mulone, 1988a). Full details of the state of the art may be found in (Mulone and Rionero, 1993; Mulone and Rionero, 2003). The MHD Bénard problem is briefly described in section 11.1. I believe that the problem of unconditional nonlinear energy stability for the full MHD Bénard problem is still open.

Another area in which little progress has been made with a nonlinear energy stability analysis is in the classical flows of Poiseuille and Couette. Many recent details are given in chapter 8 of (Straughan, 1998). Articles dealing with nonlinear energy stability studies in parallel flows are those of (Mulone, 1990; Mulone, 1991b; Mulone, 1991a), (Rionero and Mulone, 1991). We draw attention to the papers by (Kaiser and Schmitt, 2001) and (Kaiser and Tilgner, 2002) establishing interesting bounds on the energy stability limit, and to a recent paper dealing with throughflow imposed on the Couette solution, by (Doering et al., 2000).

The three classes of problem just mentioned represent fundamental problems in fluid mechanics which are relatively untouched from a nonlinear unconditional stability viewpoint. However, there are many other problems which have recently proved amenable to an unconditional analysis. We feature several of these in this book.

6.2.2 Problems solved for unconditional stability

Penetrative convection and the associated theory of convection with radiation is one area where new Lyapunov functionals have led to unconditional stability. This is covered in chapter 17.

Double diffusive convection, especially the problem of heating and salting below where effects are competing, and more generally multi-component convection is another area which has yielded promising results. This is covered in chapter 14.

For the problem of convection in a rotating porous medium (Guo and Kaloni, 1995c) and (Qin and Kaloni, 1995) have derived conditional nonlinear stability. (Vadasz, 1995; Vadasz, 1996; Vadasz, 1997; Vadasz, 1998a; Vadasz, 1998b), gives some striking results in the linearized instability and weakly nonlinear problems and explains why this problem is very important. Unlike the fluid case, the nonlinear energy stability problem is completely resolved, at least for a Darcy porous medium. This subject is covered in section 6.3.

(Zhao et al., 1995) demonstrate experimentally the important effect of temperature-dependent viscosity on thermal convection. In general, both viscosity and thermal diffusivity vary with temperature. The theory of nonlinear energy stability with temperature dependent viscosity has been analysed by (Richardson, 1993) and by (Capone and Gentile, 1994; Capone and Gentile, 1995). Through use of sophisticated Lyapunov functionals they obtained optimal Rayleigh number thresholds, although their results are conditional. More recently, very sharp unconditional results have been obtained by (Capone, 2001), (Capone and Rionero, 1999; Capone and Rionero, 2001), (Flavin and Rionero, 1998; Flavin and Rionero, 1999b; Flavin and Rionero, 1999a; Flavin and Rionero, 2001), (Rionero, 2000; Rionero, 2001) and (Payne and Straughan, 2000b). This topic is described in chapter 16.

Thawing subsea permafrost is an interesting problem involving motion of a layer of salty sediments sandwiched between the seabed and the permafrost layer which lies below, off the coast of Alaska, see chapter 7. (Galdi et al., 1987) derived a class of unconditional nonlinear stability results and recently (Budu, 2002) has obtained further sharp unconditional nonlinear stability results when the density is cubic in temperature and salt, employing a realistic density. This subject is investigated further in section 7.8.

Thermal convection in a dielectric fluid and the analogous problem in a ferromagnetic fluid yield very nonlinear equations. Yet, unconditional nonlinear stability analyses have been produced. This work is described in chapters 11 and 12.

There are a whole class of problems which may be classified as those with inclined temperature gradient, variable gravity, or time dependent boundary conditions. Stability results for such problems are reported in (Alex et al., 2001), (Capone and Rionero, 2000), (Kaloni and Qiao, 1997a; Kaloni and Qiao, 1997b; Kaloni and Qiao, 2000; Kaloni and Qiao, 2001), (Néel and Nemrouch, 2001), (Nield, 1991; Nield, 1994a; Nield, 1994b; Nield, 1998a) and (Siddheshwar and Pranesh, 1999). Some of these topics are reviewed in section 7.9.

Thus, even though we have highlighted three major areas in fluid mechanics which await new ideas to produce sharp unconditional nonlinear stability results, there are many areas where excellent progress has been achieved. I believe the energy method will continue to be a highly useful tool in thermal convection problems, but also in other areas of applied mathematics.

6.3 Nonlinear stability in rotating porous convection

In this section we study the subject of thermal convection in a layer of saturated porous material which is rotating about an axis orthogonal to the planes containing the layer. This is the analogue of the problem investigated in sections 6.1 and 6.2 for a fluid, but now we are in a porous medium. While the problem of unconditional nonlinear stability for the fluid situation is still open, the porous problem is resolved, in a sense. Thermal convection in a rotating porous layer has many applications in geophysics, in the food processing industry, and in chemical engineering industries, and such processes are described by (Nield and Bejan, 1999), (Nield, 1998a; Nield, 1999), and (Vadasz, 1995; Vadasz, 1996; Vadasz, 1997; Vadasz, 1998a; Vadasz, 1998b). (Vadasz, 1998a) studied the instability mechanisms governing convection in a rotating porous layer. He discovered the striking result that if the inertia term is left in the momentum equation then convection may commence by oscillatory convection.

We describe work of (Straughan, 2001a) which develops an unconditional nonlinear energy stability analysis for the Vadasz problem without inertia. We should point out that a sharp unconditional nonlinear energy stability result for the equivalent problem using a Brinkman porous medium is due to (Lombardo and Mulone, 2002b). Let the porous medium occupy a horizontal layer, where gravity acts in the negative $z-$direction, and the layer rotates about the $z-$axis. The nonlinear equations for convection in a saturated porous medium are, cf. (Vadasz, 1998a)

$$\begin{aligned}\frac{1}{Va}\frac{\partial u_i}{\partial t} &= -\frac{\partial \pi}{\partial x_i} + R\theta k_i - T(\mathbf{k}\times\mathbf{u})_i - u_i\,,\\ \frac{\partial u_i}{\partial x_i} &= 0\,,\\ \frac{\partial \theta}{\partial t} + u_i\frac{\partial \theta}{\partial x_i} &= Rw + \Delta\theta\,.\end{aligned} \tag{6.17}$$

The domain for (6.17) is $\{(x,y)\in\mathbb{R}^2\}\times\{z\in(0,1)\}\times\{t>0\}$. Additionally, $\mathbf{k}=(0,0,1)$, u_i,π and θ represent deviations to the velocity, pressure and temperature fields, and we denote $\mathbf{u}=(u,v,w)$. The non-dimensional numbers R,T and Va have meaning R^2 is the Rayleigh number, T^2 is the Taylor number (measuring the rate of rotation of the layer), and $Va=\phi Pr/Da$ is the Vadasz number. Here ϕ, Pr and Da are the porosity, Prandtl number and Darcy number. The Darcy number is $Da=k/H^2$, where k is the permeability and H is the depth of the layer.

For many situations there is no loss in ignoring the acceleration term in $(6.17)_1$, i.e. let $Va\to\infty$. This becomes the classical theory of Darcy, and is the case covered by (Straughan, 2001a), and in this section.

The boundary conditions considered are

$$w(\mathbf{x},t)=0,\quad \theta(\mathbf{x},t)=0,\qquad z=0,1, \tag{6.18}$$

with u_i,θ,π satisfing a plane tiling periodic boundary condition in x and y.

We now suppose $Va\to\infty$ in $(6.17)_1$. One may then show, cf. (Straughan, 2001a) that the growth rate σ for linear instability is real, i.e. $\sigma\in\mathbb{R}$.

The nonlinear stability analysis works with equations (6.17) ($Va\to\infty$) and curlcurl of $(6.17)_1$, i.e.

$$\Delta u_i + T\frac{\partial \omega_i}{\partial z} = R\big[k_i\Delta^*\theta - \theta_{,xz}\delta_{i1} - \theta_{,yz}\delta_{i2}\big]. \tag{6.19}$$

The quantity $\boldsymbol{\omega}=\operatorname{curl}\mathbf{u}$ is the vorticity. Multiply $(6.17)_3$ by θ and integrate over V to derive

$$\frac{1}{2}\frac{d}{dt}\|\theta\|^2 = R(w,\theta) - \|\nabla\theta\|^2. \tag{6.20}$$

Next, multiply (6.19) with $i=3$ by w and integrate over V to obtain

$$0 = R(\nabla^*\theta,\nabla^*w) - \|\nabla w\|^2 - T^2\|w_{,z}\|^2\,. \tag{6.21}$$

Add (6.20) to $\xi\times$(6.21) for a coupling parameter $\xi(>0)$ to obtain

$$\frac{1}{2}\frac{d}{dt}\|\theta\|^2 = RI - D, \tag{6.22}$$

where I and D are now given by

$$I = (w, \theta) + \xi(\nabla^*\theta, \nabla^* w), \qquad D = \|\nabla\theta\|^2 + \xi(\|\nabla w\|^2 + T^2\|w_{,z}\|^2). \quad (6.23)$$

The number R_E is defined as

$$\frac{1}{R_E} = \max_H \frac{I}{D}, \quad (6.24)$$

where H is the space of admissible solutions and then (6.22) leads to

$$\frac{1}{2}\frac{d}{dt}\|\theta\|^2 \leq -D\left(\frac{R_E - R}{R_E}\right). \quad (6.25)$$

Provided $R < R_E$, one may use Poincaré's inequality to see that $D \geq \pi^2\|\theta\|^2$, whence from (6.25) we conclude $\|\theta(t)\| \to 0$ at least exponentially. From $(6.17)_1$ one may deduce

$$\|\mathbf{u}\| \leq R\|\theta\| \quad (6.26)$$

from which we conclude $R < R_E$ also guarantees $\|\mathbf{u}(t)\| \to 0$ exponentially.

The Euler-Lagrange equations for (6.24) are found to be

$$\begin{aligned} R_E k_i(\theta - \xi\Delta^*\theta) + 2k_i\xi(\Delta w + T^2 w_{,zz}) &= \pi_{,i}\,, \\ R_E(w - \xi\Delta^* w) + 2\Delta\theta &= 0, \end{aligned} \quad (6.27)$$

where $\pi(\mathbf{x})$ is a Lagrange multiplier.

From (6.27) one may now deduce, (Straughan, 2001a),

$$R_E^2 = \frac{(\pi^2 + a^2)^2}{a^2} + \frac{\pi^2 T^2(\pi^2 + a^2)}{a^2}. \quad (6.28)$$

The importance of (6.28) is that it is the same as equation (16) of (Vadasz, 1998a). This shows that the linear and nonlinear boundaries are the same and the nonlinear stability threshold (6.28) is the best possible result for nonlinear, unconditional stability. Thus, inferring from the results of (Vadasz, 1998a) the stability/instability threshold is

$$Ra_{crit} = \left[1 + \sqrt{T^2 + 1}\right]^2. \quad (6.29)$$

(Vadasz, 1998a) notes, with $T^2 = Ta$, for Ta large,

$$Ra_{crit} \to Ta + O(\sqrt{Ta}). \quad (6.30)$$

6.4 Bio-Convection

We now describe an energy analysis for a suspension of swimming micro-organisms, partly because it is of interest for its own sake, but also because the choice of generalized energy is suggested directly by previous work *on the linear problem.* This illustrates an important point: if there are any special features present in the linear problem, it may be possible to exploit them to suggest an opportune generalized energy.

The continuum model of bio-convection we describe is due to (Childress et al., 1975) and (Levandowsky et al., 1975), while the energy analysis was given by (Galdi and Straughan, 1985a).

A brief explanation of the physical picture is as follows. Suppose that a suspension of micro-organisms is contained in a fluid layer, say the infinite layer between the planes $z = -H, 0$, and suppose that a gravitational field acts in the negative z direction. The organisms have a density greater than that of the containing fluid and also have a natural tendency to swim in the upward (increasing z) direction. If a sufficient number of organisms are present, eventually the situation arises where the upper layer of the fluid is dominated by micro-organisms. These in turn, being of a density greater than the fluid, will tend to fall under the action of gravity. Hence an instability somewhat akin to Rayleigh-Taylor instability may develop. The striking thing about this instability is that it does not happen in a haphazard manner; rather, the organisms tend to fall in discrete "chimneys" in an ordered pattern, although several pattern types are possible, see (Childress et al., 1975), (Levandowsky et al., 1975).

The phenomenon of swimming micro-organisms forming convection patterns is known as bio-convection. This is an active area of research. Since the original work of (Childress et al., 1975) and (Levandowsky et al., 1975) there have been new continuum models for the motion of micro-organisms swimming in reaction to effects other than gravity. The swimming motion against gravity mentioned above is now known as gravitaxis. Other effects are chemotaxis, gyrotaxis, magnetotaxis, and phototaxis. Chemotaxis corresponds to swimming along a chemical gradient, and a much studied example is swimming along an oxygen gradient (oxytaxis). Models for this are developed and studied in (Hillesdon et al., 1995) and (Hillesdon and Pedley, 1996); see also the references therein. Gyrotaxis is connected with the shape of the micro-organism and recent papers dealing with this are those of (Bees and Hill, 1999), (Ghorai and Hill, 2000b; Ghorai and Hill, 2000a); many earlier references may be found in these articles. Magnetotaxis is the response of the micro-organism to a magnetic field, (Childress, 1981). Phototaxis is the swimming of micro-organisms toward or away from a light source. A model for this is developed in (Vincent and Hill, 1996) who also provide a detailed linear instability analysis. I am unaware of any nonlinear energy stability work for the models of chemotaxis, gyrotaxis, magnetotaxis or phototaxis. Hence, in this work we focus on gravitaxis.

The basic model we employ for the motion of micro-organisms is derived by (Childress et al., 1975). Let $c(\mathbf{x}, t)$ denote the concentration of micro-organisms in the suspension. The conservation law for the organisms is easily written in terms of a flux, $\mathbf{J}$, given by

$$J_i = cU(c, z)\delta_{i3} - D_{ij}c_{,j}\,, \tag{6.31}$$

where U is the upward swimming velocity of the organisms and $\mathbf{D}$ is the diffusion tensor given by

$$\mathbf{D} = \begin{pmatrix} \kappa_1(c,z) & 0 & 0 \\ 0 & \kappa_1(c,z) & 0 \\ 0 & 0 & \kappa(c,z) \end{pmatrix} \tag{6.32}$$

in which κ and κ_1 are positive functions of the indicated arguments.

The governing equations, based on an incompressible linear viscous fluid model, are in the continuum approximation,

$$\begin{aligned} &u_{i,t} + u_j u_{i,j} = -\frac{1}{\rho} p_{,i} - g(1+\alpha c)\delta_{i3} + \nu \Delta u_i, \\ &u_{i,i} = 0, \qquad c_{,t} + u_i c_{,i} = -J_{i,i}, \end{aligned} \tag{6.33}$$

where ρ is the constant density of the suspension, u_i the velocity, p the pressure, ν the viscosity, g is gravity, and α is a positive constant that expresses the ratio of density of a micro-organism to that of the growth medium.

The linear stability of two classes of equilibrium solution is considered in (Childress et al., 1975), namely:

Case 6.4.1

$$U = U_0, \quad \kappa = \kappa_0, \quad \kappa_1 = \delta \kappa_0,$$

where U_0, κ, δ are constant.

Case 6.4.2

$$\kappa/U \text{ not explicitly dependent on } z, \kappa_1 \text{ arbitrary.}$$

(Galdi and Straughan, 1985a) analyse a subclass of the more general class 6.4.2. They need the boundary conditions:

$$J_i n_i = 0, \quad u_i = 0 \qquad \text{when} \qquad z = -H, 0.$$

The former expresses the condition that no material flows out of the bounding surfaces $z = -H, 0$, whereas the latter is the no-slip condition. To ensure uniqueness of the equilibrium solution it is also necessary to impose the following restriction on the mean concentration: $< c > \stackrel{\text{def}}{=} c_n = \text{const}$. The subclass of solutions of Case 6.4.2 investigated involves those for which κ, U depend only on z, $dU/dz \le 0$, $U/\kappa = h^{-1}$ (constant), and κ_1 is arbitrary, apart from being continuously differentiable. The basic equilibrium solution of this subclass (denoted by an overbar) is

$$\bar{u}_i = 0, \quad \bar{c} = K(z) = c_0 \exp\left(\frac{U_0 z}{\kappa_0}\right), \qquad z \in [-H, 0],$$

with $c_0 = K(0)$, and U_0, κ_0 the values of U, κ at $z = 0$.

It is important to note that the basic equilibrium solution is *nonlinear in z*, so that an energy stability analysis is likely to be different from that required for such constant gradient problems as the standard Bénard one.

To study stability we set $\mathbf{u} = (u, v, w) = (u_1, u_2, u_3)$, $c = K(z) + \phi(\mathbf{x}, t)$, $p = \bar{p}(z) + \pi$, where (u_i, ϕ, π) are perturbations to the equilibrium values $(\bar{u}_i, K, \bar{p})$. The key to the choice of energy arises from the fact that (Childress et al., 1975) observed that the growth rate of linear theory could be shown to be real by essentially using K' as a weight. In the light of this, we derive the non-dimensionalized perturbation equations to (6.33) but first divide the perturbation equation arising from $(6.33)_3$ by K' to obtain

$$\begin{aligned} &u_{i,t} + \sigma^{-1} u_j u_{i,j} = -\pi_{,i} + \Delta u_i - R\phi\delta_{i3}\,, \\ &\qquad u_{i,i} = 0, \\ &\frac{\sigma\phi_{,t} + u_i\phi_{,i}}{K'} = -Rw - \frac{(\phi U)'}{K'} + \frac{(\kappa_1\phi_x)_x + (\kappa_1\phi_y)_y}{K'} + \frac{(\kappa\phi')'}{K'}\,, \end{aligned} \tag{6.34}$$

where $R^2 = (\alpha g c_0 h^3/\nu\kappa_0)$ is like a Rayleigh number, $\sigma = \nu/\kappa_0$ is a constant Schmidt number, $h = \kappa_0/U_0$ is a constant unit of length, $(0, \lambda)$ is now the non-dimensional layer, and $'$ denotes $\partial/\partial z$.

The boundary conditions are

$$\kappa\phi' = \phi U, \quad u_i = 0 \quad \text{when} \quad z = 0, \lambda, \tag{6.35}$$

together with

$$u_i, \phi, p \text{ are periodic functions in } x, y. \tag{6.36}$$

The disturbance cell is denoted by V.

In the form (6.34), the linear operator L that acts on (u, v, w, ϕ) is given by

$$\begin{pmatrix} \Delta & 0 & 0 & 0 \\ 0 & \Delta & 0 & 0 \\ 0 & 0 & \Delta & -R \\ 0 & 0 & -R & \frac{1}{K'}\big[\partial_\alpha(\kappa_1\partial_\alpha\cdot) + (\partial_z(\kappa\partial_z\cdot) - \partial_z(U\cdot)\big] \end{pmatrix} \tag{6.37}$$

where the repeated α signifies summation over $\alpha = 1, 2$. With the aid of (6.35) and (6.36) it is easily verified that L is symmetric, thanks to the weight K'.

The natural energy to use is suggested by (6.34), namely,

$$E(t) = \frac{1}{2} < u_i u_i > + \frac{1}{2}\sigma\Big\langle \frac{\phi^2}{K'} \Big\rangle. \tag{6.38}$$

Differentiating E and substituting from (6.34), with use of the boundary conditions (6.35), (6.36), we find the energy equation is

$$\begin{aligned} \frac{dE}{dt} = &- 2R < \phi w > - < u_{i,j} u_{i,j} > - \Big\langle \frac{\kappa_1}{K'}(\phi_x^2 + \phi_y^2) \Big\rangle \\ &- \Big\langle \frac{\kappa}{K'}\Big(\phi_z - \frac{\phi}{h}\Big)^2 \Big\rangle - \frac{1}{2h}\Big\langle \frac{w\phi^2}{K'} \Big\rangle. \end{aligned} \tag{6.39}$$

Since the linear part of the system is symmetrized we may appeal to the result of section 4.3, which shows the linear and nonlinear stability boundaries are the same for a symmetric system, provided we can handle the cubic nonlinear term in (6.39). Due to the presence of the last term in (6.39), the one in $w\phi^2$, the energy decay result derived by (Galdi and Straughan, 1985a) is achieved through use of a Sobolev inequality and is only conditional.

It would appear that proving an unconditional stability result for (6.34) using the boundary conditions (6.35) is a non-trivial problem. (Buonomo and Rionero, 2001) attempted to derive unconditional stability employing an unweighted L^2 energy, rather than (6.38). They experienced severe mathematical problems, effectively due to the boundary conditions (6.35). The difficulty would appear connected to the constant swimming speed U. Physically, it does seem unrealistic that the micro-organisms are still swimming with a constant speed U in the direction of the wall when they arrive at the wall. (Hillesdon et al., 1995) propose a flux boundary condition which contains a cut-off mechanism which may remedy this problem. However, it will also complicate a nonlinear energy stability analysis and I am unaware of any such work using a modified boundary condition. (Buonomo and Rionero, 2001) do develop a nonlinear energy stability theory which is unconditional provided they prescribe the concentration of micro-organisms on the boundary. They can also handle in a natural way a diffusion coefficient which is a (nonlinear) function of the micro-organism concentration.

6.5 Convection in a porous vertical slab

We complete this chapter by looking at another problem where a standard energy may not reveal as much information as a generalized one, and, in particular, where a *natural* generalized energy is suggested by previous work on the linearized problem.

(Kassoy, 1980) describes several situations, together with relevant references, of geophysical problems where the mathematical description is provided by studying convection in a vertical porous slot. In an interesting paper (Gill, 1969) showed that when a temperature difference is imposed across the slot, no convection is possible in the sense that thc basic solution is *linearly* stable to two-dimensional disturbances no matter how large the Rayleigh number. In (Straughan, 1988) the linear analysis of (Gill, 1969) is used to motivate the choice of a generalized energy, which shows that three-dimensional nonlinear disturbances will always decay provided the initial amplitude is less than a threshold, which behaves like the inverse of the Rayleigh number.

The model uses the Boussinesq approximation and Darcy's law, see (Gill, 1969). The velocity $\mathbf{v} = (u, v, w)$ is referred to Cartesian axes x_i chosen so that the z-axis points vertically upward and such that the boundaries at temperature $T = \pm\frac{1}{2}$ are given by $x = \pm\frac{1}{2}$. The steady solution to the problem is

$$u = v = 0, \quad w = x, \quad T = x. \tag{6.40}$$

Note that the basic solution has a non-zero velocity component. The equations for a perturbation to this solution are

$$\begin{aligned} u_i = -p_{,i} + \delta_{i3}\theta, \qquad u_{i,i} = 0, \\ Ra(\theta_{,t} + u_i\theta_{,i} + u + x\theta_{,3}) = \Delta\theta, \end{aligned} \tag{6.41}$$

where Ra is the Rayleigh number. The boundary conditions are:

$$\theta = u = 0, \quad x = \pm\frac{1}{2}; \qquad u_i, \theta, p \text{ periodic in } y, z.$$

The development of a generalized energy analysis commences by multiplying $(6.41)_3$ by $\Delta\theta$ and integrating over V to obtain

$$\begin{aligned} \frac{1}{2}Ra\frac{d}{dt}\|\nabla\theta\|^2 = &- \|\Delta\theta\|^2 + Ra < u\Delta\theta > \\ &+ Ra < x\theta_{,3}\Delta\theta > + Ra < u_i\theta_{,i}\Delta\theta > . \end{aligned} \tag{6.42}$$

Next, take curlcurl of $(6.41)_1$ to see that

$$-\Delta u_i = \delta_{j3}\theta_{,ij} - \delta_{i3}\Delta\theta; \tag{6.43}$$

the first component of this is

$$-\Delta u = \theta_{,13}. \tag{6.44}$$

Integrating by parts and using the boundary conditions, we may transform the third term on the right of (6.42) to

$$\begin{aligned} < x\theta_{,3}\Delta\theta > &= - < x\theta\Delta\theta_{,3} > \\ &= \frac{1}{2} < (x\theta_{,i}\theta_{,i})_{,3} > + < \theta x_{,i}\theta_{,i3} > \\ &= < \theta\theta_{,13} > \\ &= - < \theta\Delta u > \\ &= - < u\Delta\theta >, \end{aligned}$$

substituting from (6.44) in the second last step. This relation is used to reduce (6.42) to

$$\frac{1}{2}Ra\frac{d}{dt}\|\nabla\theta\|^2 = -\|\Delta\theta\|^2 + Ra < u_i\theta_{,i}\Delta\theta > . \tag{6.45}$$

Gill's result follows easily from (6.45) since in the linear theory the last term is not present.

To deal with the nonlinearity we write

$$< u_i\theta_{,i}\Delta\theta > \leq \sup_V |\mathbf{u}| \, \|\nabla\theta\| \, \|\Delta\theta\|. \tag{6.46}$$

If we now also suppose $u_i n_i = 0$ on the lateral cell boundary (a condition that automatically holds for a finite region), then we have the inequality, see the Appendix,

$$\sup_V |\mathbf{u}| \leq C\|\Delta\mathbf{u}\|.$$

We use this inequality in (6.46) and put the result in (6.45) to find

$$\frac{1}{2}Ra\frac{d}{dt}\|\nabla\theta\|^2 \leq -\|\Delta\theta\|^2 + RaC\|\Delta\mathbf{u}\| \, \|\nabla\theta\| \, \|\Delta\theta\|. \tag{6.47}$$

We now need a bound for $\|\Delta\mathbf{u}\|$, and to this end we square (6.43) to derive

$$\|\Delta\mathbf{u}\|^2 = < \Delta\theta(\theta_{,11} + \theta_{,22}) > \leq \|\Delta\theta\|^2, \tag{6.48}$$

where we have used the boundary conditions and the fact that $\theta_{,3} = 0$ on $x = \pm\frac{1}{2}$. Inequality (6.48) applied to (6.47) leads to

$$\frac{1}{2}Ra\frac{d}{dt}\|\nabla\theta\|^2 \leq \|\Delta\theta\|^2(RaC\|\nabla\theta\| - 1). \tag{6.49}$$

We now require

$$\|\nabla\theta(\mathbf{x},0)\| < \frac{1}{RaC}, \tag{6.50}$$

and then (6.49) together with the inequality, see the Appendix, $\|\Delta\theta\| \geq \pi\|\nabla\theta\|$, allows us to deduce that $\|\nabla\theta(\mathbf{x},t)\|^2 \to 0$ at least exponentially as $t \to \infty$.

Furthermore, Poincaré's inequality allows us to deduce the same decay for $\|\theta\|^2$. Also, since from $(6.41)_1$

$$\|\mathbf{u}\|^2 = < \theta w > \leq \frac{1}{2}\Big(\|\theta\|^2 + \|w\|^2\Big),$$

we see that

$$\|\mathbf{u}\|^2 \leq \|\theta\|^2,$$

and hence $\|\mathbf{u}\|^2$ must also decay at least exponentially.

To sum up, we have demonstrated that provided the initial data satisfy the restriction (6.50), there is always nonlinear stability. Since Ra is known and C is computable, (6.50) represents a useful practical bound. However, if the initial temperature gradients exceed the bound in (6.50) then one cannot rule out the possibility of a finite amplitude instability.

6.5.1 Brinkman's equation and convection in a porous vertical slab

(Qin and Kaloni, 1993b) considered the problem just investigated but when the equations governing the porous medium are those of Brinkman rather than Darcy. The basic solution they considered has the form (6.40). For the Brinkman theory, (Kwok and Chen, 1987) show that the w variation in x is exponential for the basic solution. The pertubation equations of (Qin and Kaloni, 1993b) modify (6.41) by the addition of a Brinkman term and are

$$\begin{aligned} -\lambda\Delta u_i + u_i &= -p_{,i} + \delta_{i3}\theta, \qquad u_{i,i} = 0,\\ Ra(\theta_{,t} + u_i\theta_{,i} + u + x\theta_{,3}) &= \Delta\theta, \end{aligned} \tag{6.51}$$

where λ is a non-dimensional Brinkman coefficient.

The boundary conditions are now

$$\theta = u = v = w = 0, \quad x = \pm\frac{1}{2}, \quad u_i, \theta, p \ \text{ periodic in } y, z.$$

(Qin and Kaloni, 1993b) develop a generalized energy analysis using the functional $\|\nabla\theta\|^2$ and they derive exponential decay provided $\|\nabla\theta(\mathbf{x},0)\|$ is restricted and the Rayleigh number Ra is suitably small.

It is worth observing that one can use the L^2 analysis of (Straughan, 1988), section 2, applied to (6.51). By multiplying $(6.51)_1$ by u_i and integrating over a period cell V and multiplying $(6.51)_3$ by θ and integrating over V one finds

$$\begin{aligned} &\lambda\|\nabla\mathbf{u}\|^2 + \|\mathbf{u}\|^2 = <\theta w>, \\ &\frac{1}{2}Ra\,\frac{d}{dt}\|\theta\|^2 = -Ra<u\theta> - Ra<x\theta\theta_{,3}> - \|\nabla\theta\|^2. \end{aligned} \tag{6.52}$$

Adding these equations with a coupling parameter $\zeta(>0)$ we obtain

$$\begin{aligned} \frac{1}{2}Ra\frac{d}{dt}\|\theta\|^2 = &-\|\nabla\theta\|^2 - \zeta\|\mathbf{u}\|^2 - \zeta\lambda\|\nabla\mathbf{u}\|^2 \\ &+ \zeta<\theta w> - Ra<u\theta> - Ra<x\theta\theta_{,3}>. \end{aligned} \tag{6.53}$$

The last term is zero thanks to the boundary conditions. The arithmetic-geometric mean inequality is used on the fourth and fifth terms on the right. Poincaré's inequality is used on the gradient terms and we may show that

$$\begin{aligned} \frac{1}{2}Ra\frac{d}{dt}\|\theta\|^2 \le &-\Big(\pi^2 - \frac{\zeta}{2\alpha} - \frac{Ra}{2\beta}\Big)\|\theta\|^2 - \zeta(1+\lambda\pi^2)\|v\|^2 \\ &- \Big[\zeta(1+\lambda\pi^2) - \frac{Ra\beta}{2}\Big]\|u\|^2 - \Big[\zeta(1+\lambda\pi^2) - \frac{\zeta\alpha}{2}\Big]\|w\|^2, \end{aligned}$$

where $\alpha, \beta > 0$ are constants at our diposal. Upon selecting $\zeta = Ra$, $\alpha = \beta = 2(1+\lambda\pi^2)$ we find

$$\frac{1}{2}Ra\frac{d}{dt}\|\theta\|^2 \le -\Big[\pi^2 - \frac{Ra}{2(1+\lambda\pi^2)}\Big]\|\theta\|^2. \tag{6.54}$$

This leads to unconditional decay of $\|\theta(t)\|$ provided

$$Ra < 2\pi^2(1+\lambda\pi^2). \tag{6.55}$$

Note that this reduces to the result of (Straughan, 1988) in the Darcy case $\lambda \to 0$. Moreover, from (6.52),

$$\lambda\|\nabla\mathbf{u}\|^2 + \|\mathbf{u}\|^2 \le \frac{1}{2}\|\theta\|^2 + \frac{1}{2}\|w\|^2,$$

and so

$$2\lambda\|\nabla\mathbf{u}\|^2 + \|\mathbf{u}\|^2 \le \|\theta\|^2. \tag{6.56}$$

Thus, condition (6.55) leads also to unconditional decay of $\|\mathbf{u}\|$ and $\|\nabla\mathbf{u}\|$.

In fact, one can produce essentially the same result for the more general basic solution to Brinkman's equations as is effectively employed by (Kwok and Chen, 1987). To see this let v_i, T, p solve the Brinkman equations (6.57) with a linear density-temperature relationship. So, v_i, T, p solve

$$\begin{aligned} &-\lambda\Delta v_i + \frac{\mu}{k}v_i = -p_{,i} + g\alpha\rho_0 k_i T, \qquad v_{i,i} = 0, \\ &T_{,t} + v_i T_{,i} = \kappa\Delta T. \end{aligned} \tag{6.57}$$

The porous medium occupies the infinite three-dimensional vertical layer between the planes $x = 0$ and $x = L$. The temperatures are fixed at $T = T_L$ for $x = 0$, $T = T_R$ for $x = L$, with $T_R > T_L$, and we put $\Delta T = T_R - T_L$. Additionally, define ν, ζ, ω and the Rayleigh number, $Ra = R^2$, by

$$\nu = \frac{\mu}{\rho_0}, \quad \zeta = \sqrt{\frac{\mu}{k\lambda}}, \quad \omega = \frac{T_L}{\Delta T}, \quad Ra = \frac{Lkg\alpha\Delta T}{\kappa\nu}.$$

Then, the steady solution may be shown to have form $u = v = 0$, with $\bar{w}(x)$ and $\bar{T}(x)$ given by

$$\bar{T}(x) = \Big(\frac{T_R - T_L}{L}\Big)x + T_L, \qquad \bar{w}(x) = \frac{\kappa R^2}{L}f(x), \tag{6.58}$$

where

$$f(x) = \omega + x - \omega\cosh\hat{\lambda}x + \left(\frac{1+\omega-\omega\cosh\hat{\lambda}}{\sinh\hat{\lambda}}\right)\sinh\hat{\lambda}x,$$

and $\hat{\lambda} = \zeta L$.

With the time, velocity, and temperature scales $\mathcal{T} = L^2/\kappa$, $U = \kappa/L$, $T^\sharp = U\sqrt{(\Delta T)L\mu/\kappa k g\alpha\rho_0}$, the non-dimensional perturbation equations for the velocity, temperature, and pressure perturbations, u_i, θ, π, may be shown to be

$$\begin{aligned} &u_i - \frac{1}{\hat{\lambda}^2}\Delta u_i = -\pi_{,i} + Rk_i\theta, \\ &u_{i,i} = 0, \\ &\theta_{,t} + u_i\theta_{,i} + Ru + R^2 f(x)\theta_{,z} = \Delta\theta. \end{aligned} \tag{6.59}$$

The boundary conditions are that (u_i, θ, π) satisfy a periodic planform in y, z and $u_i = 0$, $\theta = 0$, on the non-dimensional vertical boundaries $x = 0, 1$.

From equations (6.59) we may derive the energy identities

$$\begin{aligned} 0 &= -\|\mathbf{u}\|^2 - \frac{1}{\hat{\lambda}^2}\|\nabla\mathbf{u}\|^2 + R(\theta, w), \\ \frac{d}{dt}\frac{1}{2}\|\theta\|^2 &= -R(u, \theta) - \|\nabla\theta\|^2. \end{aligned} \tag{6.60}$$

With a coupling parameter $\xi(>0)$ we then derive an energy equation

$$\frac{d}{dt}\frac{1}{2}\|\theta\|^2 = -R(u,\theta) + R\xi(\theta, w) - \xi\|\mathbf{u}\|^2 - \frac{\xi}{\hat{\lambda}^2}\|\nabla\mathbf{u}\|^2 - \|\nabla\theta\|^2. \tag{6.61}$$

We use the arithmetic-geometric mean inequality on the first two terms on the right and employ Poincaré's inequality on the gradient terms and we may show that

$$\frac{d}{dt}\frac{1}{2}\|\theta\|^2 \le -\xi\Big(1 + \frac{\pi^2}{\hat{\lambda}^2}\Big)\|v\|^2 - \left[\pi^2 - \frac{R^2}{4\xi(1+\pi^2\hat{\lambda}^{-2})} - \frac{R^2\xi}{4(1+\pi^2\hat{\lambda}^{-2})}\right]\|\theta\|^2.$$

If we now select $\xi = 1$, then we show

$$\|\theta(t)\|^2 \le \|\theta(0)\|^2 \exp\left[\left\{2\pi^2 - \frac{R^2}{(1+\pi^2\hat{\lambda}^{-2})}\right\}t\right]. \tag{6.62}$$

Thus, if $R^2 < 2\pi^2(1 + \pi^2/\hat{\lambda}^2)$ we see that $\|\theta(t)\|$ has at least exponential decay. From (6.60) a similar decay in $\|\nabla\mathbf{u}\|$ and $\|\mathbf{u}\|$ follows.

6.5.2 *Vertical porous convection and a nonlinear density law*

(Kwok and Chen, 1987) conducted an experiment with a vertical porous slab with the walls maintained at different temperatures. They backed their experiment up with a linearized instability analysis. They based their theory on a Brinkman model but also employed a cubic density-temperature relationship. Instead of discovering that the system is always (linearly) stable as has been suggested with Darcy theory, they found that convection will commence with a finite Rayleigh number (of value 308). Of course, this raises an interesting question in relation to the prediction of Darcy theory. What is responsible for convection commencing at finite Rayleigh number, is it the Brinkman theory, or the nonlinear density relationship? We here use energy theory to show, in a sense, that the Brinkman theory is responsible for convection. To do this, we commence with Darcy's law, but with a quadratic density-temperature relationship. Thus, the governing equations are

$$\begin{aligned} &\frac{\mu}{k}\,v_i = -p_{,i} - \rho_0 g k_i\big(1-\alpha[T-4]^2\big),\\ &v_{i,i}=0, \qquad T_{,t}+v_iT_{,i}=\kappa\Delta T. \end{aligned} \tag{6.63}$$

We let the water saturated porous medium occupy the three-dimensional domain between the vertical walls $x=0$ and $x=L$. The boundary $x=0$ is kept at temperature $T=0°\mathrm{C}$, while the boundary $x=L$ is held at a constant temperature $T_R \geq 4°\mathrm{C}$. There is no flow through the boundaries $x=0,L$, so $u=0$ there. The basic solution to this problem is found to be

$$\bar{u}=\bar{v}=0,\quad \bar{T}=\frac{T_Rx}{L},\quad \bar{w}=-\frac{\rho_0 gk}{\mu}\left[1-\alpha\Big(\frac{T_Rx}{L}-4\Big)^2\right]. \tag{6.64}$$

Perturbations (u_i,θ,π) are introduced to the steady solution $(\bar{v}_i,\bar{T},\bar{p})$ and these are non-dimensionalized with the time, velocity, pressure and temperature scales, $\mathcal{T},U,P$ and $T^\sharp$ given by $\mathcal{T}=L^2/\kappa$, $U=\kappa/L$, $P=LU\mu/k$, $T^\sharp=T_R$. Define variables ξ and β by $\xi=4/T_R$, $\beta=1/\alpha T_R^2$ and the Rayleigh number $Ra=R^2$ by $Ra=\alpha\rho_0 gT_R^2kL/\mu\kappa$. The non-dimensional perturbation equations are then

$$\begin{aligned} &u_i=-\pi_{,i}-2R(\xi-x)k_i\theta+Rk_i\theta^2, \qquad u_{i,i}=0,\\ &\theta_{,t}+u_i\theta_{,i}+u-Rf(x)\theta_{,z}=\Delta\theta, \end{aligned} \tag{6.65}$$

where $f(x)=\beta-(x-\xi)^2$.

We derive an equation for $\|\nabla\theta\|^2$ and this is found by multiplying $(6.65)_3$ by $-\Delta\theta$ and integrating over a period cell V,

$$\frac{d}{dt}\frac{1}{2}\|\nabla\theta\|^2=(u_i\theta_{,i},\Delta\theta)+(u,\Delta\theta)-R(f\Delta\theta,\theta_{,z})-\|\Delta\theta\|^2. \tag{6.66}$$

Next, take curlcurl of $(6.65)_1$ and then the first component of the resulting equation yields

$$-\Delta u=-2R\big[(\xi-x)\theta\big]_{,xz}+R\theta^2_{,xz}. \tag{6.67}$$

With some integrations by parts we see that

$$\begin{aligned} -R(f\Delta\theta,\theta_{,z}) =&R(f\Delta\theta_{,z},\theta)\\ =&-\frac{R}{2}<f|\nabla\theta|^2_{,z}> -R(f'\theta_{,xz},\theta)\\ =&2R\big((x-\xi)\theta,\theta_{,xz}\big)\,. \end{aligned}$$

Also, with further integrations by parts, noting $u = 0$ at $x = 0, 1$, $(u, \Delta\theta) = (\Delta u, \theta)$ and then

$$\begin{aligned}(u, \Delta\theta) - R(f\Delta\theta, \theta_{,z}) &= (\Delta u, \theta) + 2R\big((x-\xi)\theta, \theta_{,xz}\big)\\ &= \big(\theta, 2R[(\xi - x)\theta]_{,xz} - R\theta^2_{,xz}\big) + 2R\big((x-\xi)\theta, \theta_{,xz}\big)\\ &= -R(\theta, \theta^2_{,xz}),\end{aligned} \tag{6.68}$$

where (6.67) has also been employed.

Thus, using (6.68) in (6.66) we find

$$\frac{d}{dt}\frac{1}{2}\|\nabla\theta\|^2 = -\|\Delta\theta\|^2 - R(\theta, \theta^2_{,xz}) + (u_i\theta_{,i}, \Delta\theta). \tag{6.69}$$

We deduce from (6.69) that in the *linear* theory the last two terms are not present and then we have linear stability *always*. This is akin to the result described at the beginning of this section. We infer from this that the (Kwok and Chen, 1987) instability at finite Rayleigh number is due to the Brinkman effect.

6.5.3 Vertical porous convection and variable fluid properties

(Flavin and Rionero, 1999b) is a very interesting contribution to the problem of convection in a vertical porous slab. They employ Darcy's law, but allow both the viscosity and thermal diffusivity of the fluid to depend on temperature T. Their porous medium is confined in the three-dimensional vertical porous layer $-\frac{1}{2} < x < \frac{1}{2}$ with $u = 0$ at $x = \pm\frac{1}{2}$ and $T = \pm\frac{1}{2}$ at $x = \pm\frac{1}{2}$. If the thermal diffusivity is $\psi(T)$ and the viscosity is $\phi(T)$, then the governing equations are

$$\begin{aligned}\phi(T)v_i = -p_{,i} + T\delta_{i3}, \qquad v_{i,i} = 0,\\ Ra(T_{,t} + v_iT_{,i}) = \Delta\Big(\int_0^T \psi(s)ds\Big),\end{aligned} \tag{6.70}$$

where Ra is the Rayleigh number. The steady solution is now $\bar{u} = \bar{v} = 0$, with an appropriate pressure, and $\bar{T}, \bar{w}$ solving

$$\int_0^{\bar{T}} \psi(s)ds = Ax + B, \tag{6.71}$$

$$A = \int_{-\frac{1}{2}}^{\frac{1}{2}} \psi(s)ds, \tag{6.72}$$

$$B = \frac{1}{2}\int_0^{\frac{1}{2}} \big[\psi(s) - \psi(-s)\big]ds, \qquad \bar{w} = \frac{[-C + \bar{T}]}{\phi(\bar{T})},$$

where $C = 0$ or is a constant such that $\int_{-1/2}^{1/2} w\,dx = 0$. Perturbation equations are derived by (Flavin and Rionero, 1999b) as

$$\begin{aligned} &\phi(\bar{T}+\theta)u_i = -\pi_{,i} + (1-\bar{w}\phi')\theta\delta_{i3}, \qquad u_{i,i} = 0, \\ &Ra(\theta_{,t} + u_i\theta_{,i} + \bar{w}\theta_{,z} + u\bar{T}_{,x}) = \Delta\Phi_\theta, \end{aligned} \tag{6.73}$$

where $\phi' = \phi'(\bar{T}+\delta\theta)$ for some $0 < \delta < 1$, and where Φ is defined by

$$\begin{aligned} \Phi(\theta;\bar{T}) &= \int_0^\theta ds \int_{\bar{T}}^{\bar{T}+s} \psi(r)dr \\ &= \int_0^\theta ds \int_0^s \psi(\bar{T}+r)dr, \end{aligned}$$

and Φ_θ denotes $\partial\Phi/\partial\theta$.

(Flavin and Rionero, 1999b) define their generalized energy functional $E(t)$ by

$$E(t) = \int_V \Phi(\theta;\bar{T})dV$$

the integral being over the domain of the porous medium. They show that E satisfies the energy equation

$$Ra\frac{dE}{dt} = -RaA\int_V u\theta\,dV - \int_V |\nabla\Phi_\theta|^2 dV. \tag{6.74}$$

They then define the constant M by

$$M \geq |1-\bar{w}\phi'|. \tag{6.75}$$

They then show that (6.74) leads to unconditional nonlinear stability (with exponential decay) provided the Rayleigh number is such that $Ra < 2\pi^2/AM$, where A is defined by (6.72) while M is the best constant in (6.75).

Numerical studies of convection in vertical enclosures are by (Bardan et al., 2000), (Bardan and Mojtabi, 1998) for a fluid, and for a porous medium by (Charrier-Mojtabi et al., 1998), (Karimi-Fard et al., 1999), (Mercier et al., 2002), and (Rees and Lage, 1997).

7

Geophysical problems

7.1 Patterned (or polygonal) ground formation

In this chapter we primarily describe two geophysical problems where energy theory has proved very useful. Not only has an application of energy theory yielded useful information, but also, the mathematics of the problem has *necessitated* the introduction of novel generalized energies.

We begin with patterned ground formation, a subject developed analytically in the first instance by (Ray et al., 1983), (Gleason, 1984), and (Gleason et al., 1986). Early theoretical studies identified some of the processes involved, e.g., (Nordenskjold, 1909), (Low, 1925), (Gripp, 1926), and (Washburn, 1973; Washburn, 1980). A theory that involves a five step process was proposed by Professor R.D. Gunn in (George et al., 1989), and we describe this work; partly, because it would appear the process description is the most complete so far, but also, because this is where nonlinear energy stability theory was first employed in the subject.

Polygonal ground, which in this description consists of stone borders forming regular hexagons with soil centres, represents one of the most striking and interesting small scale geological phenomena. In fact, the regularity in size and shape of polygonal ground is often remarkable. However, the existence of these features is not widely known because they generally exist above the timberline in the high mountains of the temperature zone or in remote locations far beyond the treeline in the Arctic and Antarctic. The stone borders usually form regular hexagons that are all about the same size at a single location. Between different sites, polygonal stone nets vary

Site Location	USGS Quadrangle	W	D	Elev.	I
Niwot Ridge	Ward, CO	2.30	0.53	3500	a
Niwot Ridge	Ward, CO	3.80	1.00	3500	a
Caribou Mt.	Ward, CO	1.70	0.48	3650	a
Caribou Mt.	Ward, CO	1.70	0.48	3650	a
Arikaree Glacier	Monarch Lake, CO	0.86	0.18	3800	a
Arikaree Glacier	Monarch Lake, CO	0.83	0.16	3800	a
Albion Saddle	Ward, CO	0.40	0.10	3650	a
Albion Saddle	Ward, CO	0.50	0.13	3650	a
Medicine Bow Peak	Medicine Bow, WY	2.30	0.70	3650	a
Medicine Bow Peak	Medicine Bow, WY	2.90	0.70	3650	a
Trail Ridge Road	Trail Ridge, WY	4.00	1.10	3660	a
Green Lake No. 4	Ward, CO	2.30	0.71	3550	a
Chief Mountain	Franks Peak, WY	2.30	0.83	3400	b
Beartooth Plateau	Beartooth Butte & Alpine, MT	0.20	0.07	3400	b
Dana Plateau	Mono Craters, CA	0.64	0.16	3500	b
Parker Pass Creek	Mono Craters, CA	0.70	0.14	3400	b
Mt. Hare Region Richardson Mountains, Yukon Terr.	Eagle River, Canada series A 502 map 1161	0.16	0.05	1150	c

Table 7.1. Field study data for polygonal ground (hexagons) in the USA. Elev., W, D, denote elevation, width, and depth, respectively. Width, depth, and elevation are in metres. Column I denotes Investigator: a=Ray; b=Krantz & Gunn; c=Gleason & Gunn. (After (Gleason, 1984).)

from about 10cm in width to more than 4m, as may be seen from Table 7.1 of actual width to depth measurements.

In the thesis of (Gleason, 1984), pp. 117–120, he includes three tables of data. Data appropriate to hexagonal cells on land, Tables 7.1 and 7.2 are reproduced here. His second table concerns field study data for patterned ground that has formed *under water* in shallow lakes in Mono Craters, California, and in Medicine Bow, Wyoming. Finally, he includes data for sorted stripes. We include only that for hexagonal patterned ground formed on land.

Presently, theory requires three a priori conditions before stone polygons may form. The first of these is the existence of alternating freeze-thaw cycles within the soil. For the hexagonal patterns studied here these are annual cycles that follow the rhythm of the seasons throughout the year. The second condition necessitates that the soil must be saturated with water for at least part of the year. The final requirement is that an impermeable frozen soil barrier must underlie the active layer, that is, the layer of soil that

Site Location	Country	Width	Depth	Elevation	I
Macquarie Is.	Australia	0.25	0.80	300	a
Hafravatn	SW Iceland	1.46	0.37	110	b
Hraunhreppur	W Iceland	0.48	0.12	30	c
Latraheidi	NW Iceland	2.63	0.69	180	c
Thorskafjardarheidi	NW Iceland	2.03	0.50	450	c

Table 7.2. Field study data for polygonal ground (hexagons) in Australia and Iceland. Width, depth, and elevation are in metres. Column I denotes Investigator: a=Caine; b=Stingl; c=Schunke. (After (Gleason, 1984).)

alternately freezes and thaws. For the annual cycles considered here, permafrost must be present to form the impermeable frozen-soil barrier. When the three conditions outlined above are satisfied, the formation of polygonal ground follows a five step process, described completely by (George et al., 1989). Professor R.D. Gunn has, in fact, succeeded in growing stone polygons in the laboratory by reproducing these five steps. The five stages are now described.

Step 1 *Permeability enhancement as a result of the formation of needle ice and frost heaving in the soil.*

For the onset of convective motion of water in the soil, Step 2, a critical permeability must be present in the soil before polygonal ground can form. Usually, only sand or gravel possess permeabilities sufficiently high. Silty soils, such as those where patterned ground has been found, produce a large amount of frost heaving, but have low permeabilities. The necessary increase in permeability is produced by the formation of needle ice and ice lenses. Needle ice, (Washburn, 1973), is an accumulation of slender, bristle-like ice crystals found in soils subjected to freeze-thaw cycles. The elongation of these ice crystals is perpendicular to the permafrost surface, and the needles are commonly a centimetre or more in length. Since the needles consist of essentially pure ice, their growth thrusts aside soil and rocks. After a few freeze-thaw cycles, the soil volume may increase by more than a hundred percent, and the entire surface of the ground is thrust upward to accommodate this expansion, (Embleton and King, 1975). When the soil thaws, the melting of the ice crystals leaves the ground with a greatly enhanced porosity, and this in turn leads to greatly increased permeability. A silty soil thus tends to increase in permeability with each freeze-thaw cycle until the critical permeability is reached, then Step 2 of the patterned ground process commences.

During field studies by Professors Krantz and Gunn and their co-workers, they observed that soil from the central part of the stone polygons is quite silty and would not have a sufficiently large permeability to sustain patterned ground formation. Any attempt to transport these soils to the laboratory collapses the porous structure left by the needle ice and ice lenses. Laboratory measurements then confirm permeabilities well below the value required for the formation of polygonal ground. The same soils, when subjected to multiple freeze-thaw cycles in laboratory experiments, have been found to expand and frost heave upwards until after about thirty cycles they begin to form polygonal ground.

Step 2 *Onset of buoyancy driven natural convection in the water saturated soil.*

The active layer is that layer of soil above the permafrost that thaws each summer and freezes solid during the winter. In the thawed summer state, water in the active layer near the permafrost interface remains close to its freezing point, 0°C; the water in the soil nearer the surface is relatively much warmer. Water is unusual in that it has a maximum density at 4°C. Thus, in the active layer, warmer, denser water near 4°C overlies colder, less dense water near its freezing point. When the permeability is high enough and the permafrost is sufficiently melted that the active layer is deep enough, heavier water sinks with consequently lighter water rising to set up convection currents in the soil. The convection currents are responsible for fixing the size and shape of the stone polygons, which arise through Steps 3 and 4.

Step 3 *Formation of a pattern in the permafrost interface.*

The convection currents in the active layer set up a pattern of hexagonal convection cells. The water rises up the centre of the hexagon and flows down along the cell boundary. Downflow carries relatively warmer water from near the surface toward the impermeable permafrost interface, and this induces melting along the sides of the hexagons. In the centre of a cell colder water is carried toward the surface and this slows down melting. This process results in a series of isolated frozen soil peaks in the centre of the hexagon surrounded by an interconnected (hexagonal) continuous trough in the permafrost surface.

Step 4 *Formation of polygonal ground through frost heaving.*

In regions where there is vigorous frost, the process of frost heaving will push any rocks originally in the soil slowly to the surface. A well known example of frost heaving is the appearance of stones in a recently ploughed field. The type of soils in which a frost heave effect is predominant tend to be of a silty character. The experiments of (Corte, 1966) showed that in a rock-soil mixture subjected to freeze-thawing the larger stones will be pushed upwards leaving finer material below. Rocks do not necessarily move in the vertical direction, but rather, they move at ninety degrees to the freezing front, see e.g., (Washburn, 1973). In active patterned ground formations, stones are displaced perpendicularly to the frozen soil interface, and so move upward and sideways away from the peaks at the centre of a cell and congregate near the hexagonal cell boundary that moves more slowly upward. This aspect is the actual reason for the hexagonal stone pattern since the stones are following the border of the hexagonal cell formed by downflow at the permafrost interface.

One argument is that frost heaving of stones is not produced by a stationary frozen surface nor can the freezing come from above since water is present above the permafrost until the time when the soil is frozen rigid. In order that the freezing action arise from below it is necessary for the permafrost to be substantially colder than the freezing point of water and then heat will be conducted away from the ice-water interface which will then freeze in an upward direction. The permafrost temperature at 5–10m below the surface is approximately equal to the average annual temperature of the air at the surface and so this leads to the conclusion that hexagonal patterned ground formations are not likely to be found in abundance at places with average annual temperatures greater than about −5°C. These conclusions were also drawn by (Goldthwaite, 1976).

Field observations and laboratory experiments by the groups associated with Professors Gunn and Krantz evidently verify the theory above. For example, some permafrost is present in the mountains of Wyoming at heights around 3000m, where the average annual air temperature is close to 0°C. But, only a few faint and inactive stone patterns are found at these elevations and these are probably relics of the last ice age. In contrast, a considerable amount of patterned ground has been discovered at elevations of about 3600m, see Table 7.1, and here the average annual air temperature is approximately 5°C colder. In Professor Gunn's laboratory experiments, stone polygons did not form even after 500 freeze-thaw cycles when the temperature of the simulated permafrost was held close to 0°C. Stone polygon formation was, however, observed after only 20–30 freeze-thaw cycles when the lower portion of the simulated permafrost was maintained at a temperature of −5°C.

Step 5 *Perpetuation of the hexagonal pattern.*

Rocks have a higher thermal conductivity than soil, and so once stones have begun to concentrate over the troughs in the permafrost, these troughs become self-perpetuating due to accelerated melting caused by the higher thermal conductivity of rock. Eventually, the convective motion of water in the active layer may cease because the removal of the stones from the central part of the hexagonal cell causes the permeability to decrease. There still remains a tendency for further segregation of stone because of the continuing presence of an undulating permafrost surface that is now perpetuated by the higher thermal conductivity of rocks concentrated along the cell borders.

It is believed that the above five-step process, proposed by Professor R.D. Gunn, produces polygonal stone nets that are characterized by regularity in their size and shape. Previous work on the subject of patterned ground is thoroughly reviewed by (Ray et al., 1983) and (George et al., 1989).

7.2 Mathematical models for patterned ground formation

We suppose the porous material (the active layer) occupies the infinite layer $z \in (0,d)$, with the temperature of the lower plane maintained at 0°C whereas the temperature of the upper plane is kept fixed at a temperature above 4°C. Due to the fact that the porous material contains water whose density below 4°C is a decreasing function of temperature, the situation envisaged results in a gravitationally unstable layer lying below a stably stratified one. When convection occurs in the lower layer the motions will penetrate into the upper layer, as discussed in the context of a fluid in chapter 17.

The equations we employ utilize Darcy's law, although for the density in the body force term we employ the form,

$$\rho = \rho_0\{1 - \alpha(T-4)^2\}, \tag{7.1}$$

where ρ_0 is the density at 4°C and $\alpha \approx 7.68 \times 10^{-6}$ (°C^{-2}). The basic equations for the fluid motion in the active layer are then (3.61), with the body force term given by (7.1). In keeping with (George et al., 1989) we retain an inertia term in the momentum equation.

The boundary conditions we select are

$$\begin{aligned} k_iv_i &= 0 \qquad &&\text{at} \quad z = 0, d, \\ T &= 0°\text{C} &&\text{at} \quad z = 0, \\ \delta_1 k_i T_{,i} + \delta_2 T &= c &&\text{at} \quad z = d. \end{aligned} \tag{7.2}$$

The number c is a prescribed constant, and δ_1, δ_2 are constants given in terms of a *radiation parameter* a by $\delta_1 = 1/(1+a)$, $\delta_2 = a/(1+a)$. The boundary condition $(7.2)_3$ is so written, since it allows us to write the (steady) conduction solution in terms of the temperature of the upper surface in the steady state and this quantity features in the non-dimensionalization. Condition $(7.2)_3$ is only approximate since a movement of the boundary is occuring due to the phase change. We believe it is acceptable to neglect this movement since the timescale for convection is much shorter, although the phase change effect is discussed in subsection 7.2.4. Also, condition $(7.2)_3$ allows us to examine the important limiting cases of prescribed heat flux, $a = 0$, and prescribed constant temperature, $a \to \infty$.

The (steady) conduction solution is

$$v_i = 0, \qquad \bar{T} = \beta z, \tag{7.3}$$

where $\beta = T_1/d$ and the upper surface temperature T_1 is related to c by $T_1 = cd(1+a)/(1+ad)$. The steady pressure field is

$$\bar{p}(z) = p_0 - \rho_0 g z + \frac{\alpha \rho_0 g}{3\beta}(\beta z - 4)^3, \tag{7.4}$$

where p_0 is some conveniently chosen pressure reference scale.

To reflect the increased permeability due to needle ice formation, (George et al., 1989), chose a linear permeability relation of form $k(z) = k_0(1+\gamma z)$, and define $f(z) = k_0/k = 1/(1+\gamma z)$. Under a non-dimensionalization into the layer $z \in (0,1)$, defining the Prandtl number $Pr = d^2\mu/k_0\rho_0\kappa$, a parameter $\xi = 4/T_1$, and

$$R = \sqrt{\frac{g\alpha\rho_0\beta^2 d^3 k_0}{\kappa\mu}},$$

the *non-dimensionalized perturbation equations*, for a perturbation to the steady solution (7.3), (7.4), are

$$\begin{aligned} &Au_{i,t} = -p_{,i} - fu_i - 2R\theta(\xi - z)k_i + Pr\theta^2 k_i, \\ &u_{i,i} = 0, \qquad Pr(\theta_{,t} + u_i\theta_{,i}) = -Rw + \Delta\theta, \end{aligned} \tag{7.5}$$

where $w = u_3$. (George et al., 1989) define a Rayleigh number, Ra, to reflect the depth of the layer, which in the conducting state is actually destabilizing, by

$$Ra = \xi^3 R^2. \tag{7.6}$$

The boundary conditions (7.2) for the perturbation quantities become

$$\begin{aligned} w &= 0 \qquad \text{at} \qquad z = 0, 1, \\ \theta &= 0 \qquad \text{at} \qquad z = 0, \\ \frac{\partial\theta}{\partial z} + a\theta &= 0 \qquad \text{at} \qquad z = 1; \end{aligned} \tag{7.7}$$

and we also assume $\mathbf{u}, \theta, p$ have a *periodic* structure in (x, y), such as one consistent with hexagonal convection cells.

7.2.1 Linear instability

The Spiegel method of establishing exchange of stabilities, discussed in section 4.3, applies to the linearized version of (7.5), (7.7), provided $f =$ constant. Thus, due to this fact, and since the energy results are very close to the linear ones, (George et al., 1989) examine only stationary convection, which reduces to finding the smallest eigenvalue R, of the system

$$\begin{aligned} &p_{,i} = -fu_i - 2R\theta(\xi - z)k_i, \\ &u_{i,i} = 0, \qquad Rw = \Delta\theta. \end{aligned} \tag{7.8}$$

The usual normal mode method with $\theta = \Theta(z)e^{i(mx+ny)}$, $w = W(z)e^{i(mx+ny)}$, with the wave number k given by $k^2 = m^2 + n^2$ and $D = d/dz$, reduces (7.8) to

$$\begin{aligned} &(D^2 - k^2)\Theta = RW, \\ &(D^2 - k^2)W = -\frac{f'}{f}DW + \frac{2k^2 R}{f}(\xi - z)\Theta, \end{aligned} \tag{7.9}$$

where $-f'/f = \gamma/(1+\gamma z)$. This system was solved numerically by the compound matrix method, see chapter 19, and the minimum of $R^2(k^2)$ was found by golden section search. The relevant boundary conditions are:

$$\begin{aligned} W = 0 \quad &\text{at} \quad z = 0,1, \\ \Theta = 0 \quad &\text{at} \quad z = 0, \\ D\Theta + a\Theta = 0 \quad &\text{at} \quad z = 1. \end{aligned} \tag{7.10}$$

As emphasized thoroughout the book, linear theory gives a Rayleigh number boundary, which if exceeded ensures instability. It does not preclude the possibility of subcritical instabilities. Energy theory was applied to penetrative convection in a porous medium by (George et al., 1989) who find results that are very close to those obtained by linear theory. This shows that linear theory has essentially captured the physics of the onset of pore water convection, the process we believe responsible for determining the aspect ratio of the patterned ground cells.

7.2.2 Nonlinear energy stability.

To investigate the nonlinear stability of the steady solution (7.3), (7.4), we use the energy

$$E = \frac{1}{2}A\|\mathbf{u}\|^2 + \frac{1}{2}\lambda Pr\|\theta\|^2, \tag{7.11}$$

where $\lambda\,(>0)$ is a coupling parameter, $\|\cdot\|$ is the $L^2(V)$ norm, with V a disturbance cell.

We differentiate E, use (7.5) and the boundary conditions to derive the energy equation

$$\begin{aligned} \frac{dE}{dt} =& - < f|\mathbf{u}|^2 > -\lambda D(\theta) - a\int_\Gamma \theta^2 dA \\ &- R < \theta w(2\xi + \lambda - 2z) > + Pr < \theta^2 w >, \end{aligned} \tag{7.12}$$

where $D(\cdot)$ is the Dirichlet integral, $< \cdot >$ represents the integral over V, and Γ is that part of the boundary of V that lies in $z = 1$.

Define now

$$\mathcal{D} = \lambda D(\theta) + < f|\mathbf{u}|^2 > + a\int_\Gamma \theta^2 dA, \tag{7.13}$$

$$I = - < \theta w(2\xi + \lambda - 2z) >, \tag{7.14}$$

and from (7.12) we obtain

$$\frac{dE}{dt} \le -\mathcal{D}R\Big(\frac{1}{R} - \frac{1}{R_E}\Big) + Pr < \theta^2 w >, \tag{7.15}$$

where

$$\frac{1}{R_E(\lambda)} = \max_{\mathcal{H}} \frac{I}{\mathcal{D}}, \tag{7.16}$$

$\mathcal{H}$ being the space of admissible functions.

Nonlinear conditional stability follows from inequality (7.15) in a manner similar to that described in section 2.3. Hence, suppose $R < R_E$, define $b = R^{-1} - R_E^{-1}$ (> 0), and then with the aid of the Cauchy-Schwarz inequality we derive from (7.15),

$$\frac{dE}{dt} \leq -bR\mathcal{D} + Pr\|\theta^2\| \, \|w\|. \tag{7.17}$$

The Sobolev inequality $\|\theta^2\| \leq c_1 D(\theta)$, is now used in (7.17) to show

$$\frac{dE}{dt} \leq -bR\mathcal{D}\left(1 - \frac{c_1 Pr\sqrt{2}}{bR\lambda\sqrt{A}}\sqrt{E}\right). \tag{7.18}$$

Hence, if

$$\text{(i)} \quad R < R_E\,, \qquad \text{(ii)} \quad E^{1/2}(0) < \frac{bR\lambda\sqrt{A}}{c_1 Pr\sqrt{2}}\,, \tag{7.19}$$

then a calculation similar to that in section 2.3 shows that $E(t) \to 0$ as $t \to \infty$ at least exponentially.

We must now determine R_E. The Euler-Lagrange equations for the maximum in (7.16) are found after setting $\phi = \lambda^{1/2}\theta$ to remove the λ dependence from the $D(\theta)$ term in $\mathcal{D}$,

$$2\Delta\phi - R_E M(z) w = 0, \qquad 2fu_i + R_E M(z)\phi k_i = p_{,i}\,, \tag{7.20}$$

where $\operatorname{div} \mathbf{u} = 0$, p is a Lagrange multiplier, and $M(z) = (2\xi + \lambda - 2z)/\sqrt{\lambda}$. Equations (7.20) are to be solved in conjunction with the boundary conditions (7.7).

The system (7.20), (7.7) is again reduced by normal modes to an eigenvalue problem for a system of ordinary differential equations and

$$Ra_E = \xi^3 \max_{\lambda} \min_{k^2} R_E^2(\lambda, k^2) \tag{7.21}$$

is found numerically by using the compound matrix method, the optimization routine using golden section search.

For the energy stability analysis of patterned ground formation the conditional bound (7.19)(ii) is not a strong restriction. In terms of the velocity perturbation we require $\|\mathbf{u}(0)\| < bR\bar{\lambda}/c_1 Pr$, where $\bar{\lambda}$ denotes the best value of λ in (7.21). It is important to observe that this bound is independent of A, which means that the velocity condition is not as restrictive as the temperature one. Physically this is good, since we expect the fluid velocity to drive the cellular motion.

7.2.3 A cubic density law

(McKay and Straughan, 1991) extend the model just described by replacing the quadratic density-temperature relationship (7.1) by a more accurate cubic one of form

$$\rho = \rho_0(1 + AT - BT^2 + CT^3), \tag{7.22}$$

where ρ_0 is the density at 0°C and A, B, C are chosen appropriately for water. They neglect inertia in the momentum equation and so their governing equations are

$$\begin{aligned} p_{,i} &= -\rho_0 g \delta_{i3}(1 + AT - BT^2 + CT^3) - \frac{\mu}{k'} v_i, \\ v_{i,i} &= 0, \qquad T_{,t} + v_i T_{,i} = \kappa \Delta T. \end{aligned} \tag{7.23}$$

The permeability k' is linear in the vertical height z.

The nonlinear perturbation equations of (McKay and Straughan, 1991) are

$$\begin{aligned} \pi_{,i} &= -f u_i - R\theta k_i F(z) + k_i \theta^2 G(z) - k_i \frac{a_2 Pr^2}{R} \theta^3, \\ u_{i,i} &= 0, \qquad Pr(\theta_{,t} + u_i \theta_{,i}) = -Rw + \Delta\theta, \end{aligned} \tag{7.24}$$

where $f = (1 + \gamma z)^{-1}$, $F = 1 - 2a_1 z + 3a_2 z^2$, $G = Pr(a_1 - 3a_2 z)$, a_1 and a_2 being non-dimensional versions of B and C. The perturbation boundary conditions are

$$\theta = w = 0 \quad \text{on } z = 0; \qquad w = 0, \ \theta_{,z} + a\theta = 0, \quad \text{on } z = 1. \tag{7.25}$$

(McKay and Straughan, 1991) carry out a detailed linear instability analysis and perform a nonlinear energy stability (conditional) analysis. They derive nonlinear stability thresholds very close to the linear instability ones. One of the findings is that the cubic model leads to critical Rayleigh numbers which are $5 - 10\%$ lower than those predicted by the quadratic density model of (George et al., 1989). This is good from a physical point of view since it means the convection process commences more easily and thus the start of the patterned ground formation process is more likely to occur. A nonlinear unconditional energy stability theory for the cubic model has been developed using a generalized energy involving L^p integrals and a Forchheimer theory by (Carr, 2003a; Carr, 2003c).

7.2.4 A model incorporating phase change effects

Due to freeze-thawing in the active patterned ground layer with an annual cycle, one ought really to study patterned ground formation where a layer of frozen porous soil underlies an unfrozen saturated porous liquid layer, with the interface movement being accounted for. (McKay, 1996) develops a model for this. He adopts Darcy's law with a quadratic buoyancy force in the porous-liquid layer and employs the heat equation in the frozen part of the porous layer. Thus, his governing equations are Darcy's law in the region $\mathbb{R}^2 \times \{\eta < z < d\}$, with the heat equation holding in the region $\mathbb{R}^2 \times \{0 < z < \eta\}$. The interface is at $z = \eta(x, y)$ and this interface is not necessarily planar. Hence, the model of (McKay, 1996) has in the region $\mathbb{R}^2 \times \{\eta < z < d\} \times \{t > 0\}$,

$$\begin{aligned}
&\frac{\rho_0}{\phi} v_{i,t} = -p_{,i} - \rho_0 g k_i \big[1 - \alpha(T_L^2 - 8T_L)\big] - \frac{\mu}{k'} v_i,\\
&v_{i,i} = 0,\\
&(\rho_0 C_0)_L T_{,t}^L + (\rho_0 C_0)_f v_i T_{,i}^L = k_L \Delta T_L,
\end{aligned}$$

and in the region $\mathbb{R}^2 \times \{0 < z < \eta\} \times \{t > 0\}$,

$$(\rho_0 C_0)_S T_{,t}^S = k_S \Delta T_S.$$

The S and L refer to the solid and liquid regions, respectively.

The boundary conditions across the interface $z = \eta$ are non-trivial and take account of the phase change there. (McKay, 1996) linearizes all equations including the boundary conditions and develops a complete linear instability analysis. Detailed conclusions are drawn. Among these he concludes that the presence of the ice/water interface coupled with the ice region lowers the threshold for instability. This in turn predicts larger width to depth ratios for the stone polygons which form.

7.2.5 Patterned ground formation under water

(Gleason, 1984) and (Krantz et al., 1988) report striking findings and contain excellent photographs of hexagonal stone patterns on the bed of shallow alpine lakes in Colorado and Wyoming. The patterned ground typically forms in the shallow edges of the lakes. To model this phenomenon (McKay and Straughan, 1993) adopted the fundamental model of (Nield, 1977) for thermal convection in a system in which a layer of fluid overlies a layer of saturated porous material. (McKay and Straughan, 1993) analyse their model for a linear instability analysis by using a shooting technique and setting the growth rate to zero. The model of (McKay and Straughan, 1993) has been analysed in detail with an accurate spectral technique by (Carr and Straughan, 2003). The findings of (Carr and Straughan, 2003) are very revealing, especially regarding streamline formation. Their analysis shows the results are very dependent on the parameters, namely, T_U, the temperature of the upper plane of the fluid, and $\hat{d} = d/d_m$, where d is the depth of the fluid layer whereas d_m is the depth of the porous layer.

A contentious issue is what are the boundary conditions at the interface between the fluid and the saturated porous medium? This point is taken up in section 8.5. In this section we simply adopt the Beavers-Joseph interface condition.

Since the water in the region in which patterned ground is observed is in the 4°C region the model of (McKay and Straughan, 1993) adopts a quadratic density-temperature relationship. In fact, the Navier-Stokes equations are adopted in the fluid region $\mathbb{R}^2 \times \{z \in (0, d)\}$ while Darcy's law is adopted in the porous domain $\mathbb{R}^2 \times \{z \in (-d_m, 0)\}$. A steady solution is found and non-dimensional perturbation equations are derived as

$$\begin{aligned}&u_{i,t} + u_j u_{i,j} = -\pi_{,i} + \Delta u_i + 2R(z-\xi)k_i\theta + Prk_i\theta^2,\\ &u_{i,i} = 0,\\ &Pr(\theta_{,t} + u_i\theta_{,i}) = \Delta\theta - Rw\end{aligned} \tag{7.26}$$

in $\mathbb{R}^2 \times \{z \in (0,1)\} \times \{t > 0\}$, together with

$$\begin{aligned}&\delta^2\pi^m_{,i} = -u^m_i + 2R^m(z-\xi_m)k_i\theta^m + Pr_m k_i\theta^2_m,\\ &u^m_{i,i} = 0,\\ &Pr_m(G_m\theta^m_{,t^m} + u^m_i\theta^m_{,i}) = \Delta\theta^m - R^m w^m,\end{aligned} \tag{7.27}$$

in $\mathbb{R}^2 \times \{z \in (-1,0)\} \times \{t > 0\}$.

The boundary conditions on the upper and lower planes are

$$\theta = w = w_{,zz} = 0 \quad \text{on } z = 1, \qquad \theta^m = w^m = 0 \quad \text{on } z = -1, \tag{7.28}$$

whereas at the fluid-porous interface $z = 0$ we have,

$$\begin{aligned}&w^m = \frac{w}{\hat{d}}, \qquad \theta^m = \frac{\delta}{\sqrt{\hat{d}^3\epsilon_T}}\theta, \qquad \frac{\partial\theta^m}{\partial z} = \delta\sqrt{\frac{\epsilon_T}{\hat{d}^5}}\frac{\partial\theta}{\partial z},\\ &\hat{d}^2\pi^m = \pi - 2w_{,z}, \quad \frac{\partial u^\beta}{\partial z} = \frac{\hat{d}\alpha}{\delta}(u^\beta - \hat{d}u^\beta_m), \quad \beta = 1,2.\end{aligned} \tag{7.29}$$

A linear instability analysis is performed for (7.26) - (7.29) by (Carr and Straughan, 2003) employing a Chebyshev tau numerical method. The results obtained are believed to be highly accurate and yield valuable information on streamline patterns and the patterned ground formation under water model. They show that the results are very sensitive to changes of the parameters $\hat{d}$ and T_U.

7.3 Results and conclusions for patterned ground formation

We now return to discuss the model of patterned ground formation of (George et al., 1989). Extensive tabulated values of critical Rayleigh numbers of linear theory, Ra_L, and of energy theory, Ra_E, for various a values and changing permeability (varying γ) are given in (George et al., 1989).

In the context of patterned ground formation the two important dimensionless numbers are the critical Rayleigh number, $Ra = \xi^3 R^2$, and the critical wave number k. The critical Rayleigh number determines the conditions at the onset of convective motion of water in the soil. The parameters $g, \alpha, \rho_0, \kappa, \mu$ in the Rayleigh number vary little and may be treated as constant. The permeability is a separate issue since this changes greatly from one geographical location to another. Thus, in a sense, the permeability is the most important parameter in the Rayleigh number affecting stability. Attempts to obtain accurate measurements of soil permeability have proved extremely difficult. Because the soil in a stone polygon undergoes substantial lofting due to freeze-thaw cycles, it has proved impossible to measure accurate permeability values by transporting the soil back to the laboratory; the soil collapses and the effect of lofting is lost. Moreover, it is not known whether the values of permeability taken *on site* represent the state of the art when the convection process commenced. In this regard field work has not been too reliable and a comparison of Rayleigh numbers to predict the onset of instability conditions is not terribly useful, although many theoretical results are available in (George et al., 1989).

The comparison of nonlinear against linear theory is very good. The data presented in (George et al., 1989) demonstrates a maximum difference between the two approaches of less than 20% for the Rayleigh number, and less than 15% for the wave number. They demonstrate that upper surface temperatures greater than 4°C have a stabilizing influence, and it is probably realistic to consider only those upper temperatures in the range 4°C or less.

Wave number predictions have proved to be extremely useful. We take the wave number to measure the width-to-depth of a stone polygon via the equation $k = 7.664d/L$, where L is the diameter of a single circular convection cell. It is found that the wave number increases with increasing temperatures. Therefore the width-to-depth ratio of stone polygons decreases for higher upper boundary temperatures. The width-to-depth ratios from measurements in the field were found to be higher than the calculated values. This suggests that convective motion in the patterned ground cell begins in the spring as soon as the surface temperature approaches 4°C.

A vertically varying permeability was found theoretically to have almost no effect on the critical wave number. This implies that there should be little variation in the width-to-depth ratio for stone polygons at different sites or when there is a strong vertical variation in permeability. This observation has been borne out by field studies. On the other hand, the theoretical values of the Rayleigh number decrease with a linear increase in permeability, which means the convection threshold is reduced, and this is in accordance with Step 1.

The type of boundary condition that should be used at the upper surface is of importance. (George et al., 1989) show that the width-to-depth values vary by about 30% between the constant temperature and heating solely by radiation cases. The constant temperature condition is consistent with the ground surface being warmed by conduction and convection from the air, and this scenario is likely to be achieved during cloudy or foggy weather or when there are high winds. The constant temperature gradient condition is appropriate when the ground surface is heated directly and evenly by solar radiation. In practice neither condition is likely to be perfect, although it is believed that radiation heating will be predominant, see e.g., (Andersland and Anderson, 1978). The theoretical work shows the constant flux condition is consistent with a greater width-to-depth ratio and a lower Rayleigh number. Comparison of the energy and linear results with field data support the use of the constant flux condition. Indeed, the most realistic case analysed in (George et al., 1989) corresponds to heating by radiation only. The energy method yields a critical wave number of 2.456. The average width-to-depth ratio for hexagons is 3.36. (George et al., 1989) compare this theoretical value with a value of 3.61 based on data of (Ray et al., 1983) for a series of sites in the Rocky Mountains and by (Gleason et al., 1986) for additional sites in Iceland and in the Richardson Mountains of the Northern Yukon, see Tables 7.1, 7.2. The theoretical width-to-depth ratio and the least squares fit of field data differ by only 7%, which is very good for a geological phenomenon.

Certainly, the theoretical models of (Ray et al., 1983), (Gleason et al., 1986) and (George et al., 1989) predict outcomes that are in good agreement with available field studies, and this is very encouraging.

7.4 Convection in thawing subsea permafrost

During the history of the Earth the level of the sea has been dependent on the size of the glaciers and the quantity of water so held as ice. Around 18000 years ago, see (Müller-Beck, 1966), p. 1193, the level was some 100–110m lower than it presently is with the ambient air temperatures being much colder too, reaching a minimum at roughly the same time, and this led to substantial permafrost forming around some of the Earth's shores. When the sea level rose the permafrost responded to the relatively warm and salty sea, and this has created a thawing front and a layer of salty sediments beneath the sea bed. Extensive studies of this phenomenon have been made off the coast of Alaska, see (Harrison, 1982), (Harrison and Osterkamp, 1982), (Osterkamp and Harrison, 1982), (Osterkamp et al., 1989), (Swift and Harrison, 1984), and (Swift et al., 1983), According to (Harrison, 1982) a very interesting type of convection is taking place in the layer between the sea bed and permafrost, the buoyancy mechanism being one of the salty layer melting the relatively fresh ice, which being less dense then rises through the porous thawing layer, thereby creating a convective motion.

The mean sea bed temperatures measured by (Osterkamp and Harrison, 1982) are around $-1°C$ and a thermal gradient exists with colder temperatures at the downward moving permafrost interface. (Harrison, 1982) and (Swift and Harrison, 1984) employed these facts and by also assuming the convection is caused by salt effects rather than by temperature were able to produce a model to describe the subsea convection. The (Swift and Harrison, 1984) model is mathematically attractive in that it uses the salt-dominated effect together with the slow, climatic (monotonic) interface advance (2–5cm a^{-1}) to replace a moving boundary problem by one on a fixed region. (Hutter and Straughan, 1999) show this is correct to leading order on the timescale of convection. It is further interesting to observe that (McFadden et al., 1984) employ a similar approximation in crystal growth studies where they argue that, *the onset of instability is approximately the same whether or not the planar crystal melt interface is allowed to deform.* The numerical analysis of (Swift and Harrison, 1984) treats two-dimensional convection for salt Rayleigh numbers of 1750 and 17500; this is well into the convection regime, with stationary and oscillatory convection values of 40 and 400, see e.g., (Caltagirone, 1975).

The paper by (Galdi et al., 1987) determines the critical Rayleigh number for nonlinear convection by using energy stability theory: the linear instability boundary is obtained as a by-product of the nonlinear analysis. Because the parabolic salt diffusion equation is subject to a nonlinear destabilizing boundary condition at the permafrost interface, that work is unable to preclude the possibility of a large amplitude sub-critical instability. If the Rayleigh number is smaller than a critical one, R_E^α, for convection in a porous layer with a linear flux lower boundary condition then unconditional nonlinear stability holds. In the situation where the nonlinear stability is conditional upon the existence of a finite amplitude threshold, such a threshold value was calculated.

A very interesting mathematical feature of the conditional analysis of (Galdi et al., 1987) is that the two- and three-dimensional problems need different analyses. In two dimensions an L^2 energy method employing a Sobolev inequality works well: this method fails for the three-dimensional case, which evidently requires a more subtle approach in which the natural energy has to be generalized to include a controlling term that allows the nonlinearities to be dominated.

7.5 The model of Harrison and Swift

The mathematical interpretation of the geological situation is depicted in Figure 7.1. We shall assume, for the purpose of calculating the critical Rayleigh number for the onset of instability, that the permafrost boundary at $z = D(t)$ remains planar. This assumption is reasonable because appreciable boundary movement is on a timescale of years whereas the salt convective motion is on a timescale of days.

The porous layer is modelled by Darcy's law with the body force term linear in the salt field. Employing a Boussinesq approximation the equations

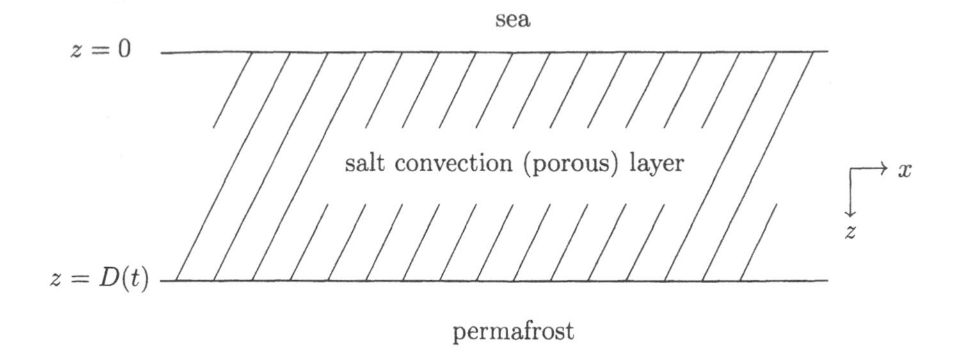

Figure 7.1. Configuration of the thawed layer of salty sediments

of motion are

$$v_i = -\mu p_{,i} + k_i g \mu \rho_0 (1 + S), \tag{7.30}$$

$$v_{i,i} = 0, \tag{7.31}$$

$$\frac{\partial S}{\partial t} + v_i S_{,i} = \kappa \Delta S, \tag{7.32}$$

where (7.32) is the equation for the diffusion of salt in the layer $z \in (0, D)$, $\mathbf{v}, S, \mu, p, \rho_0, \kappa, g$ are velocity, salt concentration, permeability divided by dynamic viscosity, pressure, (constant) density, salt diffusivity, and gravity, respectively, $\mathbf{k} = (0, 0, 1)$, and μ is here taken constant.

It is assumed there is no flow into the sea or permafrost, and so the velocity boundary conditions are

$$v_i n_i = 0, \qquad z = 0, D, \tag{7.33}$$

here n_i denotes the unit outward normal. ((Hutter and Straughan, 1999) commence with a complete system of equations for velocity, pressure, salt concentration, *and temperature* and exploit the disparity between the timescales of boundary movement and convection. They demonstrate (7.33) is correct at the lowest order in a salt dominated convection régime.) Measurements indicate the salt concentration and temperature at the sea bed are almost constant, and so we choose

$$S = S_0 \text{ (constant)}, \quad z = 0, \qquad T = T_0 \text{ (constant)}, \quad z = 0. \tag{7.34}$$

The lower boundary is moving due to a phase change occurring there; thus, appropriate forms of a Stefan condition must hold there. These are taken to be

$$L\dot{D} = -K_1 \frac{\partial T}{\partial z}\bigg|_{D^-}, \qquad S(D)\dot{D} = -\kappa \frac{\partial S}{\partial z}\bigg|_{D^-}, \qquad z = D, \tag{7.35}$$

where L is the coefficient of latent heat per unit volume of the salty layer (in units of joules m^{-3}), the subscript D^- indicates the derivative approaching D from the thawed layer, K_1 is the thermal conductivity (joules year^{-1} m^{-1} deg^{-1}), with κ the salt diffusivity coefficient (m^2 year^{-1}). The gradients of salt and temperature in the permafrost layer are neglected in (7.35).

(Swift and Harrison, 1984) simplify the above model by arguing that the driving mechanism for convection is the buoyant, relatively pure water melting at $z = D$, and the temperature input may be neglected in the sense that the temperature gradient remains constant throughout; this is in agreement with the measurements of (Osterkamp and Harrison, 1982). Thus they assume that in the porous layer

$$\frac{\partial T}{\partial z} = \frac{[T(D) - T_0]}{D}. \tag{7.36}$$

They complete the formulation of the model with a phase equilibrium condition at $z = D$, which with T measured in degrees centigrade is

$$S(D) \propto -T(D). \tag{7.37}$$

From (7.37) we deduce

$$\frac{S(D)}{S_r} = \frac{T(D)}{T_0}, \tag{7.38}$$

where S_r is the salinity of sea water that would begin to freeze at the sea bed temperature T_0. Since $T(D) < T_0 < 0$ it follows that $S(D) > S_r$. The temperature field is removed from the problem by eliminating $T(D)$ and $\dot{D}$ between (7.35), (7.36) and (7.38) to yield the nonlinear boundary condition on the salt field,

$$\frac{\partial S}{\partial z} = \frac{K_1 T_0}{L\kappa D} S\left(\frac{S}{S_r} - 1\right), \qquad z = D. \tag{7.39}$$

Thus, the governing equations (7.30)-(7.32) are to be solved in conjunction with the boundary conditions (7.33), $(7.34)_1$, and (7.39), the approximation (7.36) having the effect of fixing the moving boundary at D.

The stationary solution studied in (Galdi et al., 1987) is

$$\mathbf{v} \equiv \mathbf{0}, \quad \bar{S} = S_0 - \beta_s z, \quad \bar{p} = p_0 + g\rho_0(1 + S_0)z - \frac{1}{2} g\rho_0\beta_s z^2. \tag{7.40}$$

To investigate the stability of solution (7.40) let $\mathbf{u}$, s, and p be perturbations to the stationary values. Non-dimensionalizing the layer to $\{z \in (0,1)\}$ and introducing a salt Rayleigh number, Ra, by $Ra = R^2 = \beta_s g\mu\rho_0 D^2/\kappa$, (Galdi et al., 1987) derive the following non-dimensional perturbation equations governing the stability of the steady solution (7.40):

$$\begin{aligned} u_i &= -p_{,i} + Rsk_i, \quad u_{i,i} = 0, \\ s_{,t} + u_i s_{,i} &= Rw + \Delta s, \end{aligned} \tag{7.41}$$

with boundary conditions

$$u_i n_i = 0, \quad z = 0, 1; \tag{7.42}$$

$$s = 0, \quad z = 0; \qquad \frac{\partial s}{\partial z} = -as - bs^2 \quad \text{on} \quad z = 1. \tag{7.43}$$

In $(7.41)_3$, $w = u_3$. Moreover, we assume u_i, s, p are periodic functions of x and y with periodicities m and n, respectively. The constants a and b are given by

$$a = \Big(2\frac{\bar{S}}{S_r} - 1\Big)\frac{K_1|T_0|}{L\kappa}, \qquad b = \frac{K_1|T_0|}{LS_r}\sqrt{\frac{\beta_s}{g\mu\rho_0\kappa}}, \tag{7.44}$$

and we note that estimates of the components in (Swift and Harrison, 1984) suggest that a will always be positive.

7.6 The energy stability maximum problem

The stability analysis commences in the usual way. Multiply $(7.41)_1$ by u_i, $(7.41)_3$ by s, add and integrate over a periodicity cell V. Using the boundary conditions and integrating by parts we find

$$\frac{1}{2}\frac{d}{dt}\|s\|^2 = 2R < sw > -\mathcal{D} - b\int_\Gamma s^3\,dA, \tag{7.45}$$

where we have defined

$$\mathcal{D} = \|\mathbf{u}\|^2 + D(s) + a\int_\Gamma s^2\,dA, \tag{7.46}$$

where $< \cdot >$ and $\|\cdot\|$ denote integration and the norm on $L^2(V)$, respectively, $D(\cdot)$ denotes the Dirichlet integral, and Γ is that part of the boundary of V that lies in the plane $z = 1$.

Define further

$$I = 2 < sw >, \tag{7.47}$$

to derive from (7.45)

$$\frac{1}{2}\frac{d}{dt}\|s\|^2 \le -\Big(\frac{R_E - R}{R_E}\Big)\mathcal{D} - b\int_\Gamma s^3\,dA, \tag{7.48}$$

where R_E is defined by

$$\frac{1}{R_E} = \max_{\mathcal{H}}\frac{I}{\mathcal{D}}, \tag{7.49}$$

and where $\mathcal{H}$ is the space of admissible functions, which we here choose as

$$\begin{aligned}\mathcal{H} = \Big\{\mathbf{u}, s \big| \mathbf{u} \in (L^2(V))^3,\ u_{i,i} = 0,\ u_i n_i = 0 \text{ on } z = 0, 1;\\ s \in W^{1,2}(V),\ s = 0 \text{ at } z = 0;\\ u_i, s \text{ periodic in } x \text{ and } y\Big\}.\end{aligned} \tag{7.50}$$

The first result of (Galdi et al., 1987) is an *unconditional one* and shows that if

$$R < R_E^\alpha \equiv \max_{\mathcal{H}} \frac{I}{\mathcal{D}_\alpha}$$

with $\mathcal{D}_\alpha$ defined by

$$\mathcal{D}_\alpha \equiv \|\mathbf{u}\|^2 + D(s) + \alpha \int_\Gamma s^2\, dA,$$

then $\|s(t)\|^2$ decays at least exponentially for increasing t, *for all* $s(\mathbf{x}, 0) \in L^2(V)$. The constant α is defined by

$$\alpha = \frac{K_1|T_0|}{L\kappa}\Big(\frac{\bar{S}(D)}{S_r} - 1\Big). \tag{7.51}$$

This result hinges on the facts that $\bar{S}(D) > S_r$ and $s + \bar{S}(D) > 0$, i.e., the salinity is always positive. Using these facts they deduce from (7.48)

$$\frac{1}{2}\frac{d}{dt}\|s\|^2 \leq -\mathcal{D}_\alpha\Big(1 - \frac{R}{R_E^\alpha}\Big),$$

and this together with use of Poincaré's inequality yields rapid decay of $\|s(t)\|$ provided $R < R_E^\alpha$.

Decay for the case $R_E^\alpha < R < R_E$ is trickier and requires use of embedding inequalities. Interestingly, the two- and three-dimensional problems evidently need separate treatments. This is described in section 7.7.

An existence result for the maximizing solution to (7.49) (or the equivalent problem involving $\mathcal{D}_\alpha$) is proven in (Galdi et al., 1987). To achieve this they observe that $\mathcal{H}$ is a Hilbert space when equipped with the norm generated by $\mathcal{D}$ (where the boundary conditions are to be interpreted in the trace sense). To see this, observe that $\mathcal{H}$ is the topological product of two complete spaces and is hence itself complete with respect to the norm endowed by $\mathcal{D}$. The completeness of the space appropriate to $\mathbf{u}$ follows from, for example, (Temam, 1978), whereas the space appropriate to s is a subspace of $W^{1,2}(V)$ endowed with the standard norm; this follows because

$$\begin{aligned}\|s\|^2 + D(s) &\leq (1+4\pi^{-2})D(s)\\ &\leq (1+4\pi^{-2})\left(D(s) + a\int_\Gamma s^2\,dA\right), \qquad (7.52)\\ &\leq (1+4\pi^{-2})\left[D(s) + 2a\|s\|D^{1/2}(s)\right]\\ &\leq (1+4\pi^{-2})(1+a)\left[D(s) + \|s\|^2\right], \qquad (7.53)\end{aligned}$$

where the inequality involving the boundary integral may be found in the Appendix.

Thanks to Poincaré's inequality $I/\mathcal{D}$ is bounded above by a constant, γ, say. Hence, there exists a maximizing sequence $\{\mathbf{u}_n, s_n\}$ such that

$$\lim_{n\to\infty}\frac{I(w_n, s_n)}{\mathcal{D}(\mathbf{u}_n, s_n)} = \gamma.$$

This sequence is chosen such that $\mathcal{D}(\mathbf{u}_n, s_n) = 1$ and from this we may deduce the existence of a subsequence, again denoted by $\{\mathbf{u}_n, s_n\}$, such that

$$u_{i_n} \to u_{i_0} \quad \text{weakly in} \quad L^2(V), \qquad (7.54)$$

$$s_n \to s_0 \quad \text{weakly in} \quad W^{1,2}(V), \quad \text{strongly in} \quad L^2(V), \qquad (7.55)$$

for some $\{\mathbf{u}_0, s_0\} \in \mathcal{H}$. By using standard inequalities,

$$\begin{aligned}|I(w_n, s_n) - I(w_0, s_0)| &\leq 2| < s_0(w_n - w_0) > | + 2| < w_n(s_n - s_0) > |\\ &\leq 2| < s_0(w_n - w_0) > | + 2\|w_n\|\,\|s_n - s_0\|. \qquad (7.56)\end{aligned}$$

The first term on the right of (7.56) converges to zero by virtue of (7.54), while the second tends to zero thanks to (7.55). Therefore, $I(w_0, s_0) = \gamma$ and also one may show $\mathcal{D}(\mathbf{u}_0, s_0) = 1$. The existence of a maximizing solution to (7.49) is therefore established.

To actually solve (7.49) we find the Euler-Lagrange equations to be

$$R_E s k_i + p_{,i} = u_i\,, \qquad R_E w + \Delta s = 0, \qquad (7.57)$$

together with the boundary conditions

$$u_i n_i = 0, \quad z = 0, 1; \qquad (7.58)$$

$$s = 0, \quad z = 0; \qquad \frac{\partial s}{\partial z} + as = 0 \quad \text{on} \quad z = 1. \qquad (7.59)$$

It is very important to observe that (7.57)-(7.59)are *exactly the same as the linearized version of the full equations* (7.41)-(7.43). It transpires that this means that the critical Rayleigh number is the *same for both linear and nonlinear (conditional) stability.* Any finite amplitude instability is caused entirely by the bs^2 term in (7.43).

Critical values of R_E are given in (Galdi et al., 1987).

7.7 Decay of the energy

A nonlinear stability result follows directly from (7.48) if we consider only *two-dimensional perturbations* and employ inequality (7.60), namely,

$$\int_\Gamma s^3\,dA \le c_1\|s\|D(s), \tag{7.60}$$

see the Appendix. Suppose now, $R < R_E$, then with $h = R^{-1} - R_E^{-1}\,(>0)$, (7.48) yields

$$\frac{1}{2}\frac{d}{dt}\|s\|^2 \le -D(s)(hR - bc_1\|s\|) - hR\Big(\|\mathbf{u}\|^2 + a\int_\Gamma s^2\,dA\Big). \tag{7.61}$$

Suppose further that the initial data satisfy $\|s(0)\| < (R_E - R)/R_E c_1 b$. Then, with $\omega = hR - bc_1\|s(0)\|$, we may deduce from (7.61) that

$$\frac{1}{2}\frac{d}{dt}\|s\|^2 + \frac{1}{4}\pi^2\omega\|s\|^2 + hR\Big(\|\mathbf{u}\|^2 + a\int_\Gamma s^2\,dA\Big) \le 0. \tag{7.62}$$

Dropping the last term and integrating we arrive at

$$\|s(t)\|^2 \le \|s(0)\|^2 e^{-\frac{1}{2}\pi^2\omega t}. \tag{7.63}$$

Since from $(7.41)_1$,

$$\|\mathbf{u}\|^2 = R < sw >,$$

it is easy to see that

$$\|\mathbf{u}\|^2 \le \frac{4}{3}R^2\|s\|^2,$$

and this in (7.63) implies also decay of $\|\mathbf{u}\|^2$.

Nonlinear stability of the solution (7.40) to two-dimensional perturbations is assured if

$$\text{(i)}\quad R < R_E; \qquad \text{(ii)}\quad \|s(0)\| < \frac{hR}{3b(2 + 4/\pi m)^{1/2}}, \tag{7.64}$$

where m is the x-periodicity of u_i, s. Condition (i) determines the nonlinear critical Rayleigh number whereas (ii) is a limitation on the size of the initial amplitude.

(Galdi et al., 1987) show that inequality (7.60) is false in three dimensions (see the Appendix). Hence the above procedure to establish conditional decay fails when three-dimensional perturbations are taken into consideration. To retain the feature that the linear and nonlinear Rayleigh number boundaries are the same, the outlet adopted by (Galdi et al., 1987) is to use the generalized energy

$$E(t) = \frac{1}{2}\|s\|^2 + \frac{1}{4}\mu\|s^2\|^2, \tag{7.65}$$

for $\mu\,(>0)$ a coupling parameter. In (7.65) the L^4 piece is added wholly to control the cubic nonlinear terms, and it plays no role in determining the stability boundary.

With $m = R/R_E$ and $R < R_E$, the energy inequality for E defined by (7.65) is shown by (Galdi et al., 1987) to satisfy

$$\begin{aligned}\frac{dE}{dt} \leq &-\mathcal{D}(1-m) - b\int_\Gamma s^3 dA + \mu R < s^3 w > \\ &-\mu a\int_\Gamma s^4 dA - \mu b\int_\Gamma s^5 dA - \frac{3}{4}\mu D(s^2) \\ &-\lambda D(w) + \lambda R < s_{,\alpha} w_{,\alpha} >,\end{aligned} \tag{7.66}$$

where $\lambda\,(>0)$ is another coupling parameter and the subscript α sums over 1 and 2. The coupling parameter λ, like μ, is employed only to control and dominate the nonlinear terms; the energy stability boundary arises through the L^2 energy.

The addition of the L^4 integral of s to the energy E in (7.65) adds higher order dissipative terms to (7.66), which now allows the nonlinear boundary terms to be controlled and conditional energy decay is again achieved. The nonlinear stability criterion is still $R < R_E$, with $E(0)$ suitably small. It is, however, interesting to observe that the generalized energy (7.65) *must* be employed (as opposed to simply the L^2 energy) to achieve energy decay in three dimensions.

7.8 Other models for thawing subsea permafrost

(Payne et al., 1988) incorporated a penetrative convection effect in thawing subsea permafrost by including both temperature and salt fields directly by assuming an equation of state of form

$$\rho = \rho_0\big[1 - \alpha(T - T_R)^2 + \alpha_S(S - S_R)\big]. \tag{7.67}$$

These writers assume the convection layer has a fixed depth and so neglect the downward moving permafrost interface, arguing that the instability process occurs on a much faster time scale than that of the interface movement. (Qin et al., 1995) adopt a similar argument but they choose an equation of state of form

$$\rho = \rho_0\big[1 - \alpha_1(T - T_R) - \alpha(T - T_R)^2 + \alpha_S(S - S_R)\big]. \tag{7.68}$$

The permafrost boundary is still treated as if it were fixed. (Qin et al., 1995) allow for a temperature dependent viscosity and they additionally cater for larger fluid velocities in the thawing subsea permafrost layer by including a Forchheimer term in the momentum equation.

(Hutter and Straughan, 1999) treat the general problem of salt and temperature variation in a convection process but allow for a moving phase change boundary. Their main aim is to derive by asymptotic means the correct equations for processes such as those seen in thawing subsea permafrost. They identify three asymptotic regimes, a thermally diffusive one, a salt convective one, and an intermediate one. They concentrate on the salt convective regime and use a two time scale asymptotic analysis to derive equations governing the behaviour of the thawed layer. In this way they show that the phase boundary beneath the sea bed and below the thawing layer has a parabolic shape with respect to the shoreline. This is observed in practice.

(Hutter and Straughan, 1997) extend the Harrison-Swift model by replacing the linear density in (7.30) by the more realistic UNESCO relation given in (Marchuk and Sarkisyan, 1988), p. 167,

$$\begin{aligned}\rho = \rho_0\big[1 &+ 5.881469\times 10^{-5}T + 7.969548\times 10^{-4}S\\ &- 8.114004\times 10^{-6}T^2 - 3.252844\times 10^{-6}TS\\ &+ 1.317003\times 10^{-7}S^2\\ &+ 4.765622\times 10^{-8}T^3 + 3.891563\times 10^{-8}ST^2\\ &+ 2.879485\times 10^{-9}S^2T - 6.117825\times 10^{-11}S^3\big].\end{aligned} \tag{7.69}$$

They develop a conditional energy stability analysis. This work is further extended by (Budu, 2002) who again uses a Harrison-Swift model but with an equation of state of form

$$\rho(T,S) = \rho_0\big[f(T) + f_1(T)S + f_2(T)S^{3/2} + cS^2\big]. \tag{7.70}$$

We do not give explicitly the functions f, f_1, f_2 since they may be found in (Mellor, 1996), p. 114. However, the analysis of (Budu, 2002) employs a weighted energy and is *unconditional.*

7.8.1 Unconditional stability with a general density

We complete this section by showing how one may generalize the model of Harrison and Swift to incorporate a general equation of state and develop an unconditional nonlinear energy stability analysis. We do not need to use a weighted energy.

Thus, we adopt equations (7.31) and (7.32), but instead of the relation $\rho = \rho_0(1+S)$ used in (7.30) we employ the equation

$$\rho = \rho_0\big[1 + f_1(T)S + f_2(T,S)\big]. \tag{7.71}$$

The functions f_1 and f_2 are arbitrary but smooth. The analysis from (7.33) to (7.39) still carries over. The steady state still has form (7.40), but the nonlinear perturbation equations are modified due to (7.71). The idea is to use the Harrison-Swift relation (7.36) and replace the $T-$dependence in f_1 and f_2 by the appropriate $z-$dependence. Then, in the relation $\rho(\bar{S}+s)$

we use a Taylor series to write

$$\rho(\bar{S}+s)-\rho(\bar{S})=\big[\rho_0 f_1(z)+\rho_0 f_2'(\hat{S})\big]s. \tag{7.72}$$

Here f_2' denotes $\partial f_2/\partial S$, and $\hat{S}$ is some salt concentration between $\bar{S}$ and $\bar{S}+s$. This leads to the perturbation equations (7.41) - (7.43) being replaced by

$$\begin{aligned} u_i &= -p_{,i}+Rsk_i\big[f_1(z)+f_2'(z,\hat{S})\big],\\ u_{i,i} &= 0,\\ s_{,t}+u_i s_{,i} &= Rw+\Delta s, \end{aligned} \tag{7.73}$$

and

$$\begin{aligned} &u_i n_i = 0, \quad z=0,1;\\ &s=0, \ z=0; \qquad \frac{\partial s}{\partial z}=-as-bs^2, \ z=1. \end{aligned} \tag{7.74}$$

The constants a and b are again given by (7.44).

A linear analysis is easily developed by discarding the f_2' term. In practice we expect that $|f_2'| << |f_1|$. An unconditional nonlinear energy theory follows by multiplying $(7.73)_1$ by u_i and integrating over V, then multiplying $(7.73)_3$ by s and integrating over V. To develop an unconditional analysis we do two things. Firstly, use the fact that on $z=1$, $ss_z \le -\alpha s^2$ where α is given by (7.51). This is using the argument in (Galdi et al., 1987). We then argue as in (Budu, 2002) that $\hat{S}$ must be bounded above since S is a salt concentration and so has a saturation level. If we let m denote $m=\max|f_2'|$, then we may derive an energy inequality of form

$$\frac{d}{dt}\frac{1}{2}\|s\|^2 \le RI-D_\alpha\,, \tag{7.75}$$

where I and D_α are given by

$$\begin{aligned} I &= \Big(\big[1+\lambda f_1(z)\big]s,w\Big)+\frac{\lambda m\beta}{2}\,\|s\|^2+\frac{\lambda m}{2\beta}\,\|w\|^2,\\ D_\alpha &= \|\nabla s\|^2+\alpha\int_\Gamma s^2 dA+\lambda\|\mathbf{u}\|^2. \end{aligned}$$

In these equations $\lambda(>0)$ is a coupling parameter and $\beta(>0)$ is a constant we choose optimally. One may develop a variational theory from (7.75). The final result will yield unconditional nonlinear energy stability, and we anticipate the nonlinear stability thresholds will be extremely close to the linear instability ones for any practical choice of density in (7.71).

7.9 Other models for geophysical phenomena

There are many other models for flows which arise in geophysical situations. We here mention some which have either yielded useful results by application of energy stability theory, or we believe they may well do so. The first is Hadley flow, or convection in a layer where the lower boundary is hotter than the upper, but the temperature field varies linearly in the horizontal direction. This is interesting in that the non-uniformity in the horizontal direction induces a flow into the basic solution. This class of flows has been studied much by Nield and his co-workers. (Nield, 1998b) provides several references. However, for Hadley flow in a fluid or porous medium pertinent references are (Nield, 1990; Nield, 1991; Nield, 1994a; Nield, 1994b; Nield, 1998a), (Nield et al., 1993), and (Manole et al., 1994).

Convection with inclined temperature gradient involves a layer (of fluid or saturated porous medium) and the temperatures on the boundaries $z = \pm\frac{1}{2}$ are given by, in non-dimensional form,

$$T = \mp\frac{1}{2}R_V - R_H x, \quad z = \pm\frac{1}{2}.$$

The parameters R_V and R_H are vertical and horizontal Rayleigh numbers. In the porous medium case the steady solution has form

$$\bar{T}(z) = -R_V z + \frac{1}{24}R_H^2(z - 4z^3) - R_H x, \qquad \bar{U} = R_H z,$$

with $\bar{V} = \bar{W} = 0$. For the porous medium problem employing Darcy's law, the perturbation equations may be derived as, cf. (Nield, 1994a)

$$\begin{aligned} &\pi_{,i} = -u_i + \delta_{i3}\theta, \qquad u_{i,i} = 0, \\ &\theta_{,t} + u_i\theta_{,i} = R_H u - \bar{U}\frac{\partial\theta}{\partial x} - \frac{d\bar{T}}{dz}w + \Delta\theta. \end{aligned} \tag{7.76}$$

It is a challenge to bring out the effect of the $\bar{U}$ term using an energy theory. (Since $\bar{U} = \bar{U}(z)$ multiplication of $(7.76)_3$ by θ and integration over V yields the term $-<\bar{U}\theta_{,x}\theta> = \frac{1}{2} < \bar{U}_x\theta^2 > = 0$. Thus, a standard L^2 approach loses the effect of the $\bar{U}$ term.) Nonlinear energy stability theory for Hadley problems has been developed by (Guo and Kaloni, 1995b), (Kaloni and Qiao, 1996; Kaloni and Qiao, 1997a; Kaloni and Qiao, 1997b; Kaloni and Qiao, 2000; Kaloni and Qiao, 2001). (Alex and Patil, 2002a) derive a linear instability analysis of the problem of convective instability of an anisotropic porous layer with heat source, variable gravity, and inclined temperature gradient. The same writers (Alex and Patil, 2002b) study the onset of convection in a layer of saturated porous material with an inclined temperature gradient, vertical throughflow, and a varying gravity field. The onset of convection via linear instability techniques, in an inclined porous layer, is studied by (Rees and Bassom, 2000) and by (Rees and Postelnicu, 2001).

Another area which leads to interesting stability problems is convection where a vertical throughflow is imposed, see (Nield, 1987). Again, this has a non-zero velocity component in the steady state. Throughflow has also been combined with Hadley flow, see the cited references of Nield, and energy theory for this is due to (Qiao and Kaloni, 1998).

A very interesting geophysical process is salinization. This is where in dry regions where the rainfall is small and the water table lies relatively close to the surface, the mean flow of water through the unsaturated soil may be in the upward direction. Since groundwater is usually saline this upward flow results in salts being transported to the soil surface. Due to the heat in the dry region evaporation occurs in the soil layer and this can leave a top heavy layer of salty fluid near to the soil surface. This in turn can lead to a convective motion of salty fluid in the porous layer. Models for this are developed and analysed in (Bear and Gilman, 1995) and (Gilman and Bear, 1996). The latter contains a linearized instability analysis. Methods for efficiently dealing with the numerical eigenvalue problems which occur in the (Gilman and Bear, 1996) theory are given in (Payne and Straughan, 2000a). To my knowledge, a nonlinear energy theory for the (Gilman and Bear, 1996) problem has not been developed. The equations are highly nonlinear. However, (Payne et al., 1999) use energy like techniques to derive continuous dependence and convergence results for the basic equations arising from the (Gilman and Bear, 1996) theory. A geophysical problem which may have important environmental implications due to enhanced melting of ice shelves through increased heat transfer due to convection is that of convective motion in under ice melt ponds. Linear instability and nonlinear instability analyses for a model for this phenomena are provided in (Carr, 2003a; Carr, 2003b).

Finally in this section we mention the problem of convection in snow. If snow is lying on a surface, say a roof, then the upper boundary is cold due to the ambient air temperature, but the base may be warmer which may give rise to convective motion. The air movement can create water vapour and so convection in snow is a complicated process involving transport of air and water through snow itself. A beautiful model for this based on a Darcy porous medium and taking into account latent heat effects is due to (Powers et al., 1985). These writers solve the equations using a finite difference method to study heat transfer. I am not aware of any nonlinear energy stabilty theory applied to convection in snow, but I believe such a study on the model of (Powers et al., 1985) may prove fruitful.

8

Surface tension driven convection

8.1 Energy stability and surface tension driven convection

The topic of this section is historically important since it is now believed that in the original experiments of (Bénard, 1900) (the convection problem now bears his name), the driving mechanism was one of surface tension variation due to temperature. We point out that convective instability seems to have been first described by James Thomson, the elder brother of Lord Kelvin, (Thomson, 1882), a very clear account of this is given in (Drazin and Reid, 1981), p. 32.

The first analysis of surface tension driven convection by means of the energy method is due to (Davis, 1969a). He assumes a flat surface, a plane layer, and a linear surface tension temperature relation of the form

$$\sigma = \sigma_0 - \gamma(T - T_0), \tag{8.1}$$

σ, T being surface tension and temperature and σ_0, γ positive constants. His non-dimensional perturbation equations are

$$\begin{aligned} Pr^{-1}\Big(\frac{\partial u_i}{\partial t} + u_j u_{i,j}\Big) &= -p_{,i} + \Delta u_i + R\theta k_i, \\ \frac{\partial \theta}{\partial t} + u_i \theta_{,i} = \Delta\theta + w, &\qquad u_{i,i} = 0, \end{aligned}$$

where Pr is the Prandtl number, other variables being as before. The boundary conditions examined by (Davis, 1969a) are

$$\begin{aligned} u_i = \theta = 0, \qquad z = 0, \\ \theta_z + L\theta = w = u_z + B\theta_x = v_z + B\theta_y = 0, \qquad z = 1. \end{aligned} \tag{8.2}$$

In (8.2), $(u_1, u_2, u_3) = (u, v, w)$, a subscript x, y, or z denotes differentiation with respect to that variable, L is a positive constant, B is the Marangoni number, and the last two conditions represent the fact that the shear stress in the surface is given by a surface tension gradient.

Let V be a period cell for the motion. (Davis, 1969a) *claims* that the B terms in (8.2) act on the system like a bounded operator. He shows that for a periodic disturbance

$$\int_\Gamma (\phi_x^2 + \phi_y^2)\, dA = a^2 \int_\Gamma \phi^2\, dA,$$

where $\phi = R^{1/2}\theta$, Γ is that part of $z = 1$ intersecting ∂V, and a is the wave number, and then concludes that the B terms act like a bounded operator. His argument is based on the fact that the numerical results indicate a^2 remains bounded so he deduces p. 351: *By restricting ourselves to a finite interval of wave-numbers, due to a physical preference, we see that...behaves as would a bounded perturbation operator.* His dissipation term contains the square of the gradients of ϕ integrated over V and so it is possible to dominate the surface terms, but only by the Laplacian, i.e., the unbounded operator.

His energy theory chooses

$$E = \frac{1}{2}(Pr^{-1}\|\mathbf{u}\|^2 + \lambda\|\theta\|^2),$$

for $\lambda\,(> 0)$ a coupling parameter. Then he derives

$$\begin{aligned} \frac{dE}{dt} =& (R + \lambda) < w\theta > - \big[D(\mathbf{u}) + \lambda D(\theta)\big] \\ & - B\int_\Gamma \theta w_z\, dA - L\lambda \int_\Gamma \theta^2\, dA, \end{aligned} \tag{8.3}$$

where $D(f) = \|\nabla f\|^2$. The difficulty in establishing a variational theory of energy stability is to ensure the $B\int_\Gamma \theta w_z dA$ boundary term can be dominated by the dissipation $\|\nabla\mathbf{u}\|^2 + \lambda\|\nabla\theta\|^2 + \lambda L\int_\Gamma \theta^2 dA$. In general, one cannot control the w_z term on the boundary by $\|\nabla\mathbf{u}\|^2$. A symmetrization is introduced in (Davis, 1969a) by setting $\phi = \lambda^{1/2}\theta$, $\lambda = B\mu, R = BN_r$, and from this he uses the variational theory of energy stability to deduce that $E \to 0,\ t \to \infty$, provided $B < \rho^2$, where ρ is defined by

$$\frac{1}{\rho} = \max_{\mathcal{H}} \left\{ \frac{\mu + N_r}{\sqrt{\mu}} < w\phi > + \frac{1}{\sqrt{\mu}} \int_\Gamma (-\phi w_z)\, dA \right\}, \tag{8.4}$$

with u_i, ϕ constrained by the relation

$$D(\mathbf{u}) + D(\phi) + L\int_\Gamma \phi^2\,dA = 1.$$

The solution to (8.4) is completed by deriving the Euler-Lagrange equations in the form

$$\begin{aligned} \Delta\phi + \frac{1}{2}B_\mu\frac{\mu+N_r}{\sqrt{\mu}}\,w &= 0,\\ \Delta u_i + \frac{1}{2}B_\mu\frac{\mu+N_r}{\sqrt{\mu}}\,\phi k_i &= \pi_{,i}, \\ u_{i,i} &= 0, \end{aligned} \tag{8.5}$$

in $\{z\in(0,1)\}\times\mathbb{R}^2$, together with the boundary conditions

$$u_i = \phi = 0, \qquad z = 0, \tag{8.6}$$

$$\begin{aligned} \phi_z + L\phi + \frac{1}{2}\frac{B_\mu}{\sqrt{\mu}}\,w_z &= u_z + \frac{1}{2}\frac{B_\mu}{\sqrt{\mu}}\,\phi_x \\ &= v_z + \frac{1}{2}\frac{B_\mu}{\sqrt{\mu}}\,\phi_y = w = 0, \qquad z = 1. \end{aligned} \tag{8.7}$$

(Davis, 1969a) also derives several useful relations by parametric differentiation. He solves (8.5) - (8.7) by introducing a wave number in the x, y directions and integrating the resulting one-dimensional eigenvalue problem by using a shooting method. He compares his results against those of linear theory and shows they are very sharp, e.g., when $L = B = 0$, $R_L = R_E = 669.0$, and even when $L = 0$, $R_E, R_L \to 0$, the respective B values are still relatively close, being $B_E = 56.77$, $B_L = 79.61$. Thus, even for surface tension driven convection the energy method proves very useful.

There is an alternative way to rigorously handle the bounding of the $B\int_\Gamma \theta w_z dA$ term in (8.3). This is due to L.E. Payne and the writer (unpublished), and we now provide brief details. We use the boundary conditions and then the divergence theorem to rewrite the troublesome term as

$$-B\int_\Gamma \theta w_z dA = -B\oint_{\partial V}\theta u_{i,3}n_i dA = -B\int_V \theta_{,i}u_{i,3}dV, \tag{8.8}$$

where the ∂V integral is over all the boundary of the cell V. Upon use of (8.8) in (8.3) we may rewrite this as

$$\frac{dE}{dt} = I - D, \tag{8.9}$$

where now

$$I = (R+\lambda) < w\theta >$$

and

$$D = \|\nabla\mathbf{u}\|^2 + \lambda\|\nabla\theta\|^2 + L\lambda\int_\Gamma \theta^2 dA + B < \theta_{,i}u_{i,3} > . \tag{8.10}$$

Using the arithmetic-geometric mean inequality we may bound D below by

$$D \geq \Big(1 - \frac{B\alpha}{2}\Big)\|\nabla\mathbf{u}\|^2 + \Big(\lambda - \frac{B}{2\alpha}\Big)\|\nabla\theta\|^2 + L\lambda\int_\Gamma \theta^2 dA,$$

for $\alpha > 0$ to be selected. We now ensure λ and α are chosen so that $B\alpha/2 < 1$ and $B/2\alpha < \lambda$. Hence, provided $\lambda > B^2/4$, D defines a positive-definite functional of u_i and θ. One may now develop a variational analysis from (8.9) in which I/D is bounded above rigorously and a variational energy theory follows naturally.

There are many generalizations and extensions of the work of (Davis, 1969a). For example, (Davis and Homsy, 1980) extend the analysis to the situation where the surface is allowed to deform; this is a very technical piece of work and they find power series solutions to the Euler-Lagrange equations in the Crispation number (which measures surface deflection). (Castillo and Velarde, 1983) develop an equivalent analysis when a concentration field, C, is also present and when $\sigma = \sigma(T,C)$. (McTaggart, 1983b; McTaggart, 1983a) also considers the linear problem when $\sigma = \sigma(T,C)$ for an infinite plane layer, and for a bounded geometry. She shows that in the limit of zero buoyancy convection when both the Marangoni and solutal Marangoni numbers are positive, oscillatory instability does not occur; when, however, the Marangoni numbers have opposite signs oscillatory instability occurs. Computations are performed for an aqueous solution of magnesium sulphate. Other interesting results on surface tension convection in bounded geometries are contained in (Rosenblat et al., 1982a) and (Rosenblat et al., 1982b); these and other aspects are reviewed by (Davis, 1987). The paper by (Zeytounian, 1998) represents a comprehensive review of much recent work in surface tension driven convection; an area of much activity. (Hashim and Wilson, 1999) discusses surface tension driven convection when magnetic field effects are present and gives many references. Additionally, (Nepomnyashchy and Velarde, 1994) and (Rednikov et al., 2000) are very interesting papers dealing with solitary waves, surface waves, and internal waves, and their interaction with the process of thermal convection driven by surface tension. Interface deformations are also the subject of the paper of (Lebon et al., 2001) while Marangoni instability of an evaporating droplet is investigated by (Ha and Lai, 2001). Marangoni convection in a micropolar fluid is studied by (Ramdath, 1997), in a ferromagnetic fluid by (Rudraiah et al., 2002), and incorporating non-Boussinesq effects by (Selak and Lebon, 1997). Thermocapillary convection in models of crystal growth is analysed in detail in the book by (Kuhlmann, 1999). This demonstrates that Marangoni (surface - tension) driven convection is a subject with many applications and is a very active area of research. Many other pertinent articles may be found in the above quoted references.

8.2 Surface film driven convection

(Davis and Homsy, 1980) also briefly consider an ad hoc theory where the surface is itself a two-dimensional continuum. A more complete approach to this problem is by (McTaggart, 1983b; McTaggart, 1984) who employs the (non ad hoc) surface theory, based on a rigorous development from continuum thermodynamics, of (Lindsay and Straughan, 1979). Since the study of McTaggart is relevant to the technologically important problem of convection where a thin film of possibly different fluid overlies the convecting layer we include a description of her work.

It is known from experiments, see e.g., (Berg and Acrivos, 1965), that the presence of a surface film, overlying a layer of fluid heated from below, has a pronounced effect on the onset of convection. Results pertaining to the onset of such convection are important, for example, in the field of crystal growth where convective motion within the liquid during solidification can result in crystal defects, see e.g., (Antar et al., 1980).

The work of (McTaggart, 1983b; McTaggart, 1984) extends that of (Berg and Acrivos, 1965) who examined the effect of surface active agents on convection cells induced by surface tension in shallow layers of fluid. McTaggart examines the more realistic situation where the bulk fluid is not necessarily shallow, by adopting an alternative approach to the introduction of surface tension. The film is regarded as a two-dimensional continuum and surface tension is then introduced naturally as a combination of a surface density and the derivative of a surface free energy.

The model adopted originated with work of (Landau and Lifshitz, 1959), p. 241, on the effect of adsorbed films on the motion of a liquid. (Scriven, 1960) extended these ideas and developed the momentum equations for a two-dimensional continuum and this approach was fully developed according to the methods of modern continuum thermodynamics by (Lindsay and Straughan, 1979).

An advantage of the model of (Lindsay and Straughan, 1979) is that it enables one not only to examine the role of interfacial tension, but also the part played by other interfacial properties that pertain to the resistance of an interface to deformation. In this way (McTaggart, 1983b; McTaggart, 1984) is able to measure the stabilizing effect of the viscosity and the thermal conductivity *of the film* on the onset of convection cells induced by buoyancy forces.

The mathematical picture is now described. A thin film of fluid overlies a layer of different (immiscible) fluid, which is referred to as the bulk fluid. The bulk fluid is contained between the planes $z = 0$ and $z = d$ (> 0) and is subjected to heating from below. The equations for the bulk fluid are

$$\begin{aligned} v_{i,i} &= 0, \\ \rho\dot{v}_i &= -p_{,i} + \mu\Delta v_i + \rho g_i\big[1 - \alpha(T - T_R)\big], \\ \dot{T} &= \kappa\Delta T, \end{aligned} \qquad (8.11)$$

where v_i is the velocity in the bulk fluid, T is temperature, ρ density, p pressure, μ is the viscosity, $\mathbf{g} = (0, 0, -g)$, g being the magnitude of the acceleration due to gravity, α is the coefficient of thermal expansion, T_R is a reference temperature, and κ is the thermal diffusivity.

The film is regarded as an interface between the air (region V^+) and the bulk fluid (region V^-). Assuming the interface to be flat, equations (2.9) of (Lindsay and Straughan, 1979), being the equations of mass, momentum, and energy in the surface, reduce to

$$\begin{aligned} \dot{\gamma} + \gamma V^{\alpha}{}_{;\alpha} &= 0, \\ \gamma \dot{V}^k - S^{k\alpha}{}_{;\alpha} &= [t^{ki} n^i] + \gamma f^k, \\ \gamma\dot{\epsilon} + q^{\alpha}{}_{;\alpha} - S^{k\alpha} V^k{}_{;\alpha} - \gamma r &= -\big[Q^i n^i + t^{ki} n^k (\dot{x}^i - \dot{\xi}^i)\big], \end{aligned} \tag{8.12}$$

where γ is the surface density, $\dot{x}^i = V^i$ is the velocity of a material particle in the surface and V^{α} are the tangential components of V^i, $S^{k\alpha}$ is the surface stress, t^{ik} is the Cauchy stress, n_i denotes the unit outward normal, f^i is the surface body force, ϵ is the specific internal surface energy, q^{α} and Q^i are the surface and bulk heat flux vectors, respectively, r is the surface heat supply, and $\dot{\xi}^i = v^i$. A superposed dot denotes material differentiation, a subscript semicolon denotes differentiation with respect to the surface variables, x, y, and $[\phi] = \phi^+ - \phi^-$ denotes the jump in a quantity across the interface. The summation convention is employed with repeated Greek letters indicating summation from 1 to 2 while repeated Latin letters take values from 1 to 3. Since the interface is bounded by an inviscid fluid above and a viscous fluid below, it follows from the Clausius-Duhem inequality that

$$\big[t^{ki} n^k (\dot{x}^i - \dot{\xi}^i)\big] = 0.$$

It is additionally assumed that $f^k \equiv 0$.

By analogy with three-dimensional theories, a surface free energy ψ is introduced by $\psi = \epsilon - \eta\bar{\theta}$, where $\bar{\theta}$ and η are the temperature and specific entropy in the film and ψ, η satisfy (Lindsay and Straughan, 1979)

$$\psi = \psi(\bar{\theta}, \gamma), \qquad \eta = -\frac{\partial \psi}{\partial \bar{\theta}}.$$

A constitutive theory is then proposed for which ψ has form

$$\psi = C_A \bar{\theta}\left[1 - \log\left(\frac{\bar{\theta}}{\bar{\theta}_R}\right)\right] - \frac{s}{\gamma}(\bar{\theta} - \bar{\theta}_R),$$

where $\bar{\theta}_R$ is a reference temperature, C_A is the specific heat per unit area, and s is a positive constant. Equation (5.4) of (Lindsay and Straughan, 1979) then shows the surface tension σ has the form

$$\sigma = -\gamma^2 \frac{\partial \psi}{\partial \gamma} = -s(\bar{\theta} - \bar{\theta}_R);$$

thus it satisfies the (often assumed) condition that σ be a linear, decreasing function of $\bar{\theta}$.

The appropriate representations for $S^{\alpha\beta}$ and q^α are

$$S^{\alpha\beta} = \sigma a^{\alpha\beta} + \nu_2 a^{\alpha\beta} d^\mu_\mu + \nu_6 d^{\alpha\beta}, \qquad q^\alpha = q_0 g^\alpha,$$

with $d_{\alpha\beta} = \frac{1}{2}(V_{\alpha;\beta} + V_{\beta;\alpha})$, $g_\alpha = \bar{\theta}_{;\alpha}$, $S^{\alpha\beta}x^k{}_{;\beta} = S^{k\alpha}$, and $a_{\alpha\beta} = x^i{}_{;\alpha}x^i{}_{;\beta}$. Moreover, it is assumed that ν_2, ν_6, the coefficients of surface dilational and surface shear viscosity, and q_0, the thermal conductivity of the film, are constant. From the entropy inequality, it follows that $q_0 < 0$. (McTaggart, 1983b; McTaggart, 1984) observes that techniques for measuring ν_2 and ν_6 are available in (Briley et al., 1976). Since the bulk fluid is viscous, $v^i = V^i$ on the surface, and the temperature is assumed continuous between the bulk fluid and the film so $\bar{\theta} = T$ on $z = d$. For the assumption of a flat interface it may then be deduced that

$$v^i{}_{;\alpha} = V^i{}_{;\alpha} \quad \text{and} \quad g_\alpha = \bar{\theta}_{;\alpha} = T_{;\alpha} \quad \text{on} \quad z = d.$$

The film equations (8.12) are now rewritten

$$\begin{aligned} \gamma\dot{v}^k &= -sT_{;\alpha}x^k{}_{;\alpha} + \nu_2 d^\mu_{\mu;\alpha}x^k{}_{;\alpha} + \nu_6 d^{\alpha\beta}{}_{;\alpha}x^k{}_{;\beta} - 2\mu d^{3k}, \\ \gamma C_v\dot{T} &= -q_0 g^\alpha{}_{;\alpha} - sTv^\alpha{}_{;\alpha} - k\frac{\partial T}{\partial z} + \gamma r, \end{aligned} \tag{8.13}$$

where $d_{ij} = \frac{1}{2}(v_{i,j} + v_{j,i})$ and k is the thermal conductivity of the bulk fluid. In the derivation of (8.13) terms of second order in the velocity gradients are neglected, a procedure consistent with the derivation of $(8.11)_3$.

The steady state solution (indicated by a superposed hat) to equations (8.11) and (8.13), satisfying the boundary conditions $T(0) = T_0$ and $T(d) = T_d$, is

$$\begin{aligned} &\hat{v}_i = 0, \quad \hat{T} = T_0 - \beta z, \quad \hat{r} = -\frac{k\beta}{\gamma}, \\ &\hat{p} = p_0 - \rho g z + \alpha\rho g\big[(T_0 - T_R)z - \frac{1}{2}\beta z^2\big], \end{aligned}$$

where $\beta = (T_0 - T_d)/d$ and p_0 is a constant.

To study the stability of the above steady state solution she considers a perturbation (u_i, θ, π) to $(\hat{v}_i, \hat{T}, \hat{p})$. The resulting non-dimensional governing equations are written as:

Bulk

$$\begin{aligned} u_{i,i} &= 0, \\ \dot{u}_i &= -\pi_{,i} + \Delta u_i + R\theta\delta_{i3}, \\ Pr\dot{\theta} &= Rw + \Delta\theta; \end{aligned} \tag{8.14}$$

Film

$$
\begin{aligned}
S_1\dot{u}^k &= -\frac{B}{R}\theta_{;\alpha}x^k{}_{;\alpha} + A_1 d^\mu_{\mu;\alpha}x^k{}_{;\alpha} + A_2 d^{\alpha\beta}{}_{;\alpha}x^k{}_{;\beta} - 2d^{3k},\\
S_1 Pr\dot{\theta} &= A_3 g^\alpha{}_{;\alpha} - \frac{B}{R}A_4 u^\alpha{}_{;\alpha} - A_5\theta u^\alpha{}_{;\alpha} - \frac{\partial\theta}{\partial z},
\end{aligned} \tag{8.15}
$$

where $(u_1, u_2, u_3) = (u, v, w)$ and the Rayleigh, Prandtl, and Marangoni numbers are defined by

$$
R^2 = \frac{g\alpha\beta d^4}{\kappa\nu}, \qquad Pr = \frac{\nu}{\kappa}, \qquad B = \frac{s\beta d^2}{\rho\kappa\nu}.
$$

The non-dimensional numbers $S_1, A_1, \ldots, A_5$ are defined by

$$
\begin{aligned}
S_1 &= \frac{\gamma}{\rho d}, & A_1 &= \frac{\nu_2}{\mu d}, & A_2 &= \frac{\nu_6}{\mu d},\\
A_3 &= -\frac{q_0}{dk}, & A_4 &= \frac{T_d g\alpha}{C_v\beta}, & A_5 &= \frac{sd}{\rho\kappa\nu}.
\end{aligned}
$$

The perturbations are taken to be periodic in x and y. A typical period cell in the non-dimensional layer is denoted by V, and that part of the surface $z = 1$ that forms part of the boundary of V is denoted by Γ. The lower boundary is rigid and held at constant temperature, so

$$
u_i = \theta = 0, \qquad \text{on} \qquad z = 0.
$$

To study stability (McTaggart, 1983b; McTaggart, 1984) defines an energy $E(t)$ by

$$
E(t) = \frac{1}{2}\int_V (u^i u^i + Pr\theta^2)dx + \frac{1}{2}\int_\Gamma (u^i u^i + Pr\theta^2)da.
$$

Then using (8.14), (8.15) it is found that

$$
\begin{aligned}
\frac{dE}{dt} =& - D(\mathbf{u}) - D(\theta) + 2R < \theta w > - \frac{B}{R}(1 - A_4)\int_\Gamma \theta_{;\alpha}u^\alpha\, da\\
&- A_3\int_\Gamma \theta_{;\alpha}\theta_{;\alpha}\, da - A_1\int_\Gamma (d^\mu_\mu)^2 da\\
&- A_2\int_\Gamma d^{\alpha\beta}d_{\alpha\beta}da - A_5\int_\Gamma \theta^2 u^\alpha{}_{;\alpha}da,
\end{aligned} \tag{8.16}
$$

where $< \cdot >$ and $D(\cdot)$ denote integration over V and the Dirichlet integral on V. (McTaggart, 1983b; McTaggart, 1984) introduces the change of variables $B = R^2 N$ and $A_5 = R\bar{T}$, and then shows that

$$
\frac{dE}{dt} = -\mathcal{D} + RI \le -\mathcal{D}\Big[1 - R\max\frac{I}{\mathcal{D}}\Big],
$$

where

$$
\mathcal{D} = D(\mathbf{u}) + D(\theta) + A_3\int_\Gamma \theta_{;\alpha}\theta_{;\alpha}\, da + A_1\int_\Gamma (d^\mu_\mu)^2 da + A_2\int_\Gamma d^{\alpha\beta}d_{\alpha\beta}da,
$$

$$I = 2 < \theta w > -N(1 - A_4)\int_\Gamma \theta_{;\alpha}V^\alpha da - \bar{T}\int_\Gamma \theta^2 V^\alpha{}_{;\alpha}da\,, \qquad (8.17)$$

and where the maximum is over the space of admissible functions. She writes

$$\frac{1}{\Lambda} = \max\frac{I}{\mathcal{D}}\,, \qquad (8.18)$$

and then argues provided $R < \Lambda$ one may employ Poincaré's inequality to deduce that, for a positive constant k,

$$\frac{dE}{dt} \le -kE\Big(1 - \frac{R}{\Lambda}\Big);$$

hence global stability follows. The nonlinear stability problem, therefore, reduces to solving the maximum problem (8.18). The Euler-Lagrange equations corresponding to (8.18) are

$$R_E w + \Delta\theta = 0, \qquad R_E\theta\delta_{j3} + \Delta u_j - P_{,j} = 0, \qquad x \in V, \qquad (8.19)$$

where R_E and $P(x, y, z)$ are Lagrange multipliers, with the natural boundary conditions

$$u_i = \theta = 0, \qquad \text{on} \qquad z = 0; \qquad (8.20)$$

and

$$\begin{aligned} &w = 0,\\ &\frac{\partial\theta}{\partial z} + \bar{T}R_E\theta u^\alpha{}_{;\alpha} - \frac{1}{2}R_E N(1 - A_4)u^\alpha{}_{;\alpha} - A_3\theta_{;\alpha\alpha} = 0,\\ &\frac{\partial u}{\partial z} + R_E\Big(\frac{N}{2} - \frac{A_4}{2} - \bar{T}\theta\Big)\frac{\partial\theta}{\partial x} - A_1\frac{\partial}{\partial x}d^\mu_\mu - A_2 d^{1\beta}{}_{;\beta} = 0,\\ &\frac{\partial v}{\partial z} + R_E\Big(\frac{N}{2} - \frac{A_4}{2} - \bar{T}\theta\Big)\frac{\partial\theta}{\partial y} - A_1\frac{\partial}{\partial y}d^\mu_\mu - A_2 d^{2\beta}{}_{;\beta} = 0, \end{aligned} \qquad (8.21)$$

where (8.21) hold on the surface $z = 1$.

(McTaggart, 1983b; McTaggart, 1984) notes that (8.19) - (8.21) form a *nonlinear* system; the bulk equations (8.19) being linear, but the natural boundary conditions nonlinear. She then observes that it may be possible to solve this system using a perturbation expansion technique as in (Davis and Homsy, 1980), although she only considers the linearized version. In so doing she obtains, to first order (in $\bar{T}$) the energy limit R_E. Such a step, she argues, is a necessary preliminary to the study of the nonlinear system, and so she proceeds by neglecting the nonlinear term $\int_\Gamma \theta^2 V^\alpha{}_{;\alpha}da$. The natural boundary conditions are hence linear and reduce to

$$\begin{aligned} &u_i = \theta = 0, \qquad \text{on} \qquad z = 0, \\ &w = w_{zz} - (A_1 + A_2)\Delta^* w_z \\ &\qquad - \frac{1}{2} R_E N(1 - A_4)\Delta^*\theta = 0, \quad \text{on} \quad z = 1, \\ &\theta_z + \frac{1}{2} R_E N(1 - A_4) w_z - A_3 \Delta^*\theta = 0, \qquad \text{on} \qquad z = 1, \end{aligned} \tag{8.22}$$

with $\Delta^* \equiv \partial^2/\partial x^2 + \partial^2/\partial y^2$. It may now be verified that $R_E = \Lambda$ and hence $R < R_E$ guarantees global stability. The work of (McTaggart, 1983b; McTaggart, 1984) then carries out a numerical evaluation of R_E for various choices of parameters.

The maximum problem of (8.18) is not well set since the numerator contains a cubic term on the boundary. Hence McTaggart has essentially to remove this term by choosing $A_5 = 0$. Of course this is a strong requirement; since $A_5 \propto sd$ it means either surface tension is neglected or vanishingly thin layer theory is studied. Therefore, we now indicate another approach that allows non-zero A_5. Instead of defining I as in (8.17) we omit the cubic term and hence define

$$I = 2 < \theta w > - N(1 - A_4) \int_\Gamma \theta_{;\alpha} V^\alpha da.$$

This then yields a well-defined maximum problem that, in fact, reduces to equations (8.19) and (8.22), and hence the results for R_E derived by McTaggart continue to hold. To handle the nonlinearity, however, we observe that by use of the Cauchy-Schwarz inequality

$$\int_\Gamma \theta^2 V^\alpha{}_{;\alpha} da \le \left(\int_\Gamma \theta^4 da \right)^{1/2} \left(\int_\Gamma (d^\mu_\mu)^2 da \right)^{1/2}. \tag{8.23}$$

The d^μ_μ term is contained in $\mathcal{D}$ and to estimate the θ^4 term we additionally suppose $\int_\Gamma \theta \, da = 0$ and then use the surface Sobolev inequality (see the Appendix),

$$\left(\int_\Gamma \theta^4 da \right)^{1/2} \le c \left(\int_\Gamma \theta_{;\alpha}\theta_{;\alpha} da \right)^{1/2} \left(\int_\Gamma \theta^2 da \right)^{1/2}.$$

Employing this in the energy equation (8.16) we may derive

$$\frac{dE}{dt} \le -\mathcal{D}\left(\frac{R_E - R}{R_E} \right) + c\bar{T} \left(\frac{2}{S_1 Pr A_1 A_3} \right)^{1/2} E\mathcal{D}.$$

Hence, provided

$$R < R_E \qquad \text{and} \qquad E(0) < \left(\frac{R_E - R}{R_E c\bar{T}} \right) \left(\frac{S_1 Pr A_1 A_3}{2} \right)^{1/2},$$

a standard argument employing Poincaré's and Wirtinger's inequality (cf. chapter 4) shows there is at least exponential decay of $E(t)$ and hence global, conditional stability. Thus, the results of (McTaggart, 1983b; McTaggart, 1984) now represent a rigorous and complete nonlinear stability theory.

We point out that in the manipulation of the nonlinearity it is essential that the energy contains surface temperature terms. An attempt to dominate the nonlinearity in (8.23) by bulk terms evidently fails here.

8.3 Conclusions from surface film driven convection

The conclusions drawn by (McTaggart, 1983b; McTaggart, 1984) are worth noting. She examines how the effects of surface viscosity, $V = A_1 + A_2$, surface thermal conductivity, A_3, and surface tension, as measured by $S = N(1 - A_4)$, each affect the onset of convection in the bulk fluid. For an increase in both V and A_3 the critical Rayleigh number of energy theory increases. This agrees with (Berg and Acrivos, 1965) who found that an increase in surface viscosity leads to an increase in stability. In their work they cite the values $V = 1$ for a gaseous monolayer of stearic acid and $V = 10^3$ for a condensed monolayer.

With $V = A_3 = 0$ her results agree with the value of 669 obtained by (Davis and Homsy, 1980) in their energy stability analysis for a non-deformable free surface with the Marangoni number set equal to zero. If $A_3 \to \infty$, $V = 0$, the boundary conditions reduce to the one rigid – one free conducting boundary conditions of the classical Bénard problem, and the value of 1100.65 coincides with the result obtained there. Similarly, when $V = A_3 \to \infty$ the boundary conditions correspond to the classical problem with conducting rigid boundaries and the value of 1708 obtained is thus in agreement.

To examine the effects of surface tension she considers $S \neq 0$. The critical Marangoni number is obtained from the relation $B = R_E^2 N$. She finds there is agreement with the work of (Davis, 1969a) and (Davis and Homsy, 1980) in that increasing the Marangoni number decreases the Rayleigh number.

8.4 Energy stability and nonlinear surface tension

We now study the problem where the linear relation (8.1) is not adequate. (Cloot and Lebon, 1986) use a sort of weakly nonlinear theory to address the convection problem when

$$\sigma = \sigma_m + \frac{1}{2}b(T - T_m)^2. \tag{8.24}$$

Such a quadratic law, they point out, is appropriate to aqueous long chain alcohol solutions and some binary metallic alloys. A nonlinear energy analysis of this problem is interesting in so much as the method of (Davis, 1969a) does *not* appear to carry over. The problem arises because the perturbation surface stress now has form

$$t^{ij}n_j = Mn^i(f\theta + \frac{1}{2}\theta^2) - Ma\, x^i{}_{;\alpha}a^{\alpha\beta}\theta_{;\beta} + Mx^i{}_{;\alpha}a^{\alpha\beta}\theta\theta_{;\beta}. \qquad (8.25)$$

Here M and Ma are Marangoni numbers, $x^i_{;\alpha}$ are derivatives of x^i with respect to the surface variables u^α, and $a^{\alpha\beta}$ is the surface metric tensor. The first term on the right of (8.25) does not play an important role in energy theory. The second term on the right corresponds to the surface temperature gradients in the analysis of (Davis, 1969a), but the $\theta\theta_{;\beta}$ term introduces a problem. It gives rise to a *cubic* surface term in energy theory and evidently cannot be handled by the usual quadratic energy. One approach is to add to the energy a piece of the L^4 integral of θ; this gives rise to a three-parameter eigenvalue problem like those described in chapter 4, where minimization is in one parameter while maximization is in *two* other parameters.

A conditional nonlinear energy stability analysis for a surface tension which is a general function of temperature (including (8.24) as a special case) is given by (Lindsay and Straughan, 1991). We here restrict attention to (8.24) and the development is different from that of (Lindsay and Straughan, 1991).

We commence with the solution (v_i, T, p) satisfying the Navier-Stokes equations in the domain $\mathbb{R}^2 \times \{z \in (0,d)\} \times \{t > 0\}$ and the boundary conditions $T = T_L, v_i = 0$ on $z = 0$, with $T = T_U$, $(T_L > T_U)$, $w = 0$ on $z = d$, and the surface stress balanced by the surface tension contribution with the surface tension given by (8.24). A steady solution of form $\bar{v}_i \equiv 0$, $\bar{T} = -\beta z + T_L$, $\beta = (T_L - T_U)/d$, exists and the stability of this solution is studied. With time, velocity, pressure, and temperature scalings given by $d^2/\nu, \nu/d, \rho\nu U/d, U\sqrt{\nu\beta/\alpha g\kappa}$ the non-dimensional perturbation equations may be shown to be

$$\begin{aligned} u_{i,t} + u_j u_{i,j} &= -\pi_{,i} + \Delta u_i + R\theta k_i, \qquad u_{i,i} = 0, \\ Pr(\theta_{,t} + u_i\theta_{,i}) &= Rw + \Delta\theta, \end{aligned} \qquad (8.26)$$

where $Pr = \nu/\kappa$ is the Prandtl number and $Ra = R^2 = \alpha\beta d^4 g/\kappa\nu$ is the Rayleigh number. The boundary conditions are

$$u_i = \theta = 0, \qquad \text{on } z = 0, \qquad (8.27)$$

and

$$\begin{aligned} &w = 0, \quad \theta_z + L\theta = 0, \\ &u_z = -Ma\,\theta_x + M\theta\theta_x, \\ &v_z = -Ma\,\theta_y + M\theta\theta_y, \end{aligned} \qquad (8.28)$$

on $z = 1$. Here, L is a non-dimensional radiation parameter and $Ma = b[T_L(d) - T_m - \beta d]/\mu U$ and $M = b/2\mu U$ are non-dimensional Marangoni numbers.

To derive energy equations we multiply $(8.26)_1$ by u_i and integrate over a period cell V, use the device in (8.8), and then find

$$\begin{aligned}\frac{d}{dt}\frac{1}{2}\|\mathbf{u}\|^2 =&R < \theta w > -\|\nabla\mathbf{u}\|^2\\ &- Ma < \theta_{,i}u_{i,z} > +M < \theta\theta_{,i}u_{i,z} > .\end{aligned} \tag{8.29}$$

Then, multiplying $(8.26)_3$ by θ and integrating over V we obtain

$$\frac{d}{dt}\frac{1}{2}Pr\|\theta\|^2 = R < \theta w > -\|\nabla\theta\|^2 - L\int_\Gamma \theta^2 dA. \tag{8.30}$$

In order to handle the last term in (8.29) we multiply $(8.26)_3$ by θ^3 and integrate over V to find, with $\|\cdot\|_4$ denoting the norm on $L^4(V)$,

$$\frac{d}{dt}\frac{Pr}{4}\|\theta\|_4^4 = R < w\theta^3 > -3 < \theta^2\theta_{,i}\theta_{,i} > -L\int_\Gamma \theta^4 dA. \tag{8.31}$$

Thus, for coupling parameters $\lambda, \xi > 0$ we may derive from (8.29) - (8.31) an energy equation of form

$$\frac{dE}{dt} = RI - D + R\xi < w\theta^3 >, \tag{8.32}$$

where

$$\begin{aligned}E =&\frac{1}{2}\|\mathbf{u}\|^2 + \frac{\lambda Pr}{2}\|\theta\|^2 + \frac{\xi Pr}{4}\|\theta\|_4^4,\\ I =&(1+\lambda) < \theta w >,\\ D =&\|\nabla\mathbf{u}\|^2 + Ma < \theta_{,i}u_{i,z} > -M < \theta\theta_{,i}u_{i,z} >\\ &+ \lambda\|\nabla\theta\|^2 + 3\xi < \theta^2\theta_{,i}\theta_{,i} >\\ &+ \lambda L\int_\Gamma \theta^2 dA + \xi L\int_\Gamma \theta^4 dA.\end{aligned}$$

We next use the arithmetic-geometric mean inequality with arbitrary positive numbers α, δ to show that

$$\begin{aligned}D \geq&\|\nabla\mathbf{u}\|^2\Big(1 - \frac{Ma}{2\alpha} - \frac{M}{2\delta}\Big) + \|\nabla\theta\|^2\Big(\lambda - \frac{Ma\alpha}{2}\Big)\\ &+ \Big(3\xi - \frac{M\delta}{2}\Big) < \theta^2\theta_{,i}\theta_{,i} > +\lambda L\int_\Gamma \theta^2 dA + \xi L\int_\Gamma \theta^4 dA.\end{aligned} \tag{8.33}$$

Choose now α, δ such that $1 - Ma/2\alpha - M/2\delta > 0$ and then select λ, ξ such that $\lambda > Ma\alpha/2$ and $3\xi > M\delta/2$. Then D is positive-definite and we can show there is a positive constant c_0 such that $D \geq c_0 E$.

If it were not for the last term in (8.32) we could now develop a variational theory of energy stability and an unconditional nonlinear analysis. However, because of the presence of the $< w\theta^3 >$ term I have not seen how to do this. One can use the Cauchy-Schwarz and Sobolev inequalities to show that

$$\begin{aligned}
< w\theta^3 > &\leq \|w\| \, \|\theta\|_4 \, \|\theta^2\|_4 \\
&\leq c_1 \|w\| \, \|\nabla\theta\| \, \|\nabla\theta^2\| \\
&\leq k E^{1/2} D,
\end{aligned}$$

for some constant k. Thus, from (8.32) it is possible to develop a conditional nonlinear stability analysis.

8.5 Surface tension driven convection in a fluid overlying a porous material

A subject which has been generating interest is how to model surface tension driven convection in a fluid saturated porous material, see (Nield, 1998c) and the references therein. When the surface of the porous material is free a thin film of fluid may form and thus surface tension effects may be important. We now review work on convection which may be driven by surface tension, but when the fluid occupies a layer which in turn overlies a layer of porous medium saturated by the same fluid. This problem is not to be confused with convection in a system comprising two immiscible fluids, one overlying the other. The latter topic is reviewed in depth by (Anderdeck et al., 1998).

The classical paper which studies thermal convection in a two layer system composed of a porous layer saturated with fluid over which lies a layer of the same fluid is due to (Chen and Chen, 1988). The layer is heated from below and the bottom of the porous layer is fixed. (Chen and Chen, 1988) demonstrated that the linear instability curves for the onset of convective motion, i.e. the Rayleigh number against wavenumber curves, may be bi-modal in that they possess two local minima as opposed to convection in a single fluid where there may be only one minimum. When and how convection commences is influenced strongly by the parameter

$$\hat{d} = \frac{d}{d_m} = \frac{\text{depth of fluid layer}}{\text{depth of porous layer}}. \tag{8.34}$$

The findings of (Chen and Chen, 1988) conclude that for $\hat{d}$ small (≤ 0.13) the instability is initiated in the porous medium. For $\hat{d}$ larger than this the mechanism changes and instability is dominated by the fluid layer. Work on convection in a porous-fluid layer system is based on the fundamental original model of (Nield, 1977).

One of the fundamental problems in modelling flow of a fluid over a porous medium is that the conditions at the interface between the fluid and the porous medium are a contentious matter, see e.g. (Beavers and Joseph, 1967), (Caviglia et al., 1992),(Ciesjko and Kubik, 1999), (Jäger and

Mikelic, 1998), (Jäger et al., 1999), (Jones, 1973), (McKay, 2001), (Murdoch and Soliman, 1999), (Nield and Bejan, 1999), pp. 15, 16, (Saffman, 1971), (Taylor, 1971). Very good agreement with experiment is often achieved by employing the experimentally suggested condition proposed by (Beavers and Joseph, 1967), or its generalization by (Jones, 1973). Historically the condition of (Beavers and Joseph, 1967) precedes the one of (Jones, 1973), but there are good reasons to prefer the condition proposed by (Jones, 1973). To see this recall the form for the stress tensor, t_{ij}, of a linear viscous fluid, $t_{ij} = -p\delta_{ij} + 2\mu d_{ij}$, where $d_{ij} = (u_{i,j} + u_{j,i})/2$, with u_i being the fluid velocity. Let the interface between the porous medium and the fluid be the plane $z = C$, C constant. The stress vector, t^i, on this surface is $t_i = n_j t_{ji} = -pn_i + 2\mu d_{ij} n_j$, where $\mathbf{n} = (0, 0, 1)$ is the unit outward normal from the fluid. The tangential components of the stress vector, t_β, $\beta = 1, 2$, thus have form

$$t_{3\beta} = 2\mu d_{\beta 3} = \mu\big(u_{\beta,3} + u_{3,\beta}\big), \qquad \beta = 1, 2.$$

The (Jones, 1973) boundary condition adopts the premise that

$$t_{3\beta} \propto \big(u_\beta - u_m^\beta\big), \quad \text{on the interface,} \tag{8.35}$$

where u_m^β are the components of the (pore averaged) fluid velocity in the porous medium. The (Beavers and Joseph, 1967) condition replaces $t_{3\beta}$ by $\partial u^\beta / \partial z$. The (Jones, 1973) and the (Beavers and Joseph, 1967) boundary condition may be written

$$\frac{\partial u^\beta}{\partial z} + J \frac{\partial w}{\partial x_\beta} = \frac{\alpha}{\sqrt{K}} (u^\beta - u_m^\beta), \qquad \beta = 1, 2, \tag{8.36}$$

where α is an experimentally determined constant which varies with the fluid and porous medium, and K is the permeability. For the (Jones, 1973) condition $J = 1$ whereas $J = 0$ for the (Beavers and Joseph, 1967) condition. The condition of (Jones, 1973) is properly invariant because the left hand side is effectively the shear stress on the surface whereas the right hand side is the difference of two velocities and so is likewise invariant. (Straughan, 2002a) finds (numerically) that for the convection problems and the parameters he uses there is little difference in the results depending on whether the Beavers-Joseph or the Jones boundary condition is used. Nevertheless, due to invariance properties one should give consideration to the condition proposed by (Jones, 1973).

It is worth noting that (Payne and Straughan, 1998) demonstrate that the solution to thermal flow problems depends continuously on the parameter α in the Jones condition when Darcy's law is adopted in the porous medium and Stokes' flow holds in the fluid layer. The Beavers-Joseph condition has been successful in the slow flow of a fluid past a porous sphere (Qin and Kaloni, 1993a) If one is employing a method based on linearized instability and so is using Stokes' flow, use of condition (8.36) with $J = 0$ or $J = 1$ is probably justified. Numerical schemes are developed for the coupled fluid flow and porous flow problems by (Discacciati et al., 2002) and by (Miglio et al., 2003). Several computational simulations are reported in these papers. Another interesting numerical contribution to porous/fluid flow is by (Das et al., 2002). This paper presents a finite volume method in three-dimensions. The porous part of the domain is allowed to be anisotropic. It is shown that flow circulation may occur inside the porous medium and the direction of flow may reverse at the interface between the porous medium and fluid. (Layton et al., 2003) prove existence for weak solutions to the problem of Darcy porous media flow coupled to the Stokes equations in a fluid with the Beavers - Joseph interface boundary condition. They also analyse in detail a finite element scheme which formulates the coupled problem as uncoupled steps in the porous and fluid regions thereby allowing a user to employ some of the many existing numerical codes for the separate flow regions.

We now describe the problem of convection driven by surface tension in a fluid which overlies a layer of porous medium saturated by the same fluid, cf. (Straughan, 2001b). To describe this situation we consider the porous medium to occupy the layer $z \in (-d_m, 0)$ while the fluid is in the layer $z \in (0, d)$. The interface is at $z = 0$. The equations in the fluid are the standard equations for Bénard convection,

$$\begin{aligned} &\frac{\partial u_i}{\partial t} + u_j \frac{\partial u_i}{\partial x_j} = -\frac{1}{\rho_0} \frac{\partial p}{\partial x_i} + \nu \Delta u_i + \bar{\alpha} g T k_i, \\ &\frac{\partial u_i}{\partial x_i} = 0, \qquad \frac{\partial T}{\partial t} + u_i \frac{\partial T}{\partial x_i} = \frac{k_f}{(\rho_0 c_p)_f} \Delta T, \end{aligned} \tag{8.37}$$

in $\mathbb{R}^2 \times \{z \in (0, d)\} \times \{t > 0\}$, with $\mathbf{k} = (0, 0, 1)$. Here, u_i, p, T are velocity, pressure and temperature, and $\rho_0, \nu, \bar{\alpha}, g, k_f, c_p$ are density, kinematic viscosity, thermal expansion coefficient, gravity, thermal conductivity and specific heat at constant pressure. A subscript/superscript f or m denotes fluid or porous medium, respectively,

The equations in the porous medium are those of Darcy,

$$\begin{aligned} &\frac{1}{\phi} \frac{\partial u_i^m}{\partial t} = -\frac{1}{\rho_0} \frac{\partial p^m}{\partial x_i} - \frac{\nu}{k} u_i^m + \bar{\alpha} g T_m k_i, \\ &\frac{\partial u_i^m}{\partial x_i} = 0, \qquad (\rho_0 c_p)^* \frac{\partial T_m}{\partial t} + (\rho_0 c_p)_f u_i^m \frac{\partial T_m}{\partial x_i} = k^* \Delta T_m, \end{aligned} \tag{8.38}$$

in $\mathbb{R}^2 \times \{z \in (-d_m, 0)\} \times \{t > 0\}$. The terms u_i^m, p_m, T_m are velocity, pressure and temperature in the porous medium, ϕ is the porosity, and * denotes a quantity averaged over the porous and fluid variables, $X^* = \phi X_f + (1 - \phi) X_m$.

The steady state solution to be examined for linearized instability is one for which there is no fluid motion in either layer and the temperatures on the upper and lower boundaries are held at fixed constant temperatures, T_U and T_L, respectively, with $T_L > T_U$. So, the steady solution $(\bar{u}_i, \bar{T}, \bar{p})$,

$(\bar{u}_i^m, \bar{T}^m, \bar{p}^m)$, is

$$\begin{aligned}
&\bar{u}_i = 0, \qquad \bar{u}_i^m = 0,\\
&\bar{T} = T_0 - (T_0 - T_U)\,\frac{z}{d}, \quad 0 \le z \le d,\\
&\bar{T}_m = T_0 - (T_L - T_0)\,\frac{z}{d_m}, \quad -d_m \le z \le 0.
\end{aligned} \tag{8.39}$$

In these equations T_0 is the temperature at the interface which is given by $T_0 = (k^* dT_L + k_f d_m T_U)/(k^* d + k_f d_m)$.

At the fluid surface $z = d$ a radiation type boundary condition is assumed of form

$$\delta_1 \frac{d\bar{T}}{dz} + \delta_2 \bar{T} = c, \qquad \text{at} \quad z = d. \tag{8.40}$$

A perturbation (u_i, θ, π) to $(\bar{u}_i, \bar{T}, \bar{p})$ and (u_i^m, θ_m, π_m) to $(\bar{u}_i^m, \bar{T}_m, \bar{p}_m)$ is introduced and boundary condition (8.40) leads to a condition on the perturbation temperature field at the fluid surface, of form

$$\frac{\partial \theta}{\partial z} + L\theta = 0, \qquad \text{on } z = d. \tag{8.41}$$

Put now $u_i = \bar{u}_i + u_i$, $T = \bar{T} + \theta$, $p = \bar{p} + \pi$, $u_i^m = \bar{u}_i^m + u_i^m$, $T_m = \bar{T}_m + \theta_m$, $p_m = \bar{p}_m + \pi_m$, and derive *linearised* equations for the perturbation quantities $(u_i, \theta, \pi, u_i^m, \theta_m, \pi_m)$. (Straughan, 2001b) notes that for surface tension driven convection in a fluid convective motion may commence by oscillatory convection. Thus, all time derivatives are retained including those in the boundary conditions.

Introducing a time dependence of form $u_i = u_i(\mathbf{x})\, e^{\sigma t}$, $\theta = \theta(\mathbf{x})\, e^{\sigma t}$, $u_i^m = u_i^m(\mathbf{x})\, e^{\sigma_m t}$, $\theta_m = \theta_m(\mathbf{x})\, e^{\sigma_m t}$, the linearized perturbation equations are

$$\begin{aligned}
&\rho_0 \sigma u_i = -\frac{\partial \pi}{\partial x_i} + \mu \Delta u_i + \rho_0 \bar{\alpha} g k_i \theta,\\
&\frac{\partial u_i}{\partial x_i} = 0,\\
&\sigma \theta = \Big(\frac{T_0 - T_U}{d}\Big)\, w + \frac{k_f}{(\rho_0 c_p)_f}\, \Delta \theta,
\end{aligned} \tag{8.42}$$

and

$$\begin{aligned}
&\frac{\rho_0}{\phi}\, \sigma_m u_i^m = -\frac{\partial \pi^m}{\partial x_i} - \frac{\mu}{k}\, u_i^m + \rho_0 \bar{\alpha} g k_i \theta_m,\\
&\frac{\partial u_i^m}{\partial x_i} = 0,\\
&\sigma_m \theta_m = \Big(\frac{T_L - T_0}{d_m}\Big)\, \frac{(\rho_0 c_p)_f}{(\rho_0 c_p)_*}\, w_m + \frac{k^*}{(\rho_0 c_p)_*}\, \Delta \theta_m,
\end{aligned} \tag{8.43}$$

where $w = u_3$, $w_m = u_3^m$, and $\mu = \nu \rho_0$ is the dynamic viscosity.

With a normal mode representation $w = W(z)\,f(x,y)$, $\theta = \Theta(z)\,f(x,y)$, $w_m = W_m(z)\,f(x,y)$, $\theta_m = \Theta_m(z)\,f(x,y)$, f being the horizontal planform, the eigenvalue equations for the growth rates σ, σ_m are found to be, in terms of fluid and porous wavenumbers a and a_m,

$$
\begin{aligned}
&(D^2 - a^2)^2 W - a^2 Ra\Theta = \frac{\sigma}{Pr}(D^2 - a^2)W,\\
&(D^2 - a^2)\Theta - W = \sigma\Theta,
\end{aligned}
\tag{8.44}
$$

and

$$
\begin{aligned}
&(D^2 - a_m^2)W_m + a_m^2 Ra_m \Theta_m = -\sigma_m \frac{\delta^2}{\phi Pr_m}(D^2 - a_m^2)W_m,\\
&(D^2 - a_m^2)\Theta_m - W_m = \sigma_m G_m \Theta_m.
\end{aligned}
\tag{8.45}
$$

Equations (8.44) hold in $z \in (0,1)$ and $D = d/dz$, while equations (8.45) hold in $z_m \in (-1,0)$ with $D = d/dz_m$. The quantities Ra and Ra_m are the Rayleigh number and porous Rayleigh number given by

$$
Ra = \frac{g\bar{\alpha}\rho_0(T_U - T_0)d^3(\rho_0 c_p)_f}{\mu k_f}, \qquad Ra_m = Ra\,\frac{(\delta\epsilon_T)^2}{\hat{d}^4}. \tag{8.46}
$$

The terms Pr and Pr_m are the Prandtl and porous Prandtl numbers, δ is the Darcy number given by $\delta = \sqrt{k}/d_m$, and $G_m = (\rho_0 c_p)^*/(\rho_0 c_p)_f$, $\epsilon_T = \lambda_f/\lambda_m$, where the fluid and porous medium thermal diffusivities are defined by $\lambda_f = k_f/(\rho_0 c_p)_f$, $\lambda_m = k^*/(\rho_0 c_p)^*$. Equations (8.44) and (8.45) form a tenth order system to be solved for the eigenvalue σ with Ra, Ra_m and the other parameters fixed. Note, $\sigma_m = \hat{d}^2\sigma/\epsilon_T$, and we minimize in a or a_m.

The boundary conditions are now given, cf. (Straughan, 2001b). On $z = -1$,

$$
W_m = \Theta_m = 0, \qquad z = -1. \tag{8.47}
$$

On the surface $z = 1$ we have

$$
W = 0, \quad D\Theta + L\Theta = 0, \qquad z = 1, \tag{8.48}
$$

and at the interface we have continuity of normal velocity, of temperature, and of the heat flux which become

$$
W = \hat{d}W_m, \quad \hat{d}\Theta = \epsilon_T^2\Theta_m, \quad D\Theta = \epsilon_T D_p\Theta_m, \qquad z = 0, \tag{8.49}
$$

where $D_p = d/dz_m$.

Surface tension leads to the boundary condition

$$
D^2 W = Ma\,\Delta^*\theta, \quad \text{on } z = 1, \tag{8.50}
$$

where $\Delta^* = \partial^2/\partial x^2 + \partial^2/\partial y^2$ and Ma is the Marangoni number, $Ma = [\gamma\sigma_0(T_U - T_0)d]/\lambda_f\mu$.

The interface boundary conditions which arise from the Beavers-Joseph condition and continuity of normal stress are·

$$D^2W - \frac{\alpha\hat{d}}{\delta}DW + \frac{\alpha\hat{d}^3}{\delta}D_pW_m = 0, \quad z = 0, \tag{8.51}$$

and

$$\begin{aligned}\frac{\hat{d}^4}{\phi Pr_m}\sigma_m D_pW^m + \frac{\hat{d}^4}{\delta^2}D_pW^m =& \frac{1}{Pr}\sigma DW - D^3W \\ & - 3\Delta^* DW, \qquad \text{on } z = 0.\end{aligned} \tag{8.52}$$

Equations (8.44) and (8.45) together with the boundary conditions (8.47) - (8.52) are solved numerically by a D^2 Chebyshev tau method in (Straughan, 2001b). Many numerical results are reported. A bimodal set of neutral curves is obtained as in (Chen and Chen, 1988), although the transition from porous to fluid dominance now depends on the Marangoni number Ma in addition to the depth ratio $\hat{d}$. It is found that surface tension is important when the fluid depth is relatively large, and particularly so when the fluid depth is relatively small. For example, for a 2 cm sample with a fluid depth of 0.1 mm the surface tension effect can be 20% or more. This lends much support to employment of the two layer model of (Nield, 1998c) in interpreting surface tension driven convection in a saturated porous medium.

9
Convection in generalized fluids

9.1 The Bénard problem for a micropolar fluid

If we allow the Newtonian constitutive equation for the stress to be replaced by something for a more exotic fluid then a variety of interesting convection problems arise. Nonlinear energy stability analyses for these fluids are only relatively recent. In the first three sections we concentrate on thermal convection in a micropolar fluid. Other effects are considered in section 9.4 and section 9.5 reviews work in various other classes of generalized fluids.

The theory of micropolar fluids is due to (Eringen, 1964; Eringen, 1969; Eringen, 1972; Eringen, 1980) whose theory allows for the presence of particles in the fluid by additionally accounting for particle motion. In (Eringen, 1964) he introduced the theory for a simple microfluid and extended this in (Eringen, 1969) to allow deformation of local fluid elements (particles). This was further generalized in (Eringen, 1972) where he gave precise meaning to a thermomicropolar fluid, and in (Eringen, 1980) the theory appropriate to anisotropic fluids was elucidated.

The Bénard problem for the theory of (Eringen, 1972) was first examined by (Datta and Sastry, 1976) who studied stationary convection in the linear problem under an *assumption* on the sign of a thermal coupling term in the energy equation. (Ahmadi, 1976) used nonlinear energy stability theory on the same problem although he chose to neglect the coupling term, which gave rise to an interesting effect in (Datta and Sastry, 1976). Further use of the energy method was made by (Lebon and Perez-Garcia, 1981) who presented nonlinear stability results for the problem of (Datta and Sastry,

1976). The contribution of (Payne and Straughan, 1989) also looks at the effect of the thermal coupling term but concentrates primarily on the opposite sign to that selected by (Datta and Sastry, 1976) and (Lebon and Perez-Garcia, 1981). (Payne and Straughan, 1989) focus attention on the possibility of oscillatory convection (neglected in (Datta and Sastry, 1976)) and this yields a striking result; they also employ nonlinear energy stability but concentrate on the coefficient case not covered in (Datta and Sastry, 1976) and (Lebon and Perez-Garcia, 1981).

In (Datta and Sastry, 1976) it was shown that heating *both from above and below* could lead to stationary convection instabilities. The paper of (Payne and Straughan, 1989) provides an alternative route where stationary convection only occurs in the heated from below case. Furthermore, as the magnitude of the thermal coupling term is increased they predict a substantial decrease in the critical Rayleigh number (unlike (Datta and Sastry, 1976) and (Lebon and Perez-Garcia, 1981) where the opposite sign is chosen for the thermal coupling term). (Chandra, 1938) observed in his experiments that adding smoke particles to a layer of gas could lead to such a substantial decrease of the Rayleigh number at which convective motion commences. Since the particle spin associated with the theory of (Eringen, 1972) could possibly be appropriate to the added dust situation described by (Chandra, 1938), there may be a justification for use of the analysis of (Payne and Straughan, 1989) for convection in a fluid fairly evenly interspersed with "particles", which may be dust, dirt, ice or raindrops, or other additives. Thus we believe that heuristically the results give reason to believe that the (Eringen, 1972) micropolar convection model may be applicable to geophysical or industrial convection contexts.

It is expedient to include the relevant equations of (Eringen, 1972) for an incompressible, isotropic thermomicropolar fluid. We take the microinertia moment tensor $j_{ik} = j\delta_{ik}$ where j is constant. The continuity, momentum, moment of momentum, and balance of energy equations are then ((Eringen, 1972), pp. 489, 490)

$$v_{i,i} = 0, \tag{9.1}$$

$$\rho\dot{v}_i = \rho f_i + t_{ki,k}, \tag{9.2}$$

$$\rho j\dot{\nu}_i = \rho\ell_i + \epsilon_{ikh}t_{kh} + m_{ki,k}, \tag{9.3}$$

$$\rho\dot{\epsilon} = t_{kh}b_{kh} + m_{kh}\nu_{h,k} + q_{k,k} + \rho h, \tag{9.4}$$

where $b_{kh} = v_{h,k} - \epsilon_{khr}\nu_r$, a superposed dot denotes the material derivative, v_i, ν_i, $\epsilon(T)$ are fluid velocity, particle spin vector, and internal energy, T being temperature; $\rho, t_{ki}, f_i, \ell_i, m_{ki}, q_k$, and h are density (presumed constant except in the body force ρf_i term), stress tensor, body force, body couple, couple stress tensor, heat flux vector, and heat supply.

The constitutive equations are

$$t_{kh} = -\pi\delta_{kh} + \mu(v_{k,h} + v_{h,k}) + \bar{\kappa}(v_{h,k} - \epsilon_{kha}\nu_a), \tag{9.5}$$

$$m_{kh} = \bar{\alpha}\nu_{r,r}\delta_{kh} + \bar{\beta}\nu_{k,h} + \gamma\nu_{h,k} + \alpha\epsilon_{khm}T_{,m}, \tag{9.6}$$

$$q_k = \kappa T_{,k} + \beta\epsilon_{khm}\nu_{h,m}, \tag{9.7}$$

where π is the pressure and, in general, $\mu, \bar{\kappa}, \bar{\alpha}, \bar{\beta}, \gamma, \alpha, \kappa, \beta$ are functions of T. We treat $\mu, \bar{\kappa}, \bar{\alpha}, \bar{\beta}, \gamma, \alpha, \kappa$ as constant and in (9.1) - (9.4) ρ is assumed constant ($= \rho_0$) except in the body force term in (9.8) where

$$\rho = \rho_0\big[1 - A(T - T_0)\big], \tag{9.8}$$

A being the coefficient of thermal expansion of the fluid. It is further assumed that the body couple and heat supply are zero, i.e., $\ell_i \equiv h \equiv 0$.

The relevant equations are now obtained by inserting (9.5) - (9.8) into (9.1) - (9.4). However, some reduction of (9.4) is necessary to make the convection problem tractable. Equation (9.4) is

$$\begin{aligned} -\rho\frac{\partial\epsilon}{\partial T}\dot{T} + \Big\{&(v_{h,k} - \epsilon_{khr}\nu_r)\big[\mu(v_{k,h} + v_{h,k}) + \bar{\kappa}(v_{h,k} - \epsilon_{kha}\nu_a)\big] \\ &+ \nu_{h,k}(\bar{\alpha}\nu_{r,r}\delta_{kh} + \bar{\beta}\nu_{k,h} + \gamma\nu_{h,k})\Big\} + \kappa\Delta T \\ &+ \alpha\epsilon_{khm}T_{,m}\nu_{h,k} + \epsilon_{khm}\nu_{h,m}\beta_{,k} = 0. \end{aligned} \tag{9.9}$$

Three approaches to reducing (9.9) have been advocated in the literature. All assume the quadratic terms (in {...} parentheses) are negligible. Such an approach is consistent with the analogous reduction for a Newtonian fluid.

(Ahmadi, 1976) further assumes the coefficient β is constant and also neglects the α term. The resulting Bénard problem is then of the form

$$u_t + Lu + N(u) = 0,$$

regarded as an equation in a Hilbert space, where the nonlinear term satisfies $< u, N(u) > \geq 0$ in the appropriate inner product, and L is an unbounded, symmetric linear operator. Hence, no subcritical instabilities are possible, cf. chapter 4.

For a fluid contained in the layer $z \in (0, d)$ with gravity in the negative z-direction, the planes $z = 0, d$ kept at constant temperatures T_1, T_2 the steady solution of interest is

$$\bar{T} = -Bz + T_1, \qquad \bar{v}_i = 0, \qquad \bar{\nu}_i = 0, \tag{9.10}$$

where B is the temperature gradient.

The reduction of equation (9.9) by (Datta and Sastry, 1976) takes the coefficient β to be constant but retains the term $\alpha\epsilon_{khm}T_{,m}\nu_{h,k}$. This term gives rise to a non-zero contribution from $\partial\bar{T}/\partial z$, which makes the linear operator non-symmetric. The final approach by (Lebon and Perez-Garcia, 1981) neglects the α term, but they assume $\beta = \hat{\beta}T$ for $\hat{\beta}$ constant. This again leads to a contribution from $\partial\bar{T}/\partial z$ and hence some skew-symmetry. A richer structure is obtained by retaining a thermal coupling.

In (Datta and Sastry, 1976) and (Lebon and Perez-Garcia, 1981) the $\hat{\beta}$ term is assumed negative. However, thermodynamic arguments (Eringen, 1972) place only the following restrictions on the coefficients in (9.5) - (9.7):

$$\begin{aligned} &2\mu+\bar{\kappa}\geq 0, \quad 2\mu+\kappa\geq 0, && 3\bar{\alpha}+\bar{\beta}+\gamma\geq 0,\\ &\bar{\beta}+\gamma\geq 0, \quad \mu\geq 0, && \kappa\geq 0,\\ &\bar{\kappa}\geq 0, \quad \frac{2\kappa}{T}(\gamma-\bar{\beta})\geq\Big(\alpha-\frac{\beta}{T}\Big)^2. \end{aligned} \tag{9.11}$$

It would appear there is no reason why $\hat{\beta}=\beta T^{-1}<0$. Indeed, the theoretical results of (Payne and Straughan, 1989) might suggest it preferable to require $\hat{\beta}>0$.

The internal energy is given by $\epsilon=\psi+\eta T$ where $\eta=-\partial\psi/\partial T$. Thus,

$$\dot{\epsilon}=-\frac{\partial^2\psi}{\partial T^2}\dot{T}.$$

It is usual to assume $-\psi_{TT}=c$ (constant) and so we follow standard practice. Then, (9.9) may be reduced to

$$\rho_0 c\dot{T}=\kappa\Delta T+\hat{\beta}\epsilon_{khm}\nu_{h,m}T_{,k}. \tag{9.12}$$

To study the stability of (9.10) we define the perturbed quantities by $T=\bar{T}+\theta$, $v_i=\bar{v}_i+u_i$, $\nu_i=\bar{\nu}_i+\nu_i$. Important non-dimensional parameters are K,Γ,G, these being associated with the micropolar terms, $Ra=R^2$, the Rayleigh number, and the thermal term b, viz.,

$$K=\frac{\bar{\kappa}}{\mu},\quad \Gamma=\frac{\gamma}{\mu d^2},\quad G=\frac{\bar{\alpha}+\bar{\beta}}{\mu d^2},\quad Ra=\frac{gABd^4c\rho_0^2}{\kappa\mu},\quad b=\frac{\hat{\beta}}{d^2c\rho_0}.$$

The non-dimensionalized perturbation equations are

$$\begin{aligned} u_{i,t}+u_k u_{i,k}&=R\theta\delta_{i3}-\pi_{,i}+\Delta u_i+K(\Delta u_i-\epsilon_{kir}\nu_{r,k}),\\ u_{i,i}&=0,\\ Pr(\theta_{,t}+u_i\theta_{,i})&=Rw+\Delta\theta-bR\xi+bPr\epsilon_{khm}\nu_{h,m}\theta_{,k},\\ j(\nu_{i,t}+u_a\nu_{i,a})&=K(\epsilon_{ikh}u_{h,k}-2\nu_i)+G\nu_{a,ai}+\Gamma\Delta\nu_i, \end{aligned} \tag{9.13}$$

where $\mathbf{u}\equiv(u,v,w)$ and $\xi=(\text{curl}\,\nu)_3$.

It is important to realize that the only term that makes the linearized version of (9.13) non-symmetric is the bR term in $(9.13)_3$.

(Datta and Sastry, 1976), (Lebon and Perez-Garcia, 1981), and (Payne and Straughan, 1989) consider equations (9.13) to hold in the layer $z\in(0,1)$ with $\theta=\nu_i=0$ on $z=0,1$ together with stress free boundary conditions on the surfaces $z=0,1$. We also assume a periodic structure in the x,y-plane.

9.2 Oscillatory instability

A complete analysis of the linearized instability problem (with $b > 0$) is contained in (Payne and Straughan, 1989). We describe their findings for oscillatory convection only.

Equations (9.13) are linearized, and then a time dependence like $u_i(x_r, t) = e^{\sigma t} u_i(x_r)$ is assumed with a similar form for ν_i, θ, p. The system of equations is reduced to a single equation in w, namely,

$$\begin{aligned}(\Gamma\Delta - 2K - j\sigma)\Big\{(\Delta - Pr\sigma)\big[(1+K)\Delta^2 - \sigma\Delta\big] - R^2\Delta^*\Big\}w \\ + K\big\{\Delta K(\Delta - Pr\sigma) - bR^2\Delta^*\big\}\Delta w = 0,\end{aligned} \tag{9.14}$$

where $\Delta^* \equiv \partial^2/\partial x^2 + \partial^2/\partial y^2$.

Assume a normal mode form $w = W(z)\Phi(x, y)$, where $\Delta^*\Phi = -a^2\Phi$, and then with $D = d/dz$ the boundary conditions are

$$D^{(2n)}W = 0 \quad \text{at} \quad z = 0, 1, \ \forall n \geq 0.$$

Hence the solution to (9.14) is a half-range sine series of terms like $W = \sin n\pi z$. Denoting $\Lambda = n^2\pi^2 + a^2$, (9.14) reduces to

$$\begin{aligned}(\Gamma\Lambda + 2K + j\sigma)\Big\{(\Lambda + Pr\sigma)\big[(1+K)\Lambda^2 + \sigma\Lambda\big] - R^2a^2\Big\} \\ - K\big\{\Lambda K(\Lambda + Pr\sigma) + bR^2a^2\big\}\Lambda = 0.\end{aligned} \tag{9.15}$$

To study oscillatory convection we select $\sigma = i\sigma_1$, $\sigma_1 \in \mathbb{R}$, in (9.15). By equating real and imaginary parts the following equations are obtained:

$$\begin{aligned}\sigma_1^2 = & \frac{-1}{[j(1+Pr) + KPr(j-b)]} \\ & \times \left\{\Lambda K^2 + K\Lambda^2 b(1+K) + \left(\frac{(Kb+\Gamma)\Lambda + 2K}{j}\right)\right. \\ & \left. \times \Big(\Lambda\Gamma\big[1 + Pr(1+K)\big] + K^2Pr + 2K(1+Pr)\Big)\right\},\end{aligned} \tag{9.16}$$

$$
\begin{aligned}
R_{\rm osc}^2 = & \frac{\Lambda^2}{a^2}(1+K) \\
& + \frac{1}{ja^2}\Big\{\Gamma\Lambda^3\big[1+Pr(1+K)\big] \\
& \qquad + 2K\Lambda^2(1+Pr) + K^2\Lambda^2 Pr\Big\} \\
& + \frac{\Lambda Pr}{a^2\big[j(1+Pr)+KPr(j-b)\big]} \\
& \times\Big\{\Lambda K^2 + K\Lambda^2 b(1+K) + \Big(\frac{(Kb+\Gamma)\Lambda+2K}{j}\Big) \\
& \times\Big(\Lambda\Gamma\big[1+Pr(1+K)\big] \\
& \qquad + K^2 Pr + 2K(1+Pr)\Big)\Big\}.
\end{aligned}
\tag{9.17}
$$

Since $\sigma_1 \in \mathbb{R}$, equation (9.16) yields much useful information. If $b < 0$ then we see immediately that overstable convection will not be possible unless b is large (in an appropriate sense): certainly it would have to be necessary that $b < -\Gamma/K$. For positive b, (9.17) immediately shows that a necessary condition for overstability is that

$$
b > j\Big(1 + \frac{1+Pr}{KPr}\Big). \tag{9.18}
$$

If b satisfies (9.18), then by grouping together Λ^3, Λ^2, and Λ terms in (9.17), we see that since j, K are likely small, $R_{\rm osc}^2$ is *negatively* very large. A practical interpretation is that oscillatory convection is only possible when the layer is *heated from above.* Since the resulting critical Rayleigh numbers may well be enormous, indeed the numerical results of (Payne and Straughan, 1989) suggest $\sim 3.8 \times 10^6$, one could argue that such a situation would not be encountered in everyday life.

9.3 Nonlinear energy stability for micropolar convection

To study the nonlinear stability of (9.10) an L^2 energy, $E(t)$, is chosen, as

$$
E = \frac{1}{2}(< u_i u_i > + j < \nu_i \nu_i >) + \frac{1}{2}\lambda Pr\|\theta\|^2, \tag{9.19}
$$

where $\lambda\,(> 0)$ is a coupling parameter to be chosen. One could introduce another coupling parameter in front of the j term, but as there is symmetry between $(9.13)_1$ and $(9.13)_4$ the value of one for the second coupling parameter suffices. The functional (9.19) is also selected by (Lebon and Perez-Garcia, 1981).

We differentiate E and substitute from (9.13) to derive

$$\frac{dE}{dt} = RI - \mathcal{D}, \tag{9.20}$$

where the production term I and the dissipation $\mathcal{D}$ are in this case given by

$$I = (1+\lambda) < \theta w > + b\lambda < \xi\theta >, \tag{9.21}$$

$$\begin{aligned}\mathcal{D} = &< u_{i,j}u_{i,j} > + \lambda < \theta_{,i}\theta_{,i} > + K < \nu_i\nu_i > + \Gamma < \nu_{i,j}\nu_{i,j} > \\ &+ K < (\epsilon_{ijk}u_{k,j} - \nu_i)(\epsilon_{irs}u_{s,r} - \nu_i) > + G\|\nu_{a,a}\|^2.\end{aligned} \tag{9.22}$$

Define now

$$\frac{1}{R_E} = \max \frac{I}{\mathcal{D}}, \tag{9.23}$$

where the maximum is over the space of admissible functions. Then, from (9.20),

$$\frac{dE}{dt} \leq -\mathcal{D}\left(\frac{R_E - R}{R_E}\right),$$

and if $R < R_E$ then Poincaré's inequality together with strict forms of the thermodynamic inequalities (9.11) shows that $dE/dt \leq -kE$, for a positive constant k. Hence $R < R_E$ guarantees $E \to 0$ at least exponentially fast as $t \to \infty$, and we have nonlinear stability.

We put $\phi = \lambda^{1/2}\theta$ and calculate the Euler-Lagrange equations for R_E from (9.23), to find

$$\begin{aligned}&(1+K)\Delta u_i + K\epsilon_{ijk}\nu_{k,j} + R_E f\phi k_i = \pi_{,i}, \\ &u_{i,i} = 0, \qquad \Delta\phi + R_E(fw + \frac{1}{2}b\lambda^{1/2}\xi) = 0, \\ &\Gamma\Delta\nu_i + G\nu_{a,ai} + K(\epsilon_{ijk}u_{k,j} - 2\nu_i) - \frac{1}{2}\lambda^{1/2}b\epsilon_{3ji}R_E\phi_{,j} = 0,\end{aligned} \tag{9.24}$$

where π is a Lagrange multiplier and $f = (1+\lambda)/2\lambda^{1/2}$. Equations (9.24) are reduced to a single equation in w to find,

$$\begin{aligned}&K\left(1+\frac{1}{2}K\right)\Delta^3 w - \frac{1}{2}\Gamma(1+K)\Delta^4 w \\ &\quad + R_E^2\Big[-Kf^2\Delta^* w - \frac{1}{8}\lambda b^2(1+K)\Delta^2\Delta^* w \\ &\quad + \frac{1}{2}(K\lambda^{1/2}bf + f^2\Gamma)\Delta\Delta^* w\Big] = 0.\end{aligned} \tag{9.25}$$

By employing a normal mode analysis (9.25) reduces to

$$R_E^2 = \frac{\Lambda^3}{a^2}\left(\frac{\Gamma(1+K)\Lambda + 2K + K^2}{2Kf^2 + \Lambda(K\lambda^{1/2}bf + f^2\Gamma) + \lambda b^2(1+K)\Lambda^2/4}\right). \tag{9.26}$$

Then

$$Ra_E = \max_{\lambda} \min_{a^2} R_E^2(\lambda, a^2; \Gamma, K, b)$$

is found numerically.

(Payne and Straughan, 1989), like (Datta and Sastry, 1976) and (Lebon and Perez-Garcia, 1981), analyse the case of free-free boundaries.

The stationary convection results of (Payne and Straughan, 1989) show a drastic reduction in Ra for increasing b. Thus, an appropriate model could possibly account for the drastic reduction in Ra observed by (Chandra, 1938) when adding smoke particles to air. The energy results are useful in that they represent a *nonlinear unconditional* stability threshold and hence yield a region where *possible* subcritical instabilities may occur, although this region is quite large when $b \geq O(0.1)$. Perhaps by a more exotic choice of generalized energy one may be able to increase the nonlinear threshold substantially. The oscillatory convection results show that for Γ, K, j small, b needs to be relatively large for oscillatory convection to be possible. Even then, the numerical values of Ra are typically 10^3 larger than the ones observed for stationary convection in other parameter ranges, and this means that a very much larger *negative* temperature gradient would need to be imposed before the onset of such convection could be realized.

9.4 Micropolar convection and other effects

An important area which has seen a rapid expansion of research interest is in convection in a micropolar fluid when other effects (such as rotation) are present. Many articles have appeared during the last few years. We describe some of this work but we point out that many references to other work may be found in the articles we cite. Several of the papers we mention deal with energy stability methods, or Lyapunov functional arguments.

Thermal convection in a micropolar fluid which occupies a rotating horizontal layer has been studied by (Qin and Kaloni, 1992b) and by (Sharma and Kumar, 1994). The rotation still plays a stabilizing role, but the effect of the micro-rotation of the particles is studied in detail, especially with regard to whether the convection is via stationary or oscillatory convection. (Murty, 1999) investigates the combined effects of throughflow and rotation on thermal convection in a micropolar fluid layer. It is found that rotation may enhance the stabilizing or destabilizing effects induced by a constant vertical throughflow of fluid. (Ezzat and Othman, 2000) consider a rotating layer of dielectric micropolar fluid subject to a vertical temperature gradient. The layer is subject to a vertical ac electric field, whereby they obtain results on instability due to both factors of rotation and an applied electric field.

The paper by (Ramdath, 1997) considers the combined effects of a temperature gradient and surface tension driven convection in a layer of micropolar fluid. The microstructure effects are found to be stabilizing. A non-uniform temperature gradient is imposed on a layer of micropolar fluid in the work of (Siddheshwar and Pranesh, 1998a). Six different non-uniform temperature profiles are analysed and a detailed comparison is made. Generally, they discover that the Rayleigh number at the onset of convection is lower for a micropolar fluid than that for a Newtonian fluid, and the mechanisms which may advance or delay convection are identified. (Calmelet-Eluhu, 1996) investigates the effect of a magnetic field on thermal convection in a micropolar fluid. She derives a linear instability analysis and additionally develops energy stability theory to obtain nonlinear stability thresholds. It is found that the micro-rotation and magnetic field are stabilizing. The effect of a magnetic field on convection in a micropolar fluid is further considered by (Siddheshwar and Pranesh, 1998b). They again find the critical Rayleigh numbers for convection are lower in a micropolar fluid than in a Newtonian one, and they observe that the instability boundary is sensitive to changes in the magnetic field. (Siddheshwar and Pranesh, 1999) consider the effect of temperature and gravity modulation on the onset of convection in a micropolar fluid allowing for angular momentum effects and an applied magnetic field. The same writers in (Siddheshwar and Pranesh, 2001) study the combined effects of surface tension, magnetic field, and a non-uniform basic temperature gradient on the convection process in a fluid with suspended particles. The effect of six different non-uniform temperature profiles are analysed. The magnetic field is again found to have a strong effect. (Murty, 2003) also performs a linear instability analysis for the equations for a micropolar fluid with basic temperature profiles which deviate from being linear in the vertical coordinate. This paper also finds that the threshold for instability is reduced in the micropolar fluid as compared with the Newtonian one.

(Hung et al., 1996) and (Cheng et al., 2001) investigate the behaviour of a thin film of micropolar fluid flowing vertically downward. (Hung et al., 1996) study the situation where the film flows down a vertical plate. They find that, depending on parameters, the flow can exhibit supercritical or subcritical instability, and the micropolar parameter κ/μ plays a stabilizing role in the film as it increases. (Cheng et al., 2001) study a similar problem, but when the film flows down the outside of a circular cylinder. Again they find the flow may be subcritical or supercritical. They also find κ/μ stabilizes the film travelling downwards, and thinner cylinders are less stable.

Several papers have employed energy, or integral-like methods, to study analytical properties of micropolar flows. (Ciarletta, 1995) derives uniqueness and a variational theorem, (Chirita, 2001) establishes continuous dependence in both the forward and backward in time problems, while (Lukaszewicz, 2001) establishes results concerning the long time behaviour of two-dimensional flows. In particular, existence and uniqueness are proved, existence is proved of a global attractor, convergence to the stationary solution, continuous depedence on the micro-rotation viscosity is shown, and the global attractor is proved to be stable. (Lukaszewicz, 2003) analyses existence and uniqueness of global solutions, the associated problem of convergence to the stationary solution when the viscosity is large, existence of a global attractor and estimates of its fractal and Hausdorff dimensions, and flows when the spatial region is unbounded. (Iesan, 2002) establishes existence and uniqueness results in a theory of micromorphic continua. Questions of continuous dependence are also studied. (Ciarletta, 2001) derives spatial decay estimates for a heat-conducting micropolar fluid in a cylinder. A generalized nonlinear energy stability analysis for thermal convection in a micropolar fluid when the viscosity depends on temperature is given in (Franchi and Straughan, 1992b), and a different nonlinear energy analysis is given by (Lindsay and Straughan, 1992) for thermal convection in a micropolar fluid when the buoyancy is quadratic in temperature and so penetrative convection is allowed.

A ferromagnetic fluid (see chapter 12) is a colloidal suspension of very small particles which react when a magnetic field is applied. An important recognition of Eringen's theory of micropolar fluids via his spin effect (micro-rotation) is in the modelling of a ferromagnetic fluid. The angular momentum is of especial interest in such a theory. (Zahn and Greer, 1995) and (Zahn and Pioch, 1999) establish some interesting new effects in ferrofluids and the angular momentum equation plays a key role. (Venkatasubramanian and Kaloni, 2002) contains a complete energy stability analysis for general motions of a magnetic fluid. They specifically introduce the equation of angular momentum balance. An interesting piece of work is by (Abraham, 2002) who studies instability in the Bénard problem for a model for a ferromagnetic fluid based on Eringen's micropolar fluid theory. The equations of (Abraham, 2002) are

$$
\begin{aligned}
\rho(v_{i,t} + v_j v_{i,j}) &= -p_{,i} + \alpha\rho_0 g k_i T + (2\zeta + \eta)\Delta v_i \\
&\qquad + \zeta\epsilon_{ijk}\omega_{k,j} + (H_i B_j)_{,j}, \\
v_{i,i} &= 0, \\
\rho_0 I(\omega_{i,t} + v_j\omega_{i,j}) &= \zeta(\epsilon_{ijk}v_{k,j} - 2\omega_i) + \mu_0\epsilon_{ijk}M_j H_k \\
&\qquad + (\lambda' + \eta')\omega_{k,ki} + \eta'\Delta\omega_i, \\
T_{,t} + v_i T_{,i} &= \frac{\delta}{\rho_0 c_v}\epsilon_{ijk}T_{,i}\omega_{k,j} + K_c\Delta T, \\
B_i = \mu_0(H_i + M_i), &\quad B_{i,i} = 0, \quad \epsilon_{imn}H_{n,m} = 0.
\end{aligned}
$$

Here ω_i is the micro-rotation vector while B_i, H_i, M_i are the magnetic induction, magnetic field, and magnetization.

A steady state is found for the Bénard problem in a horizontal layer and a linear instability analysis of this solution is performed. The results of (Abraham, 2002) are very detailed.

9.5 Generalized fluids

There are other practical generalizations of a Newtonian fluid in addition to micropolar fluids. For example, within the class of viscoelastic fluids many stability/instability results are available. Energy stability theory is proving useful in classes of viscoelasticity, or non-Newtonian fluids. (Slemrod, 1978) develops a rigorous energy theory for an integral-history linear viscoelastic convection problem, while (Straughan, 1983), (Franchi and Straughan, 1988) and (Franchi and Straughan, 1993) developed energy stability analyses for thermal convection problems for fluids of second grade, third grade, and generalized second grade, respectively. The generalized second grade model had been applied previously to glacier flow by (Man and Sun, 1987). (Budu, 2002) contains a generalized energy stability analysis for a generalized second grade fluid with temperature-dependent viscosity and she also shows how an unconditional nonlinear energy stability analysis may be established for a third grade fluid with temperature-dependent viscosity. (Massoudi and Phuoc, 2001) also study flow in a generalized second grade fluid with temperature-dependent viscosity, while (Jordan and Puri, 2003) analyse Stokes' first problem for a second grade fluid in a porous half-space. Interesting stability results for Hadley flow in a Maxwell fluid are given by (Kaloni and Lou, 2002b). (Kaloni and Lou, 2002a) develop an instability analysis for the thermal convection problem with an inclined temperature gradient in an Oldroyd-B fluid. They find that the horizontal component of the temperature gradient gives rise to both oscillatory and stationary convection in the form of both transverse and longitudinal solutions. In addition they discover that the elasticity in the Oldroyd-B fluid contributes to a destabilizing effect. (Berezin et al., 1998) develop a nonlinear stability analysis for flow of a power law fluid. (Preziosi and Rionero, 1989) presents a variational calculation for a nonlinear energy stability analysis of a shear flow in a viscoelastic fluid. (Tigoiu, 2000) establishes asymptotic stability of the rest state for a fluid of third grade, while (Hayat et al., 2002) study the effect of a third order fluid on peristaltic transport in a circular cylindrical tube. (Diaz and Quintanilla, 2002) study anti-plane shear deformations in the dynamical theory of viscoelasticity. They establish existence of solutions when the domain is a semi-infinite strip. They also show that the decay rate for end effects is faster than the corresponding rate for the Laplace equation.

For the various classes of polar fluids where the stress is not symmetric, some analyses are available. Nonlinear energy stability for a dipolar fluid heated from below is investigated in (Straughan, 1987). (Budu, 2002) establishes a generalized energy stability theory for a dipolar fluid with a temperature-dependent viscosity. Several interesting results in dipolar fluid flows have been provided by (Jordan and Puri, 1999), (Jordan and Puri, 2002) and by (Puri and Jordan, 1999b; Puri and Jordan, 1999a).

9.5.1 Generalized heat conduction

There has been much recent interest in heat conduction theories which allow heat to travel with a finite wavespeed. The balance of energy equation in these theories in some ways behaves more like a hyperbolic equation than a parabolic one, see e.g. (Hetnarski and Ignaczak, 1999), (Jou et al., 1999), (Quintanilla, 2001), (Ruggeri, 2001), and the many references therein. (Metzler and Compte, 1999) have investigated a connection between one of these, the Cattaneo model, and the stochastic continuous time random walk theory. In the field of thermal convection in a fluid there has also been interest in the situation where the temperature field may be governed by an equation capable of admitting a heat wave. (Puri and Jordan, 1999b; Puri and Jordan, 1999a) adopt such a theory in connection with a dipolar fluid. When the momentum equation is that for a Newtonian fluid, the Bénard problem has been investigated by (Straughan and Franchi, 1984), (Lebon and Cloot, 1984), and by (Franchi and Straughan, 1994). (Franchi and Straughan, 1994) develop a linear instability analysis for the Bénard problem but utilize the Guyer-Krumhansl equations for the temperature and heat flux fields. (Dauby et al., 2002) study instability for the Bénard problem and the thermocapillary Bénard problem with both Maxwell-Cattaneo and Guyer-Krumhansl theories. They find convection thresholds which may be stationary or oscillatory in nature.

While linear instability has been studied for the Guyer-Krumhansl model I have not seen any work using energy stability theory. We, therefore, make some remarks here. The governing equations are (Franchi and Straughan, 1994),

$$\begin{aligned} v_{i,t} + v_j v_{i,j} &= -\frac{1}{\rho}p_{,i} + \alpha g \delta_{i3} T + \nu \Delta v_i, \qquad v_{i,i} = 0, \\ c(T_{,t} + v_i T_{,i}) &= -q_{i,i}, \\ \tau(q_{i,t} + v_j q_{i,j} - \epsilon_{ijk}\omega_j q_k) &= -q_i - \kappa T_{,i} + \hat{\tau}(\Delta q_i + 2q_{k,ki}). \end{aligned} \tag{9.27}$$

Here, τ is a relaxation coefficient, q_i is the heat flux, the $\hat{\tau}$ terms are associated with Guyer-Krumhansl theory, and ω_i is the vorticity ($\boldsymbol{\omega} = \mathrm{curl}\,\mathbf{v}$). The steady solution for prescribed constant temperatures on the horizontal boundaries $z = 0, d$, is

$$\bar{v}_i = 0, \qquad \bar{T} = -\beta z + T_0, \qquad \bar{q}_i = \kappa\beta\delta_{i3}, \tag{9.28}$$

where β is the temperature gradient. Non-dimensional perturbation equations about this steady solution may be derived as

$$\begin{aligned} &u_{i,t} + u_j u_{i,j} = -\pi_{,i} + R\theta k_i + \Delta u_i, \qquad u_{i,i} = 0, \\ &Pr(\theta_{,t} + u_i\theta_{,i}) = HRw - q_{i,i}, \\ &\tau_1(q_{i,t} + u_j q_{i,j} - \epsilon_{ijk}\omega_j q_k) = \tau_2\epsilon_{ij3}\omega_j - \tau_3 q_i - \theta_{,i} \\ &\qquad\qquad + \tau_4(\Delta q_i + 2q_{k,ki}), \end{aligned} \tag{9.29}$$

where (u_i, θ, π, q_i) are non-dimensional perturbations to velocity, temperature, pressure, and heat flux, R^2 is the Rayleigh number, τ_1 - τ_4 are positive parameters, and $H = +1$ if the layer is heated from below whereas $H = -1$ when the layer is heated from above.

The solution (9.28) is obtained with the temperatures prescribed at the upper and lower boundaries. To deal with (9.29) we specify $u_i = 0$, $q_i = 0$ on the boundaries $z = 0, 1$ with a space tiling planform periodicity in the x, y directions. We let V be a period cell for the perturbation. Then, for $\lambda > 0$ a coupling parameter, we may obtain the energy equation

$$\begin{aligned} \frac{d}{dt}\frac{1}{2}(\lambda\|\mathbf{u}\|^2 + Pr\|\theta\|^2 + \tau_1\|\mathbf{q}\|^2) &= (\lambda + H)R(\theta, w) \\ &+ \tau_2\big[(q_1, \omega_2) - (q_2, \omega_1)\big] - \lambda\|\nabla\mathbf{u}\|^2 - \tau_3\|\mathbf{q}\|^2 - 3\tau_4\|\nabla\mathbf{q}\|^2. \end{aligned} \tag{9.30}$$

When $H = +1$ (heated below) we do not have a dissipation term to control θ and it is not clear how to proceed from (9.30). If $H = -1$ (heated above) then we pick $\lambda = 1$, and since $\boldsymbol{\omega} = \operatorname{curl}\mathbf{u}$ we may produce a well defined nonlinear energy stability theory from (9.30). It is worth noting that (Straughan and Franchi, 1984) found convection may occur with heating from above in the Maxwell-Cattaneo theory, albeit for very high Rayleigh numbers.

10

Time dependent basic states

10.1 Convection problems with time dependent basic states

A physically important class of problem involves convection in a layer when the basic temperature is not simply a linear function of z, but where it depends also directly on time. For example, any attempt to grow semiconductor crystals in space will be faced with the problem of a gravity field approximately 10^{-5} times as strong as that on Earth, and hence slight movement of the spacecraft will lead to a basic state that changes with time. Also, any geophysical convection problem driven by solar radiation comes into the category under consideration here since the Sun's heating effect follows either a diurnal cycle or a yearly one, depending on the timescale involved.

The first energy theory on convection problems involving time-dependent basic states would appear to be that of (Homsy, 1973) who made some general observations and obtained numerical results for fluid layers subject to rapid heating or cooling. (Neitzel, 1982) reexamines the numerical work pertinent to the analysis of (Homsy, 1973), obtains further results for other boundary conditions, and obtains some interesting conclusions on the stability limits in time dependent problems. (Caltagirone, 1980) treats the problem of a porous layer saturated with fluid and subject to instantaneous surface heating. His analysis contains a linearized study together with a comparison by the energy method, in addition to a check by a numerical finite difference solution in two-dimensions.

(Homsy, 1974) extends his energy theory to the convection problem where:

(a) the surface temperature varies sinusoidally with time, and

(b) gravity varies sinusoidally with time.

Energy theory for sinusoidally varying convection problems is also studied by (Carmi, 1974), and for the porous convection problem with time-periodic temperature boundary conditions by (Chhuon and Caltagirone, 1979). The latter paper also discusses a linear instability theory, experimental results, and the findings of a two-dimensional finite difference simulation. (McKay, 1992a; McKay, 1992b) studies patterned ground formation taking into account the yearly freeze-thaw cycle (see section 10.3) and (Néel and Nemrouch, 2001) contains some interesting energy stability arguments when the boundary conditions are of pulsating type. Before discussing the paper of (Homsy, 1974) further we should point out that a linearized instability analysis of the gravity modulation problem is given by (Gresho and Sani, 1970). Corresponding linearized analyses for the surface temperature modulation problem are contained in (Rosenblat and Tanaka, 1971) and in (Yih and Li, 1972). These papers utilize a normal mode analysis in the horizontal x, y-directions, then assume an (Galerkin) expansion of the z- part of the perturbation into a finite set of basis functions, thereby reducing the instability problem to one of studying a finite set of coupled ordinary differential equations with time periodic coefficients. Floquet theory (see e.g., (Coddington and Levinson, 1955)) is then used to find a linear instability boundary. We do not describe the linearized analysis, although explicit details are given in the papers of (Gresho and Sani, 1970), (Rosenblat and Tanaka, 1971), and (Yih and Li, 1972). Further work on the time-dependent problem is contained in (Finucane, 1976), (Jhaveri and Homsy, 1982), (May and Bassom, 2000), and (Bardan et al., 2001) where further references may be found including those on nonlinear work employing types of truncation techniques. The paper by (Graham et al., 1992) is a very interesting one which describes a time-periodic instability in a Hele-Shaw cell, the instability having diagonal cell shape.

10.2 Time-varying gravity and surface temperatures

The paper by (Homsy, 1974) is particularly lucid and does develop a very useful theory for nonlinear energy stability for convection problems with time dependent basic states, and so we describe the contents of this work. (Homsy, 1974) considers an infinite layer of fluid bounded by surfaces $z = 0, d$, with surface temperatures

$$T = T_0 + T_s \cos \omega t, \quad z = 0, \qquad T = T_1 + \delta T_s \cos \omega t, \quad z = d.$$

The basic temperature profile $\bar{\theta} = (T - T_1)/(T_0 - T_1)$ then satisfies the diffusion equation

$$\frac{\partial \bar{\theta}}{\partial t} = \frac{\partial^2 \bar{\theta}}{\partial z^2} \tag{10.1}$$

and the boundary conditions

$$\bar{\theta} = 1 + a\cos\omega t, \quad z = 0, \qquad \bar{\theta} = \delta a\cos\omega t, \quad z = 1, \tag{10.2}$$

where $a = T_s/(T_0 - T_1)$, ω in (10.2) being a dimensionless frequency.

The solution $\bar{\theta}$ is found as

$$\bar{\theta} = 1 - z + a\,\mathrm{Re}\left\{ e^{i\omega t} \frac{\delta \sinh \beta z + \sinh[\beta(1-z)]}{\sinh \beta} \right\}, \tag{10.3}$$

where $\beta = (1 + i)\sqrt{\omega/2}$. This solution is the one appropriate to surface modulation, but since gravity modulation effects appear directly in the momentum equation it is equally valid there too.

Perturbations to solution (10.3) satisfy

$$\begin{aligned} Pr^{-1}(u_{i,t} + u_j u_{i,j}) &= -p_{,i} + \Delta u_i + R\theta k_i(1 + \epsilon \sin\omega t), \\ u_{i,i} &= 0, \\ \theta_{,t} + u_i \theta_{,i} &= \Delta\theta - Rw\frac{\partial \bar{\theta}}{\partial z}, \end{aligned} \tag{10.4}$$

where θ is the temperature perturbation, and the $\epsilon \sin\omega t$ term represents the time-dependent variation in the gravity field.

Homsy employs a coupling parameter λ and defines $\phi = \lambda^{1/2}\theta$ to obtain the energy equation

$$\frac{dE}{dt} = RI(t) - \mathcal{D}, \tag{10.5}$$

where

$$E = \frac{1}{2Pr}\|\mathbf{u}\|^2 + \frac{1}{2}\|\phi\|^2, \tag{10.6}$$

$$I(t) = \left\langle w\phi \left(\frac{1 + \epsilon \sin\omega t}{\lambda^{1/2}} - \lambda^{1/2}\frac{\partial \bar{\theta}}{\partial z} \right) \right\rangle, \qquad \mathcal{D} = \|\nabla \mathbf{u}\|^2 + \|\nabla \theta\|^2. \tag{10.7}$$

From this point (Homsy, 1974) is able to proceed interestingly in two ways. The first is to follow Joseph's variational method (chapter 4). Hence, define

$$\frac{1}{\rho_\lambda} = \max_{\mathcal{H}} \frac{I}{\mathcal{D}}. \tag{10.8}$$

The function $\rho_\lambda(t)$ is periodic in t with period $2\pi/\omega$. Define $R_{s,\lambda}$ by

$$R_{s,\lambda} = \min_{t \in [0, 2\pi/\omega)} \rho_\lambda(t).$$

Then, the usual analysis shows $R < R_{s,\lambda}$ guarantees nonlinear stability. Finally, optimize in λ to find a nonlinear stability threshold R_s given by

$$R_s = \max_{\lambda} \min_{t\in[0,2\pi/\omega)} \rho_\lambda(t). \tag{10.9}$$

The stability so obtained is referred to by (Homsy, 1974) as strong global stability. (Homsy, 1974) then argues that for time-periodic problems the condition $R < R_s$, while guaranteeing all disturbances decay rapidly, may be overly conservative. He proposes an alternative form of asymptotic stability based on an idea of (Davis and von Kerczek, 1973), which allows the amplitude of a disturbance to possibly *increase* over part of one cycle, provided the disturbance decays to zero over many cycles, i.e., as $t \to \infty$.

Basically the argument of (Homsy, 1974) is to define

$$\nu_\lambda(t) = \max_{\mathcal{H}} \left(RI - \mathcal{D}E \right), \tag{10.10}$$

and note (provided the maximum in (10.10) exists) that (10.5) yields $dE/dt \le \nu_\lambda(t)E$. This integrates to

$$E(t) \le E(0) \exp \int_0^t \nu_\lambda(s)ds.$$

The basic state is then stable provided

$$\int_0^{2\pi/\omega} \nu_\lambda(s)\, ds < 0. \tag{10.11}$$

The idea is to choose λ such that (10.11) holds for as large a number R as possible: this number is denoted by R_A. Thus the criterion for asymptotic stability is $R < R_A$.

(Homsy, 1974) treats gravity variation and surface temperature variation separately, obtaining interesting numerical results for both cases. We report only the former. For this case choose $a = 0$ and then observe that $\partial\bar{\theta}/\partial z = -1$. The strong (decay for all disturbances) limit R_s arises from

$$\frac{1}{\rho_\lambda} = \max_{\mathcal{H}} \frac{< w\phi >}{\mathcal{D}} \Lambda(t),$$

where $\Lambda(t) = (1 + \lambda + \epsilon \sin\omega t)/\sqrt{\lambda}$. The Euler-Lagrange equations for this are

$$\begin{aligned} &u_{i,i} = 0, \qquad \Delta u_i + \frac{1}{2}\rho_\lambda \phi k_i \Lambda(t) = \pi_{,i}\,, \\ &\Delta\phi + \frac{1}{2}\rho_\lambda\Lambda(t) w = 0. \end{aligned} \tag{10.12}$$

Homsy observes that these are the same as for the steady Bénard problem if R in that case is replaced by $\frac{1}{2}\rho_\lambda\Lambda(t)$. Hence, he concludes $\rho_\lambda\Lambda(t) = 2R_L$, where $R_L^2 \approx 1708$ for rigid boundaries. His global limit is found from

$$R_s = \max_\lambda \min_t \rho_\lambda(t) = \max_\lambda \frac{2R_L\lambda^{1/2}}{1+\epsilon+\lambda} = \frac{R_L}{(1+\epsilon)^{1/2}}\,.$$

Hence, if the Rayleigh number $Ra\,(=R^2)$ satisfies $Ra < R_L^2/(1+\epsilon)$ the layer is strongly stable in that disturbances decay rapidly, monotonically.

To improve this result, however, requires use of the result (10.11). The maximum problem for this is (10.10) which leads to the different Euler-Lagrange equations,

$$\begin{aligned} \nu_\lambda \frac{u_i}{Pr} &= -\pi_{,i} + \Delta u_i + \frac{1}{2}R\Lambda(t)\phi k_i,\\ \nu_\lambda \phi &= \Delta\phi + \frac{1}{2}R\Lambda(t)w, \quad u_{i,i}=0. \end{aligned} \tag{10.13}$$

(Homsy, 1974) reports he was unable to solve (10.13) in general but instead he made significant partial progress with the free-free boundary case as follows. The curlcurl of $(10.13)_2$ is taken and then the third component of this is

$$\nu_\lambda Pr^{-1}\Delta w = \Delta^2 w + \frac{1}{2}R\Lambda(t)\Delta^*\phi. \tag{10.14}$$

The system $(10.13)_3$, (10.14) is solved by taking

$$w = \hat{w}f(x,y)\sin\pi z, \qquad \phi = \hat{\phi}f(x,y)\sin\pi z,$$

where $\Delta^* f + a^2 f = 0$. For non-trivial $\hat{w}, \hat{\phi}$ he reduces $(10.13)_3$, (10.14) to

$$\nu_\lambda = \frac{1}{2}h\Big\{-(1+Pr) + \Big[(1-Pr)^2 + Pr\Big(\frac{Ra}{R_L^2}\Big)\Lambda^2\Big]^{1/2}\Big\}, \tag{10.15}$$

where $h=\pi^2+a^2$, $R_L^2 = h^3/a^2$. He again reports that it was not possible to proceed in general and apply condition (10.11), indicating the difficulty involved with time dependent basic states. Instead, he investigates the cases $Pr=1$, $Pr\to 0$, and $Pr\to\infty$. For $Pr=1$, he finds $\nu_\lambda/h = -1 + (R/R_L)[\{1+\epsilon\sin\omega t+\lambda\}/2\sqrt{\lambda}]$. Condition (10.11) is then satisfied whenever $R < (2\lambda^{1/2}/[1+\lambda])\,R_L$. By selecting $\lambda=1$ he then finds $R_A = R_L$; this is undoubtedly due to the fact that for $Pr=1$, (10.13) is a symmetric linear system.

For $Pr\to\infty$ (Homsy, 1974) finds $\nu_\lambda = (\alpha^2/h^2)\,(Ra\Lambda^2/4 - R_L^2)$, condition (10.11) yields $R^2 \le 4\lambda R_L^2/[(1+\lambda)^2 + \frac{1}{2}\epsilon^2]$, and $\lambda^2 = 1+\frac{1}{2}\epsilon^2$ maximizes, giving the asymptotic stability limit

$$R_A = \frac{2R_L}{1+(1+\frac{1}{2}\epsilon^2)^{1/2}}, \qquad Pr\to\infty. \tag{10.16}$$

By a similar argument he derives the same R_A bound as (10.16), for $Pr\to 0$.

(Homsy, 1974) observes that $R_A(Pr) = R_A(Pr^{-1})$ and $R_s \leq R_A \leq R_L$. Homsy's paper is an important one and deserves full recognition in the nonlinear energy stability literature.

Another useful contribution to energy theory in time-dependent convection is by (Gumerman and Homsy, 1975) who treat the problem of a layer of fluid impulsively cooled at its free surface, with surface tension effects taken into account. This thereby combines the analysis of (Homsy, 1973) and that of (Davis, 1969a), the surface tension theory described in chapter 8. In (Gumerman and Homsy, 1975) they also obtain estimates on the *onset time* for the instability to commence. The physical motivation for the paper of (Gumerman and Homsy, 1975) is to model evaporating liquids subject to convective instabilities induced by surface tension. In order for the surface effects to dominate the buoyancy effect, they restrict attention to thin layers, 3 mm or less. The evaporation is accounted for by a condition of constant heat flux from the upper surface. The lower surface condition is an isothermal one, thereby allowing a comparison of stability results for large times with those of (Davis, 1969a), see chapter 8. The essential difference with that of chapter 8 is that a quiescent layer of fluid, depth d, rests on a horizontal, constant temperature plate with a free upper surface. At time $t = 0$, the layer is impulsively cooled by a constant outwardly directed heat flux and this results in a *basic* (dimensionless) temperature profile of form

$$\bar{T}(z,t) = T_0 - z + 2\sum_{m=0}^{\infty} \frac{(-1)^m}{M^2} \exp\left(-M^2 t\right) \sin\left(Mz\right), \tag{10.17}$$

where $M(m) = (2m+1)\pi/2$ and T_0 is the initial (dimensionless) temperature of the layer. The nonlinear energy stability study is then akin to that at the beginning of chapter 8, *but with the basic temperature given by* (10.17). Of course, this changes the analysis considerably since the production term in the energy equation in section 8.1 now involves a time-dependent basic state. The analysis again employs Homsy's time dependent version of energy theory and the Davis-von Kerczek theory. By comparing their theoretical results with experimental measurements for propyl alcohol, methyl alcohol and acetone, they conclude that the experimental results are in broad agreement with the theoretical ones.

10.3 Patterned ground formation with time dependent surface heating

(McKay, 1992a; McKay, 1992b) studies the problem of patterned ground formation as described in section 7.1 but when the effect of freeze/thawing in the active layer is taken into account due to the surface heat flux varying over an annual cycle. He, therefore, effectively considers the model of (George et al., 1989) (see section 7.2) but where the heat flux at the ground surface is a periodic function of time. Thus, the equations of (McKay, 1992a; McKay, 1992b) are Darcy's law with a quadratic buoyancy force,

$$\begin{aligned}
&\rho_0\hat{\phi}^{-1}\frac{\partial v_i}{\partial t} = -p_{,i} - \rho_0 g k_i\big[1-\alpha(T-4)^2\big] - \frac{\mu}{k'}v_i,\\
&v_{i,i}=0,\\
&\frac{\partial T}{\partial t}+v_iT_{,i}=\kappa\Delta T,
\end{aligned} \tag{10.18}$$

where the permeability is given by $k'(z) = k'_0(1+\gamma z)$. The boundary conditions are

$$\begin{aligned}
&k_iv_i=0 \quad \text{on } z=0,d, \qquad T=0^\circ\text{C} \quad \text{on } z=0,\\
&\frac{\partial T}{\partial z}=\delta(1+\epsilon\cos\omega' t) \quad \text{on } z=d.
\end{aligned} \tag{10.19}$$

(McKay, 1992a; McKay, 1992b) finds the basic solution to (10.18), (10.19) as

$$\bar{v}_i=0, \qquad \bar{T}=\delta z+\delta\epsilon\,\text{Re}\left\{\frac{e^{i\omega' t}\sinh\beta z}{\beta\cosh\beta d}\right\}$$

where $\beta = \sqrt{i\omega'/\kappa}$. The non-dimensional perturbation equations and boundary conditions of (McKay, 1992a; McKay, 1992b) are

$$\begin{aligned}
A\frac{\partial u_i}{\partial t} &= -p_{,i}+2Rk_i\theta\left[z-\xi+\epsilon\,\text{Re}\left(\frac{e^{i\omega t}\sinh\beta dz}{\beta d\cosh\beta d}\right)\right]\\
&\qquad +\theta^2k_i-\frac{u_i}{(1+\gamma z)},\\
u_{i,i}&=0,\\
\frac{\partial\theta}{\partial t}+u_i\theta_{,i}&=-Rw\left[1+\epsilon\,\text{Re}\Big(\frac{e^{i\omega t}\cosh\beta dz}{\cosh\beta d}\Big)\right]+\Delta\theta,
\end{aligned} \tag{10.20}$$

where ω is a non-dimensional frequency, $\xi = 4/T_1$, $T_1=\delta d$, and

$$\begin{aligned}
&w=0, \quad \text{on } z=0,1,\\
&\theta=0, \quad \text{on } z=0, \qquad \frac{\partial\theta}{\partial z}=0, \quad \text{on } z=1.
\end{aligned} \tag{10.21}$$

(McKay, 1992a; McKay, 1992b) develops a complete linear instability analysis which involves Floquet theory. He also develops a nonlinear energy stability analysis along the lines of Homsy's method outlined in section 10.2. However, he also includes a weighted energy in his analysis. This is a very substantial piece of work since the numerical calculations are involved due to the coefficients in the equations depending on z and t. Extensive numerical results are presented in (McKay, 1992a; McKay, 1992b) and his analysis represents a useful model for patterned ground formation with time periodic surface heating.

(McKay, 1998b) and (McKay, 2000) studies linear instability in a saturated porous layer with a temperature field and a salt field present. The upper surface boundary condition on the temperature field is a time periodic one. Detailed Floquet analyses are presented.

11

Electrohydrodynamic and magnetohydrodynamic convection

11.1 The MHD Bénard problem and symmetry

As (Rosensweig, 1985) points out, the interaction of electromagnetic fields and fluids has been attracting increasing attention due to applications in many diverse areas. He writes that the subject may be divided into three main categories:

1. *Electrohydrodynamics* (EHD), the branch of fluid mechanics concerned with electric force effects;

2. *Magnetohydrodynamics* (MHD), the study of the interaction between magnetic fields and fluid conductors of electricity; and

3. *Ferrohydrodynamics* (FHD), which deals with the mechanics of fluid motion induced by strong forces of magnetic polarization.

Topics 1 and 3 are relatively new and are attracting increasing attention in the theoretical and engineering literature. (Rosensweig, 1985) pp. 1,2, explains succinctly the differences between the above topics. Effectively, in MHD the body force acting on the fluid is the Lorentz force. This is due to an electric current flowing at an angle to the direction of the magnetic field. In FHD there need not be electric current flowing in the fluid. The body force in this case is due to polarisation force. This requires material magnetisation and a magnetic field gradient or discontinuity. The force in EHD is usually due to free electric charge being influenced by an electric force field.

In this book we concentrate on convection-like problems in EHD and FHD, in chapters 11 and 12, respectively, although we do begin the current chapter with one result in MHD. The magnetohydrodynamic convection problem is a very important one in so much that it has intrinsic applications to the behaviour of planetary and stellar interiors, and in particular, to the behaviour inside of the Earth.

The first use of the energy method in magnetohydrodynamics, along the lines reviewed here was by (Rionero, 1967a; Rionero, 1967b; Rionero, 1968a; Rionero, 1968b; Rionero, 1970; Rionero, 1971). His contributions are fundamental, being the first to establish existence of a maximising solution in the energy variational problem. In (Rionero, 1971) he also included the Hall effect. Further recent work dealing with anisotropic MHD effects may be found in (Maiellaro and Labianca, 2002), (Rionero and Maiellaro, 1995). (Zikanov and Thess, 1998) study magnetic turbulence numerically, a situation which can have anisotropic effects induced. Linear theory for the classical MHD Bénard problem, (Chandrasekhar, 1981), shows that as the field strength is increased for the MHD convection problem with the magnetic field perpendicular to the layer, the magnetic field has a strongly inhibiting effect on the onset of convective motion. The first analysis to confirm this stabilizing effect from a nonlinear energy point of view is due to (Galdi and Straughan, 1985a). This work derives a fully unconditional nonlinear energy stability result which is optimal in that the linear instability boundary coincides with the nonlinear energy one. We now include an exposition of this material.

The situation studied by (Galdi and Straughan, 1985a) employs an approximation to the full MHD Bénard problem. They begin with Maxwell's equations (see (12.1) - (12.5)) in the form

$$\begin{aligned} &\mathrm{curl}\,\mathbf{H} = \mathbf{j}, \quad \mathrm{curl}\,\mathbf{E} = -\frac{\partial \mathbf{B}}{\partial t}, \\ &\mathrm{div}\,\mathbf{B} = 0, \quad \mathrm{div}\,\mathbf{j} = 0, \end{aligned} \tag{11.1}$$

where $\mathbf{H}$, $\mathbf{E}$ are the magnetic and electric fields, $\mathbf{j}$ is the current and $\mathbf{B} = \mu\mathbf{H}$ is the magnetic induction field. The electric field is supposed derivable from a potential ϕ so that $E_i = -\phi_{,i}$. Then, from $(11.1)_2$ it follows that $\partial B_i/\partial t = 0$ and so B_i is a function of the spatial variables x_j only. They take $\mathbf{B} = B_0\mathbf{k}$ where $\mathbf{k} = (0,0,1)$, the convection layer being $\mathbb{R}^2 \times \{z \in (0,d)\}$. The current is given by $\mathbf{j} = \sigma(-\mathbf{E} + \mathbf{v} \times \mathbf{B})$ where σ is the electrical conductivity and $\mathbf{v}$ is the fluid velocity, and then $(11.1)_4$ shows that

$$\sigma(-\Delta\phi + \mathbf{B}\cdot\nabla\times\mathbf{v} - \mathbf{v}\cdot\nabla\times\mathbf{B}) = 0. \tag{11.2}$$

It is now assumed that $\omega_3 = 0$ where $\boldsymbol{\omega}$ is the vorticity, $\boldsymbol{\omega} = \mathrm{curl}\,\mathbf{v}$. Then since $\mathbf{B} = B_0\mathbf{k}$ the term $\mathbf{B}\cdot\nabla\times\mathbf{v}$ in (11.2) vanishes. From equation $(11.1)_1$ and the expression for $\mathbf{j}$,

$$\mathrm{curl}\,\mathbf{B} = \frac{1}{\eta}(-\nabla\phi + \mathbf{v}\times\mathbf{B}), \tag{11.3}$$

where $\eta = 1/\mu\sigma$. We let $\eta \to \infty$ (which is equivalent to letting the magnetic Prandtl number $P_m = \nu/\eta \to 0$) and then from (11.3) we see that curl $\mathbf{B} = \mathbf{0}$. Therefore, the term $\mathbf{v} \cdot \nabla \times \mathbf{B} = 0$ in (11.2) and that equation reduces to $\Delta\phi = 0$. Since we suppose ϕ decays sufficiently rapidly at infinity we must have $\phi \equiv 0$ and $\mathbf{j} = \sigma(\mathbf{v} \times \mathbf{B})$ with $\mathbf{B} = B_0(z)\mathbf{k}$. (Galdi and Straughan, 1985a) then restrict attention to the case B_0 =constant. The Lorentz force, which is where the magnetic field enters the momentum equation, is then $\mathbf{j} \times \mathbf{B} = \sigma(\mathbf{v} \times B_0\mathbf{k}) \times B_0\mathbf{k}$.

The upshot of this approximation is that the non-dimensional perturbation equations for the magnetic Bénard problem may be written as

$$\begin{aligned} u_{i,t} + u_j u_{i,j} &= -\pi_{,i} + \Delta u_i + R\theta k_i - M^2\big[\mathbf{k} \times (\mathbf{u} \times \mathbf{k})\big]_i, \\ u_{i,i} = 0, \quad Pr(\theta_{,t} + u_i\theta_{,i}) &= Rw + \Delta\theta, \end{aligned} \qquad (11.4)$$

where u_i, θ are velocity and temperature perturbations, R^2 and M are the Rayleigh number and the Hartmann number (measuring the strength of the magnetic field). One may show that the linear operator associated to (11.4) is symmetric. To do this note that $[\mathbf{k} \times (\mathbf{u} \times \mathbf{k})]_i \equiv \delta_{33}u_i - u_3\delta_{i3}$ and then $\mathbf{v} \cdot [\mathbf{k} \times (\mathbf{u} \times \mathbf{k})] \equiv \delta_{33}u_i v_i - u_3 v_i \delta_{i3} = \delta_{33}u_i v_i - v_3\delta_{i3}u_i = \mathbf{u} \cdot [\mathbf{k} \times (\mathbf{v} \times \mathbf{k})]$. The conditions of the theory of section 4.3 are satisfied and we may conclude that the linear instability threshold is the same as the nonlinear energy stability one. Since the stability is unconditional the behaviour of linear theory also governs the behaviour of nonlinear theory. The linear theory shows the Rayleigh number increases with increasing M, (Chandrasekhar, 1981), e.g. for two stress free surfaces (Chandrasekhar, 1981) shows

$$R^2 = \frac{(\pi^2 + a^2)}{a^2}\left[(\pi^2 + a^2)^2 + \pi^2 M^2\right].$$

Hence, in the approximation of this section we have derived the stabilizing effect of a vertical magnetic field by using nonlinear energy stability theory.

11.2 Thermo-convective electrohydrodynamic instability

It would appear that studies of convection-like instabilities in insulating fluid layers subject to temperature gradients and electrical potential differences across the layer, first gained impetus in the 1960s. It is not the object of this book to present a critical review of the early literature on this topic. Instead we present findings of work most relevant to this book, commencing with that of (Roberts, 1969) and (Turnbull, 1968a; Turnbull, 1968b). (Roberts, 1969) reports on experiments of (Gross, 1967) in which a layer of insulating oil is confined between horizontal conducting planes and is heated from above and cooled from below. Despite the fact that this situation is gravitationally stable from a thermal point of view, Gross observed that when a vertical electric field of sufficient strength is applied across the layer, a tesselated pattern of motions is observed, in a manner similar to that of standard Bénard convection. Gross suggested that this phenomenon may be due to variation of the dielectric constant, ϵ, of the fluid with temperature.

(Roberts, 1969) performed a theoretical analysis to investigate Gross' suggestion and also investigated an alternative model that allows for the convection to be due to free charge conducted in the layer.

11.2.1 The investigations of Roberts and of Turnbull

(Roberts, 1969) first allows the dielectric constant of the fluid to vary with temperature. He considers a homogeneous insulating fluid at rest in a layer with vertical, parallel applied gradients of temperature, T, and electrostatic potential, V. He assumes the layer depth is d, and denotes by $\mathbf{D}$ the electric displacement. The body force, per unit volume, on an isotropic dielectric fluid is then given by (Landau et al., 1984), eq. (15.15)

$$\mathbf{f} = -\mathrm{grad}\, p + \frac{\rho}{8\pi}\mathrm{grad}\left\{E^2\left(\frac{\partial\epsilon}{\partial\rho}\right)_T\right\} - \frac{E^2}{8\pi}\left(\frac{\partial\epsilon}{\partial T}\right)_\rho \mathrm{grad}\, T, \tag{11.5}$$

where $p, \mathbf{E}$, and ρ denote pressure, electrical field, and density. The appropriate Maxwell equations are

$$\mathrm{div}\,\mathbf{D} = 0, \qquad \mathrm{curl}\,\mathbf{E} = 0, \tag{11.6}$$

with $\mathbf{D}$ given for an isotropic material by $\mathbf{D} = \epsilon\mathbf{E}$. Since the curl of $\mathbf{E}$ vanishes an electric potential V exists such that $\mathbf{E} = -\mathrm{grad}\, V$. (Roberts, 1969) examines a constant density, ρ_0, Newtonian fluid (apart from the thermal buoyancy term) for which the momentum and continuity equations are

$$u_{i,t} + u_j u_{i,j} = g_i + \nu\Delta u_i, \qquad u_{i,i} = 0, \tag{11.7}$$

where $\mathbf{u}$ is velocity, ν viscosity, and for a linear thermal body force,

$$g_i = -\omega_{,i} - \frac{E^2}{8\pi\rho_0}\left(\frac{\partial\epsilon}{\partial T}\right)_\rho T_{,i} - g\big[1 - \alpha(T - T_0)\big]k_i, \tag{11.8}$$

g being gravity, α thermal expansion coefficient, and $\mathbf{k} = (0, 0, 1)$, with $\omega = p/\rho_0 - (E^2/8\pi)\,(\partial\epsilon/\partial\rho)_T$. The energy equation governing the temperature field is chosen as

$$T_{,t} + u_i T_{,i} = \kappa\Delta T, \tag{11.9}$$

κ being thermal diffusivity.

The geometry of (Roberts, 1969) consists of the fluid occupying the region between the planes $z = \pm\frac{1}{2}d$, which are maintained at uniform, but different temperatures $T = \pm\frac{1}{2}\beta d$, for β constant. A uniform electric field is applied in the z–direction. The equilibrium solution is denoted by an overbar and then

$$\bar{\mathbf{u}} = \mathbf{0}, \qquad \bar{T} = \beta z, \qquad \bar{\epsilon}\bar{E}_3 = \epsilon_m E_m \,(\text{constant}), \tag{11.10}$$

or $\bar{E} \equiv \bar{E}_3 = -d\bar{V}/dz = \epsilon_m E_m/\epsilon(\bar{T}) = \epsilon_m E_m/\epsilon(\beta z)$, with $\bar{\omega}$ being determined from $(11.7)_1$. Denoting now the *perturbations* to $\bar{V}, \bar{\mathbf{E}}, \bar{\epsilon}, \bar{\omega}$, and $\bar{T}$ by $V', \mathbf{E}', \epsilon', \omega$, and θ, the *linearized* equations for instability are found to be

$$\begin{aligned}
\epsilon' &= \left(\frac{\partial \epsilon}{\partial T}\right)_{\rho_0} \theta, \\
\Delta V' &= \bar{E}\frac{\partial}{\partial z}\left(\frac{\epsilon'}{\bar{\epsilon}}\right) - \frac{1}{\bar{\epsilon}}\frac{d\bar{\epsilon}}{dz}\frac{\partial V'}{\partial z}, \\
u_{i,t} &= -\pi_{,i} + \nu\Delta u_i - B\delta_{i3} + \alpha g \delta_{i3}\theta, \\
\theta_{,t} &= -\beta w + \kappa\Delta\theta,
\end{aligned}$$

where $w = u_3$, $\pi = \omega + \epsilon'\bar{E}^2/8\pi$, $B = (\bar{E}/4\pi\rho_0)\,(d\bar{\epsilon}/dz)\,[(\epsilon'\bar{E}/\bar{\epsilon}) - V'_{,z}]$. At this point (Roberts, 1969) assumes

$$\epsilon = \epsilon_m[1 + \eta T], \tag{11.11}$$

η constant, with ϵ_m being the value of the dielectric constant at a reference temperature $T_m = 0$ deg. His Boussinesq approximation discards any term involving ηT when a similar term occurs not containing that factor.

The linearized equation for θ is then derived by (Roberts, 1969) to be

$$\Big(\frac{\partial}{\partial t} - \Delta\Big)\Big(Pr\frac{\partial}{\partial t} - \Delta\Big)\Delta^2\theta = L\Delta^{*2}\theta + Ra\Delta\Delta^*\theta, \tag{11.12}$$

where $Pr = \nu/\kappa$ is the Prandtl number and

$$Ra = \frac{\alpha\beta d^4 g}{\kappa\nu}, \qquad L = \frac{\epsilon_m E_m^2}{4\pi\rho_0}\,\frac{\beta^2\eta^2 d^4}{\kappa\nu},$$

are, respectively, the Rayleigh number and a parameter effectively measuring the potential difference between the planes. Upon seeking a normal mode representation $\theta = F(z)e^{i(lx+my)+st}$, with $D = d/dz$ and $a^2 = l^2 + m^2$ denoting the square of the wavenumber, equation (11.12) becomes

$$(\mathcal{D}^2 - s)(\mathcal{D}^2 - Pr\, s)\mathcal{D}^4 F = La^4 F - Ra\, a^2\mathcal{D}^2 F \tag{11.13}$$

where $\mathcal{D}^2 = D^2 - a^2$. For fixed surfaces the boundary conditions are

$$W = DW = 0, \qquad F = 0, \tag{11.14}$$

where $W(z)$ is the z–part of w in its normal mode representation. To interpret these as conditions on F we observe that the electric field continues outside the layer $z \in (-\frac{1}{2}, \frac{1}{2})$ and so the potential satisfies Laplace's equation there. For example, in $z > \frac{1}{2}$, if $\hat{V}'$ denotes the electric field perturbation, $\hat{V}' = G(z)\exp\left[i(lx + my) + st\right]$, then

$$(D^2 - a^2)G = 0,$$

and for $G \to 0$ as $z \to \infty$, necessarily

$$G = Ae^{-az}, \tag{11.15}$$

for some constant A. If $k = \epsilon_m/\hat{\epsilon}$, $\hat{\epsilon}$ being the dielectric constant of the solid in $z > \frac{1}{2}$, then on $z = \frac{1}{2}$,

$$\hat{V}' = V' \quad \text{and} \quad \frac{\partial \hat{V}'}{\partial z} = k\frac{\partial V'}{\partial z}.$$

From (11.15) this leads to

$$aG + kDG = 0, \quad \text{on} \quad z = \frac{1}{2}. \tag{11.16}$$

For $z < \frac{1}{2}$ composed of the same material,

$$aG - kDG = 0, \quad \text{on} \quad z = -\frac{1}{2}. \tag{11.17}$$

Roberts then shows conditions (11.14), (11.16), and (11.17) convert to the following conditions on F, on the planes $z = \pm 1/2$,

$$\begin{aligned} &F = D^2F = D(\mathcal{D}^2 - Pr\, s)F = 0, \\ &\big[\mathcal{D}^2(\mathcal{D}^2 - s)(\mathcal{D}^2 - Pr\, s) + Ra\, a^2\big](DF \pm kaF) = 0. \end{aligned} \tag{11.18}$$

Thus, the eighth-order equation (11.13) is to be solved subject to the eight boundary conditions (11.18). (Roberts, 1969) investigates only stationary convection and finds $\min_{a^2} Ra(a^2)$, for fixed L, numerically.

(Roberts, 1969) discovers that as Ra varies from -1000 to 1707.762, L decreases from 3370.077 to 0. It must be remembered that L contains β^2 and so care must be exercised with a direct comparison of critical Ra against voltage difference, the quantities measured experimentally by (Turnbull, 1968b).

(Roberts, 1969) concludes his investigation of the above model by including physical values for the various parameters that arise. For a temperature difference of 50°C across a gap of 1mm he notes that (Gross, 1967) observes a tesselated pattern for what corresponds to a value of $L \approx 2 \times 10^{-4}$. According to his theory Ra_{crit} should be near 1708 whereas Gross deliberately enforced Ra to be negative. Roberts concluded that the instability mechanism of Gross' experiments was not that of the foregoing model.

The second model studied by (Roberts, 1969) was also essentially the one investigated independently by (Turnbull, 1968a), and is the one that has effectively been used since, albeit with modification. This model assumes the variation in the dielectric constant is not important but the fluid is weakly conducting and the conductivity varies with temperature. (Turnbull, 1968b) is motivated by his experimental results, indeed on p. 2601 his graphs of conductivity, σ, against temperature indicate that for corn oil the conductivity increases about 7.5 times in a temperature change from 20°C to 50°C. For his other working fluid, castor oil, the change over the same temperature interval would appear to be four-fold. (Turnbull, 1968a), p. 2588 remarks that in liquids the density varies about 0.1% per °C, with a similar variation in the dielectric constant. The electrical conductivity exhibits a much greater variation with temperature. For the working fluids of (Turnbull, 1968b), namely corn oil and castor oil, the conductivity variation is approximately 5% per °C.

While the second model of (Roberts, 1969) and that of (Turnbull, 1968a) are essentially the same, Turnbull allows a linear variation of viscosity with temperature and a quadratic variation of electrical conductivity. Mathematically, however, there is a difference in that Roberts concentrates on stationary convection whereas (Turnbull, 1968a) does allow for oscillatory convection although he does not solve the eigenvalue problem correctly. He instead argues (p. 2592) that a Fourier sine series solution should be adequate since, ...*Ohm's law is only an approximation for poorly conducting liquids, and, therefore, it makes no sense to solve the equations exactly since the model is only an approximation.*

Graphs of the *experimental* results of (Turnbull, 1968b) on the onset of convection show that increasing the upper temperature allows a smaller voltage difference to trigger instability.

For completeness we now include a brief description of Roberts' analysis. Ohm's law and the equation of charge conservation are

$$\mathbf{j} = \sigma\mathbf{E} + Q\mathbf{u}, \tag{11.19}$$

$$\frac{\partial Q}{\partial t} + \operatorname{div}\mathbf{j} = 0, \tag{11.20}$$

where $\mathbf{j}$ is current and Q the volume space charge.

Maxwell's equations give

$$4\pi Q = \operatorname{div}\mathbf{D} = \epsilon\operatorname{div}\mathbf{E} = -\epsilon\Delta V.$$

The body force term in the momentum equation is now

$$\mathbf{g} = -\operatorname{grad}\frac{p}{\rho} + \frac{Q\mathbf{E}}{\rho},$$

together with a thermal buoyancy term, if required.

The equilibrium state has

$$\bar{E} = -\frac{d\bar{V}}{dz} = \frac{\sigma_m E_m}{\sigma(\beta z)}.$$

With a linear conductivity law of form $\sigma = \sigma_m\{1 + \bar{\eta}(T - T_m)\}$ (Roberts, 1969) develops a Boussinesq approximation and seeking instability due to stationary convection he arrives at the eigenvalue equation

$$(D^2 - a^2)^3 F = -Ra\, a^2 F - Ma^2 DF, \tag{11.21}$$

where M is a dimensionless measure of the variation of electrical conductivity with temperature, defined by $M = (\epsilon E_m^2/4\pi\rho_0)\,(\beta\bar{\eta}d^3/\nu\kappa)$. For rigid boundaries at constant temperatures the boundary conditions for (11.21) are

$$F = D^2F = D(D^2 - a^2)F = 0 \qquad \text{on} \qquad z = \pm\frac{1}{2}. \tag{11.22}$$

(Roberts, 1969) solves (11.21), (11.22) numerically, seeking the minimum over a^2. In this problem he finds that as M is increased from 0 to 1000 the Rayleigh number increases from 1707.762 to 2065.034.

11.3 Further stability analyses in thermo-convective EHD

Subsequent work would appear to have followed the second approach described above, focusing on conductivity variation with temperature and ramifications of equations (11.19), (11.20). In particular, (Takashima and Aldridge, 1976) take ϵ linear in T, σ quadratic in T and study stationary convection. (Bradley, 1978) treats a linear conductivity but resolves the discrepancy between (Roberts, 1969) stationary convection results and the oscillatory convection ones of (Turnbull, 1968a) by delimiting parameter ranges where overstable convection will be preferred, taking into account the correct boundary conditions. Useful linear instability analyses in EHD are given by (Takashima, 1976; Takashima and Ghosh, 1979; Takashima, 1980; Takashima and Hamabata, 1984) who study the additional effects of rotation, viscoelasticity, flow between coaxial cylinders, and a horizontal ac electric field, respectively. (Lietuaud and Néel, 2001) study the stability of a free shear layer with a one-directional velocity field and a superimposed electric current. They find the unconditional stability limit for the shear flow such that monotonic decay is found provided the electric current is below the threshold. (Abo-Eldahab and El Gendy, 2000) investigate free convection heat transfer in a sheet of electrically conducting fluid.

Another connected approach had been followed by P. Atten and his coworkers from the late 1960s. References to this body of work are given by (Worraker and Richardson, 1979) who point out these writers concentrated on mobility models of charge transport. In referring to the work of (Turnbull, 1968a), (Roberts, 1969), (Takashima and Aldridge, 1976), and (Bradley, 1978), (Worraker and Richardson, 1979) question the underlying mechanisms responsible for the observed fluid motions. They argue that impurities will have a profound effect on instabilities and consequently modify the model.

The equations of (Worraker and Richardson, 1979) are still (11.7) - (11.9) and (11.20). However, their constitutive theory takes

$$\mathbf{j} = Q\mathbf{u} + QK\mathbf{E}, \tag{11.23}$$

where K is a "charge carrier mobility," which they assume depends on temperature,

$$K = K_0\big[1 + k_1(T - T_0)\big]. \tag{11.24}$$

They also allow a linear variation of ϵ with temperature. Their linear analysis, which concentrates on stationary convection, is carefully carried out and they conclude, p. 43, ... *Perhaps the most striking feature of the above analysis is that it demonstrates the importance of the sign of the temperature gradient in relation to the emitting electrode. It suggests that a system with an emitter cooler than the collector is more susceptible to stationary instability than one with the opposite temperature gradient.* (Castellanos et al., 1984b; Castellanos et al., 1984a) continue in this vein analysing a variety of effects with temperature dependent mobility models, investigating both stationary and oscillatory convection.

(Martin and Richardson, 1984) return to the temperature dependent conductivity model taking an Arrhenius exponential dependence on temperature in their linear stability analysis. They conclude that a convective instability cannot be predicted by a simple conductivity model but the true picture is probably one with the instability driven by charge injection (mobility model), but modified by residual conduction.

(Rodriguez-Luis et al., 1986) further continued the above studies allowing charge injection strengths of any finite magnitude. They draw specific attention to the fact that (p. 2115), ... *it is possible to consider two well differentiated mechanisms of charge generation: (i) that resulting in the bulk of the liquid and induced by conductivity gradients and (ii) the injection of charge from one (unipolar) or both (bipolar) electrodes.*

The linear instability picture is now fairly complete for the conductivity and mobility models, covered comprehensively in the above cited works and references therein. Nonlinear studies, on the other hand and in particular those for unconditional stability, seem in relative infancy. We do report on some findings of nonlinear energy theory applied to EHD.

11.3.1 Charge injection induced instability

We describe a very interesting contribution by (Deo and Richardson, 1983) who apply a generalized energy analysis to the problem of charge injection induced instability. They study a mobility type model, but under isothermal conditions. However, in the context of this book it is important in that it is the first study of energy stability in convective electrohydrodynamics.

The object is to explain the effects of a d.c. electric field on a plane layer of dielectric liquid, and they observe that determining the onset of fluid motion and the consequent augmentation of charge transfer has led to a series of carefully controlled transient and steady-state electrochemical experiments and the application of theoretical instability analyses. By using ion-exchange membranes and electrodialytic varnishes on plane-parallel electrodes, it has been possible to investigate the consequences of almost space-charge-limited unipolar and bipolar injection into highly purified dielectric liquids and incorporate the modifying effects of residual conductivity. (Deo and Richardson, 1983) describe several experimental and linear instability results by P. Atten and his co-workers, working liquids being pyralenes, nitrobenzene, and propylene carbonate; they write p. 132, ... *because of the high degree of purity of the experimental liquids, the predictions of a diffusionless linear instability analysis that incorporates residual conductivity effects ... still do not appear to give a satisfactory explanation of the discrepancy between theory and experiment. All the theoretical estimates to date have arisen from instability analyses and have consistently exceeded the experimental values.* Presumably with this as a motivation, they chose to develop a generalized energy theory.

The precise problem studied by (Deo and Richardson, 1983) considers an incompressible dielectric liquid of constant density, ρ, electrical permittivity ϵ, and kinematic viscosity ν confined between two rigid planar perfectly conducting electrodes of infinite extent and distance d apart. They suppose an autonomous injection of unipolar charge is emitted from the electrode at $z = 0$, which is maintained at a potential Φ_0, and the collecting electrode at $z = d$ is maintained at zero potential. The governing electrodynamic field equations, neglecting magnetic effects, are

$$E_i = -\Phi_{,i}, \qquad D_{i,i} = Q, \qquad j_{i,i} = -\frac{\partial Q}{\partial t}, \tag{11.25}$$

where Φ, E_i, D_i, j_i, and Q are electrical potential, electric field, electric displacement, current density, and space-charge density, respectively. The liquid is assumed to be a linear isotropic dielectric, charge is assumed to be transported by convection, migration and diffusion, and then the additional electrical constitutive equations adopted are

$$D_i = \epsilon E_i, \qquad j_i = Qu_i + KQE_i - D_c Q_{,i}, \tag{11.26}$$

where u_i is liquid velocity, D_c is the charge diffusion coefficient, and K is an ion mobility coefficient. The remaining equations are balance of mass and of linear momentum. Taking the electrical force to be Coulombic and the liquid to be an incompressible linear viscous one, these are

$$u_{i,i} = 0, \qquad u_{i,t} + u_j u_{i,j} = -\rho^{-1} p_{,i} + \nu \Delta u_i + \rho^{-1} QE_i, \tag{11.27}$$

where p is pressure.

(Deo and Richardson, 1983) next introduce the non-dimensional parameters $T = \epsilon\Phi_0/\rho\nu K$, $C = Q_0 d^2/\epsilon\Phi_0$, $M = (1/K)\,(\epsilon/\rho)^{1/2}$, $S = \epsilon D_c/\rho\nu K^2$. Here T represents the ratio of the Coulombic forces to viscous forces, C is a measure of the injected charge, M is the ratio of hydrodynamic mobility to ionic mobility, and S, which arises because charge-diffusion effects have been retained, is typically in the range 0.04–0.4. With this non-dimensionalization the governing equations admit a steady one-dimensional hydrostatic equilibrium solution with the equilibrium electric potential $\Phi_e(z)$ solving the equation

$$\frac{S}{T}\frac{d^4\Phi_e}{dz^4}+\frac{1}{2}\frac{d^2}{dz^2}\Big(\frac{d\Phi_e}{dz}\Big)^2=0, \tag{11.28}$$

subject to the boundary conditions,

$$\Phi_e(0)=1,\quad \Phi_e(1)=0,\quad \frac{d^2}{dz^2}\Phi_e(0)=-C,\quad \frac{d^2}{dz^2}\Phi_e(1)=0. \tag{11.29}$$

Even though (Deo and Richardson, 1983) are able to solve (11.28) numerically they observe that in most dielectric fluid experiments the parameter S/T is small and so $\Phi_e(z)$ exhibits a mainstream and boundary layer character. They also worked with the basic equilibrium field of mainstream character, which is

$$E_e(z)=\Big(\frac{J^2}{C^2}+2Jz\Big)^{1/2},\qquad Q_e(z)=\frac{J}{E_e(z)}, \tag{11.30}$$

J being a real constant uniquely determined from a boundary constraint. Their energy stability analysis is carried out based on a perturbation to both base solutions (11.28) and (11.30).

The resulting perturbations u_i, e_i, and q in liquid velocity, electric field, and space-charge density satisfy the equations

$$u_{i,t}+\frac{T}{M^2}u_ju_{i,j}=-p_{,i}+\Delta u_i+T[qE_i+Qe_i+qe_i], \tag{11.31}$$

$$e_{i,t}=b_i-\frac{T}{M^2}\big[(Q+q)u_i+qE_i+(Q+q)e_i\big]+\frac{S}{M^2}q_{,i}, \tag{11.32}$$

$$q_{,t}+\frac{T}{M^2}u_j(Q+q)_{,j}=-\frac{T}{M^2}\big[qE_i+(Q+q)e_i\big]_{,i}+\frac{S}{M^2}\Delta q, \tag{11.33}$$

$$e_i=-\phi_{,i},\qquad e_{i,i}=q,\qquad u_{i,i}=0, \tag{11.34}$$

where p, ϕ are perturbations in pressure and in potential, b_i is an arbitrary solenoidal vector field, and $\mathbf{E}$, Q refer to their values in (11.28). (Deo and Richardson, 1983) rely strongly on the fact that they consider a unipolar injection of charge so that $Q+q\geq 0$, a condition, which means that nowhere in the liquid can the charge density change sign.

(Deo and Richardson, 1983) allow almost-periodic functions as perturbations and their analysis is based on an energy of the form $E=(\|\mathbf{u}\|^2+M^2[\gamma\|\mathbf{e}\|^2+\lambda\|q\|^2])/2$, for $\gamma,\lambda\,(\geq 0)$ coupling parameters.

11.3.2 Energy stability analyses for non-isothermal EHD

The writer in (Straughan, 1992) and (Straughan, 1993b) developed fully nonlinear energy stability analyses for non-isothermal convection problems in a dielectric fluid. The work in (Straughan, 1993b) is based on a model of (Stiles, 1991) and (Stiles, 1993) and this is discussed more fully in section 11.4.

The work of (Deo and Richardson, 1983) employed the energy method but no temperature field is included. Thus, they only consider the isothermal convection problem. The object of (Straughan, 1992) was to develop a nonlinear energy stability analysis for the non-isothermal theory of (Rodriguez-Luis et al., 1986). The physical problem studied is that of an incompressible dielectric fluid contained between perfectly conducting rigid planar electrodes positioned at $z = 0$ and $z = d$. The collecting electrode is at $z = d$ and is maintained at constant temperature. The lower electrode, at $z = 0$, is the emitter and is maintained at a constant temperature higher than that of the collecting electrode. A potential difference exists across the layer. Details of the equations are given in (Rodriguez-Luis et al., 1986) or (Straughan, 1992) but we here observe that the perturbation equations for a perturbation about the steady solution have form

$$\begin{aligned}
&u_{i,t} + u_j u_{i,j} = -\pi_{,i} + \Delta u_i + R\theta k_i + \frac{CT^2}{M^2}\left[(\bar{q}+q)e_i + q\bar{E}_i\right],\\
&u_{i,i} = 0,\\
&Pr(\theta_{,t} + u_i\theta_{,i}) = Rw + \Delta\theta, \qquad\qquad (11.35)\\
&q_{,t} + u_i(\bar{q}+q)_{,i} = S\Delta q - ReK\left[(\bar{q}+q)e_i + q\bar{E}_i\right]_{,i},\\
&e_{i,i} = \frac{C}{\epsilon}q, \qquad e_i = -\phi_{,i}, \qquad q = -\frac{\epsilon}{C}\Delta\phi,
\end{aligned}$$

where u_i, θ, q, e_i are perturbations to velocity, temperature, charge, and electric field, an overbar denotes the steady solution and $R, C, T, M, Pr, S, Re, K, \epsilon$ are constant parameters.

Generalized (conditional) nonlinear energy stability analyses are developed in detail for this model in (Straughan, 1992).

11.4 Unconditional nonlinear stability in thermo-convective EHD

We now present an unconditional nonlinear energy stability analysis for a model of (Stiles, 1991), see also (Stiles, 1993) where a weakly nonlinear analysis is given. The energy analysis was first given, in a different form, by (Mulone et al., 1996). (Stiles and Kagan, 1993) develop a linear instability analysis for the equivalent model for flow between two cylinders of a dielectric fluid when a radial temperature gradient is superimposed.

The physical problem we study in this section is one of a layer of dielectric fluid heated from below or above and subject to a high frequency ac electric field. In this situation the polarization body force can become dominant and enhance the instability process. This may be seen, cf. (Stiles, 1993), since in heating a 2mm layer of nitrobenzene from below a temperature difference of 3.2°K will induce convection but this is reduced to 2.6°K if a root mean square value 1kV oscillating potential difference is imposed across the layer.

There are free charges present in a dielectric fluid, but there are also dipoles densely distributed. The polarization $\mathbf{P}$ is given by $\mathbf{P} = ne\mathbf{d}$, where n is the number density of dipoles, e is the charge of the dipole, and $\mathbf{d}$ is the distance between the charges. (Landau et al., 1984), p. 66, argue that when polarization effects are important they can be included in the momentum equation via a body force term of

$$f_i = P_j E_{i,j}, \tag{11.36}$$

where E_i is the electric field. This gives rise to a momentum equation in the fluid of form

$$\rho(v_{i,t} + v_j v_{i,j}) = -p_{,i} + \eta \Delta v_i - \rho g \delta_{i3} + P_j E_{i,j}, \tag{11.37}$$

where ρ, v_i, η, and p are density, velocity, dynamic viscosity, and pressure.

(Stiles, 1991) and (Stiles, 1993) consider a dielectric fluid in a horizontal layer with constant upper and lower temperatures T_U $(z = d)$ and T_L $(z = 0)$, with $T_L > T_U$ or $T_U > T_L$. The temperature in the steady state is

$$\bar{T} = T_0 - \beta z, \tag{11.38}$$

where $\beta = (T_L - T_U)/d$ and $T_0 = (T_U + T_L)/2$. The polarization $\mathbf{P}$ is taken as a linear function of $\mathbf{E}$ and the density ρ in the buoyancy term is linear in temperature, T. The electrical permittivity ϵ is written as $\epsilon = \epsilon_r(T)\epsilon_0$ and a linear expansion is adopted for ϵ_r,

$$\epsilon_r = \epsilon_r^0 + \frac{\partial \epsilon_r}{\partial T}\bigg|_{T_0} (T - T_0). \tag{11.39}$$

The electric field may be expressed as a function of the perturbation electric potential ϕ in the form

$$E_i = E_0\Big(1 + \beta z \frac{d}{dT} \log \epsilon_r\Big)\delta_{i3} - \phi_{,i}$$

and the polarization field then becomes in terms of the temperature perturbation θ,

$$P_i = \epsilon_0 E_0 \Big[\epsilon_r - 1 - \frac{d}{dT} \log \epsilon_r \,(\beta z - \epsilon_r \theta)\Big]\delta_{i3} - \epsilon_0(\epsilon_r - 1)\phi_{,i}. \tag{11.40}$$

The non-dimensional equations for the perturbation fields of velocity u_i, pressure π, temperature θ, and electrical potential ϕ may then be derived as, cf. (Stiles, 1993), p. 3275,

$$\begin{aligned} Pr^{-1}(u_{i,t} + u_j u_{i,j}) &= -\pi_{,i} + \Delta u_i \\ &\qquad + \big[(Ra + L)\theta - L\phi_z\big]\delta_{i3} - L\theta\phi_{,iz}\,, && (11.41) \\ u_{i,i} &= 0, && (11.42) \\ \theta_{,t} + u_i\theta_{,i} &= w + \Delta\theta, && (11.43) \\ \Delta\phi &= \theta_z, && (11.44) \end{aligned}$$

where a subscript z denotes differentiation with respect to z, $w = u_3$, and Pr, Ra, L are the Prandtl, Rayleigh and Roberts numbers. The Roberts number is a measure of the electric field strength. These equations hold in the three-dimensional planar region $\mathbb{R}^2 \times \{z \in (0,1)\}$. We suppose u_i, π, θ, ϕ satisfy a plane tiling pattern in the x, y directions so they define a perturbation cell V over the lateral boundaries of which their contributions are equal and cancel out in the ensuing integrations by parts.

The boundary conditions on $z = 0, 1$ which we adopt are

$$u_i = \theta = \phi_z = 0, \quad z = 0, 1, \tag{11.45}$$

which correspond to two free surfaces, cf. (Roberts, 1969). This may be perceived to not be too serious a restriction since in some experiments a "wetting agent" is added to stop the fluid coating the wall, cf. (Rosensweig et al., 1983), and then the dielectric fluid will "see" free boundaries. The precise nature of the boundary conditions is a matter of contention as (Kaloni, 1992) points out and for now we adopt (11.45).

11.4.1 Exchange of stabilities and nonlinear stability

We firstly claim that the strong form of exchange of stabilities holds for (11.41) - (11.45), i.e. the growth rate in the linearized instability problem is real. To see this we linearize (11.41), (11.43) and assume a time dependence like $e^{\sigma t}$, then u_i, θ satisfy

$$\begin{aligned} \frac{\sigma}{Pr}\,u_i &= -\pi_{,i} + \Delta u_i + (Ra + L)\theta\delta_{i3} - L\phi_z\delta_{i3}, && (11.46) \\ \sigma\theta &= w + \Delta\theta. && (11.47) \end{aligned}$$

Multiply (11.46), (11.47) by u_i^*, θ^* (complex conjugates) and integrate over V to find

$$\begin{aligned} \frac{\sigma}{Pr}\,\|\mathbf{u}\|^2 &= -\|\nabla\mathbf{u}\|^2 + (Ra + L)(\theta, w^*) - L(\phi_z, w^*), && (11.48) \\ \sigma\|\theta\|^2 &= \|\nabla\theta\|^2 + (w, \theta^*), && (11.49) \end{aligned}$$

where $\|\cdot\|$ and $(\cdot,\cdot)$ denote the norm and inner product on $L^2(V)$. Next, we use (11.44) to see that with the aid of the boundary conditions, (11.45), and integration by parts

$$\begin{aligned}-(\phi_z, w^*) =&(\phi, w_z^*)\\ =&(\phi, \sigma^*\theta - \Delta\theta_z^*)\\ =&-(\Delta\phi, \theta_z^*) + \sigma^*(\phi, \theta_z^*)\\ =&-\|\theta_z\|^2 + \sigma^*(\phi, \Delta\phi^*)\\ =&-\|\theta_z\|^2 - \sigma^*\|\nabla\phi\|^2.\end{aligned} \tag{11.50}$$

Thus if $\sigma = \sigma_r + i\sigma_i$, (11.48) - (11.50) yield

$$\sigma_i\big(Pr^{-1}\|\mathbf{u}\|^2 + (Ra + L)\|\theta\|^2 - L\|\nabla\phi\|^2\big) = 0. \tag{11.51}$$

Now from (11.44), $\|\nabla\phi\|^2 = -(\theta_z, \phi^*) = (\theta, \phi_z^*)$, and similarly from the conjugate of (11.44), $\|\nabla\phi\|^2 = (\theta^*, \phi_z)$. Thus with the aid of the Cauchy-Schwarz inequality,

$$\|\nabla\phi\| \leq \|\theta\|.$$

Hence, (11.51) yields either $\sigma_i = 0$ or

$$Pr^{-1}\|\mathbf{u}\|^2 + Ra\|\theta\|^2 \leq 0. \tag{11.52}$$

For a non-zero solution we conclude (11.52) cannot hold and we deduce that $\sigma \in \mathbb{R}$.

This result is useful in the ensuing nonlinear energy stability theory.

To proceed with a nonlinear energy stability analysis we form the energy identities using (11.41) and (11.43)

$$\begin{aligned}\frac{1}{2}Pr^{-1}\frac{d}{dt}\|\mathbf{u}\|^2 =& \|\nabla\mathbf{u}\|^2 + (Ra + L)(\theta, w)\\ &+ L(w_z, \phi) - L < u_i\theta\phi_{,iz} >,\end{aligned} \tag{11.53}$$

$$\frac{1}{2}\frac{d}{dt}\|\theta\|^2 = (\theta, w) - \|\nabla\theta\|^2, \tag{11.54}$$

where $< \cdot >$ denotes integration over V. Additionally, from (11.42) - (11.44),

$$\begin{aligned}\frac{d}{dt}\frac{1}{2}\|\nabla\phi\|^2 =& -(\phi, \Delta\phi_t)\\ =& -(\phi, \theta_{zt})\\ =&(\phi_z, \theta_t)\\ =&(\phi_z, w + \Delta\theta - u_i\theta_{,i})\\ =&(\phi_z, w) + (\Delta\phi_z, \theta) + < u_i\theta\phi_{,iz} >\\ =&(\phi_z, w) - (\Delta\phi, \theta_z) + < u_i\theta\phi_{,iz} >\\ =&(\phi_z, w) - \|\theta_z\|^2 + < u_i\theta\phi_{,iz} > .\end{aligned} \tag{11.55}$$

For a constant $\mu(>0)$ at our disposal we form from (11.53) - (11.55) an equation for the energy

$$E(t)=\frac{1}{2}(Pr^{-1}\|\mathbf{u}\|^2+L\|\nabla\phi\|^2+\mu\|\theta\|^2),$$

namely

$$\frac{dE}{dt}=(Ra+L+\mu)(w,\theta)-\|\nabla\mathbf{u}\|^2-L\|\theta_z\|^2-\mu\|\nabla\theta\|^2. \tag{11.56}$$

It is important to realize that (11.56) has been arranged so that it does *not* contain cubic nonlinearities. By treating the indefinite term as a production (I) term we derive from (11.56)

$$\frac{dE}{dt}\leq(m-1)D. \tag{11.57}$$

In (11.57)

$$D=\|\nabla\mathbf{u}\|^2+L\|\theta_z\|^2+\mu\|\nabla\theta\|^2, \tag{11.58}$$

and m is given by

$$m=\max_H\frac{(Ra+L+\mu)(w,\theta)}{\|\nabla\mathbf{u}\|^2+L\|\theta_z\|^2+\mu\|\nabla\theta\|^2}, \tag{11.59}$$

where H is the space of admissible solutions.

The optimal stability boundary is $m=1$. We verify by calculating the Euler-Lagrange equations for (11.59) and solving the resulting max min problem numerically that the nonlinear energy critical Rayleigh - Roberts number boundary is the same as that of linear instability theory. We may see that this should be so from the following argument. Due to exchange of stabilities the linear equations are (11.46), (11.47) with $\sigma=0$ together with (11.42) and (11.44). From the resulting system we easily obtain

$$\begin{aligned}(Ra+L)(\theta,w)&=\|\nabla\mathbf{u}\|^2+L(\phi_z,w),\\ (w,\theta)=\|\nabla\theta\|^2,&\quad(\phi,\theta_z)=\|\nabla\phi\|^2.\end{aligned} \tag{11.60}$$

Now from (11.44) and (11.46),

$$\begin{aligned}(\phi_z,w)&=-(\phi,w_z)\\ &=(\phi,\Delta\theta_z)\\ &=(\Delta\phi,\theta_z)\\ &=\|\theta_z\|^2.\end{aligned} \tag{11.61}$$

By a suitable combination of (11.60) - (11.61) we find

$$(Ra+L+\mu)(\theta,w)=\|\nabla\mathbf{u}\|^2+L\|\theta_z\|^2+\mu\|\nabla\theta\|^2.$$

Thus, the solution of the linear problem satisfies the energy maximum problem at criticality, $m=1$, and the two leading eigenfunctions/eigenvalues are the same. Numerical results may be found in (Mulone et al., 1996).

12

Ferrohydrodynamic convection

12.1 The basic equations of ferrohydrodynamics

Ferrohydrodynamics (FHD) is of great interest because the fluids of concern possess a giant magnetic response. This gives rise to several striking phenomena with important applications. Among these are the spontaneous formation of a labyrinthine pattern in thin layers, the self-levitation of an immersed magnet, and of particular interest here, the enhanced convective cooling in a ferrofluid that has a temperature-dependent magnetic moment. The very well written book by (Rosensweig, 1985) is a perfect introduction to this fascinating subject. He very briefly refers to thermo-convective instability in FHD, which is what we concentrate on here. Another, more general, but again very readable account of ferromagnetism may by found in (Landau et al., 1984). We now present the relevant equations for FHD, in the forms appropriate to this chapter. Then a brief account is given of a striking convective-like instability, before embarking on the thermo-ferro convection problem.

Maxwell's equations are given below.
Faraday's law:

$$\nabla \times \mathbf{E} = -\frac{\partial \mathbf{B}}{\partial t}. \tag{12.1}$$

Ampère's law and Maxwell's correction:

$$\nabla \times \mathbf{H} = \mathbf{J} + \frac{\partial \mathbf{D}}{\partial t}. \tag{12.2}$$

Gauss' law (I):

$$\nabla \cdot \mathbf{D} = Q\,; \tag{12.3}$$

Gauss' law (II):

$$\nabla \cdot \mathbf{B} = 0\,. \tag{12.4}$$

Charge conservation:

$$\frac{\partial Q}{\partial t} + \nabla \cdot \mathbf{J} = 0\,. \tag{12.5}$$

In these equations **E**, **D**, **H**, **B** denote electric field, electric displacement, magnetic field, and magnetic induction, respectively. The quantities Q and **J** are free charge and current, respectively. In FHD, it is usual to assume the free charge Q and electric displacement **D** are absent. Hence, as (Rosensweig, 1985), p. 91, points out, the field equations (12.2), (12.4) are usually employed in the magnetostatic limit of Maxwell's equations,

$$\nabla \times \mathbf{H} = \mathbf{0}, \qquad \nabla \cdot \mathbf{B} = 0. \tag{12.6}$$

The magnetization **M** is introduced by

$$\mathbf{B} = \mu_0(\mathbf{H} + \mathbf{M}), \tag{12.7}$$

where μ_0 is the permeability of free space, $\mu_0 = 4\pi \times 10^{-7}$Henry m^{-1}.

In addition to the field equations (12.6) a constitutive law is assumed relating **B** to **H**. This has the form

$$\mathbf{B} = \mu(\mathbf{H}; T, \rho)\mathbf{H}, \tag{12.8}$$

where the T and ρ dependence are often suppressed.

The fluid is assumed incompressible and then with **v** denoting the velocity field, the continuity and momentum equations are

$$\nabla \cdot \mathbf{v} = 0, \qquad \frac{\partial \mathbf{v}}{\partial t} + (\mathbf{v}.\nabla)\mathbf{v} = -\frac{1}{\rho}\nabla p + \nu\Delta\mathbf{v} + \mathbf{f}, \tag{12.9}$$

where p is the pressure, ρ is the (constant) density, and **f** is the total force. In this work we allow **f** to be composed of two pieces, one due to gravity, while the other is due to the magnetic field. The gravity contribution may be written $-g\,\mathbf{f}_g$, where, for example, in a thermal convection problem,

$$\mathbf{f}_g = \big[1 - \alpha(T - T_R)\big]\mathbf{k}, \tag{12.10}$$

α being the thermal expansion coefficient, T temperature, T_R being a reference temperature, and $\mathbf{k} = (0, 0, 1)$.

The magnetic body force $\mathbf{f}_m$ has several forms and these are discussed in (Rosensweig, 1985), pp. 110–119, and in (Landau et al., 1984), p. 127. For now we note that

$$\mathbf{f}_m = -\nabla\left[\mu_0 \int_0^H \left(\frac{\partial M v}{\partial v}\right)_{H,T} dH + \frac{1}{2}\mu_0 H^2\right] + \nabla(\mathbf{B}\mathbf{H}) \tag{12.11}$$

or

$$\mathbf{f}_m = -\nabla\left[\mu_0 \int_0^H \left(\frac{\partial M v}{\partial v}\right)_{H,T} dH\right] + \mu_0 M \nabla H, \qquad (12.12)$$

where $v = 1/\rho$ is the specific volume and $M = |\mathbf{M}|$, $H = |\mathbf{H}|$. The above forms are useful in section 12.2.

For convection problems, incorporating (12.10) and either (12.11) or (12.12) in $(12.9)_2$, the momentum equation is conveniently written

$$\frac{d\mathbf{v}}{dt} = -\frac{1}{\rho}\nabla\tilde{p} + \nu\Delta\mathbf{v} - g(1 - \alpha[T - T_R])\mathbf{k} + \nabla(\mathbf{BH}) \qquad (12.13)$$

or

$$\frac{d\mathbf{v}}{dt} = -\frac{1}{\rho}\nabla\hat{p} + \nu\Delta\mathbf{v} - g(1 - \alpha[T - T_R])\mathbf{k} + \mu_0 M \nabla H, \qquad (12.14)$$

where the modified pressures $\tilde{p}, \hat{p}$ have the forms

$$\tilde{p} = p - \mu_0 \int_0^H \left(\frac{\partial M v}{\partial v}\right)_{H,T} dH + \frac{1}{2}\mu_0 H^2, \qquad (12.15)$$

$$\hat{p} = p - \mu_0 \int_0^H \left(\frac{\partial M v}{\partial v}\right)_{H,T} dH. \qquad (12.16)$$

Thus, the system of equations governing the ferrofluid motion are (12.6) - (12.9), together with *either* (12.13), (12.15) *or* (12.14), (12.16). A further equation for the temperature field must be added and this is discussed in section 12.2.

This section is completed by reviewing a striking isothermal instability of (Cowley and Rosensweig, 1967), analysed nonlinearly by (Gailitis, 1977). (Cowley and Rosensweig, 1967) consider a horizontal layer of ferromagnetic fluid with a free surface. A magnetic field passes through the fluid at right angles, in the vertical direction. On the basis of a static linear instability theory they deduce that the free surface cannot remain flat and an instability must develop when the magnetic field strength exceeds a critical value, H_c. They also analysed the situation experimentally and verified their theoretical findings. Experimentally, the instability is truly striking. It consists of a very regular hexagon pattern of crests on the free surface, with spikes at the centre of the hexagon. Their working ferromagnetic fluid was prepared by grinding magnetite to particles of submicron size and adding to kerosene to which 5.8% by weight of oleic acid had been added. In a paper that demonstrates a very practical example of secondary bifurcation, (Gailitis, 1977) further investigates the (Cowley and Rosensweig, 1967) problem. He used a potential energy argument and expanded the free surface and magnetic induction in a Fourier series. By examining various modes he discovered that a second critical field strength, H_2, exists, $H_2 > H_c$, such that the hexagonal shape exists for $H_c < H < H_2$, but after this square surface waveforms are the stable ones. The phenomenon of (Gailitis, 1977) exhibits hysteresis. The papers of (Cowley and Rosensweig, 1967) and (Gailitis, 1977) are fundamental in this field.

12.2 Thermo-convective instability in FHD

(Shliomis, 1974) writes that the temperature dependence of the magnetization is important in thermo-convective instability in FHD. This point is, in fact, taken up in (Finlayson, 1970), (Curtis, 1971), (Lalas and Carmi, 1971), and (Shliomis, 1974). As we pointed out in the last section, the energy equation for the temperature field must be decided upon. For linear instability we neglect viscous dissipation in the energy equation and then the equation adopted by (Finlayson, 1970), and essentially by (Curtis, 1971) and (Shliomis, 1974) is

$$\left[\rho C_{VH} - \mu_0 \mathbf{H}.\left(\frac{\partial \mathbf{M}}{\partial T}\right)_{V,\mathbf{H}}\right]\frac{dT}{dt} + \mu_0 T\left(\frac{\partial \mathbf{M}}{\partial T}\right)_{V,\mathbf{H}} \cdot \frac{d\mathbf{H}}{dt} = k\Delta T, \quad (12.17)$$

where C_{VH} and k are heat capacity at constant volume and magnetic field and (constant) thermal conductivity, respectively. Thus, the complete system of equations is (12.17), (12.6) - (12.9) and *either* (12.13), (12.15) *or* (12.14), (12.16).

(Finlayson, 1970) takes

$$\mathbf{M} = \frac{\mathbf{H}}{H} M(H,T) \quad (12.18)$$

and

$$M = M_0 + \chi(H - H_0) - K(T - T_a), \quad (12.19)$$

where H_0 is a constant and T_a (constant) is an average temperature field. The coefficients χ and K are called the susceptibility and the pyromagnetic coefficient, respectively, and are defined by

$$\chi = \left(\frac{\partial M}{\partial H}\right)_{H_0,T_a}, \qquad K = -\left(\frac{\partial M}{\partial T}\right)_{H_0,T_a}. \quad (12.20)$$

The fluid is assumed to occupy the layer $z \in (-d/2, d/2)$, and the magnetic field, $\mathbf{H} = H_0^{ext}\mathbf{k}$, acts outside the layer.

(Finlayson, 1970) assumes

$$T = T_0 \quad \text{at} \quad z = \frac{1}{2}d, \qquad T = T_1 \quad \text{at} \quad z = -\frac{1}{2}d, \quad (12.21)$$

T_0, T_1 constants, and $T_a = (T_0 + T_1)/2$. He investigates the instability of the solution

$$\begin{aligned} &\mathbf{v} \equiv \mathbf{0}, && T = T_a - \beta z, \\ &\beta = \frac{T_1 - T_0}{d}, && H_0 + M_0 = H_0^{ext}, \\ &\mathbf{H}_0 = \mathbf{k}\left(H_0 - \frac{K\beta z}{1+\chi}\right), && \mathbf{M}_0 = \mathbf{k}\left(M_0 + \frac{K\beta z}{1+\chi}\right). \end{aligned} \tag{12.22}$$

The analysis used is a linearized one based on stationary convection with the eigenvalue (critical Rayleigh number) being found numerically by a Galerkin method. In his work he introduces a constant C by $C = C_{VH} + \mu_0 K H_0/\rho$, and his Rayleigh number is then $Ra = \alpha g \beta d^4 \rho C/\nu k$, so the Rayleigh number includes the temperature dependencc of the magnetization. His results are certainly interesting. In particular, he deduces that magnetic forces are only appreciably dominant in very thin layers; for example, in a 1mm layer of kerosene a temperature difference of 51°C is required to induce convection in a zero magnetic field, whereas in the same situation with the kerosene at saturation magnetization instability occurs for a temperature difference of only 19°C. For a depth of the order of 5mm, there is no difference.

The results of (Curtis, 1971) and (Shliomis, 1974) continue in this vein. (Lalas and Carmi, 1971) do attempt a different approach. They assume

$$M = M_0\big[1 - \gamma(T - T_R)\big], \tag{12.23}$$

and argue that for a constant magnetic field gradient $\nabla H > 100$ Gauss/cm, the terms involving the derivatives of $\mathbf{M}$ in (12.17) may be neglected thus reducing the energy equation to

$$\rho C_{VH}\frac{dT}{dt} = k\Delta T. \tag{12.24}$$

Their momentum equation becomes, using (12.23), (12.14), and $(12.9)_2$,

$$\rho\frac{d\mathbf{v}}{dt} + \nabla\hat{p} = M_0\nabla H\big[1 - \gamma(T - T_R)\big] - \rho g\mathbf{k}\big[1 - \alpha(T - T_R)\big] + \rho\nu\Delta\mathbf{v}.$$

At this point they argue that the $M_0\nabla H$ term can be regarded as constant and reduce the system to one equivalent to a standard Bénard one and can thus apply energy stability theory to derive *nonlinear* stability. To apply energy stability theory correctly it is surely necessary to include all perturbations in the $T\nabla H$ term.

12.3 Unconditional nonlinear stability in FHD

(Stiles et al., 1992), (Stiles et al., 1993) and (Blennerhassett et al., 1991) have studied heat transfer and thermal convection in a layer of ferromagnetic fluid. (Blennerhassett et al., 1991) develop linear and weakly nonlinear analyses for a strongly magnetized ferrofluid. (Straughan, 1992) developed a conditional nonlinear energy stability analysis for the (Blennerhassett et al., 1991) model. Here we show how one can derive a completely unconditional analysis for the same problem. We now describe the (Blennerhassett et al., 1991) model and use their notation.

The ferromagnetic fluid is contained in a horizontal layer $\mathbb{R}^2 \times \{0 < z < d\}$. The limit form (12.6) of Maxwell's equations is used. The body force in the momentum equation has form

$$f_i = \mu_0 M_j H_{i,j}, \tag{12.25}$$

where μ_0 is the permeability of free space. (Blennerhassett et al., 1991) consider a strong applied magnetic field and then take $\mathbf{M}$ to have form

$$M_i = \big[M_0 - K(T - T_0)\big]\,\frac{H_i}{H}\,, \tag{12.26}$$

where M_0, T_0, K are constants.

The steady state whose stability is investigated is

$$\bar{T} = T_0 - \beta z d, \qquad \beta = \frac{T_L - T_U}{d}\,, \qquad T_0 = \frac{1}{2}(T_L + T_U), \tag{12.27}$$

where T_L, T_U are the temperatures of the boundaries $z = 0, z = d$, respectively. Both heating above ($\beta < 0$) and heating below ($\beta > 0$) is allowed. The steady magnetization, $\bar{\mathbf{M}}$, and steady magnetic field, $\bar{\mathbf{H}}$, have form

$$\bar{M}_i = (M_0 + K\beta z d)\delta_{i3}, \qquad \bar{H}_i = (H_0 - K\beta z d)\delta_{i3}.$$

The Rayleigh number, Ra, and a non-dimensional magnetic number, N, are defined as $Ra = g\alpha\beta d^4/\kappa\nu$, $N = \mu_0 K^2\beta^2 d^4/\kappa\nu\rho_0$. Then, the non-dimensional fully nonlinear perturbation equations may be derived as

$$\begin{aligned} Pr^{-1}(u_{i,t} + u_j u_{i,j}) &= -\pi_{,i} + \Delta u_i + (Ra + N)\theta\delta_{i3} \\ &\qquad - N\phi_{,z}\delta_{i3} - N\theta\phi_{,zi}, \\ u_{i,i} &= 0, \\ \theta_{,t} + u_i\theta_{,i} &= w + \Delta\theta, \\ \Delta\phi &= \theta_{,z}, \end{aligned} \tag{12.28}$$

which are defined on $\mathbb{R}^2 \times \{-1/2 < z < 1/2\} \times \{t > 0\}$. The boundary conditions are

$$u_i = 0, \quad \theta = 0, \quad \phi_{,z} \pm a\phi = 0, \qquad \text{on } z = \pm\frac{1}{2}, \tag{12.29}$$

where a effectively plays the same role as the constant in (11.16), (11.17). These are the perturbation equations of (Blennerhassett et al., 1991) and this is the model studied also by (Straughan, 1992).

For the present section we take $a = 0$ in (12.29), i.e. assume $\phi_{,z} = 0$ on $z = \pm 1/2$. This is effectively assuming free surfaces. However, when we do this equations (12.28) and the boundary conditions (12.29) are mathematically the same as (11.41) - (11.45) for the dielectric fluid problem. Thus, we may proceed as in section 11.4 to establish a sharp unconditional nonlinear energy stability result. In fact, we may show that the nonlinear unconditional energy stability boundary coincides with the linear instability one. The energy measure used to obtain decay is now $E = (1/2)(Pr^{-1}\|\mathbf{u}\|^2 + N\|\nabla\phi\|^2 + \lambda\|\theta\|^2)$, where λ is a positive coupling parameter.

12.4 Other models and results in FHD

(Stiles and Blennerhassett, 1993) have applied linear instability theory to the problem of flow between two concentric cylinders. The ferromagnetic fluid is subject to a radial temperature gradient and the magnetization is also in the radial direction. They discover that with no gravity and a strong enough magnetic field, then a very small temperature gradient can lead to instability.

(Venkatasubramanian and Kaloni, 1994) treat the Bénard problem for a ferromagnetic fluid in a rotating layer. Their analysis is a linearized one taking into account oscillatory convection. It is found that rotation stabilizes but less so than the non-ferromagnetic case. Various types of boundary conditions are analysed in detail. (Shivakumara et al., 2002) investigate the effect of changing the steady temperature profile on thermal convection in a ferrofluid. Both gravitational and surface tension effects are allowed. The physical interpretation is discussed at length.

(Zahn and Greer, 1995) and (Zahn and Pioch, 1999) are very interesting contributions which examine instability problems where the magnetic field has the effect of rendering the viscosity essentially zero or negative, depending on the field strength.

(Qin and Kaloni, 1994b) present a nonlinear energy stability (variational) analysis for the Bénard problem for a ferromagnetic fluid with a free surface allowing for surface tension effects in the surface. A detailed comparison is made of the nonlinear stability and linear instability boundaries.

(Venkatasubramanian and Kaloni, 2002) develop a nonlinear energy stability for a class of ferromagnetic fluids in an isothermal state, but allowing for angular momentum balance, i.e. spin of ferromagnetic particles. (Rosensweig, 1985), chapter 8, considers the effects of asymmetric stress and polar fluids on ferrohydrodynamics, in detail. The equations of (Venkatasubramanian and Kaloni, 2002) are those of balance of momentum, balance of angular momentum, the magnetization relaxation equation, the continuity equation, and suitable forms of Maxwell's equations, namely

$$\begin{aligned}
&\rho_0(v_{i,t} + v_j v_{i,j}) = -p_{,i} + \mu\Delta v_i + 2\mu_{vv}\epsilon_{ijk}\omega_{k,j} + \mu_0 M_j H_{i,j},\\
&v_{i,i} = 0,\\
&\rho_0 I(\omega_{i,t} + v_j\omega_{i,j}) = (\lambda' + \eta')\omega_{j,ji} + \eta'\Delta\omega_i + \mu_0\epsilon_{ijk}M_j H_k\\
&\qquad\qquad + 2\mu_{vv}(\epsilon_{ijk}v_{k,j} - 2\omega_i),\\
&M_{i,t} + v_j M_{i,j} = \epsilon_{ijk}\omega_j M_k - \frac{1}{\tau_B}(M_i - \chi H_i),\\
&v_{i,i} = 0, \quad \epsilon_{ijk}H_{k,j} = 0, \quad B_{i,i} = 0, \quad B_i = \mu_0(M_i + H_i).
\end{aligned}$$

In these equations $v_i, p, \omega_i, M_j, H_i, B_i$ are velocity, pressure, spin, magnetization, magnetic field, and the magnetic flux density, and the various coefficients are constants. (Venkatasubramanian and Kaloni, 2002) study the stability and uniqueness questions for this system with the fluid in a bounded domain, allowing for the fact that the magnetic field permeates out into the region beyond the fluid. They develop general stability and uniqueness results somewhat akin to the famous results of (Serrin, 1959a) for a viscous fluid.

(Rosensweig, 1985), chapter 9, draws attention to the very important class of problems involving magnetized fluidized solids. Into this class we add fluidized beds, and porous media. There are many applications of magnetic field effects applied to such media. However, one very important aspect is to know how to incorporate the magnetic interaction into the equations of motion. (Rosensweig, 1985) discusses averaging procedures which may be used.

(Qin and Chadam, 1995) develop a nonlinear energy stability analysis for the ferrofluid Bénard problem in a porous medium allowing for Darcy and Forchheimer effects. This is certainly an interesting analysis. (Vaidyanathan et al., 1991) also study the ferromagnetic Bénard problem in a porous medium, while (Sekar et al., 1993) and (Sekar and Vaidyanathan, 1993) include effects of rotation for Darcy and Brinkman porous media, respectively. All of these papers include the magnetic field interaction in the momentum equation by adapting the same form as what is true in a ferrofluid when no porous medium is present.

(Nield, 1999) and (Nield, 2001) draws attention to the modelling of magnetic field effects in convection problems in porous media. The magnetic field may well affect the ferrofluid differently from the solid matrix, and indeed, may induce anisotropy. This would appear to be a potentially fruitful area of future research.

13
Convective instabilities for reacting viscous fluids far from equilibrium

13.1 Chemical convective instability

The phenomenon of double-diffusive convection in a fluid layer, where two scalar fields (such as heat and salinity concentration) affect the density distribution in a fluid, has become increasingly important in recent years. The behaviour in the double-diffusive case is much more diverse than for the Bénard problem. In particular, linear stability theory, cf. (Baines and Gill, 1969), predicts that the first occurrence of instability may be via oscillatory rather than stationary convection if the component with the smaller diffusivity is stably stratified. Finite amplitude convection in the double-diffusive context was investigated by (Veronis, 1965; Veronis, 1968a) whose results suggested steady finite amplitude motion could occur at critical values of a Rayleigh number much less than that predicted by linearized theory. Several later papers confirmed this, usually by weakly nonlinear theory, see e.g., (Proctor, 1981) and the references therein. The boundary layer analysis of (Proctor, 1981) is an interesting one and provides some explanation for the energy results of (Shir and Joseph, 1968). The phenomenon of double diffusive convection and even multi-diffusive convection is examined in detail in the next chapter.

When temperature and one or more species are present and interactions between species are allowed, then the system becomes increasingly richer. Furthermore, reaction-diffusion equations for mixtures of viscous fluids play an important role in everyday life. We mention specifically acid rain effects, (Pandiz, 1989); the nuclear winter phenomenon, (Giorgi, 1989); warming of the stratosphere, (Rood, 1987), (Kaye and Rood, 1989); and enzyme recovery from reacting mixtures, (Duong and Weiland, 1981), (Malikkides and Weiland, 1982). However, the equations are usually written down in an ad hoc manner and often vary considerably. Therefore, a rational derivation of a relevant system of equations would seem appropriate. We therefore develop such a theory here and commence an analysis of linearized stability, following (Morro and Straughan, 1990). Before doing this, however, we examine a previous analysis of nonlinear energy stability in a dissociating gas.

13.2 Chemical convective instability and quasi-equilibrium thermodynamics

(Wollkind and Frisch, 1971) developed a linear analysis for a chemical instability problem and deduced that for large enough Rayleigh number the Bénard problem involving a chemically dissociating fluid is unstable, in the situation that is stable for a non-reactive fluid, when the fluid layer is heated from above. The nonlinear analyses described for the Wollkind & Frisch system were given by (Straughan et al., 1984).

The model with which Wollkind & Frisch commence considers a dissociating fluid contained in the infinite layer $0 < z < d$ and attention is restricted to essentially isochoric motions by adopting a Boussinesq-type approximation. The mass flux through the boundaries $z = 0, d$ is zero, and the prescribed temperatures there are T_0, T_1, with $T_1 > T_0$. The relevant equations admit a steady solution in which the velocity is zero, the fraction, α, of free atoms is constant, and the temperature, T, is linear in z across the layer.

In non-dimensional form the equations for the perturbation to the constant concentration solution are

$$\begin{aligned} \dot{u}_i &= -p_{,i} + \delta_{i3}(R\theta + \phi) + \Delta u_i, \\ Sc\,\dot{\phi} &= \Delta\phi - X_1\phi, \\ Pr\,\dot{\theta} &= -Rw + \Delta\theta + \Big(S + \frac{PrA}{ScR}\Big)\Delta\phi - \frac{X_1 PrA}{ScR}\,\phi, \end{aligned} \tag{13.1}$$

where $\mathbf{u}, \theta, \phi, p$ are the perturbation fields of velocity (solenoidal), temperature, fraction of free atoms, and pressure, R^2 is the Rayleigh number, Δ is the three-dimensional Laplacian, a superposed dot denotes the material derivative, $w = u_3$, and where for completeness we include below the non-dimensionalization appropriate to the notation of (Wollkind and Frisch, 1971),

$$\begin{aligned} &Sc = \nu/D_{12}^0, &&X_1 = d^2/\tau D_{12}^0, &&\kappa = \lambda^0/\rho_0 c_1, \\ &Pr = \nu/\kappa, &&S = D_{12}^0 D_0/2mc_1\kappa, &&A = c_2\alpha/c_1 b. \end{aligned}$$

The functions $\mathbf{u}, \theta, \phi, p$ are assumed periodic in x, y and satisfy the boundary conditions

$$\mathbf{u} = \mathbf{0}, \qquad \frac{\partial \phi}{\partial z} = 0, \qquad \theta = 0, \tag{13.2}$$

on $z = 0, d$.

To investigate nonlinear stability we now define, for $\lambda\,(> 0)$ to be chosen, an energy $E(t)$ by

$$E(t) = \frac{1}{2}\big(\|\mathbf{u}\|^2 + Pr\|\theta\|^2 + \lambda Sc\|\phi\|^2\big), \tag{13.3}$$

where for V a period cell of the perturbed solution, $\|\cdot\|$, $<\cdot>$, and $D(\cdot)$, denote again the L^2-norm, integral, and Dirichlet integral on V.

The independence of equation $(13.1)_2$ allows us to use λ in a different way from the normal theory of the best λ. From (13.1) we derive

$$\begin{aligned} \frac{dE}{dt} &= <\phi w> - D(\mathbf{u}) - \lambda D(\phi) - D(\theta) - \lambda X_1\|\phi\|^2 \\ &\quad - \left(S + \frac{PrA}{ScR}\right) <\nabla\phi . \nabla\theta> - \frac{X_1 PrA}{ScR} <\phi\theta> . \end{aligned} \tag{13.4}$$

From use of the arithmetic-geometric mean and Poincaré inequalities we next establish the following estimates:

$$<\phi w> \leq \frac{1}{2}\lambda_1 \|w\|^2 + \frac{1}{2}\lambda_1^{-1}\|\phi\|^2,$$

$$-\left(S + \frac{PrA}{ScR}\right) <\nabla\phi . \nabla\theta> \leq \frac{1}{3}D(\theta) + \frac{3}{4}\left(S + \frac{PrA}{ScR}\right)^2 D(\phi),$$

$$\frac{X_1 PrA}{ScR} <\phi\theta> \leq \frac{1}{3}D(\theta) + \frac{3}{4\lambda_1}\left(\frac{X_1 PrA}{ScR}\right)^2 \|\phi\|^2,$$

where λ_1 is the constant in Poincaré's inequality. These estimates are used in (13.4), and we then select λ so large that

$$\lambda \geq \frac{3}{4}\left(S + \frac{PrA}{ScR}\right)^2$$

and

$$\lambda X_1 \geq \lambda_1^{-1} + \frac{3}{2\lambda_1}\left(\frac{X_1 PrA}{ScR}\right)^2,$$

and from the resulting inequality and further use of Poincaré's inequality we obtain

$$\frac{dE}{dt} \leq -ME,$$

where $M = \min\{\lambda_1, 2\lambda_1/3Pr, X_1/Sc\}$.

Clearly then, $E \to 0$ as $t \to \infty$, and so there is *no* instability when the fluid layer is heated from above. (Such a conclusion was suggested from a different viewpoint by (Wollkind and Bdzil, 1971).)

We now include an analysis of another system suggested by (Wollkind and Frisch, 1971).

A perturbation to the constant concentration equilibrium solution for the *modified* system of (Wollkind and Frisch, 1971) (using *their* chemical quasi-equilibrium approximation) satisfies the equations

$$\begin{aligned}\dot{u}_i &= -p_{,i} + g\alpha k_i\theta + gbk_i\phi + \nu\Delta u_i,\\ \dot{\theta} &= \beta w + \kappa\Delta\theta,\\ \dot{\theta} &= \beta w - M\phi + D_{12}^0\Delta\theta,\end{aligned} \tag{13.5}$$

for divergence free $\mathbf{u}$. The constant coefficents are in the notation of (Wollkind and Frisch, 1971) and we do not include them explicitly as we give a non-dimensional version, which corresponds to our notation below. A key factor is that from $(13.5)_{2,3}$ $\phi = \Delta\theta\,[(D_{12}^0 - \kappa)/M]$, and hence ϕ may be eliminated to yield the following (non-dimensional) equations,

$$\begin{aligned}\dot{u}_i &= -p_{,i} + \delta_{i3}R(\theta + \epsilon B\Delta\theta) + \Delta u_i,\\ Pr\dot{\theta} &= -Rw + \Delta\theta,\end{aligned} \tag{13.6}$$

in which ϵB is a reaction term of small magnitude, introduced by (Wollkind and Frisch, 1971).

To investigate the stability of the zero solution to (13.6) we choose

$$E(t) = \frac{1}{2}\|\mathbf{u}\|^2 + \frac{1}{2}Pr\|\theta\|^2. \tag{13.7}$$

The energy equation appropriate to (13.7) is determined to be

$$\frac{dE}{dt} = RI - \mathcal{D}, \tag{13.8}$$

where

$$\mathcal{D} = D(\mathbf{u}) + D(\theta), \qquad I = -\epsilon B < \nabla\theta.\nabla w > .$$

Define now

$$\frac{1}{R_E} = \max \frac{I}{\mathcal{D}} \quad (= \lambda), \tag{13.9}$$

where the maximum is over the space of admissible solutions and from (13.8)

$$\frac{dE}{dt} \le -\mathcal{D}R\Big(\frac{1}{R} - \frac{1}{R_E}\Big). \tag{13.10}$$

If now $R < R_E$, then (13.10) and Poincaré's inequality show that $E \to 0$ at least exponentially as $t \to \infty$.

The problem is then to find R_E, or equivalently λ, as in (13.9). To this end we derive the Euler-Lagrange equations for this maximum as

$$\begin{aligned} \epsilon B\delta_{i3}\Delta\theta + 2\lambda\Delta u_i &= 2p_{,i}, \\ \epsilon B\Delta w + 2\lambda\Delta\theta &= 0. \end{aligned} \tag{13.11}$$

These equations are linear and so we may use a normal mode technique to obtain

$$4\lambda^2(D^2 - a^2)^2 W = -a^2(\epsilon B)^2(D^2 - a^2)W, \tag{13.12}$$

where $D = d/dz$, a^2 is the wave number, and $W(z)$ is the z-part of w. For the two free boundaries situation covered in (Wollkind and Frisch, 1971) the boundary conditions allow W to be composed of $\sin m\pi z$, $m = 1, 2, ...,$ and (13.12) yields

$$\lambda^2 = \frac{(\epsilon B)^2 a^2}{4(m^2\pi^2 + a^2)}. \tag{13.13}$$

Obviously, as a function of m, λ is maximum for $m = 1$. We then see that the maximum of λ is achieved asymptotically as $a^2 \to \infty$. We may, therefore, conclude that $R_E = 2/\epsilon B$.

If we denote by R_L^2 the critical Rayleigh number of linear theory, the asymptotic expression given in (Wollkind and Frisch, 1971), eq. (36), is

$$R_L^2 = \frac{4}{(\epsilon B)^2} + \frac{2\pi^2}{\epsilon B} + O(1), \tag{13.14}$$

which compares with the energy limit

$$R_E^2 = \frac{4}{(\epsilon B)^2}. \tag{13.15}$$

Estimates (13.14) and (13.15), which agree to leading order, determine quantitatively a band of Rayleigh numbers where subcritical bifurcation may occur.

13.3 Basic equations for a chemically reacting mixture

In this and the next two sections we describe a theory of a mixture of chemically reacting viscous fluids due to (Morro and Straughan, 1990) that uses one velocity field.

Consider a mixture of $N + 1$ fluids, with each constituent labelled by a Greek index. The notation we employ is standard; a repeated Greek index signifies summation from 1 to N whereas a repeated Roman index denotes summation over 1 to 3. We denote by ρ_α the mass density of the αth constituent, and by $\rho\,(= \sum_{\alpha=1}^{N+1} \rho_\alpha)$ the total density; $c_\alpha = \rho_\alpha/\rho$ is the concentration of the αth species, $\mathbf{h}_\alpha$ is the relative αth mass flux, and m_α is the mass supply.

The equation of balance of mass for each constituent is

$$\rho \dot{c}_\alpha = -\nabla \cdot \mathbf{h}_\alpha + m_\alpha, \qquad \alpha = 1, ..., N, \tag{13.16}$$

where a superposed dot denotes the material time derivative; in (13.16) the $N+1$th components are determined from the relations

$$\sum_{\alpha=1}^{N+1} c_\alpha = 1, \qquad \sum_{\alpha=1}^{N+1} \mathbf{h}_\alpha = 0, \qquad \sum_{\alpha=1}^{N+1} m_\alpha = 0.$$

The equations of balance of mass, linear momentum, and energy for the total mixture are, respectively,

$$\dot{\rho} + \rho \nabla \cdot \mathbf{v} = 0, \tag{13.17}$$

$$\rho \dot{\mathbf{v}} = \nabla \cdot \mathbf{T} + \rho \mathbf{b}, \tag{13.18}$$

$$\rho \dot{e} = -\nabla \cdot \mathbf{q} + \mathbf{T} \cdot \mathbf{D} + \rho r, \tag{13.19}$$

where $\mathbf{v}$ is the velocity, $\mathbf{T}$ is the stress tensor, $\mathbf{b}$ is the body force, e is the internal energy, $\mathbf{q}$ is the heat flux, $\mathbf{D}$ is the symmetric part of the velocity gradient $\mathbf{L}$, and r is the heat supply.

Define η to be the entropy and θ the temperature. (Morro and Straughan, 1990) employ the entropy inequality in the form

$$\rho \dot{\eta} \geq -\nabla \cdot \left(\frac{\mathbf{q}}{\theta} + \mathbf{k} \right) + \frac{\rho r}{\theta}, \tag{13.20}$$

where $\mathbf{k}$ is Müller's entropy extra flux. For application, this inequality is rewritten in terms of the free energy ψ, as

$$-\rho(\dot{\psi} + \eta \dot{\theta}) + \mathbf{T} \cdot \mathbf{D} - \frac{\mathbf{q} \cdot \mathbf{g}}{\theta} - \theta \nabla \cdot \mathbf{k} \geq 0, \tag{13.21}$$

$\mathbf{g}$ denoting the temperature gradient.

Consequences from the relations (13.16) - (13.21) are now derived for an incompressible mixture, that is a mixture of $N+1$ fluids with ρ constant and so equation (13.17) becomes

$$\nabla \cdot \mathbf{v} = 0. \tag{13.22}$$

(Morro and Straughan, 1990) adopt the following constitutive theory:

$$\psi, \mathbf{T}, \eta, \mathbf{q}, \mathbf{k}, \mathbf{h}_\alpha, \text{ and } m_\alpha$$

are functions of

$$\theta, C, \mathbf{g}, \mathbf{S}, \mathbf{D},$$

where $C = (c_1, ..., c_N)$, $\mathbf{S} = (\mathbf{s}_1, ..., \mathbf{s}_N)$, with $\mathbf{s}_\alpha = \nabla c_\alpha$. Then setting $\mu_\alpha = \partial\psi/\partial c_\alpha$ and $\mathbf{J} = \theta\mathbf{k} + \mu_\alpha \mathbf{h}_\alpha$ they write the entropy inequality as

$$\begin{aligned} &-\rho(\psi_\theta + \eta)\dot{\theta} - \rho\psi_{\mathbf{g}} \cdot \dot{\mathbf{g}} - \rho\psi_{\mathbf{s}_\alpha} \cdot \dot{\mathbf{s}}_\alpha - \rho\psi_{\mathbf{D}} \cdot \dot{\mathbf{D}} \\ &+ \mathbf{T} \cdot \mathbf{D} + \Big(\mathbf{J}_\theta - \frac{\mathbf{q}}{\theta} - \mathbf{k} - \mathbf{h}_\alpha \frac{\partial \mu_\alpha}{\partial \theta}\Big) \cdot \mathbf{g} \\ &+ \Big(\mathbf{J}_{c_\alpha} - \mathbf{h}_\beta \frac{\partial \mu_\beta}{\partial c_\alpha}\Big) \cdot \mathbf{s}_\alpha + \mathbf{J}_{\mathbf{g}} \cdot (\nabla\nabla\theta) \\ &+ \mathbf{J}_{\mathbf{s}_\alpha} \cdot (\nabla\nabla c_\alpha) + \mathbf{J}_{\mathbf{D}} \cdot (\nabla\mathbf{D}) - \mu_\alpha m_\alpha \geq 0. \end{aligned} \tag{13.23}$$

They make use of standard thermodynamic arguments and the fact that $\dot{\theta}, \dot{\mathbf{g}}, \dot{\mathbf{s}}_\alpha, \dot{\mathbf{D}}$ appear linearly in (13.23) to deduce the facts below.

The free energy ψ is independent of $\mathbf{g}$, $\mathbf{S}$, $\mathbf{D}$, and so,

$$\psi = \psi(\theta, C).$$

The entropy η is related to the free energy ψ by

$$\eta = -\psi_\theta. \tag{13.24}$$

The quantity $\mathbf{J}$, which is the energy flux due to diffusion, satisfies the restrictions

$$\mathrm{sym}\, \mathbf{J}_{\mathbf{g}} = 0, \qquad \mathrm{sym}\, \mathbf{J}_{\mathbf{s}_\alpha} = 0, \quad \alpha = 1, ..., N, \qquad \mathbf{J}_{\mathbf{D}} = 0, \tag{13.25}$$

where sym denotes the symmetric part.

The entropy inequality remaining from (13.23) is

$$\begin{aligned} \mathbf{T} \cdot \mathbf{D} &+ \left(\mathbf{J}_\theta - \frac{\mathbf{q}}{\theta} - \mathbf{k} - \mathbf{h}_\alpha \frac{\partial \mu_\alpha}{\partial \theta}\right) \cdot \mathbf{g} \\ &+ \left(\mathbf{J}_{c_\alpha} - \mathbf{h}_\beta \frac{\partial \mu_\beta}{\partial c_\alpha}\right) \cdot \mathbf{s}_\alpha - m_\alpha \mu_\alpha \geq 0, \end{aligned} \tag{13.26}$$

and this must be true for every motion of the incompressible mixture.

(Morro and Straughan, 1990) appeal to a result of (Gurtin, 1971), to conclude that for $\mathbf{J}$ isotropic, (13.25) implies it vanishes. Thus when $\mathbf{J} = 0$ the entropy inequality finally becomes

$$\mathbf{T} \cdot \mathbf{D} - \left[\frac{\mathbf{q}}{\theta} + \Big(\frac{\partial \mu_\alpha}{\partial \theta} - \frac{\mu_\alpha}{\theta}\Big)\mathbf{h}_\alpha\right] \cdot \mathbf{g} - \mathbf{h}_\beta \frac{\partial \mu_\beta}{\partial c_\alpha} \cdot \mathbf{s}_\alpha - \mu_\alpha m_\alpha \geq 0, \tag{13.27}$$

and the entropy flux $\mathbf{k}$ has form $\mathbf{k} = -\mu_\alpha \mathbf{h}_\alpha/\theta$.

13.4 A model for reactions far from equilibrium

To apply the general theory above it is necessary to be more specific about the form of constitutive equations. Before doing this, however, it is instructive to the understanding of the problem to review some basic properties of chemistry connecting the chemical affinity of a reaction and the chemical reaction rates.

13.4.1 Chemical affinities and mass supplies

Let the $N+1$ molecular constituents involved in the chemical reactions be composed of A atomic substances, where $A \leq N$. For $a = 1, ..., A$ and $\beta = 1, ..., N+1$, let $T_{a\beta}$ represent the number of atoms of the ath species in the molecule of the βth species. For the case under consideration the atomic substances are assumed indestructabile, and so

$$\sum_{\beta=1}^{N+1} T_{a\beta} \frac{m_\beta}{M_\beta} = 0, \qquad a = 1, ..., A, \tag{13.28}$$

M_β being the molecular weight. As a consequence of (13.28), and since $\varrho = \text{rank}\,(T_{a\beta}) \leq A$, the reaction rates j_r, $r = 1, ..., N+1-\varrho$, determine the mass supplies m_β through the equations

$$m_\beta = \rho M_\beta \sum_r P_{\beta r} j_r, \tag{13.29}$$

where $P_{\beta r}$ are the stoichiometric coefficients. These coefficients are not totally independent since the total mass supply is zero, i.e. $\sum_{\beta=1}^{N+1} m_\beta = 0$. This follows because if $\nu_{\beta r} = \rho M_\beta P_{\beta r}$, we may sum (13.29) over β and then because of the arbitrariness of j_r one sees that

$$\sum_{\beta=1}^{N+1} \nu_{\beta r} = 0, \qquad r = 1, ..., N+1-\varrho. \tag{13.30}$$

It may be shown that the chemical affinity of reaction r, A_r, is given in terms of the reduced chemical potentials μ_α by

$$A_r = \nu_{\alpha r} \mu_\alpha, \tag{13.31}$$

where summation is over $\alpha = 1, ..., N$. This then leads to the useful relation connecting mass supplies and reaction rates,

$$m_\alpha \mu_\alpha = \sum_r j_r A_r. \tag{13.32}$$

13.4.2 Thermodynamic deductions

(Morro and Straughan, 1990) assume that the nonlinearities of the model occur only through the mass supplies. They follow the approach of (Loper and Roberts, 1978; Loper and Roberts, 1980) in compositional convection and introduce the explicit (linear) constitutive assumptions:

$$\begin{aligned} \psi &= \psi(\theta, C), & \mathbf{q} &= -\kappa \mathbf{g} - \omega_\alpha \mathbf{s}_\alpha, \\ \mathbf{h}_\alpha &= \chi_\alpha \mathbf{g} - \lambda_{\alpha\beta} \mathbf{s}_\beta, & \mathbf{T} &= -p\mathbf{I} + 2\mu \mathbf{D}, \end{aligned} \tag{13.33}$$

where $\kappa, \omega_\alpha, \chi_\alpha, \lambda_{\alpha\beta}, p, \mu$ are functions of θ, C; the mass supplies $m_\alpha (= m_\alpha(\theta, C))$ are still general functions of their arguments but they may be written in terms of the reaction rates j_r according to (13.32). Upon substitution into the entropy inequality (13.27) they deduce that $\mu > 0$, and the residual entropy inequality yields

$$\frac{\mathbf{q}\cdot\mathbf{g}}{\theta} + \mu_\alpha m_\alpha + \zeta_\alpha \mathbf{h}_\alpha \cdot \mathbf{g} + a_{\alpha\beta}\mathbf{s}_\alpha \cdot \mathbf{h}_\beta \leq 0, \tag{13.34}$$

where $\zeta_\alpha = \partial\mu_\alpha/\partial\theta - \mu_\alpha/\theta$, $a_{\alpha\beta} = \partial\mu_\alpha/\partial c_\beta$, and the matrix $(a_{\alpha\beta})$ is symmetric. Upon employing equations $(13.33)_{2,3}$ in (13.34) one obtains

$$\Big(-\frac{\kappa}{\theta}+\zeta_\alpha\chi_\alpha\Big)\mathbf{g}\cdot\mathbf{g} - a_{\gamma\beta}\lambda_{\gamma\alpha}\mathbf{s}_\alpha\cdot\mathbf{s}_\beta + \Big(\chi_\alpha a_{\alpha\beta} - \zeta_\alpha\lambda_{\alpha\beta} - \frac{\omega_\beta}{\theta}\Big)\mathbf{s}_\beta\cdot\mathbf{g} + \mu_\alpha m_\alpha \leq 0.$$

In view of the fact that $\mu_\alpha m_\alpha$ is independent of $\mathbf{g}$, $\mathbf{s}_\beta$, (Morro and Straughan, 1990) then conclude that this inequality can hold if and only if $\mu_\alpha m_\alpha \leq 0$, or, alternatively, due to (13.32) $\sum_r j_r A_r \leq 0$ and, in addition, the matrix

$$\begin{pmatrix} \kappa/\theta - \chi_\gamma\zeta_\gamma & \frac{1}{2}(\zeta_\gamma\lambda_{\gamma\alpha} - \chi_\gamma a_{\gamma\alpha} + \omega_\alpha/\theta) \\ \frac{1}{2}(\zeta_\gamma\lambda_{\gamma\beta} - \chi_\gamma a_{\gamma\beta} + \omega_\beta/\theta) & a_{\gamma\alpha}\lambda_{\gamma\beta} \end{pmatrix}$$

is positive semidefinite.

13.4.3 Reduction of the energy equation

The energy equation (13.19) is still too general for practical purposes and some reduction is necessary. This is done by recalling the definition of μ_α and using (13.16) and (13.24) to rewrite equation (13.19) as

$$\rho\theta\eta_\theta\dot{\theta} - \rho\theta\frac{\partial\mu_\alpha}{\partial\theta}\dot{c}_\alpha = -\nabla\cdot(\mathbf{q} - \mu_\alpha\mathbf{h}_\alpha) + \mathbf{T}\cdot\mathbf{D} - \mathbf{h}_\alpha\cdot\nabla\mu_\alpha - m_\alpha\mu_\alpha. \tag{13.35}$$

Without loss of physical content, we now set r equal to zero. The terms $\mathbf{T}$ and $\mathbf{h}_\alpha$ are expressed using $(13.33)_{3,4}$ to find

$$\mathbf{T}\cdot\mathbf{D} - \mathbf{h}_\alpha\cdot\nabla\mu_\alpha = 2\mu\mathbf{D}\cdot\mathbf{D} + \Big(\frac{\partial\mu_\alpha}{\partial\theta}\mathbf{g} + \frac{\partial\mu_\alpha}{\partial c_\omega}\mathbf{s}_\omega\Big)\cdot(\lambda_{\alpha\beta}\mathbf{s}_\beta - \chi_\alpha\mathbf{g}). \tag{13.36}$$

We now neglect the terms of (13.36) in (13.35), which all involve products of gradients. This leads to a reduced energy equation from (13.35) of form

$$\rho\theta\eta_\theta\dot{\theta} - \rho\theta\frac{\partial\mu_\alpha}{\partial\theta}\dot{c}_\alpha = -\nabla\cdot(\mathbf{q} - \mu_\alpha\mathbf{h}_\alpha) - m_\alpha\mu_\alpha. \tag{13.37}$$

13.4.4 Selection of m_α and ψ

Up to this point linear constitutive equations (in $\mathbf{g}$ and $\mathbf{s}_\alpha$) have been prescribed but m_α has not been specified. When we wish to analyze a specific problem it is then necessary to prescribe equations for m_α and ψ.

This approach is not unlike that of (Loper and Roberts, 1980), p. 91, who derive an exact model to describe compositional convection.

(Morro and Straughan, 1990) choose the free energy ψ to be

$$\psi = \tilde{c}\theta\left\{1 - \log\left(\frac{\theta}{\tilde{\theta}}\right)\right\} + \epsilon_\alpha c_\alpha \theta + \frac{1}{2}\phi_{\alpha\beta}c_\alpha c_\beta, \tag{13.38}$$

where $\tilde{c}, \epsilon_\alpha, \phi_{\alpha\beta}, \tilde{\theta}$ are constants. They further assume a linear relationship between the mass supplies and the temperature and concentrations so that for constants $\tau_{\alpha\beta}, \delta_\alpha$,

$$m_\alpha = -\tau_{\alpha\beta}c_\beta - \delta_\alpha\theta. \tag{13.39}$$

Since the entropy inequality (13.34) requires $\mu_\alpha m_\alpha \le 0$, this implies $\tau_{\alpha\beta}, \delta_\alpha, \epsilon_\alpha, \phi_{\alpha\beta}$ must comply with the restriction

$$(\tau_{\alpha\beta}c_\beta + \delta_\alpha\theta)(\epsilon_\alpha\theta + \phi_{\alpha\beta}c_\beta) \ge 0, \tag{13.40}$$

for all admissible temperatures and concentrations θ, c_α.

The energy balance equation (13.37) and the balance of mass laws (13.16) are then succinctly written with the aid of (13.33) and (13.38) as

$$\rho c\dot{\theta} - \rho\epsilon_\alpha\theta\dot{c}_\alpha = \nabla\cdot(k\mathbf{g}) + \nabla\cdot(l_\alpha\mathbf{s}_\alpha) - m_\alpha\mu_\alpha, \tag{13.41}$$

$$\rho\dot{c}_\alpha = \nabla\cdot(\lambda_{\alpha\beta}\mathbf{s}_\beta) - \nabla\cdot(\chi_\alpha\mathbf{g}) + m_\alpha, \tag{13.42}$$

where $k = \kappa + \mu_\alpha\chi_\alpha$, and $l_\alpha = \lambda_\alpha - \mu_\gamma\lambda_{\gamma\alpha}$.

To complete the model it is necessary to prescribe a form for the buoyancy term in the momentum equation. (Morro and Straughan, 1990) employ a Boussinesq approximation and the momentum equation then becomes

$$\dot{v}_i = -\frac{1}{\rho}p_{,i} + \nu\Delta v_i - g\delta_{i3}\big[1 - \alpha(\theta - \bar{\theta}) + \beta_\omega(c_\omega - \bar{c}_\omega)\big], \tag{13.43}$$

where g is the gravity acceleration, $\rho, \nu, \alpha, \beta_\omega$ are constants, and $\bar{\theta}, \bar{c}_\omega$ are conveniently chosen, constant reference values.

The model of convection in a chemically reacting mixture, not necessarily near equilibrium, is then composed of the system of equations (13.41) - (13.43) together with the incompressibility condition (13.22). For any model to be practically useful it must allow the determination of the solution or certainly it must allow deduction of useful quantitative information. The model described above is still highly nonlinear and numerical computation is evidently necessary.

To investigate convective instability, (Morro and Straughan, 1990) resort to a yet simpler version of (13.41) - (13.43). They suppose $\lambda_{\alpha\beta}$ and k are constant and take $l_\alpha, \chi_\alpha = 0$, and thereby neglect, respectively, the Dufour and Soret effects. (Within the context of a non-reacting mixture such effects have been intensely investigated, see e.g., (Ybarra and Velarde, 1979), (Straughan and Hutter, 1999) and the references therein.) The inclusion of such effects may, however, be desirable in a detailed study of a specific reacting gaseous atmosphere.

13.5 Convection in a layer

Suppose now the reacting mixture is confined to the layer $z \in (0, d)$, with the boundaries $z = 0, d$ retained at fixed temperatures, namely,

$$\theta = \theta_1, \quad z = 0; \qquad \theta = \theta_2, \quad z = d, \tag{13.44}$$

with $\theta_2 < \theta_1$. In addition, assume there is no mass flux out of the boundaries, which using $(13.33)_3$ is equivalent to

$$\mathbf{k} \cdot \lambda_{\alpha\beta} \nabla c_\alpha = 0, \quad \text{on} \quad z = 0, d. \tag{13.45}$$

(Morro and Straughan, 1990) choose to omit the $m_\alpha \mu_\alpha$ term in (13.41), since its presence requires a numerically determined base state. Instead they seek a steady solution of form

$$\bar{\theta} = -\gamma z + \theta_1, \qquad \bar{\mathbf{v}} = \mathbf{0}, \tag{13.46}$$

where $\gamma = (\theta_1 - \theta_2)/d$ is the temperature gradient. The steady concentrations $\bar{c}_\alpha(z)$ satisfy the equations

$$\lambda_{\alpha\beta} \bar{c}_\beta'' - \tau_{\alpha\beta} c_\beta = \delta_\alpha \bar{\theta}(z), \tag{13.47}$$

where a prime indicates differentiation with respect to the variable z. In (Morro and Straughan, 1990) investigation is confined to the case of one constituent, $c = c_1$: this has applications to a dissociating gas.

For simplicity put $\lambda = \lambda_{11}$, $\tau = \tau_{11}$ $\delta = \delta_1$, $\beta = \beta_1$, $\epsilon = \epsilon_1$. Equation (13.47) becomes for $\bar{c}(z)$

$$\lambda \bar{c}'' - \tau \bar{c} = \delta(\theta_1 - \gamma z),$$

and the boundary conditions (13.45) are

$$\bar{c}' = 0, \qquad z = 0, d. \tag{13.48}$$

The solution is found to be

$$\bar{c} = \frac{\delta\beta}{\tau} \left\{ -\frac{1}{\epsilon} \sinh \epsilon z + \cosh(\epsilon d - 1) \cosh \epsilon z + z \right\} - \frac{\delta\theta_1}{\tau}, \tag{13.49}$$

where $\epsilon = (\tau/\lambda)^{1/2}$. It is noteworthy to compare (13.49) with the non-reacting case which has $\bar{c}$ constant: here the nonlinear dependence on z is a direct consequence of the reaction.

A linear instability analysis of the steady solution (13.46), (13.49) commences with the introduction of perturbations $\mathbf{u}, \vartheta, \phi$, and π to $\bar{\mathbf{v}}, \bar{\theta}, \bar{c}$, and the steady pressure $\bar{p}$. From (13.41) - (13.43) the perturbations are found to satisfy

$$\vartheta_{,t} - \frac{\epsilon}{\tilde{c}}\bar{\theta}\phi_{,t} = -\bar{\theta}' w + \frac{\epsilon}{\tilde{c}}\bar{\theta}\bar{c}' w + K\Delta\vartheta, \tag{13.50}$$

$$\rho\phi_{,t} = -\rho w\bar{c}' + \lambda\Delta\phi - \tau\phi - \delta\vartheta, \tag{13.51}$$

$$\mathbf{u}_{,t} = -\frac{1}{\rho}\nabla\pi + \nu\Delta\mathbf{u} + \alpha g\vartheta\mathbf{k} - g\beta\phi\mathbf{k}, \tag{13.52}$$

where $K = \kappa/\rho\tilde{c}$, $w = u_3$, and g is the acceleration due to gravity.

If one restricts attention to stationary convection and employs normal modes in (13.50) - (13.52) in such a way that the x, y dependence of $\mathbf{u}, \vartheta, \phi, \pi$ has a periodic shape with wavenumber ε, then after elimination of $\vartheta, \phi, u_1, u_2$, and π one may produce the following non-dimensional equation which w must satisfy on $z \in (0, 1)$,

$$\begin{aligned} &\Delta^4 w - R(1 - fS\Upsilon)\Delta\Delta^* w + \Lambda R(1 - fS\Upsilon)\Delta^* w \\ &\quad - \Lambda\Delta^3 w + R\Upsilon A\Delta(\mathcal{F}w) - \Upsilon R(B - fC)\varepsilon^2 w = 0, \end{aligned} \tag{13.53}$$

where Λ, Υ are concentration and thermal reaction terms, respectively, and

$$\Delta^* \equiv \partial^2/\partial x^2 + \partial^2/\partial y^2, \qquad R = \frac{\alpha\gamma g d^4}{K\nu} \quad \text{(Rayleigh number)},$$

$$\Lambda = \frac{\tau d^2}{\lambda}, \qquad \Upsilon = \frac{\gamma d\delta}{\tau}, \qquad S = \frac{\epsilon}{\tilde{c}} \qquad A = \frac{\beta K\rho}{\alpha\lambda\gamma d},$$

$$B = \frac{\beta d\tau}{\alpha\gamma\lambda}, \qquad C = \frac{d^2\rho\epsilon K\beta}{\lambda\alpha}.$$

The function $\mathcal{F}$, which is the non-dimensional z-part of $\bar{c}'$, is defined by

$$\mathcal{F} = 1 - \sinh\Lambda^{1/2}z + \cosh(\Lambda^{1/2} - 1)\cosh\Lambda^{1/2}z$$

and in (13.53) $f = z\mathcal{F}$.

(Morro and Straughan, 1990) present an asymptotic solution for the approximate free boundary conditions

$$\frac{d^{2n}W}{dz^{2n}} = 0, \quad z = 0, 1; \qquad n = 0, 1, \dots,$$

where $W(z)$ is the z-part of w. This allows one to see the first order effects of the reaction term Υ and the concentration reaction term Λ. Indeed the asymptotic analysis is probably valuable, since the reaction rates of many atmospheric reactions are typically of order 10^{-12}, see e.g. (Kaye and Rood, 1989). By writing

$$W = W_{00} + \Upsilon W_{10} + \Lambda W_{01} + \cdots$$

and

$$R = R_{00} + \Upsilon R_{10} + \Lambda R_{01} + \cdots,$$

(Morro and Straughan, 1990) find

$$R = \frac{27\pi^4}{4}\left\{1 + \Upsilon\left(\frac{S}{4} - \frac{A}{\pi^2} - \frac{B}{3\pi^2}\right) + O(\Upsilon^2, \Upsilon\Lambda, \Lambda^2)\right\},$$

which shows the thermal reaction term may have an important effect.

We conclude this chapter by mentioning work of (McKay, 1998a) who studies the effect of a chemical reaction term on convection in a fluid overlying a porous layer. The chemical reaction is included in the model by a temperature - dependent heat source in both the fluid energy balance and in the energy balance of the porous temperature equation.

14
Multi-component convection diffusion

14.1 Convection with heating and salting below

In the standard Bénard problem the instability is driven by a density difference caused by a temperature difference between the upper and lower planes bounding the fluid. If the fluid layer additionally has salt dissolved in it then there are potentially two destabilizing sources for the density difference, the temperature field and the salt field. A similar scenario could be witnessed in isothermal conditions but with two dissolved salts such as sodium and potassium chloride. When there are two effects such as this the phenomenon of convection which arises is called *double diffusive convection*. For the specific case involving a temperature field and sodium chloride it is frequently referred to as *thermohaline convection*. There are many recent studies involving three or more fields, such as temperature and two salts such as NaCl, KCl. For the three or greater field case we shall refer to *multi-component convection*.

The driving force for many studies in double diffusive or multi-component convection is largely physical applications. For instance modelling geothermal reservoirs, e.g. in the Imperial valley in California, cf. (Cheng, 1978) the Wairakei system in New Zealand, cf. (Griffiths, 1981) near Lake Kinnert in Israel, cf. (Rubin, 1973) the Floridan aquifer, cf. (Kohout, 1965) and in the Salton sea geothermal system in California, cf. (Fournier, 1990), (Helgeson, 1968), (Oldenburg and Pruess, 1998), (Williams and McKibben, 1989), (Younker et al., 1982): a recent review of numerical techniques and their application in geothermal reservoir simulation may be found in (O'Sullivan et al., 2001). The Salton sea geothermal system in southern California is particularly interesting in that it involves convection of hypersaline fluids. For example, to model the Salton sea geothermal system (Oldenburg and Pruess, 1998) develop a model for convection in a Darcy porous medium where the mechanism involves temperature, NaCl, $CaCl_2$, and KCl. Other applications include the oceans, cf. (Stern, 1960), (Cathles, 1990), (Mellor, 1996), (Pedlosky, 1996), the Earth's magma, e.g. (Hansen and Yuen, 1989), (Carrigan and Cygan, 1986), (Huppert and Sparks, 1984). Drainage in a mangrove system, (van Duijn et al., 2001) is yet another area encompassing double diffusive flows. Solar ponds are a particularly promising means of harnessing energy from the Sun by preventing convective overturning in a thermohaline system by salting from below, cf. (Rothmeyer, 1980), (Tabor, 1980), (Zangrando, 1991). Bio-remediation, involving the introduction of micro-organisms to change the chemical composition of contaminants is a very important area, cf. (Celia et al., 1989), (Chen et al., 1994), (Suchomel et al., 1998). Contaminant movement or pollution transport is a further area of multi-component flow in porous media which is of much interest in environmental engineering, cf. (Curran and Allen, 1990), (Ewing et al., 1997), (Ewing, 1996), (Ewing, 1997), (Ewing and Weekes, 1998), (Franchi and Straughan, 2001). Other very important areas of double diffusive transport occur in oil reservoir simulation, cf. (Allen, 1984), (Allen, 1986), (Allen et al., 1988), (Allen et al., 1992), (Ludvigsen et al., 1990), and salinization, cf. (Gilman and Bear, 1996).

To describe nonlinear energy stability results in double-diffusive convection we begin by introducing the relevant equations. For a fluid, these have already been encountered in section 4.6. With no Soret effect the equations are

$$\dot{v}_i = -\frac{1}{\rho_0}p_{,i} + \nu\Delta v_i - gk_i\big(1 - \alpha[T - T_0] + \gamma[C - C_0]\big), \qquad (14.1)$$

$$v_{i,i} = 0, \qquad (14.2)$$

$$\dot{T} = \kappa\Delta T\,, \qquad (14.3)$$

$$\dot{C} = \kappa_C\Delta C\,. \qquad (14.4)$$

If we suppose equations (14.1-14.4) are defined on the spatial region $\mathbb{R} \times \{z \in (0,d)\}$ with $t > 0$, then with boundary conditions of fixed temperatures and salt concentrations, T_L, T_U, C_L, C_U, and no-slip velocity boundary conditions the steady solution in whose stability we are interested is

$$\bar{v}_i \equiv 0, \qquad \bar{T} = -\beta z + T_L, \qquad \bar{C} = -\beta_C z + C_L, \qquad (14.5)$$

where the temperature and concentration gradients are given by

$$\beta = \frac{T_L - T_U}{d}, \qquad \beta_C = \frac{C_L - C_U}{d}. \qquad (14.6)$$

Perturbations u_i, π, θ, ϕ to the steady fields $\bar{v}_i, \bar{p}, \bar{T}$ and $\bar{C}$ then satisfy the equations

$$u_{i,t} + u_j u_{i,j} = -\frac{1}{\rho_0} p_{,i} + \nu \Delta u_i + g\alpha k_i \theta - g\gamma k_i \phi, \tag{14.7}$$

$$u_{i,i} = 0, \tag{14.8}$$

$$\theta_{,t} + u_i \theta_{,i} = \beta w + \kappa \Delta \theta, \tag{14.9}$$

$$\phi_{,t} + u_i \phi_{,i} = \beta_C w + \kappa_C \Delta \phi. \tag{14.10}$$

In general, the thermal and salt diffusivities κ, κ_C are very different and this gives rise to interesting effects. We non-dimensionalize equations (14.7-14.10) with the time, velocity, temperature, concentration, length, scales of $\mathcal{T} = d^2/\kappa$, $U = \kappa/d$, $T^\sharp = (\beta\nu/\alpha g\kappa)^{1/2} U$, $C^\sharp = (\beta_C \nu/\gamma g \kappa_C)^{1/2} U$, and $L = d$. We introduce the Lewis number, Le, (the ratio of diffusivities), the thermal and salt Prandtl numbers, Pr, P_C, and the Rayleigh number and salt Rayleigh numbers, $Ra = R^2$, $Ra_S = R_C^2$, by

$$Le = \frac{\kappa}{\kappa_C}, \quad Pr = \frac{\nu}{\kappa_C}, \quad P_C = \frac{\nu}{\kappa_C}, \quad R^2 = \frac{g\alpha\beta d^4}{\kappa\nu}, \quad R_C^2 = \frac{\beta_C g\gamma d^4}{\kappa_C \nu}.$$

Then equations (14.7-14.10) assume the non-dimensional form

$$\frac{1}{Pr}(u_{i,t} + u_j u_{i,j}) = -\pi_{,i} + \Delta u_i + R k_i \theta - R_C k_i \phi, \tag{14.11}$$

$$u_{i,i} = 0, \tag{14.12}$$

$$\theta_{,t} + u_i \theta_{,i} = \hat{H} R w + \Delta \theta, \tag{14.13}$$

$$Le(\phi_{,t} + u_i \phi_{,i}) = \tilde{H} R_C w + \Delta \phi, \tag{14.14}$$

where $k_i = \delta_{i3}$ and (14.11-14.14) hold on the spatial region $\mathbb{R}^2 \times (0,1)$ with $t > 0$. The constants $\hat{H}$ and $\tilde{H}$ take values ± 1. If the fluid is heated from below, $T_L > T_U$, and then $\hat{H} = 1$, otherwise $\hat{H} = -1$, whereas if the fluid is salted below, $C_L > C_U$, so $\tilde{H} = +1$ with salting above yielding $\tilde{H} = -1$. The boundary conditions are

$$u_i = 0, \quad \theta = 0, \quad \phi = 0, \qquad z = 0, 1, \tag{14.15}$$

and u_i, π, θ, ϕ satisfy a periodic plane tiling pattern in x, y.

In this section we pay particular attention to the case of infinite Prandtl number, Pr. While this is an unphysical case it is very useful example for illustration of what can be achieved with an energy method. The equations for this case are

$$\pi_{,i} = \Delta u_i + R k_i \theta - R_C k_i \phi, \tag{14.16}$$

$$u_{i,i} = 0, \tag{14.17}$$

$$\theta_{,t} + u_i \theta_{,i} = \hat{H} R w + \Delta \theta, \tag{14.18}$$

$$Le(\phi_{,t} + u_i \phi_{,i}) = \tilde{H} R_C w + \Delta \phi. \tag{14.19}$$

The equivalent perturbation equations for a Darcy porous medium may be derived from e.g. (Nield and Bejan, 1999), (Lombardo et al., 2001b) and are

$$\pi_{,i} = -u_i + Rk_i\theta - R_C k_i\phi, \qquad u_{i,i} = 0, \tag{14.20}$$

$$\theta_{,t} + u_i\theta_{,i} = \hat{H}Rw + \Delta\theta, \tag{14.21}$$

$$Le(\epsilon\phi_{,t} + u_i\phi_{,i}) = \tilde{H}R_C w + \Delta\phi, \tag{14.22}$$

where the constant ϵ is given by $\epsilon = \varepsilon M$ with ε the porosity and $M = (\rho_0 c_p)_f/(\rho_0 c)_m$, this being the ratio of heat capacities in the fluid and the porous matrix.

We now return to equations (14.11-14.14). In equations (14.11-14.14) if the layer is salty above and heated below then both effects are destabilizing, $\tilde{H} = -1$, $\hat{H} = 1$ in (14.11-14.14) and the linearized system is symmetric, therefore, the linear and nonlinear boundaries coincide and no sub-critical instabilities can occur. This result was first established by (Shir and Joseph, 1968). If, however, the layer is salted below, which is a stabilizing effect, while the layer is simultaneously heated from below, which is a destabilizing effect, then the two opposing effects make an energy analysis decidely more complicated, to achieve sharp results.

The really interesting situation from both a geophysical and a mathematical viewpoint arises when the layer is simultaneously heated from below and salted from below. In this situation heating expands the fluid at the bottom of the layer and this in turn wants to rise thereby encouraging motion due to thermal convection. On the other hand, the heavier salt at the lower part of the layer has exactly the opposite effect and this acts to prevent motion through convective overturning. Thus, these two physical effects are competing against each other. Due to this competition, it means that the linear theory of instability does not always capture the physics of instability completely and (sub-critical) instabilities may arise before the linear threshold is reached, cf. (Hansen and Yuen, 1989), (Proctor, 1981), (Veronis, 1965). Due to the possibility of sub-critical instabilities occurring, it is very important to obtain (unconditional) nonlinear stability thresholds which guarantee bounds below which convective overturning will not occur. For the heated and salted below situation the equations are (14.11-14.14) with $\tilde{H} = \hat{H} = +1$. For precisely this problem (Joseph, 1970) introduced a new twist into the theory of energy stability; the idea of a *generalized energy*. He sets $\tau = Pc/Pr$, $R_C = \alpha R$ and then chooses

$$E(t) = \frac{1}{2}\|\mathbf{u}\|^2 + \frac{1}{2}\frac{Pr}{1+\tau}\big[\|\gamma\|^2 + \|\psi\|^2\big],$$

where

$$\gamma = \lambda_1\theta - \lambda_2\phi, \qquad \psi = \lambda_1\theta - \tau\lambda_2\phi, \tag{14.23}$$

and where λ_1, λ_2 are coupling parameters linked by the equation

$$\lambda_1 + \frac{1}{\lambda_1} - \frac{2\alpha\lambda_2}{(1+\tau)} = \alpha\Big(\frac{1}{\lambda_2} - \lambda_2\Big) + \frac{2\lambda_1\tau}{1+\tau}.$$

The above choice of energy is evidently necessary to produce a sharp result. Indeed, (Joseph, 1970) finds a stability boundary that is very close to the linear instability one. He indicates that sub-critical instabilities arise in precisely the region delimited by his energy analysis.

From the point of view of Lyapunov functionals, or generalized energies, in fluid mechanics, the paper of (Joseph, 1970) is a very important one. It is the first one I know of where such a *generalized energy* is employed *very effectively* to achieve sharp results evidently not attainable by a standard energy analysis. Joseph's paper has unquestionably had a major influence on the work in energy theory which has subsequently developed.

We now demonstrate some unconditional nonlinear energy stability results for the $Pr = \infty$ case described by equations (14.16-14.19). While we begin with $Le = 1$ we note that this is an instructive case. In this section as throughout much of the book, we concentrate on unconditional nonlinear stability results.

Suppose $Le = 1$ in (14.16-14.19) and take $\hat{H} = \tilde{H} = 1$, i.e. this is the tricky case of heating below and salting below. It is instructive to write the right hand side of (14.16-14.19) as a matrix system in (u_i, θ, ϕ), for then we see it has form

$$\begin{pmatrix} \Delta & 0 & 0 & 0 & 0 \\ 0 & \Delta & 0 & 0 & 0 \\ 0 & 0 & \Delta & R & 0 \\ 0 & 0 & R & \Delta & 0 \\ 0 & 0 & 0 & 0 & \Delta \end{pmatrix} \begin{pmatrix} u \\ v \\ w \\ \theta \\ \phi \end{pmatrix} + \begin{pmatrix} 0 & 0 & 0 & 0 & 0 \\ 0 & 0 & 0 & 0 & 0 \\ 0 & 0 & 0 & 0 & -R_C \\ 0 & 0 & 0 & 0 & 0 \\ 0 & 0 & R_C & 0 & 0 \end{pmatrix} \begin{pmatrix} u \\ v \\ w \\ \theta \\ \phi \end{pmatrix}$$

By writing the right hand side in this way the symmetric effect of R is evident as is the anti-symmetric effect of the salt field via R_C.

To progress with a nonlinear energy stability analysis we introduce a *natural variable* $a = R\theta - R_C\phi$. The introduction of this natural variable and its usefulness in the heated/salted below problem is due to (Mulone, 1994). Equations (14.16-14.19) may be rearranged to combine θ and ϕ into a as

$$\pi_{,i} = \Delta u_i + k_i a, \qquad u_{i,i} = 0, \tag{14.24}$$

$$a_{,t} + u_i a_{,i} = (R^2 - R_C^2)w + \Delta a. \tag{14.25}$$

The boundary conditions are

$$u_i = 0, \quad a = 0, \qquad z = 0, 1$$

with u_i, π and a satisfying the periodicity condition with period cell V.

We first suppose $R^2 < R_C^2$. Then we put $\mu = R_C^2 - R^2 > 0$. Multiply (14.24) by u_i and integrate over V and likewise multiply (14.25) by a and integrate over V to obtain the energy identities

$$0 =(a, w) - \|\nabla \mathbf{u}\|^2 \tag{14.26}$$

$$\frac{d}{dt}\frac{1}{2}\|a\|^2 = - \mu(w, a) - \|\nabla a\|^2. \tag{14.27}$$

By adding this with μ as a coupling parameter one finds

$$\frac{d}{dt}\frac{1}{2}\|a\|^2 = -(\|\nabla a\|^2 + \mu\|\nabla \mathbf{u}\|^2).$$

From this we can clearly deduce that

$$\|a(t)\|^2 \leq \|a(0)\|^2 \exp(-2\pi^2 t) \tag{14.28}$$

and it follows a decays to zero at least exponentially fast in L^2 measure.

Then, from (14.26), using the arithmetic - geometric mean inequality and Poincaré's inequality we have

$$\begin{aligned}\pi^2\|\mathbf{u}\|^2 \leq \|\nabla \mathbf{u}\|^2 = (a, w) &\leq \frac{1}{2\pi^2}\|a\|^2 + \frac{\pi^2}{2}\|\mathbf{u}\|^2 \\ &\leq \frac{1}{2\pi^2}\|a\|^2 + \frac{1}{2}\|\nabla \mathbf{u}\|^2 .\end{aligned} \tag{14.29}$$

Hence,

$$\pi^4\|\mathbf{u}\|^2 \leq \pi^2\|\nabla \mathbf{u}\|^2 \leq \|a\|^2. \tag{14.30}$$

Employing (14.28) we then find that $R^2 < R_C^2$ also guarantees $\|\mathbf{u}\|$ and $\|\nabla \mathbf{u}\|$ decay exponentially fast in addition to $\|a\|$. Physically this is what we expect. The heavier salting at the bottom of the layer is dominating the heating from below and preventing convective overturning. These results are unconditional, i.e. for all initial data.

Suppose now $R^2 > R_C^2$ and put $\gamma = R^2 - R_C^2 > 0$. By multiplying (14.24) by γ this system becomes

$$\gamma\pi_{,i} = \gamma\Delta u_i + \gamma k_i a, \qquad u_{i,i} = 0, \tag{14.31}$$

$$a_{,t} + u_i a_{,i} = \gamma w + \Delta a. \tag{14.32}$$

The right hand side is equivalent to

$$\begin{pmatrix} \gamma\Delta & 0 & 0 & 0 \\ 0 & \gamma\Delta & 0 & 0 \\ 0 & 0 & \gamma\Delta & \gamma \\ 0 & 0 & \gamma & \Delta \end{pmatrix}\begin{pmatrix} u \\ v \\ w \\ a \end{pmatrix}$$

In this manner the symmetry of the system is evident and so we can conclude that the linear instability boundary is the same as the (unconditional) nonlinear energy stability one and hence no sub-critical instabilities are possible. Nevertheless, it is very instructive to prove this directly. To do this multiply (14.31) by u_i, (14.32) by a and integrate each over V to derive the energy identities

$$\begin{aligned} 0 =& \gamma(a,w) - \gamma\|\nabla\mathbf{u}\|^2 \\ \frac{d}{dt}\frac{1}{2}\|a\|^2 =& \gamma(w,a) - \|\nabla a\|^2. \end{aligned}$$

Adding these yields the equation

$$\frac{dE}{dt} = I - D \tag{14.33}$$

where

$$E(t) = \frac{1}{2}\|a(t)\|^2, \quad I(t) = 2\gamma(w,a), \quad D(t) = \|\nabla a\|^2 + \gamma\|\nabla\mathbf{u}\|^2.$$

Thus if we define

$$\frac{1}{R_E} = \max_{\mathcal{H}} \frac{I}{D} \tag{14.34}$$

then we derive

$$\frac{dE}{dt} \le -D\Big(1 - \frac{1}{R_E}\Big). \tag{14.35}$$

Nonlinear stability ensues when $R_E > 1$. The Euler-Lagrange equations from (14.34) are easily seen to be

$$\begin{aligned} &R_E 2\gamma a k_i + 2\gamma\Delta u_i = 2\lambda_{,i} \\ &R_E 2\gamma w + 2\Delta a = 0. \end{aligned} \tag{14.36}$$

For the linear theory it is not difficult to show $\sigma \in \mathbb{R}$ and then the equations governing linear instability are

$$\begin{aligned} &\pi_{,i} = \Delta u_i + a k_i, \quad u_{i,i} = 0, \\ &0 = \gamma w + \Delta a. \end{aligned} \tag{14.37}$$

Since λ is a Lagrange multiplier we divide the first of (14.36) by γ and then for the threshold case $R_E = 1$ it is easily seen that (14.36) are the same as (14.37) and so the linear instability threshold is identical to the nonlinear energy stability one.

To complete the proof we argue from (14.35) as in the case $R_C^2 > R^2$ to show $\|a\|, \|\mathbf{u}\|, \|\nabla\mathbf{u}\|$ are bounded by a decreasing exponential function of time.

We now abandon the restriction $Le = 1$ and derive an unconditional stability result valid for any Lewis number. To achieve this we form energy identities from (14.16-14.19) as

$$0 = -\|\nabla\mathbf{u}\|^2 + R(\theta,w) - R_C(\phi,w) \tag{14.38}$$

$$\frac{d}{dt}\frac{1}{2}\|\theta\|^2 = R(w,\theta) - \|\nabla\theta\|^2 \tag{14.39}$$

$$\frac{d}{dt}\frac{1}{2}Le\|\phi\|^2 = R_C(w,\phi) - \|\nabla\phi\|^2. \tag{14.40}$$

Form now the combination (14.38)+μ(14.40) +λ(14.39) where μ, λ are now positive constants we are using as coupling parameters. This leads to the energy equation (14.33) where now

$$\begin{aligned} E &= \frac{\lambda}{2}\|\theta\|^2 + \frac{\mu Le}{2}\|\phi\|^2 \\ I &= R(1+\lambda)(\theta, w) + R_C(\mu - 1)(\phi, w) \\ D &= \|\nabla \mathbf{u}\|^2 + \lambda\|\nabla\theta\|^2 + \mu\|\nabla\phi\|^2. \end{aligned} \tag{14.41}$$

We may now establish at least exponential decay of E provided $R_E^{-1} < 1$ where $R_E^{-1} = \max_{\mathcal{H}}(I/D)$. From (14.38) we also find

$$\begin{aligned} \pi^2\|\mathbf{u}\|^2 \le \|\nabla\mathbf{u}\|^2 &\le \frac{R}{2\alpha}\|\theta\|^2 + \frac{R_C}{2\beta}\|\phi\|^2 + \left(\frac{R\alpha}{2} + \frac{R_C\beta}{2}\right)\|w\|^2 \\ &\le \frac{R}{2\alpha}\|\theta\|^2 + \frac{R_C}{2\beta}\|\phi\|^2 + \frac{1}{2\pi^2}(R\alpha + R_C\beta)\|\nabla\mathbf{u}\|^2 \end{aligned}$$

for positive constants α, β. Select $\alpha = \pi^2/2R$, $\beta = \pi^2/2R_C$ and then we deduce

$$\pi^2\|\mathbf{u}\|^2 \le \|\nabla\mathbf{u}\|^2 \le \frac{2R^2}{\pi^2}\|\theta\|^2 + \frac{2R_C^2}{\pi^2}\|\phi\|^2.$$

This relation shows that R_E^{-1} guarantees in addition to decay of $\|\theta\|$ and $\|\phi\|$, also decay of $\|\mathbf{u}\|$ and $\|\nabla\mathbf{u}\|$.

To calculate the nonlinear stability boundary it remains to calculate the Euler-Lagrange equations for the maximum of I/D and this we do in the threshold case $R_E = 1$. The relevant Euler-Lagrange equations for a Lagrange multiplier ζ are

$$\begin{aligned} &R(1+\lambda)\theta k_i + R_C(\mu-1)\phi k_i + 2\Delta u_i = \zeta_{,i} \\ &R(1+\lambda)w + 2\lambda\Delta\theta = 0 \\ &R_C(\mu-1)w + 2\mu\Delta\phi = 0. \end{aligned}$$

We may eliminate ζ to derive the equation

$$2\Delta^2 w + R(1+\lambda)\Delta^*\theta + R_C(\mu-1)\Delta^*\phi = 0,$$

where Δ^* is the horizontal Laplacian. While we could solve this eigenvalue system numerically we assume we are dealing with two surfaces free of tangential stress. In this way we may look for solutions of form $w = W \sin n\pi z\, f(x,y)$, $\theta = \Theta \sin n\pi z\, f(x,y)$, $\phi = \Phi \sin n\pi z\, f(x,y)$, where W, Θ, Φ are constant amplitudes and f is a horizontal planform satisfying $\Delta^* f = -a^2 f$, a being the wavenumber. This allows us to deduce the relation below

$$\frac{2(\pi^2n^2 + a^2)^3}{a^2} = \frac{R^2(1+\lambda)^2}{2\lambda} + \frac{R_C^2(\mu-1)^2}{2\mu}.$$

We see that for R^2 minimized in n we need $n = 1$. Since μ is a coupling parameter we choose it to make R^2 as large as possible and so take $\mu = 1$.

Thus

$$R^2 = \frac{4\lambda}{(1+\lambda)^2} \frac{(\pi^2+a^2)^3}{a^2}.$$

The critical Rayleigh number of this energy theory is found as

$$Ra_E = \max_{\lambda} \min_{a^2} R^2$$

for which we find $a^2 = \pi^2/2$ and $\lambda = 1$, then $Ra_E = 27\pi^4/4$.

Thus, for system (14.16-14.19) with stress free velocity boundary conditions we have shown that $R^2 < 27\pi^4/4$ guarantees exponential decay of $\|\mathbf{u}\|, \|\nabla\mathbf{u}\|, \|\theta\|$ and $\|\phi\|$ for all initial data regardless of the values of R_C and Le.

One can prove much more than the above results for the $Pr = \infty$ problem, at the expense of increasing the technical nature of the calculations, by introducing generalized energies. The idea is to use the methods of (Mulone, 1994), (Mulone and Rionero, 1998), (Lombardo et al., 2000), and (Lombardo et al., 2001a).

Detailed linear instability theory for the problem of convection with temperature and salt fields has been worked out by (Nield, 1968) for a porous medium and by (Baines and Gill, 1969) in a fluid.

As we have already pointed out (Joseph, 1970) is the first to use a generalized energy method to incorporate the stabilizing effect of salting below in a nonlinear energy stability analysis. (Joseph, 1970) treats system (14.11-14.14) with $\hat{H} = \tilde{H} = 1$ and works with the variables γ and ψ given by (14.23). His energy uses two functionals,

$$\Psi(t) = \frac{Pr}{2(1+\tau)}\|\psi\|^2 \qquad \text{and} \qquad E(t) = \frac{1}{2}\|\mathbf{u}\|^2 + \frac{Pr\,\tau}{(1+\tau)}\|\phi\|^2.$$

He derives a sharp stabilty threshold with stability guaranteed in the sense that

$$\lim_{t\to\infty} \int_0^t E(s)ds < \infty. \tag{14.42}$$

(Rionero and Mulone, 1987) provide a rigorous linearization principle for double-diffusive convection problems using energy ideas. (Mulone, 1991b) develops a conditional nonlinear energy stability analysis for the heated and salted below problem allowing the fluid also to be undergoing Poisueille or Couette flow. This work is extended by (Mulone, 1994) but he now develops a sharp unconditional result for the heated and salted below problem. The analysis of (Mulone, 1994) also derives exponential decay for the energy instead of the condition (14.42). It is in this paper where the use of the "natural" variable suggested by the density, namely

$$a = R\theta - R_C\phi$$

is used with great effect. (Mulone, 1994) develops a nonlinear energy stability theory based on the generalized energy functional

$$E(t) = \frac{1}{2}\|\mathbf{u}\|^2 + \frac{1}{2}\lambda_1 P_C\|a\|^2 + \frac{1}{2}\lambda_2 Pr\|\theta\|^2.$$

Very sharp nonlinear stability thresholds are derived and compared carefully with the linear instability boundaries. (Mulone and Rionero, 1998), (Mulone, 1998), (Lombardo et al., 2000), (Xu, 2000), and (Lombardo et al., 2001a) derive further sharp nonlinear stability boundaries for the heated and salted below problem for a fluid by employing other kinds of generalized energies. The decay obtained is always exponential in time. (Lombardo et al., 2001b) and (Lombardo and Mulone, 2002a) use not dissimilar generalized energies to derive sharp nonlinear stability boundaries in the analogous problem for a porous medium.

Further interesting nonlinear (conditional) energy stability results in double-diffusive convection are due to (Guo et al., 1994), (Guo and Kaloni, 1995a), (Guo and Kaloni, 1995c), (Qin et al., 1995). These writers derive nonlinear stability bounds for double-diffusive convection in a Brinkman porous medium, (Guo and Kaloni, 1995a), in a rotating layer in (Guo et al., 1994) and (Guo and Kaloni, 1995c), and incorporating penetrative convection in (Qin et al., 1995). (Guo and Kaloni, 1995b) develop a nonlinear unconditional analysis for the interesting problem in a porous medium when the convection is induced by temperature and concentration gradients which have a horizontal as well as a vertical component. This nonlinear energy theory complements the linearized instability theory of (Nield et al., 1993). In addition (Kaloni and Qiao, 1997a), (Kaloni and Qiao, 2000), (Qiao and Kaloni, 1997), (Qiao and Kaloni, 1998), have developed a series of interesting articles in which they establish *unconditional* nonlinear energy stability results for double diffusive problems but when the temperature and concentration gradients vary in the horizontal as well as the vertical directions. The analogous problems in a fluid and in a porous medium are investigated and throughflow effects are additionally incorporated. (Bardan and Mojtabi, 1998) and (Bardan et al., 2000) study a nonlinear double diffusive problem in a two-dimensional enclosure. The temperature and concentration gradients are horizontal and opposing. Subcritical instabilities are found and a careful bifurcation analysis is performed. (Charrier-Mojtabi et al., 1998) and (Karimi-Fard et al., 1999) investigate double diffusive convection in a rectangular container filled with saturated porous material. The driving forces are horizontal temperature and concentration gradients and in the case of (Karimi-Fard et al., 1999) the rectangular cavity is tilted. These are both very interesting articles and go into much detail about the bifurcation patterns which develop.

14.2 Convection with three components

We have already seen that the solution behaviour in the double-diffusive convection problem is more interesting than that of the single component situation in so much as new instability phenomena may occur which are not present in the classical Bénard problem. Nonlinear energy stability theory is correspondingly more complicated and mathematically richer. When temperature and two or more component agencies, or three different salts, are present then the physical and mathematical situation becomes increasingly richer. Very interesting results in triply diffusive convection have been obtained by (Pearlstein et al., 1989). The results of (Pearlstein et al., 1989) are in some ways remarkable. They demonstrate that for triple diffusive convection linear instability can occur in discrete sections of the Rayleigh number domain with the fluid being linearly stable in a region in between the linear instability ones. This is because for certain parameters the neutral curve has a finite isolated oscillatory instability curve lying below the usual unbounded stationary convection one. The shape of the oscillatory convection curve is topologically complex being approximately heart shaped which in turn leads to the possibility of complex dynamical behaviour. (Pearlstein et al., 1989) cite specific examples of convection in a triply diffusive system, in crystal growth, (Coriell et al., 1987), and the chemical application of (Noulty and Leaist, 1989).

(Straughan and Walker, 1997) derive the equations for non-Boussinesq convection in a multi - component fluid [1] and investigate the situation analogous to that of (Pearlstein et al., 1989) but allowing for a density nonlinear in the temperature field. In this way they also encompass penetrative convection.

(Pearlstein et al., 1989) treat triply diffusive convection with a density linear in all fields and they employ stress free boundary conditions. They find that the oscillatory instability curve has a perfectly symmetric shape about a vertical axis. In (Lopez et al., 1990) they study the equivalent problem with fixed boundary conditions and show that the effect of the boundary conditions breaks the perfect symmetry. In reality the density of a fluid is never a linear function of temperature, and so the work of (Straughan and Walker, 1997) applies to the general situation where the equation of state is one of the density being quadratic in temperature, T. This is important, since they find that departure from the linear Boussinesq equation of state changes the perfect symmetry of the heart shaped neutral curve of (Pearlstein et al., 1989). This means that one is never likely to actually see the interesting dynamics predicted by (Pearlstein et al.,

[1] Some of the material in this section is reprinted from Fluid Dynamics Research, Vol. 19, B. Straughan and D.W. Walker, Multi component diffusion and penetrative convection, pp. 77-89, Copyright (1997), with permission from Elsevier Science.

1989). In fact, (Pearlstein et al., 1989) find that one can have a "heart shaped" curve lying below the usual unbounded stationary convection one and it is isolated in that below the stationary convection curve and above the lobes of the heart shaped curve the system is linearly stable. The lobes of the heart shape have the same Rayleigh number but different wavenumbers which suggests instability could occur at a given Rayleigh number for *two* distinct values of the wavenumber. The breaking of symmetry by rigid boundary conditions, (Lopez et al., 1990), or due to a non-Boussinesq equation of state, (Straughan and Walker, 1997), does not allow this equal Rayleigh number lobe behaviour. (Straughan and Walker, 1997) do, however, find that in the non-Boussinesq situation one can have equal *minima* of the stationary convection curve and the isolated oscillatory convection curve. This is further described after (14.51). While (Lopez et al., 1990) also find the perfect symmetry of (Pearlstein et al., 1989) is broken, they do not find the dynamical behaviour of (Straughan and Walker, 1997).

The work of (Straughan and Walker, 1997) restricts attention to two stress free surfaces since this is likely to be a physically relevant case in astrophysics. However, this still does not permit the use of the analytical method of (Pearlstein et al., 1989) who also employed surfaces free of tangential stress. The coefficients of the differential equations for the instability eigenvalue problem in (Straughan and Walker, 1997) are functions of the spatial variables. This necessitates a numerical approach and so they employ a Chebyshev tau method. This appraoch yields as many eigenvalues as are desired and so the structure of the growth rate may be studied for more than the leading eigenvalue. In fact, (Straughan and Walker, 1997) find a change from a complex conjugate eigenvalue to a real one as the eigenvalues change positions in Rayleigh number parameter space.

(Straughan and Walker, 1997) develop a linearized instability analysis for the problem of penetrative thermal convection in a three-component fluid. They study the situation of a layer of fluid occupying the three - dimensional region $\mathbb{R}^2 \times \{0 < z < d\}$. The lower plane of the fluid layer is held at the fixed temperature $T = 0°$C while the upper boundary is maintained at a constant temperature $T_1 \geq 4°$C. Thus for a working fluid of water which has a density maximum at approximately 4°C, penetrative convection can occur and convection which commences in the gravitationally unstable lower section of the layer may penetrate into the upper part. (Straughan and Walker, 1997) assume that the fluid occupying the layer has dissolved in it two or more different species of chemicals and so they have a three or more component fluid with the solvent being one component. The analysis of (Straughan and Walker, 1997) proceeds by developing the general theory for $5 + A$ equations, A being the number of species components. We here present equations for the case of two species of chemicals. The mathematical picture is described by 7 partial differential equations for the velocity, pressure, temperature, and 2 species components, of concentrations C_1 and C_2. The governing partial differential equations are,

$$
\begin{aligned}
&\rho_0(v_{i,t} + v_j v_{i,j}) = -p_{,i} + \rho_0 \nu \Delta v_i - g\rho(T, C_1, C_2)k_i, \qquad v_{i,i} = 0,\\
&T_{,t} + v_i T_{,i} = \kappa \Delta T,\\
&C_{1,t} + v_i C_{1,i} = \kappa_1 \Delta C_1,\\
&C_{2,t} + v_i C_{2,i} = \kappa_2 \Delta C_2,
\end{aligned}
$$

where ρ_0 is a constant, $\mathbf{k} = (0,0,1)$, and where $v_i, p, \nu, g, \kappa, \kappa_\alpha$ denote, respectively, velocity, pressure, viscosity, gravity, thermal diffusivity, and solute diffusivity of component α. The density is quadratic in the temperature field and linear in the two concentrations, i.e.

$$
\rho = \rho_0\big[1 - \alpha_T(T-4)^2 + A_1(C_1 - \hat{C}_1) + A_2(C_2 - \hat{C}_2)\big], \tag{14.43}
$$

where $\hat{C}_1, \hat{C}_2$, are constants.

The boundary conditions equivalent to those of (Straughan and Walker, 1997) are

$$
\begin{aligned}
T = 0, \quad C^1 = C^1_\ell, \quad C^2 = C^2_\ell, \qquad z = 0,\\
T = T_1, \quad C^1 = C^1_u, \quad C^2 = C^2_u, \qquad z = d,
\end{aligned} \tag{14.44}
$$

for prescribed constants $T_1, C^\alpha_\ell, C^\alpha_u$, $\alpha = 1,2$. They adopt stress free velocity boundary conditions.

The steady (conduction) solution whose stability is under investigation is

$$
\bar{v}_i = 0, \quad \bar{T} = \beta z, \quad \bar{C}^1 = C^1_\ell - \frac{\Delta C^1}{d}\,z, \quad \bar{C}^2 = C^2_\ell - \frac{\Delta C^2}{d}\,z, \tag{14.45}
$$

with $\beta = T_1/d$, $\Delta C^\alpha = C^\alpha_\ell - C^\alpha_u$, $\alpha = 1,2$. With the scalings $t = t^* d^2/\kappa, \mathbf{x} = \mathbf{x}^* d, Pr = \nu/\kappa, U = \nu/d$, $P = \nu U \rho_0/d, \xi = 4/T_1$, $T^\sharp = U\sqrt{\nu/\kappa\alpha_T d g}$, $C^\sharp_\alpha = U\sqrt{|\Delta C^\alpha|/\kappa g d A_\alpha}$, $\tau_\alpha = \kappa_\alpha/\kappa$, $H_\alpha = \operatorname{sgn}(\Delta C^\alpha)$, we define the Rayleigh number, R^2, and the solutal Rayleigh numbers R_1^2 and R_2^2 as

$$
R^2 = T_1^2\left(\frac{\alpha_T g d^3}{\nu\kappa}\right), \qquad R_1^2 = \frac{|\Delta C^1| d^3 g A_1}{\nu^2\kappa}, \qquad R_2^2 = \frac{|\Delta C^2| d^3 g A_2}{\nu^2\kappa}.
$$

The quantities $T^\sharp$ and $C^\sharp_\alpha$ are temperature and concentration scales. The non-dimensional perturbation equations to the conduction solution (14.45) become, with π being the pressure perturbation,

$$
\begin{aligned}
Pr^{-1}u_{i,t} + u_j u_{i,j} &= -\pi_{,i} + \Delta u_i - 2R\theta k_i(\xi - z)\\
&\qquad - (R_1 c_1 + R_2 c_2)k_i + Pr\theta^2 k_i,\\
u_{i,i} &= 0,\\
\theta_{,t} + Pr\, u_i\theta_{,i} &= -Rw + \Delta\theta,\\
c^1_{,t} + Pr\, u_i c^1_{,i} &= H_1 R_1 w + \tau_1 \Delta c_1,\\
c^2_{,t} + Pr\, u_i c^2_{,i} &= H_2 R_2 w + \tau_2 \Delta c_2.
\end{aligned} \tag{14.46}
$$

(Straughan and Walker, 1997) concentrate on cases where the H_α have different signs and they develop a systematic linear analysis. They do cite one nonlinear energy stability result. This shows that when all the salts in the steady state are "salting" from above, and the layer is heated from below, then the nonlinear energy stability boundary coincides with the linear instability one in the *non - penetrative convection case*, where the density is linear in T. Thus sub-critical bifurcation is not possible in that particular scenario. We can illustrate this in the case of two salt fields and then instead of (14.43) one considers the linear equation of state of (Pearlstein et al., 1989),

$$\rho = \rho_0\big[1 - \alpha_T(T - \hat{T}_0) + A_1(C_1 - \hat{C}_1) + A_2(C_2 - \hat{C}_2)\big]. \qquad (14.47)$$

Thus, assume we have equations (14.46) but with the density relation (14.43) replaced by (14.47). The nonlinear, non-dimensional perturbation equations which one derives become

$$\begin{aligned}
&Pr^{-1}u_{i,t} + u_j u_{i,j} = -\pi_{,i} + \Delta u_i + R\theta k_i - (R_1 c_1 + R_2 c_2)k_i,\\
&u_{i,i} = 0,\\
&\theta_{,t} + Pr\, u_i \theta_{,i} = Rw + \Delta\theta, \qquad\qquad (14.48)\\
&c^1_{,t} + Pr\, u_i c^1_{,i} = H_1 R_1 w + \tau_1 \Delta c_1,\\
&c^2_{,t} + Pr\, u_i c^2_{,i} = H_2 R_2 w + \tau_2 \Delta c_2.
\end{aligned}$$

When $H_1 = H_2 = -1$, equations (14.48) are symmetric and an unconditional nonlinear energy stability analysis in the energy

$$E(t) = \frac{1}{2Pr}\|\mathbf{u}\|^2 + \frac{1}{2}\|\theta\|^2 + \frac{1}{2} < c^1 c^1 > + \frac{1}{2} < c^2 c^2 >$$

leads to the same critical Rayleigh number as a linearized instability analysis. One shows the growth rate of linearized instability in this case is real and then the equations to be solved for the linear instability critical Rayleigh number threshold are

$$\begin{aligned}
&\Delta^2 w = -R\Delta^*\theta + \sum_{\alpha=1}^{2} R_\alpha \Delta^* c_\alpha,\\
&-Rw = \Delta\theta, \qquad R_\alpha w = \tau_\alpha \Delta c_\alpha, \quad \alpha = 1, 2.
\end{aligned}$$

One shows from this that

$$\frac{(\pi^2 + a^2)^3}{a^2} = R^2 + \frac{R_1^2}{\tau_1} + \frac{R_2^2}{\tau_2}.$$

Hence, due to symmetry in equations (14.48) the threshold for linearized instability and for nonlinear stability is the boundary

$$\frac{27\pi^4}{4} = R^2 + \frac{R_1^2}{\tau_1} + \frac{R_2^2}{\tau_2}. \qquad (14.49)$$

We have just demonstrated this coincidence for stress free boundaries, but the method works and the result is also true in the fixed velocity boundary condition case (although the left hand side of (14.49) changes from $27\pi^4/4$).

(Straughan and Walker, 1997) develop a linearized instability analysis in the general case where (14.46) have A concentration equations rather than 2. They concentrate on the case of $A = 2$ where they have a temperature field and two concentrations, C_1, C_2, only when a numerical investigation of the instability boundary is sought. We here restrict attention to the problem of $H_1 = +1, H_2 = -1$, which means salting below (bottom heavy) in C_1 with salting above in C_2. Thus, the C_1 effect is stabilizing whereas C_2 is destabilizing. In addition, the temperature field destabilizes. Thus, there is competition between two destabilizing effects and one stabilizing effect. The linearized equations which one finds from (14.46) are, after eliminating π and seeking a time-dependence like $e^{\sigma t}$,

$$\begin{aligned}
&\sigma Pr^{-1}\Delta w = \Delta^2 w - 2R(\xi - z)\Delta^*\theta - R_1\Delta^* c_1 - R_2\Delta^* c_2\,,\\
&\sigma\theta = -Rw + \Delta\theta,\\
&\sigma c_1 = R_1 w + \tau_1\Delta c_1,\\
&\sigma c_2 = -R_2 w + \tau_2\Delta c_2,
\end{aligned}\tag{14.50}$$

where $\Delta^* = \partial^2/\partial x^2 + \partial^2/\partial y^2$.

The normal mode form of equations arising from (14.50) is

$$\begin{aligned}
&(D^2 - a^2)\big[(D^2 - a^2) - \sigma Pr^{-1}\big]W + 2R(\xi - z)a^2\Theta\\
&\qquad\qquad\qquad\qquad + R_1 a^2 C_1 + R_2 a^2 C_2 = 0,\\
&\big[(D^2 - a^2) - \sigma\big]\Theta - RW = 0,\\
&\big[(D^2 - a^2) - \sigma\tau_1^{-1}\big]C_1 + \frac{R_1}{\tau_1}W = 0,\\
&\big[(D^2 - a^2) - \sigma\tau_2^{-1}\big]C_2 - \frac{R_2}{\tau_2}W = 0,
\end{aligned}\tag{14.51}$$

where $D = d/dz$, a is the wavenumber, and W, Θ, C_1, C_2, are the $z-$dependent parts of w, θ, c_1 and c_2. The above presentation is different from that of (Straughan and Walker, 1997) and care must be taken in comparing the Rayleigh numbers above with those of (Straughan and Walker, 1997).

The numerical findings of (Straughan and Walker, 1997) are based on a Chebyshev tau analysis of (14.51) although the Rayleigh number interpretations are different. Many results are reported in this paper for various upper temperatures in the range 4°C - 8°C. We only report one result. This is highlighted in figure 14.1 which illustrates the neutral stability curves for a special case. The upper temperature is 7°C and the salt Rayleigh numbers are $R_1^2 = 950.964$, $R_2^2 = 814.1119$. The salt diffusivities have values $\tau_1 = 0.22$, $\tau_2 = 0.21$. It is important to note that in the steady state the system is bottom heavy in the species with the larger diffusion coefficient

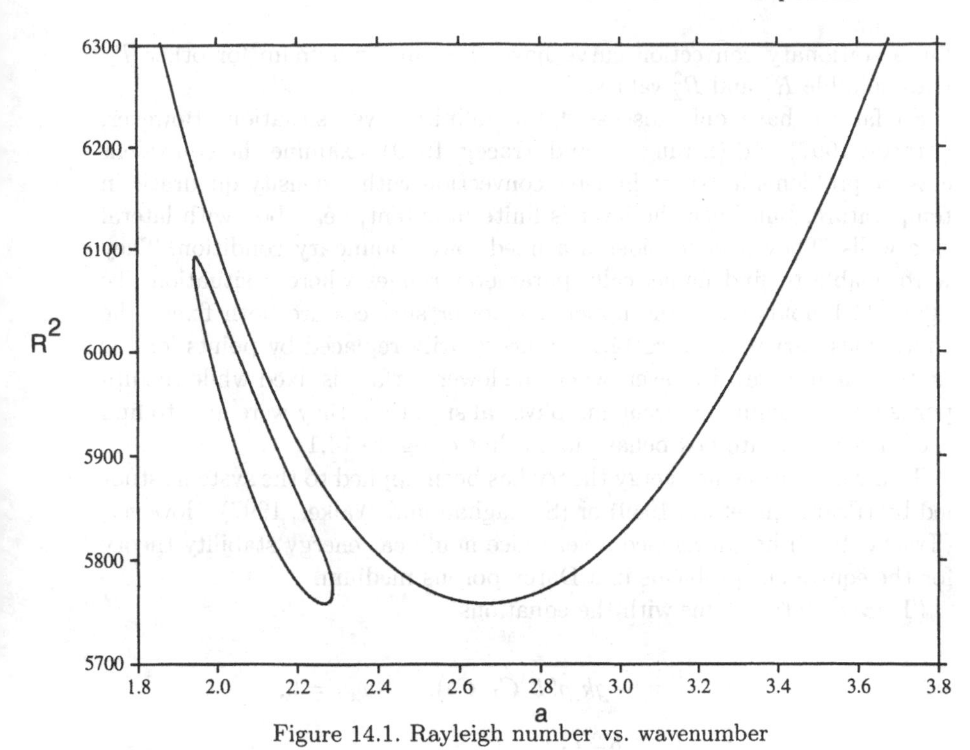

Figure 14.1. Rayleigh number vs. wavenumber

whereas it is top heavy in the species with the smaller diffusion coefficient, although the diffusion coefficients are close together in numerical value. Thus, this is essentially a stable configuration. However, temperature is destabilizing in the conduction state and this together with the competing concentration fields yields the complex dynamical behaviour depicted by figure 14.1. This allows the possibility for convection to commence simultaneously as stationary convection with a wavenumber $a = 2.66$ and as oscillatory convection with wavenumber $a = 2.27$. If R_1^2 is decreased, i.e. salting below decreased in species 1, then the oscillatory curve drops and the first occurrence of linear instability is via oscillatory convection. When R_1^2 is increased the opposite happens, the isolated closed curve rises and convection commences via stationary convection.

We have only shown the situation where the isolated oscillatory convection curve and the infinite stationary convection curve have the same minimum for one set of Rayleigh number parameters and one upper temperature, $T_1 = 7°$C. However, by careful numerical searching one can find other cases. For example, (Straughan and Walker, 1997) report a similar situation when the upper temperature T_1 is 8°C. The closed oscillatory convection curve becomes much thinner as T_1 increases. We could produce other examples where the isolated oscillatory convection curve and the infinite stationary convection curve have the same minimum for other T_1, with suitable R_1^2 and R_2^2 values.

So far we have only discussed the infinite layer situation. However, (Tracey, 1997) and (Straughan and Tracey, 1999), examine the equivalent class of problems in triply diffusive convection with a density quadratic in temperature, but when the layer is finite in extent, i.e. a box with lateral side walls. They also considered a fixed lower boundary condition. They were unable to find numerically parameter ranges where a situation like figure 14.1 holds when the upper and lower surfaces are both fixed (the continuous curves of figure 14.1 are necessarily replaced by points for the finite domain case). However, when the lower surface is fixed while the upper is free, a commonly occurring physical situation, they were able to find the analogous finite box behaviour to that of figure 14.1.

To my knowledge no energy theory has been applied to the systems studied by (Pearlstein et al., 1989) or (Straughan and Walker, 1997). However, (Tracey, 1997) has developed a very nice nonlinear energy stability theory for the equivalent problems in a Darcy porous medium.

(Tracey, 1997) begins with the equations

$$\begin{aligned}
&p_{,i} = -\frac{\mu}{k}\, v_i - g k_i \rho(T, C_1, C_2), \qquad v_{i,i} = 0,\\
&T_{,t} + v_i T_{,i} = \kappa \Delta T,\\
&C^1_{,t} + v_i C^1_{,i} = \kappa_1 \Delta C^1,\\
&C^2_{,t} + v_i C^2_{,i} = \kappa_2 \Delta C^2.
\end{aligned} \tag{14.52}$$

The density ρ has form, either,

$$\rho = \rho_0 \big[1 - A(T - T_0) + A_1(C^1 - C^1_0) + A_2(C^2 - C^2_0)\big] \tag{14.53}$$

or, in the penetrative convection case with water as the saturating fluid

$$\rho = \rho_0 \big[1 - A(T - 4)^2 + A_1(C^1 - C^1_0) + A_2(C^2 - C^2_0)\big] \tag{14.54}$$

where $A, A_1, A_2, T_0, C^1_0, C^2_0$ are constants.

We concentrate first on the density given by (14.53). For boundary conditions $T(0) = T_L, T(d) = T_U$, $C^\alpha(0) = C^\alpha_L, C^\alpha(d) = C^\alpha_U$, $\alpha = 1, 2$, with $v_3 = 0$ at $z = 0, d$ the conduction solution to (14.52) with density given by (14.53) is

$$\bar{T} = T_L - \beta z, \qquad \bar{C}^1 = C^1_L - \beta_1 z, \qquad \bar{C}^2 = C^2_L - \beta_2 z, \tag{14.55}$$

with $\beta = (T_L - T_U)/d$, $\beta^\alpha = (C^\alpha_L - C^\alpha_U)/d$, $\alpha = 1, 2$. The non-dimensional perturbation equations arising from this steady solution are (Tracey, 1997),

p. 14,

$$\pi_{,i} = -u_i + (R\theta - R_1\phi^1 - R_2\phi^2)k_i, \tag{14.56}$$

$$u_{i,i} = 0, \tag{14.57}$$

$$\theta_{,t} + u_i\theta_{,i} = HRw + \Delta\theta, \tag{14.58}$$

$$P_1(\phi^1_{,t} + u_i\phi^1_{,i}) = H_1R_1w + \Delta\phi^1, \tag{14.59}$$

$$P_2(\phi^2_{,t} + u_i\phi^2_{,i}) = H_2R_2w + \Delta\phi^2, \tag{14.60}$$

these equations holding on $\mathbb{R}^2 \times \{z \in (0,d)\} \times \{t > 0\}$ with the boundary conditions

$$w = \theta = \phi^1 = \phi^2 = 0, \qquad \text{on } z = 0, 1, \tag{14.61}$$

together with a plane tiling periodicity with period cell V.

(Tracey, 1997) develops a comprehensive linear instability analysis for (14.56-14.60). The linearized analysis is very detailed and yields many interesting effects including calculating the location of isolated oscillatory convection curves which lie below the standard unbounded stationary convection one. Our interest here centres mainly on his energy stability work. For this he first develops an L^2 analysis based on multiplying (14.56) by u_i, (14.58) by θ, (14.59) by ϕ^1, (14.60) by ϕ^2 and with coupling parameters λ, ξ, μ he derives

$$\frac{dE}{dt} = I - D \tag{14.62}$$

where

$$\begin{aligned} E =& \frac{1}{2}\lambda\|\theta\|^2 + \frac{1}{2}\xi P_1\|\phi^1\|^2 + \frac{1}{2}\mu P_2\|\phi^2\|^2 \\ I =& (\lambda H + 1)R(w,\theta) \\ &+ (\xi H_1 - 1)R_1(w,\phi^1) + (\mu H_2 - 1)R_2(w,\phi^2) \\ D =& \|\mathbf{u}\|^2 + \lambda\|\nabla\theta\|^2 + \xi\|\nabla\phi^1\|^2 + \mu\|\nabla\phi^2\|^2. \end{aligned} \tag{14.63}$$

This leads him to unconditional exponential decay results in $E(t)$ together with similar unconditional nonlinear stability in $\|\mathbf{u}(t)\|$. The Rayleigh number thresholds depend on the values of H, H_1, H_2, which are ± 1 depending on whether there is heating or salting above or below.

(Tracey, 1997) selects two cases. The first is $H = 1$, $H_1 = H_2 = -1$, so that there is heating below and salting above in both C_1 and C_2. All three effects are destabilizing and (14.56-14.60) defines a symmetric system. Thus, he derives the optimal result that the linear instability and nonlinear stability boundaries coincide and no sub-critical instabilities occur.

He then studies the very interesting case of $H = H_1 = 1$ with $H_2 = -1$. This is the competitive case where there is heating below, salting below in component 1 but salting above in component 2. For this situation (14.56-14.60) does not define a symmetric system. His energy theory based on

(14.63) for this particular case leads to the nonlinear stability boundary

$$R^2 + R_2^2 = 4\pi^2. \tag{14.64}$$

To improve on this threshold (Tracey, 1997) develops a very interesting generalized energy analysis. To do this he introduces the variables

$$\psi = \phi_1 + \phi_2 \qquad \text{and} \qquad \phi = \phi_1 - \phi_2. \tag{14.65}$$

He works with the generalized energy

$$E(t) = \frac{1}{2}\gamma_0\|\theta\|^2 + \frac{1}{2}\gamma_1\|\psi\|^2 + \frac{1}{2}\gamma_2\|\phi\|^2 \tag{14.66}$$

for coupling parameters $\gamma_0, \gamma_1, \gamma_2$. (Tracey, 1997) deduces exponential decay in $E(t)$ and $\|\mathbf{u}(t)\|$ from his energy equation and calculates the appropriate Euler-Lagrange equations for the Rayleigh number stability boundary.

The way the nonlinear energy stability boundary is computed and the improvement obtained over the standard energy boundary (14.64) is very interesting indeed. Many results are given in (Tracey, 1997), pp. 56-58, and the nature of these is depicted in the schematic figure 14.2. In figure 14.2 curve 3 represents the linear instability curve. Above this curve the steady solution is unstable. The kink is due to the occurrence of an isolated oscillatory convection curve. Curve 1 is the standard L^2 energy stability threshold given by (14.64). Between curves 1 and 3 would be the region of possible sub-critical instabilities. (Tracey, 1997) fixes γ_1, γ_2 and computes a nonlinear stability curve which gives a point on curve 2. By varying γ_1 and γ_2 one obtains the curve 2. Below this curve no instability can arise. Between curves 2 and 3 is now the region of possible sub-critical instabilities. This is a very strong energy stability result and is a major improvement over standard L^2 theory.

When the density is given by (14.54) (Tracey, 1997) adopts constant concentration boundary conditions and fixes $T_L = 0°\text{C}$ with $T_U = T_1°\text{C}$ where $T_1 \geq 4°\text{C}$. This gives him the steady (conduction) solution $\bar{v}_i = 0$, $\bar{C}^\delta = C_L^\delta - \beta_\delta z$, $\delta = 1, 2$, and $\bar{T} = T_1 z/d$. The non-dimensional, nonlinear perturbation equations for this problem are

$$\begin{aligned}
&\pi_{,i} = -u_i - 2R(\xi - z)\theta k_i - R_1\phi^1 k_i - R_2\phi^2 k_i + \theta^2 k_i,\\
&u_{i,i} = 0,\\
&\theta_{,t} + u_i\theta_{,i} = -Rw + \Delta\theta,\\
&P_1(\phi^1_{,t} + u_i\phi^1_{,i}) = H_1 R_1 w + \Delta\phi^1,\\
&P_2(\phi^2_{,t} + u_i\phi^2_{,i}) = H_2 R_2 w + \Delta\phi^2.
\end{aligned} \tag{14.67}$$

A careful linear instability investigation is carried out by (Tracey, 1997) who finds parameters where the minimum of the isolated oscillatory convection curve is the same as that of the unbounded stationary convection one, in a manner similar to that depicted in figure 14.1.

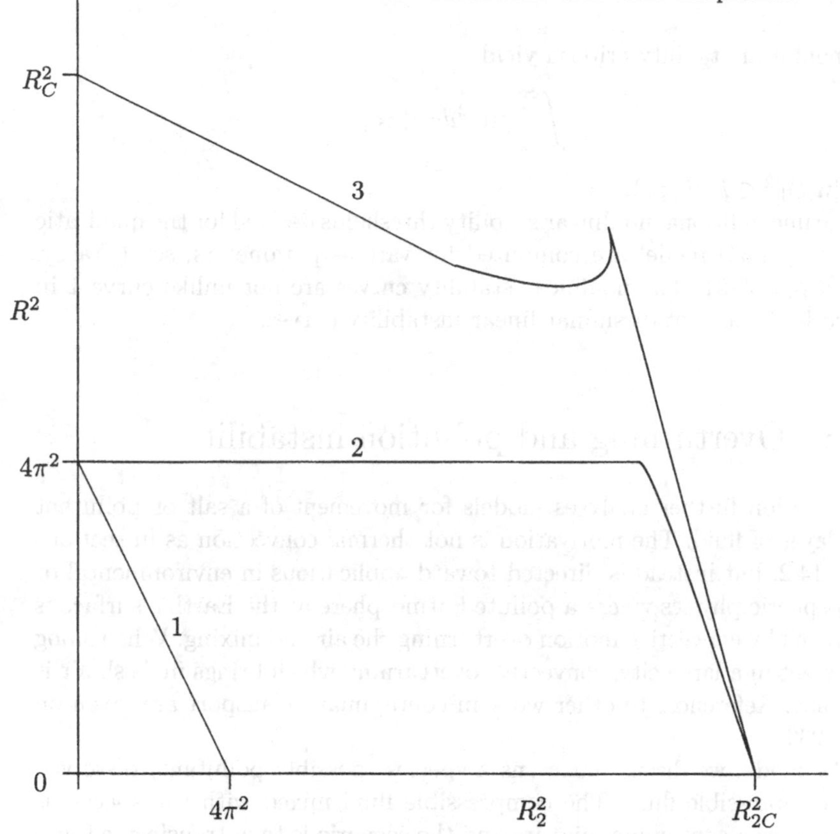

Figure 14.2. Stability and instability boundaries

A standard energy identity may be formed from (14.67) and we see that

$$\|\mathbf{u}\|^2 = -2R < (\xi - z)\theta w > -R_1(\phi^1, w) - R_2(\phi^2, w) + < \theta^2 w > . \quad (14.68)$$

The cubic term $\theta^2 w$ is troublesome if one desires to have an unconditional nonlinear stability theory. Hence, (Tracey, 1997) employs a weight in the velocity part of the energy, cf. section 17.1. He works with the energy

$$E(t) = \frac{1}{2} < \hat{\mu}\theta^2 > + \frac{1}{2}\lambda_1 P_1 \|\phi_1\|^2 + \frac{1}{2}\lambda_2 P_2 \|\phi_2\|^2,$$

for $\hat{\mu} = \mu - 2z$ with $\mu > 2$, λ_1, λ_2 coupling parameters. This choice of energy removes the cubic term of (14.68) and leads to a quadratic energy stability theory with equation (14.62) where now

$$I = -2R < (\xi - z)\theta w > + (\lambda_1 H_1 - 1)R_1(w, \phi_1) + (\lambda_2 H_2 - 1)R_2(w, \phi_2),$$
$$D = \|\mathbf{u}\|^2 + < \hat{\mu}|\nabla\theta|^2 > + \lambda_1 \|\nabla\phi_1\|^2 + \lambda_2 \|\nabla\phi_2\|^2.$$

He then derives conditions for exponential decay in the measure $E(t)$. Interestingly, a similar decay in u_i is not proven. Instead, it is shown that the nonlinear stability criteria yield

$$\int_0^\infty \|\mathbf{u}\|^2 ds < \infty,$$

i.e. $\|\mathbf{u}(t)\|^2 \in L^1(0,\infty)$.

The unconditional nonlinear stability thresholds derived for the quadratic density (14.54) model are computed for various parameters, see (Tracey, 1997), pp. 88-91. The nonlinear stability curves are not unlike curve 1 in figure 14.2 with not dissimilar linear instability curves.

14.3 Overturning and pollution instability

This section further analyses models for movement of a salt or pollutant in a layer of fluid. The motivation is not thermal convection as in sections 14.1, 14.2, but instead is directed toward applications in environmental or atmospheric physics where a polluted atmosphere at the Earth's surface is improved by convective motion overturning the air and mixing. When smog is present in a large city, convective overturning which brings in fresh air is welcome. References to other work in contaminant transport are given on page 239.

The model we discuss concerns a species (possibly pollutant) dissolved in a compressible fluid. The compressible fluid mixed with the species is regarded as a continuum mixture and the scenario is then transformed into one in which the fluid is incompressible. The models are due to (Graffi, 1955), (Kazhikhov and Smagulov, 1977), and (Beirao da Veiga, 1983). Various questions of existence, regularity and uniqueness are investigated by (Beirao da Veiga, 1983), (Beirao da Veiga et al., 1982), (Graffi, 1955), (Kazhikhov and Smagulov, 1977), (Prouse and Zaretti, 1987) and (Secchi, 1982). These papers place the Graffi, Kazhikhov - Smagulov, and Beirao da Veiga models on a sound mathematical footing. (Franchi and Straughan, 2001) develop a Graffi - Kazhikhov - Smagulov model to be applicable to convective overturning in a plane layer.

To describe the Graffi, Kazhikhov - Smagulov, and Beirao da Veiga models [2] we begin with a horizontal layer containing a mixture of two incompressible miscible fluids. Before mixing the densities of fluids 1 and 2 are ρ_{10} and ρ_{20}, respectively, while at point $\mathbf{x}$ and time t in the mixture they are $\rho_1(\mathbf{x},t)$ and $\rho_2(\mathbf{x},t)$. It is assumed that upon mixing the volume of the two individual incompressible fluids at the outset does not change

[2]Some of the material in this section is reprinted from Advances in Water Resources, Vol. 24, F. Franchi and B. Straughan, A comparison of the Graffi and Kazhikhov-Smagulov models for top heavy pollution instability, pp. 585-594, Copyright (2001), with permission from Elsevier Science.

and mathematically this is represented as,

$$\frac{\rho_1}{\rho_{10}} + \frac{\rho_2}{\rho_{20}} = 1 . \tag{14.69}$$

The total (mixture) density $\rho(\mathbf{x},t) = \rho_1 + \rho_2$, and v_i^1 is the velocity of constituent 1 in the mixture while v_i^2 is that of constituent 2. The individual balance of mass equations are

$$\rho_{1,t} + (\rho_1 v_i^1)_{,i} = 0, \qquad \rho_{2,t} + (\rho_2 v_i^2)_{,i} = 0. \tag{14.70}$$

The models alluded to earlier require the introduction of w_i, which is the mean *mass* velocity, or barycentric velocity, viz.

$$\rho w_i = \rho_1 v_i^1 + \rho_2 v_i^2 . \tag{14.71}$$

By using equations (14.70) one can see that ρ and w_i satisfy the equation

$$\rho_{,t} + (\rho w_i)_{,i} = 0. \tag{14.72}$$

A further velocity, v_i, the mean *volume* velocity is defined by

$$v_i = \frac{\rho_1}{\rho_{10}} v_i^1 + \frac{\rho_2}{\rho_{20}} v_i^2 . \tag{14.73}$$

Then, using (14.69) and (14.72) we derive the important relation that v_i is divergence free, i.e. $\partial v_i / \partial x_i = 0$.

(Franchi and Straughan, 2001) include a derivation of a fundamental relation of (Kazhikhov and Smagulov, 1977). This is stated in terms of the mass concentration, c, and volume concentration, α, of constituent 1 in the mixture given by

$$c = \frac{\rho_1}{\rho}, \qquad \alpha = \frac{\rho_1}{\rho_{10}} .$$

The fundamental relation of (Kazhikhov and Smagulov, 1977) is

$$w_i = v_i - \lambda \frac{\rho_{,i}}{\rho} \tag{14.74}$$

which follows from Fick's law of diffusion

$$v_i^1 = w_i - \lambda \frac{c_{,i}}{c} , \tag{14.75}$$

where λ is a (constant) diffusion coefficient, as described by (Franchi and Straughan, 2001).

Equation (14.74) is fundamental to further development since it allows the equations to be written in terms of the mixture velocities w_i and v_i rather than the velocity of individual constituents which appear in Fick's law.

We now describe the derivation of the Graffi, Kazhikhov-Smagulov, and Beirao da Veiga equations using relation (14.74).

The starting point involves the equations for flow of a compressible fluid with density ρ, velocity w_i, body force f_i and pressure p, under isothermal conditions, cf. section 15.2,

$$\rho(w_{i,t} + w_k w_{i,k}) = -p_{,i} + (\mu + \mu_2)w_{m,mi} + \mu\Delta w_i + \rho f_i, \tag{14.76}$$
$$\rho_{,t} + (\rho w_m)_{,m} = 0. \tag{14.77}$$

The quantities μ and μ_2 are viscosities of the compressible fluid and satisfy the restrictions $\mu \geq 0$ and $3\mu_2 + 2\mu \geq 0$, c.f. (Truesdell and Toupin, 1960), p. 719. In a mixture of two constituents w_i is interpreted as the barycentric velocity with ρ being the total density. The idea of the Graffi, Kazhikhov-Smagulov, and Beirao da Veiga models is to transform equations (14.76) and (14.77) with the aid of relation (14.74) into equations which involve the density ρ and mean volume velocity v_i so that the velocity field which appears, v_i, is divergence free.

One substitutes in equation (14.77) for w_i from (14.74) to derive

$$\rho_{,t} + \rho_{,i}v_i = \lambda\Delta\rho. \tag{14.78}$$

The momentum equation is a little more involved but the method is to use (14.74) to eliminate w_i from (14.76) and then use (14.78) on a $\rho_{,t}$ term. In this manner one derives

$$\begin{aligned} \rho(v_{i,t} + v_j v_{i,j}) - \underline{\lambda\big[\rho_{,j}v_{i,j} + v_j\rho_{,ij}\big]} &= -P_{,i} + \mu\Delta v_i \\ &+ \rho f_i + \underline{\lambda^2\left[\frac{\rho_{,i}|\nabla\rho|^2}{\rho^2} - \frac{\rho_{,j}\rho_{,ij}}{\rho} - \frac{\rho_{,i}\Delta\rho}{\rho}\right]} \end{aligned} \tag{14.79}$$

where P is a generalized pressure given by

$$P = -\lambda\rho_{,t} + p + \lambda(\mu_2 + 2\mu)\Delta(\log\rho).$$

The theory of (Kazhikhov and Smagulov, 1977) is derived by discarding the underlined $O(\lambda^2)$ terms in (14.79), while the (Graffi, 1955) theory discards both the $O(\lambda^2)$ and $O(\lambda)$ underlined terms. The full equation (14.79)gives rise to the (Beirao da Veiga, 1983) model.

For the present discussion we are interested in instability in a horizontal layer under the influence of a downward vertical gravitational field. Hence let $f_i = -gk_i$ where g is gravity and $\mathbf{k}$=(0,0,1). The equations for gravity driven convective motion according to the three theories are now given. For the Graffi theory we have

$$\begin{aligned} &\rho(v_{i,t} + v_j v_{i,j}) = -p_{,i} + \mu\Delta v_i - \rho g k_i \\ &v_{i,i} = 0 \\ &\rho_{,t} + v_i\rho_{,i} = \lambda\Delta\rho. \end{aligned} \tag{14.80}$$

The Kazhikhov-Smagulov theory gives rise to

$$\begin{aligned} &\rho v_{i,t} + \rho v_j v_{i,j} - \lambda\big[\rho_{,j}v_{i,j} + \rho_{,ij}v_j\big] = -P_{,i} + \mu\Delta v_i - \rho g k_i\,, \\ &v_{i,i} = 0, \\ &\rho_{,t} + \rho_{,i}v_i = \lambda\Delta\rho\,. \end{aligned} \tag{14.81}$$

Finally the Beirao da Veiga theory leads to the system

$$
\begin{aligned}
\rho(v_{i,t}+v_jv_{i,j})-\lambda\big[\rho_{,j}v_{i,j}+v_j\rho_{,ij}\big] &= -P_{,i}+\mu\Delta v_i\\
&\quad -\rho g k_i+\lambda^2\left[\frac{\rho_{,i}|\nabla\rho|^2}{\rho^2}-\frac{\rho_{,j}\rho_{,ij}}{\rho}-\frac{\rho_{,i}\Delta\rho}{\rho}\right],\\
v_{i,i}&=0,\\
\rho_{,t}+\rho_{,i}v_i&=\lambda\Delta\rho.
\end{aligned}
\tag{14.82}
$$

The work of (Franchi and Straughan, 2001) concentrates on the Kazhikhov-Smagulov model, equations (14.81). They study a layer of a mixture governed by equations (14.81) occupying the domain $\{z\in(0,d),(x,y)\in\mathbb{R}^2\}$, with gravity in the negative $z-$direction. The density is supposed known at the upper and lower surfaces with $\rho=\rho_L$, $z=0$, $\rho=\rho_U$, $z=d$, where $\rho_L<\rho_U$, ρ_L,ρ_U constants. Their objective is to determine a critical "Rayleigh" number, $Ra_c\propto d^3(\rho_U-\rho_L)$, for which the fluid will convectively overturn once the Rayleigh number exceeds Ra_c.

The boundary conditions of (Franchi and Straughan, 2001) keep the lower surface, $z=0$, fixed, and since the fluid is viscous, $v_i=0$ there. On $z=d$ the surface is free of tangential stress and remains horizontal. The free surface boundary conditions correspond to $t_1=t_2=0$ on $z=d$, where t_α are components of the stress vector $t_i=n_jt_{ij}$ where t_{ij} denotes the stress tensor. The steady solution which satisfies the boundary conditions is

$$
\bar\rho=\left(\frac{\rho_U-\rho_L}{d}\right)z+\rho_L\,,\qquad \bar v_i\equiv 0. \tag{14.83}
$$

The steady pressure $\bar P$ is a quadratic function of z, given by

$$
\bar P=-g\left(\frac{\rho_U-\rho_L}{2d}\right)z^2+\rho_Lz+p_0,
$$

where p_0 is a constant reference pressure. Perturbations (u_i,ρ,π) to $(\bar v_i,\bar\rho,\bar P)$ are introduced. (Franchi and Straughan, 2001) investigate linearized instability of the Graffi and Kazhikhov-Smagulov models. These writers introduce the Rayleigh number, Ra, the Graffi number, G, and Prandtl number, Pr, by

$$
Ra=\frac{gd^3(\rho_U-\rho_L)}{\mu\lambda}\qquad G=\frac{\mu\lambda}{gd^3\rho_L}\qquad Pr=\frac{\mu}{\lambda\rho_L}
$$

and for convenience define $R=\sqrt{Ra}$. The perturbation equations for linearized instability in the Kazhikhov-Smagulov model, about the steady solution (14.83), in terms of non-dimensional variables become

$$
\begin{aligned}
&\sigma(1+GRa\,z)u_i+R\rho k_i-\frac{GRa}{Pr}\frac{\partial u_i}{\partial z}=-\frac{\partial\pi}{\partial x_i}+\Delta u_i,\\
&u_{i,i}=0,\\
&\sigma Pr\rho+Rw=\Delta\rho,
\end{aligned}
\tag{14.84}
$$

these equations holding on $\{(x,y) \in \mathbb{R}^2\} \times \{z \in (0,1)\}$. The free surface boundary condition is

$$\frac{\partial^2 w}{\partial z^2} + \frac{2GR}{(1+GRa)} \Delta^* \frac{\partial \rho}{\partial z} = 0. \tag{14.85}$$

The analogous equations which one finds from the Graffi theory are

$$\sigma(1 + GRa\, z)u_i + R\rho k_i = -\frac{\partial \pi}{\partial x_i} + \Delta u_i, \tag{14.86}$$

$$u_{i,i} = 0, \tag{14.87}$$

$$\sigma Pr \rho + Rw = \Delta \rho, \tag{14.88}$$

in $\{(x,y) \in \mathbb{R}^2\} \times \{z \in (0,1)\}$, with the free surface boundary condition

$$\frac{\partial^2 w}{\partial z^2} = 0, \qquad \text{on } z = 1. \tag{14.89}$$

(Franchi and Straughan, 2001) show that for the Graffi theory $\sigma \in \mathbb{R}$. Hence, the instability problem for the Graffi theory reduces to the classical one for Bénard convection with the lower surface being fixed but the upper surface is free of tangential stress. For this case the critical Rayleigh number, Ra_c, for linear instability is given by $Ra_c = 1100.65$. One should, however, observe that the Graffi equations (14.86) - (14.89) are *not* the same as those of the classical Bénard problem,

For the Kazhikhov-Smagulov model equations (14.84) may be reduced to the following equations in w and ρ,

$$\begin{aligned} &\sigma\Big[(1 + GRa\, z)\Delta + GRa \frac{\partial}{\partial z}\Big] w + R\Delta^* \rho = \Delta \Big[\frac{G\,Ra}{Pr}\frac{\partial}{\partial z} + \Delta\Big] w, \\ &\sigma Pr \rho + Rw = \Delta \rho. \end{aligned} \tag{14.90}$$

In (Franchi and Straughan, 2001) the Prandtl number is kept fixed at the value 6. However, we note that equations (14.90) are, *a priori*, strongly dependent on the Prandtl number. In this book we investigate the quantitative effect of the Prandtl number.

The eigenvalue problem (14.90), (14.85) is solved numerically by Chebyshev tau and compound matrix methods in (Franchi and Straughan, 2001) for the value of $Pr = 6$. They find that for small diffusion, i.e. small Graffi number, the Rayleigh number increases and thus the effect of the Kazhikhov-Smagulov terms is one of inhibiting instability. However, as G increases further (Franchi and Straughan, 2001) find a dramatic effect. They discover that the instability threshold remains essentially constant for $G \leq 0.035$. Then, Ra rises rapidly for increasing G, and indeed, they find that Ra_{crit} apparently becomes infinite for $G \to G_{crit} \approx 0.0487^+$. They interpret this phenomenon of Ra_{crit} becoming infinite for $G \to G_{crit}$ as meaning that for $G > G_{crit}$ there is never convective overturning with the model developed from the Kazhikhov-Smagulov theory. This conclusion in (Franchi and Straughan, 2001) is based on calculation with $Pr = 6$.

In this book we find that G_{crit} is strongly dependent on the value of the Prandtl number Pr.

We now present new numerical computations which investigate the behaviour of the Rayleigh number against the Graffi number, but for various values of Prandtl number other than 6. We select the value of 0.72 since this is appropriate for the working fluid being air, see (Batchelor, 1967), p. 597. We also employ the values $Pr = 9.5$ and $Pr = 11.2$, these being appropriate to water at temperatures of 10°C and 5°C, respectively. Since we might interpret overturning instability results in air as being physically important we expect the curves displayed here to be useful. Likewise, many oceans and lakes possess temperatures in the range $5 - 10$°C and so our density driven instability results should be of practical value.

In figures 14.3 - 14.8 we present numerical findings for the instability curves with the Rayleigh number, Ra, graphed against the Graffi number, G. Figures 14.3, 14.4 are for $Pr = 9.5$ which corresponds to water at 10°C. Figures 14.5, 14.6 are for $Pr = 11.2$ which corresponds to water at 5°C, while figures 14.7, 14.8 have $Pr = 0.72$ which is a value appropriate to air. The first figure of each pair, figures 14.3, 14.5, 14.7, displays the instability curve for small values of G. In these figures there is instability above the curve. In each of these we see a similar trend. The Rayleigh number rises then dips slightly before rising again. This is consistent with the behaviour found in (Franchi and Straughan, 2001) for $Pr = 6$. Of course, the rate of increase as a function of G for G close to 0 depends very strongly on Pr. When $Pr = 0.72$ this increase is small whereas it is relatively rapid for $Pr = 11.2$. The second figure of each pair, figures 14.4, 14.6, 14.8, show the instability curves as far as we have been able to compute. In each of these we see a similar trend, again confirming what (Franchi and Straughan, 2001) discovered for $Pr = 6$. There is instability above the curve in each of these figures but in each there is a critical value of G, $G_{crit}(Pr)$, such that instability does not appear to be witnessed for $G > G_{crit}$. The value of G_{crit} is strongly dependent on the Prandtl number. For $Pr = 0.72$ we find $G_{crit} \approx 0.0058457^+$, for $Pr = 9.5$ we find $G_{crit} \approx 0.07708^+$, while for $Pr = 11.2$, $G_{crit} \approx 0.09087^+$. This is a more comprehensive study than that of (Franchi and Straughan, 2001) and certainly underlines the dependence of the Kazhikhov-Smagulov equations (14.84) on the Prandtl number. Of course, the Prandtl number itself is defined in terms of the diffusion coefficient λ.

To explain the apparent singular behaviour Ra displays at G_{crit}, (Franchi and Straughan, 2001) argue that the Kazhikhov-Smagulov theory neglects $O(\lambda^2)$ diffusion terms. Hence, the Kazhikhov-Smagulov model is likely to be realistic only for small G. Our computations bear this out for a range of Prandtl numbers.

To overcome the infinite Rayleigh number behaviour observed with the Kazhikhov-Smagulov model we believe one must return to the full Beirao da Veiga equations (14.79) and retain all λ and λ^2 terms in the ensuing

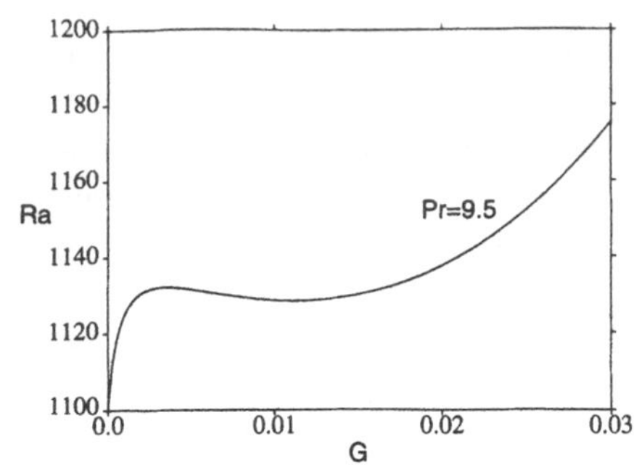

Figure 14.3. Rayleigh number, Ra, vs. Graffi number, G. Small G, $Pr = 9.50$.

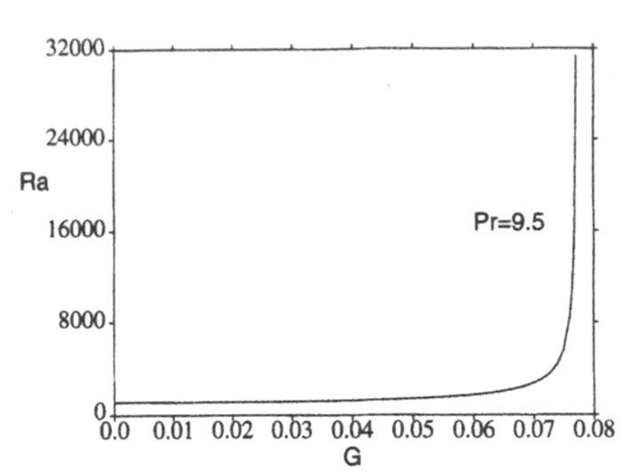

Figure 14.4. Rayleigh number, Ra, vs. Graffi number, G. Larger G, $Pr = 9.5$.

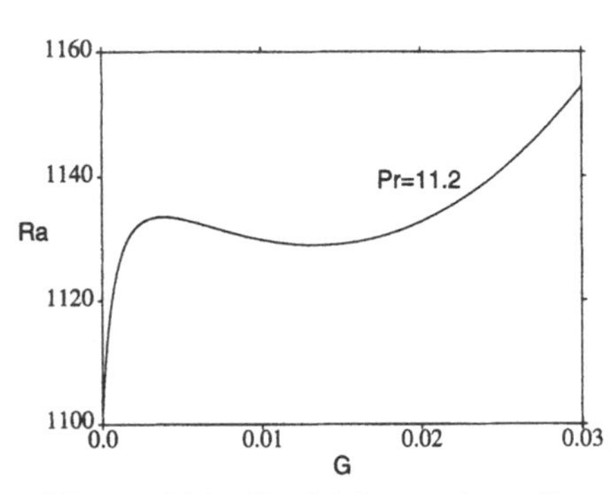

Figure 14.5. Rayleigh number, Ra, vs. Graffi number, G. Small G, $Pr = 11.2$.

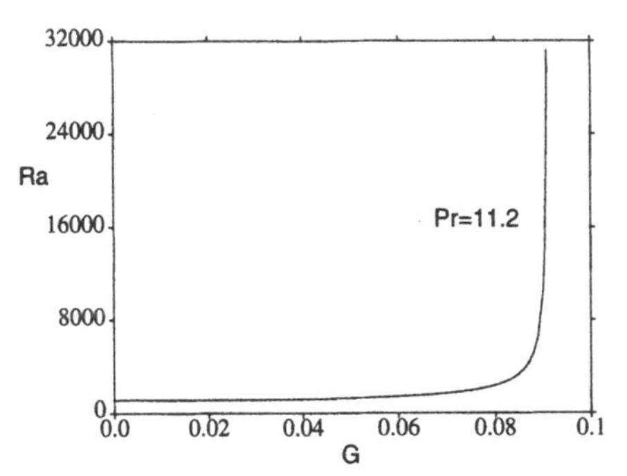

Figure 14.6. Rayleigh number, Ra, vs. Graffi number, G. Larger G, $Pr = 11.2$.

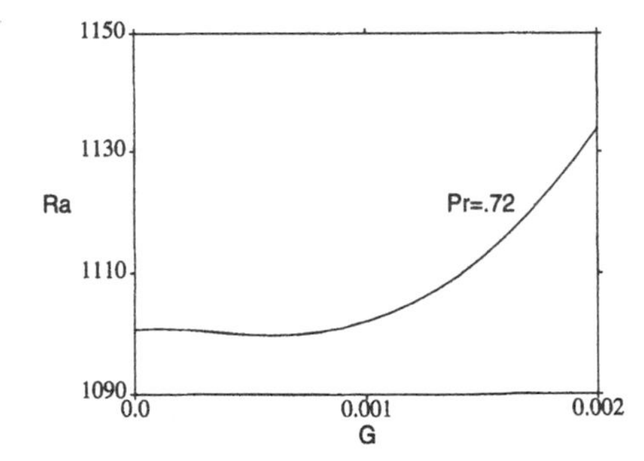

Figure 14.7. Rayleigh number, Ra, vs. Graffi number, G. Small G, $Pr = 0.72$.

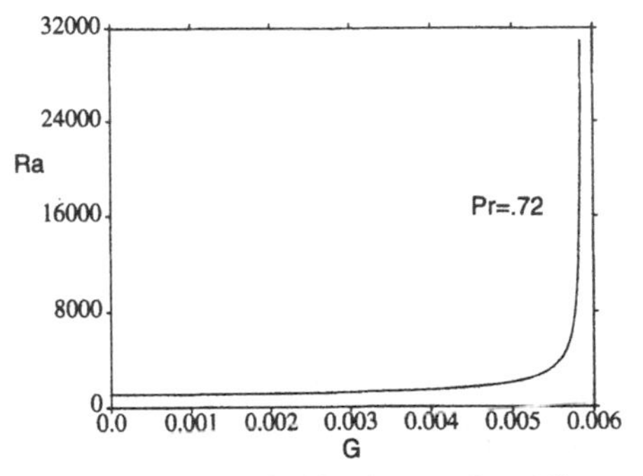

Figure 14.8. Rayleigh number, Ra, vs. Graffi number, G. Larger G, $Pr = 0.72$.

development. The linearized equations for top heavy pollution instability according to the Beirao da Veiga model (14.79) may be shown to be

$$\sigma\left[(1+G\,Ra\,z)\Delta w+GRa\,\frac{\partial w}{\partial z}+\frac{G^2R^3\Delta^*\rho}{(1+GRa\,z)}\right]+R\Delta^*\rho-\frac{GRa}{Pr}\,\Delta\frac{\partial w}{\partial z}$$
$$-\frac{G^3R^5}{Pr[1+GRa\,z]^2}\Delta^*\frac{\partial\rho}{\partial z}+\frac{G^2Ra^2}{Pr(1+GRa\,z)}\,\Delta^*w=\Delta^2w,$$
$$\sigma Pr\rho+Rw=\Delta\rho.$$

The free surface boundary condition still has form

$$\frac{\partial^2 w}{\partial z^2}+\frac{2GR}{(1+GRa)}\,\Delta^*\,\frac{\partial\rho}{\partial z}=0,\qquad \text{on } z=1.$$

(Bresch et al., 2002) propose new Kazhikhov-Smagulov models and investigate questions of global existence of weak solutions.

14.4 Chemical convection

(Pons et al., 2000) review various chemistry experiments in which patterns are created in a layer of fluid due to chemical reactions inducing density gradients. They concentrate on a "blue bottle experiment" where the instability mechanism is gravity driven with the density being increased in the upper part of the fluid layer due to a chemical reaction. The reaction is an akaline oxidation of glucose by oxygen in which methylene blue acts as a catalyst and an indicator. The main oxidation product is gluconic acid and it is the accumulation of this in the upper layer of the fluid which initiates the convective overturning instability. The patterns which arise are very interesting.

A mathematical model for the experiment just outlined has been developed by (Bees et al., 2001). These writers refer to the instability thus arising as chemoconvection. Their model employs the Navier-Stokes equations with a Boussinesq approximation involving the density being a linear function of the concentration, A, of gluconic acid. Thus, they have

$$v_{i,t}+v_jv_{i,j}=-\frac{1}{\rho_0}\,p_{,i}-g\alpha Ak_i+\nu\Delta v_i,\qquad v_{i,i}=0 \tag{14.91}$$

where $\alpha=(\partial\rho/\partial A)|_{A_0}\rho_0^{-1}$ is the coefficient of expansion due to the dependence of ρ on gluconic acid. They study the situation of a free surface bounded below by a fixed surface corresponding to an experiment in an open Petri dish. (The effect of surface tension is neglected in (Bees et al., 2001).)

In addition to the balances of momentum and mass, equations (14.91), there are partial differential equations which describe the conservation of chemical species. These are given in the manner of interaction-diffusion equations of general form

$$C_{,t} + v_j C_{,j} = (D_C C_{,i})_{,i} + I_C \tag{14.92}$$

where C is the species concentration, allowance is made for variable diffusivity, D_C, and I_C represent the source terms for the reactions. The model of (Bees et al., 2001) involves four equations of type (14.92) for four chemical species believed dominant in the blue bottle reaction, namely the concentrations of the colourless reduced form of methylene blue, the blue oxidised form of methylene blue, gluconic acid, and oxygen, and these are denoted by W, B, A and Ω, respectively. Reasons for the precise forms of the interaction terms I_C which appear are given in (Bees et al., 2001). However, the relevant equations are

$$A_{,t} + v_i A_{,i} = (D_A A_{,i})_{,i} + k_2 B, \tag{14.93}$$
$$B_{,t} + v_i B_{,i} = (D B_{,i})_{,i} + 2k_1 \Omega W - k_2 B, \tag{14.94}$$
$$W_{,t} + v_i W_{,i} = (D W_{,i})_{,i} - 2k_1 \Omega W + k_2 B, \tag{14.95}$$
$$\Omega_{,t} + v_i \Omega_{,i} = (D_\Omega \Omega_{,i})_{,i} - k_1 \Omega W, \tag{14.96}$$

where k_1 and k_2 are reaction constants. (Bees et al., 2001) simplify their system by assuming that if $W + B$ is constant at some time, say $t = 0$, then by adding (14.94) and (14.95) one observes $W + B$ remains constant for all time. With this they eliminate W.

The seven equations (14.91) and what remains from (14.93) - (14.96) are non-dimensionalized with the groups $\kappa = 2k_1\Omega_0/k_2$, $\lambda = k_1 W_0/k_2$, $\delta = D_\Omega/D$, $\delta_A = D_A/D$, $Ra = g\alpha W_0 \bar{H}^3/\nu D$, $Sc = \nu/D$, where W_0 and Ω_0 are reference values of W and Ω and $\bar{H} = \sqrt{D/k_2}$ is a depth reflecting the depth of build up of gluconic acid near the open surface. The diffusion coefficients D, D_Ω and D_A are taken to be constant. The resulting non-dimensional equations are, (Bees et al., 2001),

$$Sc^{-1}(v_{i,t} + v_j v_{i,j}) = -P_{,i} - Ra\, A k_i + \Delta v_i, \tag{14.97}$$
$$v_{i,i} = 0, \tag{14.98}$$
$$A_{,t} + v_i A_{,i} = \delta_A \Delta A + B, \tag{14.99}$$
$$B_{,t} + v_i B_{,i} = \Delta B + \kappa\Omega(1 - B) - B, \tag{14.100}$$
$$\Omega_{,t} + v_i \Omega_{,i} = \delta \Delta\Omega - \lambda\Omega(1 - B), \tag{14.101}$$

holding on $\{(x, y) \in \mathbb{R}^2\} \times \{z \in (-d, 0)\} \times \{t > 0\}$. Here d is a scaled depth, so that $z = 0$ represents the upper (open) surface while $z = -d$ is the lower (fixed) surface. The boundary conditions are

$$v_i = 0, \qquad z = -d, 0, \tag{14.102}$$
$$\frac{\partial A}{\partial z} = \frac{\partial B}{\partial z} = \frac{\partial \Omega}{\partial z} = 0, \qquad z = -d, \tag{14.103}$$
$$\frac{\partial A}{\partial z} = \frac{\partial B}{\partial z} = 0, \quad \Omega = 1, \qquad z = 0, \tag{14.104}$$

together with free surface velocity boundary conditions at $z = 0$. The conditions on the chemical species are commensurate with no flux of A, B at the upper and lower surfaces, no flux of oxygen at the lower, and a continuous supply of oxygen at the upper surface, $z = 0$.

At this point (Bees et al., 2001) observe that the gluconic acid concentration never reaches steady state. They thus write

$$A(\mathbf{x}, t) = < A > (t) + a(\mathbf{x}, t)$$

where $< A >$ is a spatial average of A and $< a >= 0$. If we assume a spatial periodicity of the solution horizontal planforms then $< \cdot >$ may be interpreted as integration over a period cell V. The spatial average of (14.99) gives

$$\frac{\partial < A >}{\partial t} = < B >$$

and leads to

$$a_{,t} + v_i a_{,i} = \delta_A \Delta a + B - < B > .$$

The next step is to seek a steady solution to (14.97) - (14.104). (Bees et al., 2001) denote this steady solution by $v^i_{ss}(z), P_{ss}(z), a_{ss}(z), B_{ss}(z), \Omega_{ss}(z)$. They concentrate on a linearized instability analysis paying particular attention to stationary convection. The perturbations to this steady solution may be written as u_i, π, a, b, ω, where $v_i = v^i_{ss}(z) + u_i(\mathbf{x}, t)$, etc. We find it necessary to impose the constraint $< b >= 0$ to proceed. (This is, however, consistent with no flux boundary conditions.)

The linearized equations for stationary convection are presented in (Bees et al., 2001). However, if one goes through the requisite analysis we believe the following nonlinear perturbation equations may be deduced,

$$Sc^{-1}(u_{i,t} + u_j u_{i,j}) = -\pi_{,i} - Ra\, k_i a + \Delta u_i, \qquad (14.105)$$

$$u_{i,i} = 0, \qquad (14.106)$$

$$a_{,t} + u_i a_{,i} = -a'_{ss} w + \delta_A \Delta a + b, \qquad (14.107)$$

$$b_{,t} + u_i b_{,i} = -B'_{ss} w + \Delta b + \kappa\big[\omega(1 - B_{ss}) - \Omega_{ss} b\big] - \underline{\kappa b\omega}, \qquad (14.108)$$

$$\omega_{,t} + u_i \omega_{,i} = -\Omega'_{ss} w + \delta \Delta \omega - \lambda\big[\omega(1 - B_{ss}) - \Omega_{ss} b\big] + \underline{\lambda b\omega}, \qquad (14.109)$$

where $'$ denotes d/dz. The boundary conditions are

$$\begin{aligned} &u_i = a_{,z} = b_{,z} = 0, \quad z = -d, 0; \\ &\omega = 0, \quad z = 0; \qquad \omega_{,z} = 0, \quad z = -d. \end{aligned} \qquad (14.110)$$

As we have already stated, (Bees et al., 2001) develop a detailed linearized instability analysis of (14.105)-(14.110). They restrict attention to stationary convection, i.e. the growth rate σ in the linearized analysis is *a priori* set equal to zero. Oscillatory instabilities may be important in such a system. We add that it is not difficult to develop a *conditional* nonlinear energy stability analysis by multiplying (14.105) by u_i, (14.107) by a, (14.108) by

b, (14.109) by ω. The convective nonlinearities do not cause any problem and the conditional aspect arises only due to the $b\omega$ terms underlined in (14.108) and (14.109). Use of the Cauchy-Schwarz and Sobolev inequalities allows one to overcome these, but I have only seen how to obtain *conditional* nonlinear stability. One can form an equation for the variable $\phi = b + \kappa\omega/\lambda$ which removes the underlined quadratic terms. Whether a generalized energy analysis may be developed utilizing this variable, and which leads to *unconditional* nonlinear stability, is not clear to me.

We now mention recent work on swarming of bacteria in thin fluid films. This swarming behaviour may be regarded as an instability and since such effects are important in areas like food contaminantion the whole field is topical. (Bees et al., 2000) develop a very interesting model for the behaviour of the bacteria Serratia liquefaciens in a thin layer of growth culture. By studying the biology carefully they write down a model involving biological interactions and the fluid is described by a thin film approximation to the Navier-Stokes equations. The work of (Bees et al., 2000) is a stimulating paper in a very promising research area. They develop a numerical simulation using finite differences and to my knowledge no stability theory for their model has been developed.

To finish the present section we point out that nonlinear energy stability arguments, or Lyapunov functional arguments, have been effectively applied to Keller-Segel type biological models, see e.g. (Diaz and Nagai, 1995), (Diaz et al., 1989), (Gajewski and Zacharias, 1998), (Nagai et al., 1997), (Nagai et al., 2000), (Senba and Suzuki, 2000), (Senba and Suzuki, 2001). This represents an active and exciting area of research using energy stability arguments in bio-chemical systems and future work along these lines applied to models such as those of (Bees et al., 2000) will undoubtedly prove rewarding.

15

Convection in a compressible fluid

15.1 The Boussinesq approximation and convection in a deep layer

In section 3.2 we have discussed the Boussinesq approximation for the equations of thermal convection in a linear viscous fluid. The equations arising from the Boussinesq approximation, (3.41), have been extensively employed in this book. However, there are situations in which compressibility effects are important, or one wishes to study convection in a very deep layer. In these circumstances, the Boussinesq approximation is generally believed to be inappropriate. In this chapter we examine some of the models which have been employed to take account of compressibility effects or convection in a deep layer. We begin in this section with an energy analysis of two models for thermal convection in a deep layer.

Justifications of the Boussinesq approximation and a presentation of the conditions under which it is likely to be valid have been given by (Spiegel and Veronis, 1960), (Mihaljan, 1962), (Roberts, 1967b), and (Hills and Roberts, 1991). (Velarde and Perez Cordon, 1976) and (Perez Cordon and Velarde, 1976) look closely at derivations of the Boussinesq approximation and (Velarde and Perez Cordon, 1976) additionally derive a model for convection in a deep layer. Starting with the full nonlinear system of equations for convection in a deep layer of heat conducting compressible viscous fluid (Velarde and Perez Cordon, 1976) show that a reference state (an adiabatic hydrostatic field), $(\bar{T},\bar{p},\bar{\rho})$, from which to study convection in a deep layer is

$$\bar{T} = T_0 - \beta z$$
$$\bar{p} = p_0 + \frac{1}{\chi}\Big[\Big(1 - \frac{\alpha\beta}{\chi g \rho_0}\Big)(e^{-g\rho_0\chi z} - 1) - \alpha\beta z\Big] \tag{15.1}$$
$$\bar{\rho} = \rho_0\Big[\Big(1 - \frac{\alpha\beta}{\chi g\rho_0} e^{-g\rho_0\chi z} + \frac{\alpha\beta}{\chi g \rho_0}\Big)\dots\Big].$$

Here β is the temperature gradient, $\bar{T}, \bar{p}, \bar{\rho}$, are temperature, pressure, and density, and z is the vertical height coordinate. The other quantities are constant and the expansion in ρ is presented to leading order. These writers develop a perturbation procedure about the stationary solution (15.1) expanding in the parameters $\epsilon_1 = \alpha\Delta T$ and $\eta = \chi\rho_0\nu\kappa/\alpha^3(\Delta T)^3 d^2$ where $\alpha, \Delta T, \chi, \nu, \kappa$ and d are the thermal expansion coefficient, temperature difference across the layer, isothermal compressibility, kinematic viscosity, thermal diffusivity, and depth of the layer. For example, for convection in the Earth's mantle they estimate $\epsilon_1 \approx 6 \times 10^{-2}$, $\eta \approx 2 \times 10^{-3}$ in c.g.s. units. However, by this method, (Velarde and Perez Cordon, 1976) derive the following nonlinear equations which a perturbation to (15.1) satisfies for *convection in a deep layer*,

$$\epsilon_2 w = u_{i,i}\,, \tag{15.2}$$
$$e^{-\epsilon_2 z}\dot{u}_i = -Pr\,\pi_{,i} + Pr\,\tau_{ji,j} + (PrRa\Omega\,\theta - Pr\epsilon_2\,\pi)\delta_{i3}\,, \tag{15.3}$$
$$\Omega\dot{\theta} - w + Di\Big(\frac{1}{\phi} - z + \Omega\theta\Big)\,w = e^{\epsilon_2 z}\Omega\Delta\theta\,. \tag{15.4}$$

In these equations $\epsilon_2 = \chi\rho_0 g d$ is a constant, Pr and Ra are the Prandtl and Rayleigh numbers, $Di = \alpha g d/c_p$ (a dissipation number), $\Omega = \theta_{\max}/\beta d$, where $\theta_{\max}$ is the maximum the perturbation temperature can achieve, g is gravity, c_p is the specific heat at constant pressure, and τ_{ji} is the perturbation stress tensor, so

$$\tau_{ji} = \mu_1(u_{i,j} + u_{j,i}) + \frac{1}{3}\,(\mu_2 - 2\mu_1)u_{r,r}\,\delta_{ij}\,. \tag{15.5}$$

This is the standard form of stress tensor for a compressible linear viscous fluid, see e.g. (Serrin, 1959a; Serrin, 1959d).

The convection layer has been scaled to $z \in (0,1)$ and so (15.2-15.4) hold for $t > 0$ on the spatial layer $\{(x,y) \in \mathbb{R}^2\} \times \{z \in (0,1)\}$. Notice that the compressibility and depth effects are directly visible in (15.2-15.4) because ϵ_2 is proportional to both the isothermal compressibility χ and the layer depth d, while Di is also linearly proportional to d. (In fact, the usual nonlinear perturbation equations which arise from the Boussinesq approximation are recovered from (15.2-15.4) in the limit $\epsilon_2, Di \to 0$.)

The boundary conditions to be imposed on u_i and θ are

$$u_i = 0, \quad \theta = 0, \qquad z = 0, 1 \tag{15.6}$$

together with a plane tiling periodicity in (x, y) defining a period cell V. It is noteworthy that we do not require surfaces free of tangential stress, we can allow fixed boundary conditions. This is in contrast to other nonlinear energy stability analyses reported in this chapter where stress free conditions are imposed.

A nonlinear energy analysis of the (Velarde and Perez Cordon, 1976) model commences by multiplying (15.3) by u_i and integrating over V, and (15.4) by θ and performing a similar integration. In this way we derive the two equations

$$\begin{aligned}\frac{d}{dt}\frac{1}{2} < e^{-\epsilon_2 z}u_iu_i > =& PrRa\Omega < \theta w > + Pr < \tau_{ji,j}u_i > - < e^{-\epsilon_2 z}u_iu_ju_{i,j} > \\ &- Pr < \pi_{,i}u_i > - Pr\epsilon_2 < \pi w >, \end{aligned} \tag{15.7}$$

$$\begin{aligned}\frac{d}{dt}\frac{1}{2}\Omega\|\theta\|^2 =& \Big\langle \Big(1 - \frac{Di}{\phi} + zDi\Big)\theta w \Big\rangle + \frac{1}{2}\Omega\epsilon_2^2 < e^{\epsilon_2 z}\theta^2 > \\ &- \Omega < e^{\epsilon_2 z}\theta_{,i}\theta_{,i} > - Di\,\Omega < \theta^2 w > . \end{aligned} \tag{15.8}$$

In deriving (15.8) we have performed some integrations by parts. The last two terms in (15.7) are zero as may be seen by multiplying equation (15.2) by π and integrating. To handle the stress term we use (15.5) and find

$$\begin{aligned}< \tau_{ji,j}u_i > =& \mu_1 < u_i(u_{i,j} + u_{j,i})_{,j} > + \frac{1}{3}(\mu_2 - 2\mu_1) < u_iu_{j,ji} > \\ =& -\mu_1\|\nabla\mathbf{u}\|^2 - \frac{1}{3}(\mu_1 + \mu_2) < (u_{i,i})^2 > \\ =& -\mu_1\|\nabla\mathbf{u}\|^2 - \frac{1}{3}(\mu_1 + \mu_2)\epsilon_2^2\|w\|^2 . \end{aligned} \tag{15.9}$$

Expression (15.9) is used in (15.7) and then $\lambda(15.8)$ is added to derive an energy equation of form

$$\frac{dE}{dt} = I - D + N, \tag{15.10}$$

where

$$E(t) = \frac{1}{2} < e^{-\epsilon_2 z}u_iu_i > + \frac{1}{2}\lambda\Omega\|\theta\|^2, \tag{15.11}$$

$$I(t) = \Big\langle \Big[PrRa\Omega + \lambda\Big(1 - \frac{Di}{\phi}\Big) + Di\lambda z\Big]\theta w \Big\rangle + \frac{1}{2}\lambda\Omega\epsilon_2^2 < e^{\epsilon_2 z}\theta^2 >, \tag{15.12}$$

$$D(t) = \mu_1 Pr\|\nabla\mathbf{u}\|^2 + \frac{1}{3}Pr\epsilon_2^2(\mu_1 + \mu_2)\|w\|^2 + \lambda\Omega < e^{\epsilon_2 z}\theta_{,i}\theta_{,i} >, \tag{15.13}$$

$$N(t) = \frac{1}{2}\epsilon_2 < wu_iu_i > + \lambda\Omega Di < \theta^2 w > . \tag{15.14}$$

Unfortunately, due to the cubic N term all we can develop from (15.10) is a conditional nonlinear analysis. It is, however, sharp in the limit $\epsilon_2, Di \to$

0. The method is to set

$$\frac{1}{R_E} = \max_{\mathcal{H}} \frac{I}{D}, \tag{15.15}$$

where $\mathcal{H}$ is the space of admissible solutions, and then derive

$$\frac{dE}{dt} \leq -D\left(1 - \frac{1}{R_E}\right) + N. \tag{15.16}$$

The space $\mathcal{H}$ is not constrained by the solenoidal condition on u_i but instead by (15.2). We require

$$R_E^{-1} < 1, \tag{15.17}$$

and then put $a = (R_E - 1)/R_E > 0$, to find

$$\frac{dE}{dt} \leq -aD + N. \tag{15.18}$$

We can estimate the terms comprising N as follows, with the aid of the Sobolev inequality $\|\phi\|_4^2 \leq c_1\|\phi\|^{1/2}\|\nabla\phi\|^{3/2}$, and the Poincaré inequality $\pi\|\phi\| \leq \|\nabla\phi\|$,

$$\begin{aligned} < wu_iu_i > &\leq \left(\int_V w^2 dV\right)^{1/2} \left(\int_V |\mathbf{u}|^4 dV\right)^{1/2} \\ &\leq c_1 \left(\int_V w^2 dV\right)^{1/2} \left(\int_V u_iu_i dV\right)^{1/4} \left(\int_V |\nabla\mathbf{u}|^2 dV\right)^{3/4} \\ &\leq \frac{c_1 e^{\epsilon_2/2}}{\pi^{1/2}} < e^{-\epsilon_2 z} u_iu_i >^{1/2} \|\nabla\mathbf{u}\|^2, \end{aligned} \tag{15.19}$$

$$\begin{aligned} < \theta^2 w > &\leq \left(\int_V w^2 dV\right)^{1/2} \left(\int_V \theta^4 dV\right)^{1/2} \\ &\leq c_1\|w\|\|\theta\|\|\nabla\theta\|^{3/2} \\ &\leq \frac{c_1}{\pi^{1/2}} \|w\|\|\nabla\theta\|^2 \\ &\leq \frac{c_1 e^{\epsilon_2/2}}{\pi^{1/2}} < e^{-\epsilon_2 z} u_iu_i >^{1/2} < e^{\epsilon_2 z}|\nabla\theta|^2 > . \end{aligned} \tag{15.20}$$

Combining these we obtain

$$N \leq \omega E^{1/2} D, \tag{15.21}$$

where

$$\omega = \frac{c_1}{\sqrt{2\pi}} e^{\epsilon_2/2} \max\left\{\frac{\epsilon_2}{\mu_1 Pr}, Di\right\}. \tag{15.22}$$

Thus, returning to (15.18) we may derive

$$\frac{dE}{dt} \leq -D(a - \omega E^{1/2}). \tag{15.23}$$

Since $D \geq 2\min\{1, \mu_1\pi^2 Pr\}E$ we establish unconditional nonlinear stability from (15.23) provided that

$$E^{1/2}(0) < a/\omega\,. \tag{15.24}$$

It is worth observing that as $\epsilon_2, Di \to 0$, $\omega \to 0$, and so the bound in (15.24) increases.

Taking into account the constraint $u_{i,i} - \epsilon_2 w = 0$ and introducing the Lagrange multiplier $\pi(\mathbf{x})$ the Euler-Lagrange equations for (15.15) in the stability threshold case of $R_E = 1$ are found to be

$$\begin{aligned}&\Big[PrRa\Omega + \lambda\Big(1 - \frac{Di}{\phi} + zDi\Big)\Big]\theta + 2\mu_1 Pr\Delta u_i \\ &\qquad - \frac{2}{3}Pr\epsilon_2^2(\mu_1 + \mu_2)w\delta_{i3} - Pr(\pi_{,i} + \epsilon_2\pi\delta_{i3}) = 0,\end{aligned} \tag{15.25}$$

$$\begin{aligned}&\Big[PrRa\Omega + \lambda\Big(1 - \frac{Di}{\phi} + zDi\Big)\Big]w + \lambda\Omega\epsilon_2^2 e^{\epsilon_2 z}\theta \\ &\qquad + 2\Omega\lambda\frac{\partial}{\partial x_i}\Big(e^{\epsilon_2 z}\frac{\partial\theta}{\partial x_i}\Big) = 0\,.\end{aligned} \tag{15.26}$$

The nonlinear energy stability Rayleigh number threshold is determined by solving these equations. We do not do this here, but we do note that for $\epsilon_2, Di \to 0$ these are the same as the equations one obtains from (15.2-15.4) in the same limit. The limit equations are those of standard Bénard convection with the Boussinesq approximation and in this limit we recover the strong result of coincidence of the linear instability and nonlinear stability boundaries. In fact, it is not difficult to develop an asymptotic analysis from (15.25,15.26) by expanding Ra as a series in ϵ_2 and Di. The derivation of the nonlinear energy stability Rayleigh number Ra_E and the linear instability one Ra_L could then be obtained to leading order. Since the (Velarde and Perez Cordon, 1976) equations are for deep convection for which ϵ_2 and Di are not small we do not do this. A nonlinear stability boundary could be found by solving (15.25,15.26) numerically.

15.1.1 The Zeytounian model

We complete this section with an investigation into the nonlinear stability characteristics of a second system of equations governing thermal convection in a deep fluid layer. The model we study is that of (Zeytounian, 1989). The mathematical well posedness of the Zeytounian model is studied by (Charki, 1995) and (Charki, 1996), while (Charki and Zeytounian, 1994) and (Charki and Zeytounian, 1995) investigate the model in the region where the fluid is undergoing a convective motion. As with the work of (Velarde and Perez Cordon, 1976) the analysis of (Zeytounian, 1989) begins with the full system of compressible, heat-conducting equations for a viscous fluid and reduces in systematic fashion to a system appropriate to thermal convection in a deep layer. A linearized instability analysis for the conduction solution found by (Zeytounian, 1989) is performed by (Errafiy and Zeytounian, 1991). The present section describes work of (Franchi and Straughan, 1992a) and shows how to construct a suitable generalized energy theory to give rigorous stability estimates for the full *nonlinear* equations of (Zeytounian, 1989).

The nonlinear perturbation equations derived by (Zeytounian, 1989) to be applicable to convection in a deep layer are given below. (These may be compared with equations (15.2-15.4), those of the (Velarde and Perez Cordon, 1976) model.) Instead of writing directly Zeytounian's equations we first facilitate a symmetry between the linear interaction terms of equations (4.6) of (Zeytounian, 1989), by setting $\theta = R\Xi$. Here, $R = \sqrt{Ra}$, Ra being the Rayleigh number, Ξ is the temperature perturbation field of Zeytounian, and so θ is the temperature perturbation field in this work. In our notation the nonlinear *perturbation* equations of (Zeytounian, 1989), become

$$Pr^{-1}(u_{i,t} + u_j u_{i,j}) = -\pi_{,i} + \Delta u_i + R\theta\delta_{i3}, \tag{15.27}$$

$$u_{i,i} = 0, \tag{15.28}$$

$$\theta_{,t} + u_i\theta_{,i} = Rw + \mu(z)\Delta\theta + 2\frac{\delta}{R}\mu(z)d_{ij}d_{ij}, \tag{15.29}$$

where u_i, π, Pr are velocity field, pressure, Prandtl number, and as elsewhere in the book $w = u_3$, $d_{ij} = \frac{1}{2}(u_{i,j} + u_{j,i})$. The number δ is a constant which represents a depth parameter. It satisfies $0 < \delta < 1$. The function μ in (15.29) is given by

$$\mu(z) = \frac{1}{1 + \delta(1 - z)}. \tag{15.30}$$

Equations (15.27-15.29) are defined for $t > 0$ on the three-dimensional spatial layer $z \in (0, 1)$ subject to stress free boundary conditions (although we assume the surface itself remains fixed), i.e.

$$\theta = w = \frac{\partial u}{\partial z} = \frac{\partial v}{\partial z} = 0, \qquad \text{on} \qquad z = 0, 1. \tag{15.31}$$

In addition (u_i, θ, π) are assumed to have an (x, y) dependence with a periodic shape in the plane. As throughout the book we denote by V a period cell for the perturbation (u_i, θ, π). Note that the analysis for the Zeytounian system appears to require the stress free boundary conditions, unlike the work earlier in this section on the (Velarde and Perez Cordon, 1976) model.

To analyse the *nonlinear* stability of the conduction state of (Zeytounian, 1989), we begin with the governing perturbation equations (15.27-15.29) and multiply (15.27) by u_i, (15.29) by θ, and integrate over V. After some calculation we obtain

$$\frac{d}{dt}\frac{1}{2Pr}\|\mathbf{u}\|^2 = R < \theta w > -\|\nabla\mathbf{u}\|^2, \tag{15.32}$$

and

$$\begin{aligned}\frac{d}{dt}\frac{1}{2}\|\theta\|^2 =& R < w\theta > - < \mu|\nabla\theta|^2 > \\ &+ \delta^2 < \mu^3\theta^2 > + 2\frac{\delta}{R} < \mu\theta d_{ij}d_{ij} > .\end{aligned} \tag{15.33}$$

We now add (15.32) and (15.33) to deduce

$$\frac{dE}{dt} = RI - D + 2\frac{\delta}{R} < \mu(z)\theta d_{ij}d_{ij} >, \tag{15.34}$$

where the energy E, the production term I, and the dissipation term D are given by

$$E(t) = \frac{1}{2Pr}\|\mathbf{u}\|^2 + \frac{1}{2}\|\theta\|^2, \qquad I(t) = 2 < \theta w >, \tag{15.35}$$

$$D(t) = \|\nabla\mathbf{u}\|^2 + < \mu|\nabla\theta|^2 > -\delta^2 < \mu^3\theta^2 > . \tag{15.36}$$

The dissipation term D must be positive-definite. This is certainly true provided

$$\pi^2 > \delta^2(1+\delta). \tag{15.37}$$

We assume this to be the case. Then, one can show with the aid of Poincaré's inequality that

$$D \geq \|\nabla\mathbf{u}\|^2 + k\|\nabla\theta\|^2, \tag{15.38}$$

with $k = [1/(1+\delta)] - (\delta^2/\pi^2)$. If (15.37) holds then $k > 0$ and so D is positive-definite.

The main difficulty to proceed beyond (15.34) is the nonlinear term $< \mu\theta d_{ij}d_{ij} >$. Unlike equations (15.7) and (15.8) we see that the gradients of u_i appear as squared terms. This makes an analysis like that for the (Velarde and Perez Cordon, 1976) model difficult. The dissipation term, D, would not appear to be strong enough to dominate the nonlinearity and so it appears necessary to develop a generalized energy theory.

The generalized energy analysis requires the introduction of an L^2 norm of ∇u_i and begins by multiplying (15.27) by $-\Delta u_i$ and integrating over V to obtain

$$\frac{d}{dt}\frac{1}{2Pr}\|\nabla\mathbf{u}\|^2 = \frac{1}{Pr} < u_j u_{i,j}\Delta u_i > -\|\Delta\mathbf{u}\|^2 - R < \theta\Delta w > . \tag{15.39}$$

The analysis of (Franchi and Straughan, 1992a) adds a multiple of (15.39) to (15.34) and then estimates the nonlinear terms using the inequality

$\sup_V |\mathbf{u}| \le c\|\Delta\mathbf{u}\|$. They show that

$$\begin{aligned}\frac{d\mathcal{E}}{dt} \le &-\Big(\frac{R_E - R}{R_E}\Big)D - \lambda\|\Delta\mathbf{u}\|^2 - \lambda R < \theta\Delta w > \\ &+ \frac{\lambda c}{Pr}\|\nabla\mathbf{u}\|\|\Delta\mathbf{u}\|^2 + 2\frac{\delta^2 c}{R}\|\theta\|\|\nabla\mathbf{u}\|\|\Delta\mathbf{u}\| \\ &+ \frac{\delta c}{R}\|\theta\|\|\Delta\mathbf{u}\|^2 + 2\frac{\delta c}{R}\|\nabla\theta\|\|\nabla\mathbf{u}\|\|\Delta\mathbf{u}\|, \end{aligned} \tag{15.40}$$

where the generalized energy $\mathcal{E}$ is given by

$$\mathcal{E}(t) = E + \frac{\lambda}{2Pr}\|\nabla\mathbf{u}\|^2, \tag{15.41}$$

and R_E is the quantity:

$$\frac{1}{R_E} = \max_{\mathcal{H}} \frac{I}{D}, \tag{15.42}$$

$\mathcal{H}$ being the space of admissible solutions.

Next suppose $R < R_E$ and put $a = (R_E - R)/R_E$ (> 0). Select now λ by $\lambda = (a\pi^2/R)[(1+\delta)^{-1} - (\delta^2/\pi^2)]$ (> 0). Further, define a generalized dissipation function $\mathcal{D}$ by

$$\mathcal{D} = \frac{1}{2}aD + \frac{1}{2}\lambda\|\Delta\mathbf{u}\|^2. \tag{15.43}$$

From inequality (15.40) (Franchi and Straughan, 1992a) then derive the estimate

$$\frac{d\mathcal{E}}{dt} \le -\mathcal{D} + A\mathcal{E}^{1/2}\mathcal{D}, \tag{15.44}$$

where A is given by

$$A = \frac{2^{3/2}c}{R}\Big[\frac{R}{\sqrt{Pr\lambda}} + \sqrt{\frac{2}{k}\frac{\delta^2}{\lambda}} + \delta + \frac{2\delta}{\lambda\sqrt{ak}}\Big]. \tag{15.45}$$

Thus provided

$$R < R_E \qquad \text{and} \qquad \mathcal{E}^{1/2}(0) < \frac{1}{A}, \tag{15.46}$$

$\mathcal{E}(t) \to 0$ as $t \to \infty$, and we have rigorously established nonlinear stability.

While we do not include a numerical calculation of $R_E(\delta)$, the nonlinear stability threshold, an asymptotic analysis is revealing and now developed.

The Euler-Lagrange equations arising from (15.42) are

$$\begin{aligned} &\Delta u_i + R_E\delta_{i3}\theta = \varpi_{,i}, \qquad u_{i,i} = 0, \\ &\nabla(\mu(z)\nabla\theta) + R_E w - \delta^2\mu^3\theta = 0, \end{aligned} \tag{15.47}$$

where ϖ is now a Lagrange multiplier introduced because $\mathbf{u}$ is solenoidal.

These equations are reduced to a system in w and θ and then normal mode forms $w = W(z)G(x,y)$, $\theta = \Theta(z)G(x,y)$, are chosen where $\Delta^* G =$

$-a^2G$, G being the spatial planform of the solution and a the horizontal wavenumber.

Hence, the eigenvalue system for R_E is

$$\begin{aligned}&(D^2-a^2)^2W-R_Ea^2\Theta=0,\\&(D^2-a^2)\Theta+\delta D\Theta+\delta^2\mu^2a^2\Theta-\frac{R_E}{\mu}a^2W=0,\end{aligned}\tag{15.48}$$

which is to be solved subject to the boundary conditions for two free surfaces

$$\Theta=W=D^2W=0,\tag{15.49}$$

on $z=0,1$. Since μ depends on z a complete analysis would proceed numerically to find

$$Ra_E=\min_{a^2}R_E^2(a^2;\delta),\tag{15.50}$$

Ra_E being the critical Rayleigh number of energy stability theory. In this book we include only an asymptotic result for small δ; this is, however, in line with work of (Errafiy and Zeytounian, 1991).

(Franchi and Straughan, 1992a) expand $Ra=R_E^2$ and W in δ, and derive the asymptotic formula

$$Ra_E=\frac{27}{4}\pi^4\Big(1-\frac{1}{2}\delta+O(\delta^2)\Big),\tag{15.51}$$

which agrees exactly with the critical Rayleigh number of linear instability theory, (Errafiy and Zeytounian, 1991), to $O(\delta)$. Thus, to order δ the linear instability and nonlinear energy stability critical Rayleigh numbers are the same.

15.2 Slightly compressible convection

An interesting approach to the Boussinesq approximation is given in (Hills and Roberts, 1991) who adopt the principles of modern continuum thermodynamics, a philosophy to which we subscribe. A brief review of their approach, which also includes effects of compressibility in a precise manner, is now given.

(Hills and Roberts, 1991) begin with the full equations of motion for a compressible fluid in the form

$$\dot\rho=-\rho v_{i,i}\,,\tag{15.52}$$

$$\rho\dot v_i=\rho f_i+\sigma_{ji,j}\,,\tag{15.53}$$

$$\rho\dot U=\sigma_{ji}d_{ij}-q_{i,i}+\rho q,\tag{15.54}$$

where q is the heat supply per unit mass per unit time and U is the specific internal energy. Their entropy production inequality is

$$\rho(T\dot{S} - \dot{U}) + \sigma_{ji}d_{ij} - \frac{q_i T_{,i}}{T} \geq 0. \tag{15.55}$$

Their interest is in materials whose density can be changed by variations in the temperature, T, but not in the thermodynamic pressure P, and so they formulate the constitutive theory in terms of P and T. They argue that the natural thermodynamic potential is the Gibbs energy

$$G = U - ST + \frac{P}{\rho}.$$

The entropy inequality (15.55) may thus be rewritten

$$-\rho\big(\dot{G} + S\dot{T}\big) + \dot{P} + (\sigma_{ji} + P\delta_{ji})d_{ij} - \frac{q_i T_{,i}}{T} \geq 0. \tag{15.56}$$

The constitutive theory of (Hills and Roberts, 1991) chooses

$$G = G(T, P), \qquad S = S(T, P), \qquad \rho = \rho(T), \tag{15.57}$$

$$\sigma_{ij} = -p\delta_{ij} + \lambda d_{mm}\delta_{ij} + 2\mu d_{ij}, \qquad q_i = -\kappa T_{,i}, \tag{15.58}$$

where p is the mechanical pressure, and the coefficients λ, μ, κ depend on P and T.

For the form of ρ in (15.57), the continuity equation (15.52) becomes

$$\alpha\dot{T} = v_{i,i}, \tag{15.59}$$

where $\alpha(= -\rho^{-1}d\rho/dT)$ is the thermal expansion coefficient of the fluid. Equation (15.59) is regarded as a constraint and then included in (15.56) via a Lagrange multiplier, Γ. By using the arbitrariness of the body force and heat supply they then deduce from (15.56) that

$$S = -\left(\frac{\partial G}{\partial T} + \frac{\Gamma\alpha}{\rho}\right), \qquad \frac{\partial G}{\partial P} = \frac{1}{\rho},$$

$$p = P + \Gamma, \qquad \Gamma = \Gamma(T, P),$$

$$\lambda + \frac{2}{3}\mu \geq 0, \qquad \mu \geq 0, \qquad \kappa \geq 0.$$

They then work with another Gibbs energy, $\hat{G} = G + \Gamma/\rho$, which allows them to replace G, P, Γ by $\hat{G}, p$, for which

$$\hat{G} = \hat{G}(T, p) = G_0(T) + \frac{p}{\rho}, \qquad \frac{\partial \hat{G}}{\partial T} = -S, \qquad \frac{\partial \hat{G}}{\partial p} = \frac{1}{\rho},$$

where now the material parameters depend on p and T. The governing equations for what (Hills and Roberts, 1991) term a *generalized* incompressible linear viscous fluid become

$$\alpha\dot{T} = v_{i,i}, \tag{15.60}$$

$$\rho\dot{v}_i = \rho f_i - p_{,i} + (\lambda v_{m,m}\delta_{ij} + 2\mu d_{ij})_{,j}, \tag{15.61}$$

$$\rho c_p\dot{T} - \alpha T\dot{p} = \rho q + (\kappa T_{,i})_{,i} + \lambda(d_{ii})^2 + 2\mu d_{ij}d_{ij}, \tag{15.62}$$

where $c_p = T(\partial S/\partial T)_p$ is again the specific heat at constant pressure.

The boundary conditions they adopt are continuity of temperature and the normal component of the heat flux. At a stationary, no-slip boundary they have $v_i = 0$, while at a free surface, $\mathcal{S}$, at an ambient pressure π the stress vector is continuous, which results in

$$-\pi = -p + \lambda v_{i,i} + 2\mu n_i n_j v_{i,j}, \qquad \epsilon_{ijk}n_j\omega_k = 0, \qquad \text{on } \mathcal{S}, \tag{15.63}$$

n_i being the unit normal to $\mathcal{S}$ with ω_i the vorticity vector. In addition to (15.63) they have a kinematic condition

$$n_i v_i = V, \qquad \text{on } \mathcal{S}, \tag{15.64}$$

V being the velocity of $\mathcal{S}$ along its normal n_i.

This system must be further reduced to arrive at the usual equations, and so they consider a horizontal layer of fluid contained between a fixed lower boundary $z = L$ and a free top surface $z = f(x_1, x_2, t)$. The condition (15.64) is then

$$v_3 = \dot{f}, \qquad \text{on} \qquad z = f. \tag{15.65}$$

They assume gravity acts in the z-direction so $f_i = g\delta_{i3}$ and expand about a reference state (T_r, p_r) to obtain the expansion for the density

$$\rho(T) = \rho_r\big[1 - \alpha_r(T - T_r) + \cdots\big],$$

where $\rho_r = \rho(T_r)$, etc.

The key philosophy of (Hills and Roberts, 1991) (cf. (Roberts, 1967b), pp. 196–197) is that *typical accelerations promoted in the fluid by variations in the density are always much less than the acceleration of gravity.* The resulting equations from the Boussinesq approximation, the so-called Oberbeck-Boussinesq equations, arise by taking the simultaneous limits, $g \to \infty$, $\alpha_r \to 0$, with the restriction that $g\alpha_r$ remains finite.

(Hills and Roberts, 1991) expand the pressure, velocity, and temperature fields in g^{-1}, viz.

$$p = p^0 g + p^1 + \frac{p^2}{g} + \cdots,$$

$$v_i = v_i^1 + \frac{v_i^2}{g} + \cdots,$$

$$T - T_r = T^1 - T_r + \frac{T^2 - T_r}{g} + \cdots,$$

together with an expansion of the free surface

$$f = \frac{f^2}{g} + \cdots .$$

(Hills and Roberts, 1991) derive the Oberbeck-Boussinesq equations at the $O(1)$ level *with the limit*

$$\epsilon = \frac{g\alpha_r L}{(c_p)_r} \to 0.$$

In fact, under a suitable non-dimensionalization, with

$$Ra = \frac{g\alpha_r \beta L^4}{\nu_r k_r}$$

being the Rayleigh number, and $Pr = \nu_r/k_r$ the Prandtl number, their $O(1)$ equations are

$$v^1_{i,i} = 0, \tag{15.66}$$

$$\dot{v}^1_i = -RaT^1\delta_{i3} - p^1_{,i} + \Delta v^1_i, \tag{15.67}$$

$$Pr\Big(\dot{T}^1 - \epsilon\big[T_r + T^1\big]v^1_3\Big) = q + \Delta T + 2\frac{Pr}{Ra}\epsilon d^1_{ij}d^1_{ij}. \tag{15.68}$$

The standard equations of the Boussinesq approximation are found when $\epsilon \to 0$ with $q \equiv 0$ in (15.66)-(15.68). It is important to observe that (15.66)-(15.68) do contain first order effects of compressibility via the ϵ terms.

15.2.1 Stability analysis for slightly compressible convection

(Richardson, 1993) studies instability and stability for the Bénard problem according to the Hills-Roberts equations (15.66)-(15.68). She considers a layer of heat conducting viscous fluid contained between the planes $z = 0$ and $z = d$. The temperature on the boundaries is supposed fixed such that

$$T = T_0, \quad \text{on } z = 0, \qquad T = T_U, \quad \text{on } z = d, \tag{15.69}$$

where T_0 and T_U are constants with $T_0 > T_U$. The steady solution of (Richardson, 1993) satisfies $\bar{v}_i = 0$, $\bar{T} = \beta z + T_0$ where $\beta = (T_U - T_0)/d\,(<0)$ is the temperature gradient, and the steady pressure $\bar{p}$ is quadratic in z.

She analyses the stability of this steady solution by introducing perturbations (u_i, θ, π) as

$$v_i = \bar{v}_i + u_i, \quad T = \bar{T} + \theta, \quad p = \bar{p} + \pi. \tag{15.70}$$

With the non-dimensional variables ϵ, Γ defined by

$$\epsilon = \frac{g\alpha_r d}{(c_p)r} \qquad \Gamma = z + Pr\,T_0 \tag{15.71}$$

she derives equations (15.72,15.73) for the nonlinear non-dimensional perturbations. The quantity ϵ is a dissipation parameter such that by assuming that the layer is adiabatically thin in the sense that $\beta_{\rm ad} << \beta$, then equivalently $\epsilon \to 0$. Thus the parameter ϵ is effectively a measure of the compressibility of the fluid layer, and $Ra = R^2$ is the Rayleigh number.

The non-dimensional perturbation equations of (Richardson, 1993) are

$$u_{i,t} + u_j u_{i,j} = -p_{,i} + \Delta u_i - R\theta k_i, \qquad u_{i,i} = 0, \tag{15.72}$$

$$Pr(\theta_{,t} + u_i\theta_{,i}) = Pr\epsilon\theta w - R(1-\epsilon\Gamma)w + \Delta\theta + \frac{2Pr\epsilon}{R}\, d_{ij}d_{ij}\,. \tag{15.73}$$

Note that the three ϵ terms in equation (15.73) represent changes arising from compressibility as compared with the usual equations for the Boussinesq approximation model.

The boundary conditions employed by (Richardson, 1993) are those appropriate to two stress free surfaces on $z = 0, 1$ and so

$$w = \frac{\partial^2 w}{\partial z^2} = \Delta w = 0, \qquad \theta = \Delta\theta = 0, \qquad \frac{\partial u}{\partial z} = \frac{\partial v}{\partial z} = 0\,.$$

(Richardson, 1993) considers stationary convection in the linearized instability theory for (15.72,15.73) so she sets

$$u_i(\mathbf{x}, t) = u_i(\mathbf{x})e^{\sigma t}, \qquad \theta(\mathbf{x}, t) = \theta(\mathbf{x})e^{\sigma t}$$

in the linearized version of (15.72,15.73) and takes $\sigma = 0$. The linearized equations for stationary instability are then

$$\begin{aligned} \Delta u_i - R\theta k_i - p_{,i} &= 0, \\ \Delta\theta - R(1-\epsilon\Gamma)w &= 0, \end{aligned} \tag{15.74}$$

with u_i being a solenoidal field. By introducing normal modes

$$w = W(z)G(x,y), \qquad \theta = \Theta(z)G(x,y)$$

equations (15.74) yield

$$\begin{aligned} (D^2 - a^2)^2 W &= -Ra^2\Theta, \\ (D^2 - a^2)\Theta &= R(1-\epsilon\Gamma)W. \end{aligned} \tag{15.75}$$

The critical Rayleigh number of stationary convection is found by searching numerically for

$$Ra_L = \min_{a^2} R^2\,.$$

Numerical results are given in (Richardson, 1993).

In addition, (Richardson, 1993) develops a fully nonlinear stability analysis for the Hills-Roberts model. To do this she multiplies the first of (15.72) by u_i, (15.73) by θ and integrates over a period cell V. The individual energy equations thus obtained are

$$\frac{d}{dt}\frac{1}{2}\|\mathbf{u}\|^2 = -\|\nabla\mathbf{u}\|^2 - R < \theta w >, \tag{15.76}$$

and

$$\frac{d}{dt}\frac{1}{2}Pr\|\theta\|^2 = - \|\nabla\theta\|^2 - R < (1-\epsilon\Gamma)\theta w > \\ + Pr\epsilon < \theta^2 w > + \frac{2Pr\epsilon}{R} < \theta d_{ij}d_{ij} > . \tag{15.77}$$

By adding equations (15.76) and (15.77) with a coupling parameter $\lambda > 0$ there results an energy equation of form

$$\frac{dE}{dt} = RI - D + Pr\epsilon\lambda < \theta^2 w > + \frac{2Pr\epsilon\lambda}{R} < \theta d_{ij}d_{ij} > . \tag{15.78}$$

The energy E, production I, and dissipation D are in this situation

$$E(t) = \frac{1}{2}\|\mathbf{u}\|^2 + \frac{1}{2}\lambda Pr\|\theta\|^2, \tag{15.79}$$

$$I(t) = - < \theta w\big[1 + \lambda(1-\epsilon\Gamma)\big] >, \tag{15.80}$$

$$D(t) = \|\nabla\mathbf{u}\|^2 + \lambda\|\nabla\theta\|^2. \tag{15.81}$$

(Richardson, 1993) remarks that the $< \theta^2 w >$ term is not difficult to bound in terms of $DE^{1/2}$ by using a Sobolev inequality. However, the treatment of the term $< \theta d_{ij}d_{ij} >$ appears to require the introduction of a generalized energy containing higher derivative terms.

The analysis of (Richardson, 1993) additionally requires the equation, obtained from (15.72)

$$\frac{d}{dt}\frac{1}{2}\|\nabla\mathbf{u}\|^2 = R < \theta\Delta w > - \|\Delta\mathbf{u}\|^2 + < u_j u_{i,j}\Delta u_i > . \tag{15.82}$$

For another coupling parameter μ (Richardson, 1993) defines a generalized energy $\mathcal{E}(t)$ to be

$$\mathcal{E}(t) = \frac{1}{2}\|\mathbf{u}\|^2 + \frac{1}{2}Pr\lambda\|\theta\|^2 + \frac{1}{2}\mu\|\nabla\mathbf{u}\|^2. \tag{15.83}$$

Her generalized energy identity is then

$$\frac{d\mathcal{E}}{dt} = RI - D + Pr\epsilon\lambda < \theta^2 w > + \mu < u_j u_{i,j}\Delta u_i > \\ + \frac{2Pr\epsilon\lambda}{R} < \theta d_{ij}d_{ij} > - \mu\|\Delta\mathbf{u}\|^2 + R\mu < \theta\Delta w > . \tag{15.84}$$

By defining

$$\frac{1}{R_E} = \max_{\mathcal{H}} \frac{I}{D}, \tag{15.85}$$

assuming $R < R_E$ and putting $a = (R_E - R)/R_E \; (> 0)$ she finds

$$\frac{d\mathcal{E}}{dt} \leq - aD + Pr\epsilon\lambda < \theta^2 w > + \mu < u_j u_{i,j}\Delta u_i > \\ + \frac{2Pr\epsilon\lambda}{R} < \theta d_{ij}d_{ij} > - \mu\|\Delta\mathbf{u}\|^2 + R\mu < \theta\Delta w > . \tag{15.86}$$

She defines a generalized dissipation $\mathcal{D}$ by the quantity $\mathcal{D} = a\|\nabla\mathbf{u}\|^2 + a\lambda\|\nabla\theta\|^2 + \mu\|\Delta\mathbf{u}\|^2$, the idea being to bound the cubic nonlinearities in inequality (15.86) in terms of $\mathcal{D}^{7/8}\mathcal{E}^{5/8}$. To achieve this aim she employs the Sobolev inequalities $\sup_V|\theta| \le c\|\Delta\theta\|^{3/4}\|\theta\|^{1/4}$ and $\|\theta\|_4^2 \le c_1\|\nabla\theta\|^{3/2}\|\theta\|^{1/2}$. In this manner she arrives from (15.86) to the inequality

$$\frac{d\mathcal{E}}{dt} \le -\frac{1}{2}\mathcal{D} + A\mathcal{D}^{7/8}\mathcal{E}^{5/8}, \tag{15.87}$$

for a constant A given explicitly in (Richardson, 1993). Provided $R < R_E$ and $\mathcal{E}(0) < \xi^{1/4}/(4A^2)$, (Richardson, 1993) shows that $\mathcal{E} \to 0$ at least exponentially as $t \to \infty$, and so (conditional) nonlinear stability follows.

To complete her analysis she derives the Euler-Lagrange equations for (15.85) as

$$\begin{aligned} &2\Delta u_i - R_E M\theta k_i = \varpi_{,i}\\ &2\Delta\theta - R_E M w = 0 \end{aligned} \tag{15.88}$$

where $M = [1+\lambda(1-\epsilon\Gamma)]/\sqrt{\lambda}$. Normal modes are introduced and equations (15.88) are reduced to

$$\begin{aligned} &(D^2-a^2)^2W = -\frac{1}{2}R_E M a^2\Theta\\ &(D^2-a^2)\Theta = \frac{1}{2}R_E M W. \end{aligned} \tag{15.89}$$

She then finds

$$Ra_E = \max_{\lambda}\min_{a^2} R_E^2$$

numerically.

In addition to presenting numerical results (Richardson, 1993) also develops an asymptotic analysis of (15.75) for small ϵ. She takes the Prandtl number to be 0.72 in accordance with that for air. For the linear theory she shows

$$Ra_L = \frac{27\pi^4}{4}\left[1 + \epsilon(T_0Pr + \frac{1}{2}) + \cdots\right]. \tag{15.90}$$

(Richardson, 1993) develops a similar small ϵ analysis for the nonlinear stability eigenvalue problem (15.89). She now finds

$$Ra_E = \frac{27\pi^4}{4}\left[1 + \epsilon(T_0Pr + \frac{1}{2}) + \cdots\right]. \tag{15.91}$$

It is important to realize that (Richardson, 1993) shows that to $O(\epsilon)$, $Ra_L = Ra_E$, in other words, to this order of precision the linear instability boundary is the same as the nonlinear stability one.

15.2.2 The Berezin-Hutter model

To complete this section on effects of slight compressibility we note that (Berezin and Hutter, 1997) also argue that the pressure dependence can be discarded in the equation of state $\rho = \rho(p, T)$. They argue that for subsonic conditions or for example in lakes, $\rho = \rho(T)$ is adequate. In particular, they adopt a linear relationship

$$\rho = \rho_0\big[1 - \beta(T - T_0)\big]$$

where $\beta = \rho_0^{-1}(\partial\rho/\partial T)|_p$ is a constant. They write the temperature T as $T = T_0(z) + T_1(\mathbf{x}, t)$ where $T_0(z) = T_0 - Az$, A constant, and then

$$\rho = \rho_0(1 + \beta Az - \beta T_1).$$

The mass balance equation $\rho_{,t} + (\rho v_i)_{,i} = 0$ then becomes

$$(1 + \beta Az - \beta T_1)\, v_{i,i} = \beta\Big(\frac{\partial T^1}{\partial t} + v_i T^1_{,i} - Aw\Big) \tag{15.92}$$

where $w = v_3$. These writers show that the heat conduction equation $T_{,t} + v_i T_{,i} = \kappa\Delta T$ becomes

$$\frac{\partial T_1}{\partial t} + v_i T^1_{,i} = Aw + \kappa\Delta T_1 \,. \tag{15.93}$$

Equations (15.92) and (15.93) effectively allow (Berezin and Hutter, 1997) to remove the density from the system of equations. In fact, they use (15.92) and (15.93) to obtain

$$(1 + \beta Az - \beta T_1)\, v_{i,i} = \beta\kappa\Delta T_1$$

and then argue that for small density variations $\beta|Az - T_1| << 1$ so that this equation reduces to

$$v_{i,i} = \beta\kappa\Delta T_1. \tag{15.94}$$

(Berezin and Hutter, 1997) proceed to study thermal convection under rotation and turbulent conditions. By writing $v_i = u_i + v_i'$, $T_1 = \theta + T'$, where v_i' and T' are fields with a zero ensemble average they derive the following linearized non-dimensional system of equations

$$\begin{aligned}\frac{\partial u_i}{\partial t} =& - p_{,i} + Ra\,\theta\delta_{i3} - 2(\boldsymbol{\Omega}\times\mathbf{u})_i + \Delta u_i\\ &+ \Omega s\big[\mathbf{k}(\mathbf{k}\cdot\mathrm{curl}\,\mathbf{u}) - (\mathbf{k}\cdot\nabla)(\mathbf{k}\times\mathbf{u})\big]_{,i}\,, \qquad (15.95)\\ u_{i,i} =& B\Delta\theta \qquad (15.96)\\ Pr\,\frac{\partial\theta}{\partial t} =& \mathbf{k}\cdot\mathbf{u} + \Delta\theta \qquad (15.97)\end{aligned}$$

where Ra is a Rayleigh number, $\mathbf{k} = (0, 0, 1)$, s is a helicity coefficient, B is a measure of the departure from incompressibility, and $\boldsymbol{\Omega}$ is the angular velocity of the rotating frame. (Berezin and Hutter, 1997) study their system in some detail numerically. To the best of my knowledge an energy stability theory has not been developed for the (Berezin and Hutter, 1997) model.

15.3 The low Mach number approximation

The full equations for compressible convection are (15.52) - (15.54). We rewrite them here using (3.34) for the balance of energy, so the equations of conservation of mass, conservation of linear momentum, and the balance of energy are

$$\dot\rho + \rho v_{i,i} = 0\,, \tag{15.98}$$

$$\rho\dot v_i = \sigma_{ji,j} - gk_i\rho\,, \tag{15.99}$$

$$\rho T\dot S = \sigma_{ji}d_{ij} - q_{i,i}, \tag{15.100}$$

where $\mathbf{k} = (0,0,1)$ and g is gravity. We use equation (3.35) to rewrite the left hand side of (15.100) and then use the representation (15.5) for the stress so

$$\sigma_{ij} = -p\delta_{ij} + \mu_1 d_{ij} + \frac{1}{3}(\mu_2 - 2\mu_1)d_{rr}\delta_{ij} \tag{15.101}$$

where $d_{ij} = (v_{i,j} + v_{j,i})/2$.

In this manner equations (15.98) - (15.100) become

$$\dot\rho + \rho v_{i,i} = 0\,, \tag{15.102}$$

$$\rho\dot v_i = -p_{,i} + \mu_1\Delta v_i + \underline{\frac{1}{3}(\mu_1 + \mu_2)d_{rr,i}} - gk_i\rho\,, \tag{15.103}$$

$$\rho c_p\dot T - \underline{\alpha T\dot p} = -\underline{p\,d_{ii}} + \underline{2\mu_1 d_{ij}d_{ij}} + \underline{\frac{1}{3}(\mu_2 - 2\mu_1)d_{ii}^2} + k\Delta T. \tag{15.104}$$

In deriving (15.104) we have assumed Fourier's heat law $q_i = -kT_{,i}$ with k constant. In addition to (15.102) - (15.104) one needs an equation of state for $p = p(\rho, T)$.

(Mlaouah et al., 1997) is an interesting paper which analyses thermal convection numerically in a cavity filled with a compressible fluid. This paper also contains other pertinent references to numerical studies of compressible convection and, in particular, to articles using a low Mach number approximation. These writers discard the underlined term in (15.103) and the four underlined terms in (15.104) on the grounds that they are several orders of magnitude smaller than the other terms in the relevant equations, for their application. The (Mlaouah et al., 1997) system is reviewed again in the next section. However, (Mlaouah et al., 1997) also deal with a system of equations they term the low Mach number approximation. This approximation is achieved by expanding all variables in a series of small Mach numbers. The low Mach number approximation uses the equation of state

$$\rho T = c, \tag{15.105}$$

for a constant c. If we were to employ a perfect gas law $p = R\rho T$ then (15.105) is effectively looking at convection under constant pressure conditions.

The equations for convection in a low Mach number approximation are then

$$\dot{\rho} + \rho v_{i,i} = 0\,, \tag{15.106}$$

$$\rho \dot{v}_i = -p_{,i} + \mu_1 \Delta v_i - g k_i \rho\,, \tag{15.107}$$

$$\rho c_p \dot{T} = k \Delta T, \tag{15.108}$$

together with (15.105). If we eliminate ρ using (15.105) we are then left with the system of equations

$$\frac{c}{T}\dot{v}_i = -p_{,i} + \mu_1 \Delta v_i - \frac{cg}{T} k_i\,, \tag{15.109}$$

$$\frac{cc_p}{T}\dot{T} = k\Delta T, \tag{15.110}$$

$$v_{i,i} = \frac{k}{c_p}\Delta T. \tag{15.111}$$

It is interesting to note that equation (15.111) is mathematically the same as equation (15.94), arising in the theory of (Berezin and Hutter, 1997). However, the derivations are different.

I am unaware of any nonlinear energy stability calculations on the low Mach number model but, (Frölich et al., 1992) do present an appealing linear instability analysis. (Mlaouah et al., 1997) find that the low Mach number model does agree very well with predictions made using their full compressible model. If one recalls the physical constraints under which the low Mach number approximation applies then a nonlinear stability analysis might be worthwhile especially if a meaningful comparison can be made with the linear analysis of (Frölich et al., 1992). Equations (15.109) - (15.111) admit the steady (conduction) solution in the layer $\mathbb{R}^2 \times \{z \in (0,d)\}$,

$$\bar{v}_i = 0, \qquad \bar{T} = -\beta z + T_L\,, \qquad \bar{p} = \int_a^z \left(\frac{gc}{\beta s - T_L}\right) ds\,,$$

which is consistent with no motion and constant upper and lower temperatures T_U and T_L.

15.4 Stability in compressible convection

In this section we examine some studies which have looked at convection and stability for the full system of compressible flow equations.

(Mlaouah et al., 1997) develop a two-dimensional numerical simulation of convection in a square cavity. Their boundary conditions are that on the top and bottom planes $z = 0, d$ the walls are subject to adiabatic conditions whereas the convection is driven by differentially heated sidewalls. The sidewalls $x = 0, d$ have a temperature difference of ΔT°C. They consider ΔT up to 34°C which is almost certainly beyond the realm of the Boussinesq approximation. This is a very interesting paper which presents results for the square cavity problem found by solving three models. The Boussinesq approximation model, the one employing a low Mach number approximation as seen in the last section, and their system for the full compressible fluid model (which we denote by the MTN system). In fact, the MTN system is composed of equations (15.106) - (15.108) together with the equation of state for a perfect gas. Thus, the system of equations representing convection in a compressible fluid given by (Mlaouah et al., 1997) is

$$\dot{\rho} + \rho v_{i,i} = 0\,, \tag{15.112}$$

$$\rho \dot{v}_i = -p_{,i} + \mu_1 \Delta v_i - g k_i \rho\,, \tag{15.113}$$

$$\rho c_p \dot{T} = k \Delta T, \tag{15.114}$$

$$p = R^* \rho T, \tag{15.115}$$

where R^* is the gas constant.

We are not aware of stability calculations based on this precise model, although we next detail work on a very similar theory by (Spiegel, 1965) and then by (Padula, 1986). It is noteworthy that for the thermal convection problem with the fluid between the planes $z = 0, d$, gravity in the negative z-direction, and constant boundary temperatures $T_L(z = 0)$, $T_U(z = d)$, $T_L > T_U$, then equations (15.112) - (15.115) admit the conduction solution

$$\bar{v}_i = 0, \quad \bar{T} = -\beta z + T_L\,, \quad \bar{\rho} = \frac{\rho_0}{T_L^m}\,(T_L - \beta z)^m, \quad \bar{p} = \frac{\rho_0 R^*}{T_L^m}\,(T_L - \beta z)^{m+1},$$

where $\beta = (T_L - T_U)/d$, $\rho_0 \equiv \rho(z = 0)$, and m is the polytropic index $m = (g/R^*\beta) - 1$.

(Spiegel, 1965) is historically an important contribution. He investigates convection for the compressible system which retains the $\mu_1 d_{rr,i}/3$ term in (15.103) and the $p\, d_{ii}$ term in (15.104). In addition, he employs the specific heat at constant volume, c_v, in (15.104) in place of c_p. Thus, Spiegel's system for compressible convection is

$$\dot{\rho} + \rho v_{i,i} = 0\,, \tag{15.116}$$

$$\rho \dot{v}_i = -p_{,i} + \mu_1 \Delta v_i + \frac{1}{3}\,\mu_1 v_{j,ji} - g k_i \rho\,, \tag{15.117}$$

$$\rho c_v \dot{T} = -p v_{i,i} + k \Delta T, \tag{15.118}$$

$$p = R^* \rho T. \tag{15.119}$$

(Spiegel, 1965) studies instability of the situation where the fluid is contained between the planes $z = z_0$ and $z = z_0 + d$ with constant boundary temperatures T_L and T_U, $T_L > T_U$ (he actually treats $z = z_0$ as the upper boundary). His steady solution is

$$\bar{v}_i = 0, \quad \bar{T} = \beta z, \quad \bar{\rho} = \frac{P}{R^* \beta} z^m, \quad \bar{p} = P z^{m+1}$$

where P is a constant of integration.

The work of (Spiegel, 1965) centres around an analysis of linearized instability. He remarks that for a compressible atmosphere the problems of the onset of convection and of overstability can be treated as separate and his interest centres on stationary convection. He shows that the equations for stationary convection, in terms of a stretched wavenumber and vertical coordinate $\zeta = z/d$ are

$$\mathcal{L}W = -\frac{PR^*(m+1)^2 d^{m+1}}{g\mu} a^2 \zeta^m \Theta$$
$$(D^2 - a^2)\Theta = \frac{\beta P c_p (m+1) d^{m+2}}{gk} \zeta^m W$$

where $D = d/d\zeta$ and the operator $\mathcal{L}$ has form

$$\begin{aligned}\mathcal{L} =& \zeta(D^2 - a^2)^2 + \zeta(D^2 - a^2) D \frac{m}{\zeta} \\ &- (m+1)(D^2 - a^2)\Big(D + \frac{m}{\zeta}\Big) + \frac{m(m+1)}{3\zeta} a^2 .\end{aligned}$$

(Spiegel, 1965) develops his results in terms of two Rayleigh numbers, one defined in terms of $z = z_0 + d$, the other in terms of the mean depth of the layer. He interestingly points out that the boundary conditions in compressible convection may be a matter of contention since in stellar convection the boundaries are not well defined. This is because motions can penetrate from unstable zones into stable ones and the phenomenon of penetrative convection (see chapter 17) ought to be considered. For the study of (Spiegel, 1965) boundary conditions free of tangential stress are considered. Thus,

$$w = \theta = \frac{\partial u}{\partial z} = \frac{\partial v}{\partial z} = 0, \qquad \zeta = \zeta_0, \zeta_0 + 1.$$

Spiegel observes that in astrophysical convection the convecting layer in the thick outer zones of late-type stars is highly important and for this it is likely to be important to study departures from the Boussinesq approximation. With this in mind he develops a perturbation analysis of (15.116) - (15.119) (his thin layer theory). He employs the non-dimensional thickness Z as an expansion parameter, where $\zeta = \zeta_0(1 + Zx)$ and then solves the W equation at each expansion level, putting

$$W = W_0 + ZW_1, \qquad R = R_0 + ZR_1.$$

At $O(1)$ level he finds $R_0 = (a^2 + n^2\pi^2)^2/a^2$ and then $R_1 = -(m - 1/2)R_0$. Thus, in terms of the Rayleigh number evaluated at the top of the layer the first order effect of compressibility is to reduce the critical Rayleigh number. (Spiegel, 1965) continues a systematic asymptotic analysis of his compressible convection problem, including use of a WKB approximation.

15.4.1 Nonlinear energy stability in compressible convection

It is worth pointing out that energy methods were employed to obtain uniqueness theorems for compressible flows in the 1950's. (Graffi, 1953) (see also (Graffi, 1999), p. 229) and (Serrin, 1959d) are notable papers which primarily focussed on bounded spatial domains. (Graffi, 1959; Graffi, 1960) (see also (Graffi, 1999), p. 251 and p. 263) also presents a generalized energy method which he adapts for uniqueness for a compressible fluid on an unbounded domain without requiring stringent decay conditions at infinity. Graffi's unbounded domain energy technique is quite beautiful. Nonlinear energy stability analyses of the Spiegel model are given by (Padula, 1986) and by (Coscia and Padula, 1990). These analyses both yield conditional stability results and are very technical.

(Padula, 1986) considers the thermal convection problem with stress free boundary conditions and her analysis uses the generalized energy

$$\mathcal{E}(t) = E(t) + \sum_{i=1}^{4} \lambda_i E_i(t),$$

for suitable coupling parameters λ_i. The functions E_i are effectively introduced to control the nonlinearities and the "energy" $E(t)$ is given by

$$\begin{aligned} 2E(t) =& (a-1) < z^{2m} u_i u_i > + Pr(c-1) < z^{2m+1} \rho^2 > + Pr < z^{2m-1} \theta^2 > \\ & + \gamma b < \rho_{,i} \rho_{,i} > + \frac{2R\gamma}{\mathcal{V}} < z^m u_i \rho_{,i} > . \end{aligned}$$

She develops an energy analysis leading to an equation of form $dE/dt = I - D + N$, where her reduced dissipation function $D' (\leq D)$ has form

$$D' = (a-1) < z^m u_{i,j} u_{i,j} > + \frac{1}{2} < z^{m-1} \theta_{,i} \theta_{,i} > + \frac{R^2 \gamma}{2(m+1)} < z^{m+1} \rho_{,i} \rho_{,i} > .$$

(The actual dissipation function D contains sign indefinite terms and needs to be bounded below.) Her (conditional) nonlinear stability result is expressed in terms of the maximum

$$\mathcal{M} = \max_{\mathcal{H}} \frac{I}{D} .$$

Provided $\mathcal{M}$ is suitably small she shows that $\mathcal{E}(t)$ decreases exponentially if $\mathcal{E}(0) < A$. The maximum condition is reduced to a polynomial inequality in R and then she deduces conditions on the Rayleigh number for nonlinear stability. These are

$$R^2 < \frac{\pi^4 2g(c_p\beta_0 - g)(1-\eta)^2}{(c_p\beta_0 - g)^2 + g^2} \qquad \text{when} \quad Pr > \frac{gc_v\beta_0}{2g},$$

$$R^2 < \frac{\pi^4 2Prg(c_p\beta_0 - g)(1-\eta)^2}{2Pr(c_p\beta_0 - g)^2 + g^2 c_v\beta_0} \qquad \text{when} \quad Pr \leq \frac{gc_v\beta_0}{2g}.$$

No comparison of these thresholds with the linear instability results is given, nor is a detailed analysis of the conditional restriction provided. However, it is remarkable that any progress at all has been possible with a nonlinear stability analysis when one considers the complexity of the compressible flow equations.

(Bormann, 2001) is a recent contribution to the compressible Bénard problem. He treats equations (15.98) and (15.99) with μ_2 absent from (15.101) and employs (15.54) for the balance of energy. The steady state temperature is linear in z and (Bormann, 2001) considers an equation of state $p = p(\rho, T)$ which allows him to reduce the steady state equation

$$\frac{\partial \bar{p}}{\partial z} = -g\bar{\rho}(z)$$

to

$$\frac{\partial \bar{\rho}}{\partial z} + \alpha\bar{\rho}\frac{\partial \bar{T}}{\partial z} = -\kappa_T\bar{\rho}^2 g \tag{15.120}$$

where the coefficients α and κ_T are constant. He calls α the thermal expansion coefficient and κ_T the isothermal compressibility.

Bormann develops a linearized instability analysis for the Bénard problem with fixed boundary conditions and because of his choice of equation of state the steady solution for $\bar{\rho}$ from (15.120) has form

$$\bar{\rho}(z) = \frac{k_1}{k_2 \exp[A(z-1)] + k_3}$$

for constants A, k_1, k_2, k_3. Various conclusions are drawn based entirely on a linearized instability analysis for stationary convection. With compressible convection there may well be strongly nonlinear subcritical instabilities which entirely change the predictions drawn from linear analyses. Thus, nonlinear stability analyses, and particularly unconditional nonlinear stability analyses, would be particularly welcome. To my knowledge there is no nonlinear energy stability work available for the (Bormann, 2001) model.

16

Convection with temperature dependent fluid properties

16.1 Depth dependent viscosity and symmetry

The instability of the thermal conduction solution when a layer of fluid is heated from below and convection cells form is treated extensively in this book. Up to this point we have mostly assumed the properties of the fluid are constant. It may, however, be argued that viscosity should always be treated as a function of temperature in the Bénard problem, since it is one of the fluid properties which does exhibit a considerable change with varying temperature. To appreciate this variation we simply look at values for mundane fluids. For example, (Rossby, 1969) quotes that the kinematic viscosity of water varies from 0.01008 cm^2sec^{-1} at 20°C to 0.00896 cm^2sec^{-1} at 25°C, this being approximately a 10% change. Over the same temperature range the thermal conductivity only exhibits a 1.5% change. (Rossby, 1969) also quotes values for the viscosity of a 20 cSt silicone oil, of 0.2137 cm^2sec^{-1} at 20°C and 0.1904 cm^2sec^{-1} at 25°C, i.e. approximately a 20% variation. Over this temperature range the thermal conductivity of the same silicone oil is constant. The viscosity values just quoted are for typical room temperatures over a range in which the Boussinesq approximation may be expected to hold. If the temperature range is greater the viscosity variation is also typically much larger. For example (Lide, 1991) states that the viscosity of olive oil varies from 138.0 centipoise at 10°C to 12.4 centipoise at 70°C.

In this chapter we investigate various nonlinear stability analyses which have been performed for the Bénard problem when the viscosity varies with temperature. However, we begin with a simpler model in which the viscosity variation is a function of the depth coordinate. This means that the viscosity is still varying with temperature, but really only with the steady state temperature. There are geophysical situations where this is perfectly reasonable. For example, (Torrance and Turcotte, 1971) argue that in the earth's mantle the pressure increases strongly with depth due to the hydrostatic head and this increase should lead to a corresponding increase in viscosity. They propose a viscosity dependence of form

$$\nu = \nu_0 h(z), \tag{16.1}$$

where ν is the kinematic viscosity, ν_0 is constant, and

$$h(z) = \exp\{c(\frac{1}{2} - z)\}. \tag{16.2}$$

In (16.2) c is a constant and the fluid occupies the dimensionless layer $0 < z < 1$.

The equations which govern convection with the viscosity given by (16.1) are essentially (3.41) but where the stress tensor $t_{ij} = -p\delta_{ij} + 2\nu\rho_0 D_{ij}$ for ν constant is replaced by $t_{ij} = -p\delta_{ij} + 2\nu(z)\rho_0 D_{ij}$ where $D_{ij} = (v_{i,j} + v_{j,i})/2$. One again shows that the steady solution to the relevant equations together with boundary conditions (3.42) is given by (3.43). The non-dimensional equations for a perturbation to the steady state may be derived in the usual manner and are, cf. (Straughan, 1986),

$$\begin{aligned} &u_{i,t} + u_j u_{i,j} = -\pi_{,i} + R\theta k_i + 2(hd_{ij})_{,j}, \qquad u_{i,i} = 0, \\ &Pr(\theta_{,t} + u_i\theta_{,i}) = Rw + \Delta\theta. \end{aligned} \tag{16.3}$$

The perturbation boundary conditions for the two fixed surfaces are

$$u_i = 0, \quad \theta = 0, \qquad \text{on } z = 0, 1 \tag{16.4}$$

with u_i, θ, π satisfying a plane tiling periodic planform in the x, y directions.

The system (16.3), (16.4), is such that the linear operator is symmetric and the conditions of section 4.3 are satisfied. Thus, there is coincidence of the linear instability critical Rayleigh number and the nonlinear energy one. The nonlinear stability obtained with this model in which the viscosity is a function only of z is unconditional. Thus, using the theory of section 4.3, the *nonlinear* critical Rayleigh number is given by the linearized equations derived from (16.3) with the growth rate σ set equal to zero.

Numerical results for this problem are presented in (Straughan, 1986). This paper also considers the equivalent convection problem in a Darcy porous medium when the viscosity is given by (16.1) and the permeability is also a function of z. The resulting system is again symmetric and the coincidence of the linear and nonlinear Rayleigh number thresholds is found. We add that the symmetry arguments applied in this section work for an arbitrary function of z, not just that given by (16.1).

16.2 Stability when the viscosity depends on temperature

Fundamental early articles on convection in temperature-dependent viscosity fluids are those of (Tippelskirch, 1956) and (Palm et al., 1967). The viscosity-temperature relationship is generally a nonlinear one but for many practical studies it is adequate to adopt the linear relationship of (Palm et al., 1967),

$$\nu(T) = \nu_0\big[1 - \gamma(T - T_0)\big], \tag{16.5}$$

where ν_0, γ, T_0 are positive constants, and T is the temperature field.

When (16.5) is adopted the equations derived from the Navier-Stokes equations for convective motion in the planar region between $z = 0$ and $z = d$ are

$$\begin{aligned} v_{i,t} + v_j v_{i,j} &= -\frac{1}{\rho_0}\, p_{,i} + \nu_0\Big\{\big(1 - \gamma[T - T_0]\big)2D_{ij}\Big\} + \alpha g k_i T, \\ v_{i,i} &= 0, \\ T_{,t} + v_i T_{,i} &= \kappa \Delta T, \end{aligned} \tag{16.6}$$

where $v_i, p, \rho_0, \alpha, g, \kappa$ are velocity, pressure, constant density, thermal expansion coefficient, gravity and thermal conductivity, respectively, and $D_{ij} = (v_{i,j} + v_{j,i})/2$. We study the problem with constant boundary temperatures

$$T = T_1, \quad z = 0, \qquad T = T_2, \quad z = d, \qquad (T_1 > T_2),$$

which gives rise to the steady solution

$$\bar{v}_i = 0, \qquad \bar{T} = -\beta z + T_1, \qquad \beta = \frac{T_1 - T_2}{d}.$$

The perturbations u_i, θ, π to $\bar{v}_i, \bar{T}, \bar{p}$ are introduced and non-dimensionalized and then the non-dimensional perturbation equations are, cf. (Richardson, 1993), (Richardson and Straughan, 1993),

$$\begin{aligned} u_{i,t} + u_j u_{i,j} &= -\pi_{,i} + 2\frac{\partial}{\partial x_j}\Big\{\big[1 + \Gamma(z - \xi)\big]d_{ij}\Big\} \\ &\quad + R\theta k_i - 2\frac{\Gamma Pr}{R}\,\frac{\partial}{\partial x_j}(\theta d_{ij}), \\ u_{i,i} &= 0, \\ Pr(\theta_{,t} + u_i\theta_{,i}) &= Rw + \Delta\theta, \end{aligned} \tag{16.7}$$

where $\Gamma = \gamma\beta d$ with $\xi = (T_1 - T_0)/\beta d$.

(Richardson, 1993) developed a nonlinear energy stability analysis for (16.7). This is highly non-trivial due to the presence of the $(\theta d_{ij})_{,j}$ term. This is easily recognised by forming a standard energy equation from (16.7) of form

$$\frac{dE}{dt} = RI - D + 2\frac{\Gamma Pr}{R} < \theta d_{ij} d_{ij} >, \tag{16.8}$$

where

$$E = \frac{1}{2}\|\mathbf{u}\|^2 + \frac{1}{2}Pr\|\theta\|^2, \qquad I = 2(\theta, w),$$
$$D = \|\nabla \mathbf{u}\|^2 + 2\Gamma < (z - \xi)d_{ij}d_{ij} > + \|\nabla\theta\|^2.$$

The difficulty is to dominate the cubic term $< \theta d_{ij} d_{ij} >$. Indeed, I believe within Navier-Stokes theory an unconditional nonlinear stability analysis which yields a sharp Rayleigh number threshold has not yet been developed.

(Richardson, 1993) instead develops a conditional nonlinear analysis for two stress free surfaces. Her analysis is very intricate and relies on developing a theory for a generalized energy of form

$$\mathcal{E}(t) = E(t) + \frac{1}{2}\eta\|\nabla \mathbf{u}\|^2 + \frac{1}{2}\zeta Pr\|\Delta\theta\|^2,$$

for coupling parameters η and ζ. Although the result she finds is only a conditional one it is very sharp in that it derives coincidence of the nonlinear energy and linear instability Rayleigh number thresholds.

(Capone and Gentile, 1994; Capone and Gentile, 1995) also develop intricate conditional nonlinear energy stability analyses for the variable viscosity problem. In (Capone and Gentile, 1994) they consider the viscosity relation

$$\nu(T) = \nu_0 \exp\big[-\eta(T - T_0)\big]$$

whereas in (Capone and Gentile, 1995) they treat a very general viscosity of form

$$\nu(T) = \nu_0 f(T),$$

in which f is a convex non-increasing function. (Capone and Gentile, 1995) provide specific results for glycerol and golden syrup which they model using the viscosity fits $\nu = \nu_0 \exp(a - bT + cT^2 - dT^3 + eT^4)$ and $\nu = \nu_0 \exp(1/[pT^2 + qT + r])$, respectively, where values for the constants may be found in (Capone and Gentile, 1995).

16.3 Unconditional nonlinear stability

As we have stated in section 16.2 (Richardson, 1993) used a generalized energy analysis to establish nonlinear stability bounds which are very sharp when compared to those of linearised instability theory. She chose a viscosity which decreased linearly with temperature. Richardson's work employed a non-trivial Lyapunov function and is an intricate analysis. The extension of (Capone and Gentile, 1994; Capone and Gentile, 1995) allowed for a more general viscosity-temperature relationship. These articles are important since they are the first to derive nonlinear energy stability bounds for Bénard convection with temperature-dependent viscosity. Nevertheless they have two deficiencies. The first is that the analysis holds only for two surfaces free of tangential stress. The second deficiency is that the stability thresholds derived are conditional. (Straughan, 2002c) removed these restrictions by working with alternative models of fluid behaviour to the Navier-Stokes equations, namely those of (Ladyzhenskaya, 1967; Ladyzhenskaya, 1968; Ladyzhenskaya, 1969). In this section we effectively describe the results of (Straughan, 2002c) but in the context of a viscoelastic fluid.

It is worth noting that there has been much recent mathematical work dealing with convection problems when the viscosity or conductivity is a function of temperature or concentration, cf. (Abo-Eldahab and El Gendy, 2000), (Abo-Eldahab and Salem, 2001), (Capone and Gentile, 1994; Capone and Gentile, 1995), (Diaz and Galiano, 1997; Diaz and Galiano, 1998), (Flavin and Rionero, 1998; Flavin and Rionero, 1999a; Flavin and Rionero, 1999b), (Galiano, 2000), (Kafoussias and Williams, 1995), (Payne et al., 1999), (Payne and Straughan, 2000b), (Qin and Chadam, 1996), (Richardson, 1993), (Straughan, 2002c), (Zhao et al., 1995). (Diaz and Galiano, 1997; Diaz and Galiano, 1998) develop an existence theory for solutions to the fluid equations with temperature varying viscosity and thermal conductivity.

We now denote the kinematic viscosity by ν_0 and we let this be a function of temperature of form

$$\nu_0(T) = \nu_L\big(1 - \gamma[T - T_L]\big), \tag{16.9}$$

for a constant $\gamma > 0$, where T_L is a reference temperature, chosen here as the temperature of the lower plane of the Bénard layer.

The basic viscoelastic fluid model we select is

$$\begin{aligned} \frac{\partial v_i}{\partial t} + v_j \frac{\partial v_i}{\partial x_j} &= -\frac{1}{\rho}\frac{\partial p}{\partial x_i} + \frac{\partial}{\partial x_k}\Big[\Big(\nu_0(T) + \nu_1|\mathbf{D}|^{2\mu}\Big) D_{ik}\Big] \\ &\quad + g\alpha k_i T, \end{aligned} \tag{16.10}$$

$$\frac{\partial v_i}{\partial x_i} = 0, \qquad \frac{\partial T}{\partial t} + v_i \frac{\partial T}{\partial x_i} = \kappa \Delta T,$$

where $\nu_1 > 0$ is a constant, $D_{ik} = (v_{i,k} + v_{k,i})/2$, and $|\mathbf{D}| = \sqrt{D_{ik}D_{ik}}$, the other coefficients being as in the rest of the chapter. We concentrate on the cases $\mu = 1/2$ and $\mu = 1$. Note that (16.10) is a generalization of a well known model in viscoelasticity, cf. (Antontsev et al., 2001), for ν_0 a function of T. Similar models are also used in glaciology (Man and Sun, 1987) and in applications of generalized second and third grade fluids, cf. (Massoudi and Phuoc, 2001), (Tigoiu, 2000).

In (16.10) we assume ν_0 is given by (16.9) and we employ a Boussinesq approximation in the sense that the density is constant except in the buoyancy term which is linear in temperature.

The physical picture is one where the fluid behaviour is governed by the viscoelastic equations (16.10) and the fluid occupies the layer $\mathbb{R}^2 \times \{z \in (0,d)\}$. Gravity is in the negative $z-$direction and the planes $z = 0, d$ are, respectively, held at fixed temperatures T_L and T_U, with $T_L > T_U$.

For any $\mu > 0$ (16.10) admits the steady solution

$$\bar{T} = T_L - \beta z, \qquad \bar{v}_i \equiv 0, \qquad \beta = \frac{T_L - T_U}{d}, \tag{16.11}$$

with the steady pressure $\bar{p}(z)$ found from the momentum equation.

We introduce perturbations (u_i, θ, π) by $v_i = \bar{v}_i + u_i, T = \bar{T} + \theta, p = \bar{p} + \pi$ and rescale the perturbation equations to make them non-dimensional. The time scale $\mathcal{T}$, velocity scale U, pressure scale P, and temperature scale $T^\sharp$ chosen are $\mathcal{T} = d^2/\nu_L$, $U = \nu_L/d$, $P = \rho\nu_L U/d$, $T^\sharp = (\beta\nu_L/\kappa g\alpha)^{1/2}\, U$. In addition we have non-dimensional expressions of the viscosity coefficient γ, Prandtl number, Pr, Rayleigh number, R^2, and the coefficient ν_1, namely, $\Gamma = \gamma\beta d$, $Pr = \nu_L/\kappa$, $R^2 = d^4\beta g\alpha/\kappa\nu_L$, $\omega = \nu_L\nu_1/d^4$. Using these variables, one may derive the following non-dimensional equations for the perturbations to the steady solution (16.11),

$$\begin{aligned} \frac{\partial u_i}{\partial t} + u_j\frac{\partial u_i}{\partial x_j} &= -\frac{\partial \pi}{\partial x_i} + \big[(1+\Gamma z)\, d_{ij}\big]_{,j} - \frac{\Gamma Pr}{R}\,(\theta d_{ij})_{,j} \\ &\quad + \omega\big(|\mathbf{d}|^{2\mu} d_{ij}\big)_{,j} + R\theta k_i, \\ \frac{\partial u_i}{\partial x_i} = 0, \qquad & Pr\Big(\frac{\partial \theta}{\partial t} + u_i\frac{\partial \theta}{\partial x_i}\Big) = Rw + \Delta\theta. \end{aligned} \tag{16.12}$$

The boundary conditions are

$$u_i = 0, \quad z = 0, 1, \qquad \theta = 0, \quad z = 0, 1, \tag{16.13}$$

and u_i, θ, π satisfy a plane tiling periodic shape in the (x, y) plane. The cell which arises due to this plane tiling form is denoted by V.

16.3.1 Unconditional stability for $\mu = 1$

The analysis commences by deriving two energy equations. Multiply $(16.12)_1$ by u_i, and $(16.12)_3$ by θ and intergrate over V. One finds

$$\frac{d}{dt}\frac{1}{2}\|\mathbf{u}\|^2 = R(\theta, w) - < f|\mathbf{d}|^2 > -\omega < |\mathbf{d}|^4 > + \frac{\Gamma Pr}{R}\int_V \theta|\mathbf{d}|^2 dx, \tag{16.14}$$

$$\frac{d}{dt}\frac{1}{2}Pr\|\theta\|^2 = R(\theta, w) - \|\nabla\theta\|^2, \tag{16.15}$$

where $w = u_3$, $f = 1 + \Gamma z$, and $< \cdot >$ denotes integration over V.

The idea is to use the ω term in (16.14) to control the cubic nonlinearity in the same equation. To this end, we employ the arithmetic - geometric mean inequality on the last term in (16.14) and add the result to (16.15) multiplied by a positive coupling parameter λ, to see that

$$\frac{dE}{dt} \leq I - D, \tag{16.16}$$

where the energy, E, dissipative term, D, and the production term, I, are defined by

$$E(t) = \frac{1}{2}\|\mathbf{u}\|^2 + \frac{1}{2}\lambda Pr\|\theta\|^2,$$

$$D(t) = < f|\mathbf{d}|^2 > +\lambda\|\nabla\theta\|^2, \qquad I(t) = R(1+\lambda)(\theta, w) + \frac{\Gamma^2 Pr^2}{4R^2\omega}\|\theta\|^2.$$

Next set

$$\frac{1}{R_E} = \max_H \frac{I}{D}, \tag{16.17}$$

where H is the space of admissible solutions. The existence of a maximising solution is ensured since $f \geq 1$. From (16.16) we now deduce

$$\frac{dE}{dt} \leq -D\Big(1 - \frac{1}{R_E}\Big). \tag{16.18}$$

For $R_E > 1$ global nonlinear (exponential) stability follows because $D \geq cE$ for a constant $c > 0$ thanks to Poincaré's inequality (since $f \geq 1$). The nonlinear stability threshold is given by $R_E = 1$.

The Euler-Lagrange equations obtained from (16.17) with $R_E = 1$ are

$$\begin{aligned} &(fd_{ij})_{,j} + \frac{1}{2}R(1+\lambda)\theta k_i = -\pi_{,i}\,, \qquad u_{i,i} = 0,\\ &\lambda\Delta\theta + \frac{1}{2}R(1+\lambda)w + \frac{\Gamma^2 Pr^2}{4\omega R^2}\,\theta = 0, \end{aligned} \tag{16.19}$$

where $\pi(\mathbf{x})$ is a Lagrange multiplier. The solution of (16.19) yields the critical Rayleigh number threshold for R^2 for *unconditional* nonlinear stability. Note that with $\lambda = 1$ equations (16.19) are identical to the linearised instability equations obtained from (16.12), apart from the last term in $(16.19)_3$. The coefficient of this term is $O(\Gamma^2)$ and so the linearised instability critical Rayleigh numbers will be very close to the nonlinear ones determined from (16.19), certainly when Γ is small. However, previous experience with numerical calculations for eigenvalue problems of this type suggests that these boundaries will be close even for larger Γ.

16.3.2 Unconditional stability for $\mu = 1/2$.

For $\mu = 1/2$ we again commence with equations (16.14) and (16.15) but now add them directly (i.e. use a coupling parameter $\lambda = 1$) to see that

$$\begin{aligned}\frac{d}{dt}\left(\frac{1}{2}\|\mathbf{u}\|^2 + \frac{1}{2}Pr\|\theta\|^2\right) =& 2R(\theta, w) - < f|\mathbf{d}|^2 > - \|\nabla\theta\|^2 \\ & - \omega < |\mathbf{d}|^3 > + \frac{\Gamma Pr}{R} < \theta|\mathbf{d}|^2 > .\end{aligned} \tag{16.20}$$

It transpires that to control the last term in (16.20) we need to incoporate an equation for the L^3 norm of θ. Thus, let $\|\cdot\|_3$ denote the norm on $L^3(V)$ and then from $(16.12)_3$ we find

$$\frac{d}{dt}\frac{1}{3}Pr\|\theta\|_3^3 = R\int_V w\theta^2(\text{sign}\,\theta)dx - \frac{8}{9}\|\nabla|\theta|^{3/2}\|^2. \tag{16.21}$$

By employing Poincaré's inequality we then arrive at

$$\frac{d}{dt}\frac{1}{3}Pr\|\theta\|_3^3 \leq R\int_V w\theta^2(\text{sign}\,\theta)dx - \frac{8\pi^2}{9}\|\theta\|_3^3. \tag{16.22}$$

We now use Young's inequality for positive α and β to obtain

$$R\int_V w\theta^2(\text{sign}\,\theta)dx \leq \frac{R\beta^2}{3}\|w\|_3^3 + \frac{2R}{3\beta}\|\theta\|_3^3, \tag{16.23}$$

and

$$\frac{\Gamma Pr}{R}\int_V \theta|\mathbf{d}|^2 dx \leq \frac{2\Gamma Pr\alpha}{3R}\int_V |\mathbf{d}|^3 dx + \frac{\Gamma Pr}{3\alpha^2 R}\|\theta\|_3^3. \tag{16.24}$$

We need to bound $\|w\|_3$ in terms of $\|\mathbf{d}\|_3$ and this is achieved by using Hölder's inequality and the Sobolev inequality as follows

$$\begin{aligned}\|w\|_3^3 \leq \int_V |\mathbf{u}|^3 dx \leq & V^{1/2}\left(\int_V |\mathbf{u}|^6 dx\right)^{1/2} \\ \leq & V^{1/2}c_S^3\left(\int_V |\nabla\mathbf{u}|^2 dx\right)^{3/2} \\ =& 2^{3/2}V^{1/2}c_S^3\left(\int_V |\mathbf{d}|^2 dx\right)^{3/2} \\ \leq & 2^{3/2}Vc_S^3\int_V |\mathbf{d}|^3 dx,\end{aligned}$$

where V is the volume of V and c_S is the optimal constant in the Sobolev inequality $\|\mathbf{u}\|_{L^6} \leq c_S\|\mathbf{u}\|_{H_0^1}$. We denote $c_1 = 2^{3/2}Vc_S^3$ and then we have

$$\|w\|_3^3 \leq c_1\int_V |\mathbf{d}|^3 dx. \tag{16.25}$$

Using (16.25), inequality (16.23) may be replaced by

$$R\int_V w\theta^2(\text{sign}\,\theta)dx \le \frac{R\beta^2 c_1}{3}\int_V |\mathbf{d}|^3 dx + \frac{2R}{3\beta}\|\theta\|_3^3. \tag{16.26}$$

We next employ (16.26) in (16.22) to find

$$\frac{d}{dt}\frac{1}{3}Pr\|\theta\|_3^3 \le \frac{R\beta^2 c_1}{3}\int_V |\mathbf{d}|^3 dx - \|\theta\|_3^3\Big(\frac{8\pi^2}{9} - \frac{2R}{3\beta}\Big). \tag{16.27}$$

Now let a denote a positive constant and from (16.20) and (16.27) we may derive the inequality

$$\begin{aligned}\frac{dE}{dt} \le & 2R(\theta, w) - < f|\mathbf{d}|^2 > - \|\nabla\theta\|^2 \\ & - \Big\{\omega - \frac{Ra\beta^2 c_1}{3} - \frac{2\Gamma Pr\alpha}{3R}\Big\} < |\mathbf{d}|^3 > \\ & - \Big\{\frac{8\pi^2 a}{9} - \frac{2Ra}{3\beta} - \frac{\Gamma Pr}{3R\alpha^2}\Big\}\|\theta\|_3^3,\end{aligned} \tag{16.28}$$

where E is an energy given by

$$E = \frac{1}{2}\|\mathbf{u}\|^2 + \frac{1}{2}Pr\|\theta\|^2 + \frac{a}{3}\,Pr\,\|\theta\|_3^3\,. \tag{16.29}$$

We now define I and D by

$$I = 2(w, \theta), \qquad D = \int_V f|\mathbf{d}|^2 dx + \|\nabla\theta\|^2. \tag{16.30}$$

Next, set $a = a' + k\epsilon$ for a number k to be selected and an arbitrarily small $\epsilon(> 0)$. We choose α, β such that $(8\pi^2 a'/9) - (\Gamma Pr/3R\alpha^2) - (2Ra'/3\beta) = 0$, and then minimise $f(\alpha, \beta) = (Rc_1/3)\beta^2 a' + (2\Gamma Pr\alpha/3R)$. This is to have ω as small as possible. To minimize f we first differentiate with respect to α to find $\alpha = \alpha(\beta)$. Then we find $df/d\beta$. This leads to the choices $\alpha = 9Rc_1^{1/3}/8\pi^2$, $\beta = 9R/8\pi^2$, $a' = 8\Gamma Pr\pi^2/9R^3c_1^{2/3}$. Next select $k = 27/8\pi^2$ and from (16.28) we deduce

$$\frac{dE}{dt} \le RI - D - \hat{\omega}\|\mathbf{d}\|_3^3 - \epsilon\|\theta\|_3^3\,, \tag{16.31}$$

where $\hat{\omega}$ is given by $\hat{\omega} = \omega - (9\Gamma Prc_1^{1/3}/8\pi^2) - (27c_1R^3k/64\pi^4)\,\epsilon$.

To utilize the $\|\mathbf{d}\|^3$ term we require that $\hat{\omega} \ge 0$. Since $\epsilon > 0$ is arbitrary this essentially implies the restriction

$$\omega > \frac{9\Gamma Prc_1^{1/3}}{8\pi^2}\,. \tag{16.32}$$

Inequality (16.32) is the restriction on ω by comparison with the viscosity coefficient γ. Define R_E by

$$\frac{1}{R_E} = \max_H \frac{I}{D}\,. \tag{16.33}$$

Inequality (16.31) allows us to deduce

$$\frac{dE}{dt} \leq -D\left(\frac{R_E - R}{R_E}\right) - \epsilon\|\theta\|_3^3. \qquad (16.34)$$

We require $R < R_E$ and then use of the Poincaré inequality in (16.34) allows us to conclude that

$$\frac{dE}{dt} \leq -\pi^2\left(\frac{R_E - R}{R_E}\right)\left(\frac{1}{2}\|\mathbf{u}\|^2 + \|\theta\|^2\right) - \epsilon\|\theta\|_3^3. \qquad (16.35)$$

Hence, employing the definition of $E(t)$ we may show

$$\frac{dE}{dt} \leq -cE \qquad (16.36)$$

where c is the constant

$$c = \min\left\{\pi^2\left(\frac{R_E - R}{R_E}\right), \frac{2\pi^2}{Pr}\left(\frac{R_E - R}{R_E}\right), \frac{3\epsilon}{aPr}\right\}.$$

From inequality (16.36), we see that the energy decays exponentially and unconditional nonlinear stability follows provided

$$R < R_E \qquad \text{and} \qquad \omega > \frac{9\Gamma Prc_1^{1/3}}{8\pi^2}. \qquad (16.37)$$

If we calculate the Euler-Lagrange equations from (16.33) we find that these are the same as those we obtain for the linearised instability problem from (16.12) with $\mu = 1/2$. This is an optimal result because it shows that the critical Rayleigh number for nonlinear unconditional stability is precisely the same as that for linearised instability. Hence sub-critical instabilities cannot arise. The restriction on the viscoelastic coefficient ω (i.e. ν_1) is a very weak requirement since for practical fluids and many situations Γ is small.

16.4 Quadratic viscosity dependence

In section 16.3 we derived unconditional nonlinear stability using a viscoelastic fluid model when the viscosity is a linear function of temperature. One of the important developments in thermal convection is the recognition that increasing or decreasing of viscosity with temperature was shown experimentally to give rise to decreasing or increasing, respectively, fluid motion in the centre of the Bénard cell. This important work is due to (Tippelskirch, 1956) who performed an experiment with liquid sulphur to verify his conclusion. Sulphur is special in that the viscosity has a maximum in temperature. In fact, not all fluids may be adequately modelled by a linear viscosity-temperature relationship. Examples are liquid sulphur and bismuth both of which have a viscosity which achieves a maximum in temperature, cf. (Lide, 1991). For such fluids a quadratic relation of form

$$\nu(T) = \nu_0\big[1 - \gamma(T - T_0)^2\big] \tag{16.38}$$

is necessary to reflect the maximum behaviour in temperature. The coefficients ν_0, T_0 and γ are constants. (Richardson, 1993) developed a nonlinear, conditional energy method to handle the Bénard problem when the viscosity has the form of (16.38). In this section we describe work of (Diaz and Straughan, 2002) who show how to construct an unconditional nonlinear energy stability theory for the Bénard problem when the viscosity is given by (16.38). To do this they employ a viscoelastic model as in section 16.3. Thus, they introduce a constitutive equation composed of two parts. One part is the Newtonian one with the viscosity being a function of temperature. The second component involves a nonlinear function of the symmetric part of the velocity gradient.

For the viscoelastic model under consideration the equations of momentum, continuity, and energy balance are

$$\begin{aligned}
\frac{\partial v_i}{\partial t} + v_j\frac{\partial v_i}{\partial x_j} &= -\frac{1}{\rho}\frac{\partial p}{\partial x_i} + \frac{\partial}{\partial x_k}2\nu_0\Big\{\big[1 - \gamma(T - T_0)^2\big]D_{ik}\Big\}\\
&\quad + 2\nu_1\frac{\partial}{\partial x_k}(|\mathbf{D}|^{2\mu}D_{ik}) + g\alpha k_i T, \qquad\qquad (16.39)\\
\frac{\partial v_i}{\partial x_i} = 0, \qquad &\frac{\partial T}{\partial t} + v_i\frac{\partial T}{\partial x_i} = \kappa\Delta T.
\end{aligned}$$

The notation is as in section 16.3 but we note that the coefficient ν_1 is positive and we restrict attention to $\mu = 1$.

The fluid occupies the layer $\mathbb{R}^2 \times \{0 < z < d\}$ with gravity in the negative $z-$direction. The boundaries $z = 0, d$ are, respectively, held at fixed temperatures T_L and T_U, with $T_L > T_U$.

The steady solution is

$$\bar{T} = T_L - \beta z, \qquad \bar{v}_i \equiv 0, \qquad \beta = \frac{T_L - T_U}{d}, \tag{16.40}$$

with the steady pressure $\bar{p}(z)$ determined from the momentum equation.

To study nonlinear stability we introduce perturbations (u_i, θ, π) by $v_i = \bar{v}_i + u_i$, $T = \bar{T} + \theta$, $p = \bar{p} + \pi$, and rescale the perturbation equations to make them non-dimensional. The scalings employed by (Diaz and Straughan, 2002) are those of (Richardson, 1993), p. 58, who works in terms of an average viscosity, $\nu_m = d^{-1}\int_0^d \nu(\bar{T})dz$. With ω denoting a non-dimensional form of ν_1, the non-dimensional perturbation equations are,

$$\begin{aligned}\frac{\partial u_i}{\partial t} + u_j\frac{\partial u_i}{\partial x_j} = -\frac{\partial \pi}{\partial x_i} &+ R\theta k_i + (fd_{ij})_{,j} + h(z\theta d_{ij})_{,j}\\ &- \zeta(d_{ij}\theta^2)_{,j} + \omega(|\mathbf{d}|^{2\mu}d_{ij})_{,j}\,, \qquad (16.41)\\ \frac{\partial u_i}{\partial x_i} = 0, \qquad Pr\Big(\frac{\partial \theta}{\partial t} &+ u_i\frac{\partial \theta}{\partial x_i}\Big) = Rw + \Delta\theta,\end{aligned}$$

in which $d_{ij} = (u_{i,j} + u_{j,i})/2$, $f(z) = 6(1-\Gamma z^2)/(3-\Gamma)$, $h = 12\Gamma Pr/R(3-\Gamma)$, and $\zeta = 6\Gamma Pr^2/R^2(3-\Gamma)$. In addition Γ satisfies $0 < \Gamma < 1$.

These equations hold on the domain $\{(x,y)\in\mathbb{R}^2\}\times\{z\in(0,1)\}\times\{t>0\}$ and the appropriate boundary conditions are

$$u_i = 0, \quad z = 0,1, \qquad \theta = 0, \quad z = 0,1, \qquad (16.42)$$

with u_i, θ, π satisfying a plane tiling periodic shape in the (x,y) plane. As elsewhere in this book the cell which arises due to this plane tiling form is denoted by V.

16.4.1 Unconditional nonlinear stability

The analysis commences by constructing two "energy" identities. Multiply $(16.41)_1$ by u_i, and $(16.41)_3$ by θ and integrate over V. Note that $\mu = 1$, so this procedure yields

$$\begin{aligned}\frac{d}{dt}\frac{1}{2}\|\mathbf{u}\|^2 =& R(\theta,w) - \int_V f|\mathbf{d}|^2 dx - h < z\theta|\mathbf{d}|^2 >\\ &- \omega\int_V |\mathbf{d}|^4 dx + \zeta < \theta^2|\mathbf{d}|^2 >, \qquad (16.43)\end{aligned}$$

$$\frac{d}{dt}\frac{1}{2}Pr\|\theta\|^2 = R(\theta,w) - \|\nabla\theta\|^2. \qquad (16.44)$$

We next estimate the indefinite term $< z\theta|\mathbf{d}|^2 >$ and the positive term $< \theta^2|\mathbf{d}|^2 >$ by using the arithmetic - geometric mean inequality to derive for constants $\alpha, \beta > 0$ to be selected,

$$\begin{aligned}\frac{d}{dt}\frac{1}{2}\|\mathbf{u}\|^2 \le& R(\theta,w) - < f|\mathbf{d}|^2 > - < |\mathbf{d}|^4 > \Big(\omega - \frac{\alpha h}{2} - \frac{\zeta\beta}{2}\Big)\\ &+ \frac{h}{2\alpha} < z^2\theta^2 > + \frac{\zeta}{2\beta} < \theta^4 > . \qquad (16.45)\end{aligned}$$

By introducing a coupling parameter $\lambda > 0$ (Diaz and Straughan, 2002) derive the following energy inequality,

$$\begin{aligned}\frac{d}{dt}\Big(\frac{1}{2}\|\mathbf{u}\|^2 + \frac{1}{2}\lambda Pr\|\theta\|^2\Big) \leq & R(1+\lambda)(\theta, w) - < f|\mathbf{d}|^2 > -\lambda\|\nabla\theta\|^2 \\ & + \frac{h}{2\alpha} < z^2\theta^2 > + \frac{\zeta}{2\beta} < \theta^4 > \\ & - < |\mathbf{d}|^4 > \Big(\omega - \frac{1}{2}\alpha h - \frac{1}{2}\zeta\beta\Big).\end{aligned} \tag{16.46}$$

There does not seem to be a way of handling the $< \theta^4 >$ term directly and so (Diaz and Straughan, 2002) derived an extra equation for $< \theta^4 >$, namely,

$$\frac{d}{dt}\frac{aPr}{4} < \theta^4 >= aR(\theta^3, w) - \frac{3a}{4}\|\nabla\theta^2\|^2. \tag{16.47}$$

The number $a > 0$ is another coupling parameter at our disposal. The inequalities of Poincaré, Sobolev, and Young are now used and a new energy inequality is derived of form

$$\begin{aligned}\frac{dE}{dt} \leq & I - D - \|\theta\|_4^4\Big(\frac{3\pi^2 a}{4} - \frac{3aR\epsilon^{4/3}}{4} - \frac{\zeta}{2\beta}\Big) \\ & - < |\mathbf{d}|^4 > \Big(\omega - \frac{\alpha h}{2} - \frac{\zeta\beta}{2} - \frac{ac_1 R}{4\epsilon^4}\Big).\end{aligned} \tag{16.48}$$

The Lyapunov function E, dissipation D, and production term I, are defined by

$$E(t) = \frac{1}{2}\|\mathbf{u}\|^2 + \frac{1}{2}\lambda Pr\|\theta\|^2 + \frac{aPr}{4}\|\theta\|_4^4, \tag{16.49}$$

$$\begin{aligned} I(t) &= R(1+\lambda)(\theta, w) + \frac{h}{2\alpha} < z^2\theta^2 >, \\ D(t) &= < f|\mathbf{d}|^2 > +\lambda\|\nabla\theta\|^2. \end{aligned} \tag{16.50}$$

By employing a procedure not dissimilar to that in section 16.3, (Diaz and Straughan, 2002) select the constants in (16.48) optimally to minimize the size of ω while retaining a part of the $\|\theta\|_4^4$ term.

The restriction ω must satisfy is found to be

$$\omega > \frac{64\sqrt{c_1}Pr^2}{3\pi^4(3-\Gamma)}\Gamma, \tag{16.51}$$

and the energy inequality is reduced to

$$\frac{dE}{dt} \leq -D\Big(1 - \frac{1}{R_E}\Big) - \delta\|\theta\|_4^4, \tag{16.52}$$

for a suitable positive constant δ, where

$$\frac{1}{R_E} = \max_H \frac{I}{D}. \tag{16.53}$$

The requirement is made that $R_E > 1$ and f is bounded by $f \geq 6(1 - \Gamma)/(3 - \Gamma) \equiv f_0 > 0$. (Diaz and Straughan, 2002) then show that (16.52) leads to

$$\frac{dE}{dt} \leq -KE, \tag{16.54}$$

where the coefficient K is given by $K = [(R_E - 1)/R_E] \min\{[48\pi^2(1 - \Gamma)/(3 - \Gamma)], 2\pi^2/Pr, 4\delta/aPr\}$. The above inequality yields exponential decay and hence unconditional nonlinear stability. The viscoelastic coefficient ω must satisfy (16.51), but in practice Γ is relatively small and so this is a weak requirement.

The Euler-Lagrange equations for the determination of the nonlinear critical Rayleigh number R^2 which follow from (16.53) are

$$\begin{aligned} &R(1+\lambda)\theta k_i + 2(fd_{ij})_{,j} = -\pi_{,i}\,, \qquad u_{i,i} = 0, \\ &R(1+\lambda)w + \frac{27\pi^4\Gamma}{4R^2(3-\Gamma)\sqrt{c_1}}\, z^2\theta + 2\lambda\Delta\theta = 0. \end{aligned} \tag{16.55}$$

The linear instability equations follow easily from (16.41) and because these are symmetric, the equations which determine R_L^2, the linear critical Rayleigh number, are

$$\begin{aligned} &R\theta k_i + (fd_{ij})_{,j} = -\pi_{,i}\,, \\ &u_{i,i} = 0, \quad \Delta\theta + Rw = 0. \end{aligned} \tag{16.56}$$

The Rayleigh numbers produced from the above two sets of equations are close and show that there is at most a small range where sub-critical instabilities may possibly be found.

16.5 Unconditional nonlinear stability and temperature dependent diffusivity

While it is usually the viscosity which changes more rapidly with temperature than the thermal conductivity, there are cases where the conductivity variation may be significant. (Flavin and Rionero, 1999a) developed a very interesting way of obtaining unconditional nonlinear stability when the viscosity is constant, but the thermal conductivity is a nonlinear function of temperature. This is very much of interest, because when the conductivity is a temperature-dependent function, the basic temperature profile is no longer a linear function of z. The method of (Flavin and Rionero, 1999a) is further developed and applied to other problems in (Flavin and Rionero, 1995; Flavin and Rionero, 1997; Flavin and Rionero, 1998; Flavin and Rionero, 1999b) and (Rionero and Maiellaro, 1995).

The basic equations governing the (Flavin and Rionero, 1999a) problem are

$$\begin{aligned} v_{i,t} + v_j v_{i,j} &= -\frac{1}{\rho_0} p_{,i} + \nu \Delta v_i + k_i \alpha g T, \\ v_{i,i} = 0, \qquad T_{,t} + v_i T_{,i} &= \big(\kappa(T) T_{,i}\big)_{,i}, \end{aligned} \tag{16.57}$$

where the notation is as in the rest of the chapter. The key difference in this section is the dependence on temperature of κ. They specifically write $\kappa(T) = \kappa_0 \psi(T/T_r)$ where κ_0 is a positive constant, T_r is a reference temperature, and the function ψ satisfies $\psi(\cdot) \geq 1$. By introducing ψ in this manner they are able to rewrite (16.57) in the form

$$\begin{aligned} v_{i,t} + v_j v_{i,j} &= -\frac{1}{\rho_0} p_{,i} + \nu \Delta v_i + k_i \alpha g T, \\ v_{i,i} = 0, \qquad T_{,t} + v_i T_{,i} &= \kappa_0 T_r \Delta \bigg(\int_0^{T/T_r} \psi(s) ds \bigg). \end{aligned} \tag{16.58}$$

The fluid is contained in the horizontal layer $\mathbb{R}^2 \times \{z \in (0, d)\}$ and the boundary conditions are

$$T = T_0, \quad z = 0, \qquad T = T_1, \quad z = d, \tag{16.59}$$

with T_0, T_1 constants and $T_0 > T_1$.

The steady solution to (16.58), (16.59), is given by $\bar{v}_1 \equiv 0$ and $\bar{T}(z)$ where $\bar{T}(z)$ solves the equation

$$\int_{T_0/T_r}^{\bar{T}/T_r} \psi(s) ds = -\frac{\beta}{T_r} z, \tag{16.60}$$

where $\beta = (T_r/d) \int_{T_1/T_r}^{T_0/T_r} \psi(s) ds$, see (Flavin and Rionero, 1999a). Thus, the form $\bar{T}$ takes depends entirely on what ψ is, i.e. what form $\kappa(T)$ has. (Flavin and Rionero, 1999a) also show that

$$\frac{d\bar{T}}{dz} = -\frac{\beta}{\psi(\bar{T}/T_r)}. \tag{16.61}$$

It is convenient to introduce the notation $\mathcal{T}(z) = \bar{T}(z)/T_r$. (Flavin and Rionero, 1999a) derive the perturbation equations about the steady solution and for perturbations u_i, θ, π they derive non-dimensional equations of form

$$\begin{aligned} & u_{i,t} + u_j u_{i,j} = -\pi_{,i} + \Delta u_i + k_i R \theta, \\ & u_{i,i} = 0, \\ & Pr(\theta_{,t} + u_i \theta_{,i}) = \frac{R}{\psi(\mathcal{T})} w + \Delta \bigg[\int_{\mathcal{T}}^{\mathcal{T}+\theta} \psi(s) ds \bigg]. \end{aligned} \tag{16.62}$$

The key ingredient in the (Flavin and Rionero, 1999a) approach is the introduction of the functional $\Phi(\theta;\mathcal{T})$ by

$$\Phi(\theta;\mathcal{T})=\int_0^\theta\int_{\mathcal{T}}^{\mathcal{T}+\eta}\psi(s)ds\,d\eta=\int_0^\theta\int_0^\eta\psi(s+\mathcal{T})ds\,d\eta.$$

The perturbation boundary conditions are $\theta=0$ on $z=0,1$ and $u_i=0$ for fixed boundaries or the usual stress free ones. The introduction of Φ allows (16.62) to be rewritten as

$$\begin{aligned}&u_{i,t}+u_ju_{i,j}=-\pi_{,i}+\Delta u_i+k_iR\theta,\\&u_{i,i}=0,\\&Pr(\theta_{,t}+u_i\theta_{,i})=\frac{R}{\psi(\mathcal{T})}w+\Delta\Phi_\theta(\theta;\mathcal{T}),\end{aligned}\tag{16.63}$$

where $\Phi_\theta=\partial\Phi/\partial\theta$. If V denotes a period cell for the disturbance then the development of (Flavin and Rionero, 1999a) in the u_i energy equation is standard, so that

$$\frac{d}{dt}\frac{1}{2}\|\mathbf{u}\|^2=R(\theta,w)-\|\nabla\mathbf{u}\|^2.\tag{16.64}$$

For the temperature field they work with the thermal energy functional $Pr\int_V\Phi(\theta,\mathcal{T})dV$. For this functional they show

$$\begin{aligned}\frac{d}{dt}Pr\int_V\Phi(\theta;\mathcal{T})dV=&Pr\int_V\Phi_\theta\theta_{,t}dV\\=&-Pr\int_V\Phi_\theta\theta_{,i}u_idV+R\int_V\frac{\Phi_\theta w}{\psi(\mathcal{T})}dV-\|\nabla\Phi_\theta\|^2\\=&R(\theta,w)-\|\nabla\Phi_\theta\|^2.\end{aligned}\tag{16.65}$$

Equations (16.64) and (16.65) now lead to an energy equation of form

$$\frac{dE}{dt}=RI-D\tag{16.66}$$

where

$$I=2(\theta,w),\qquad D=\|\nabla\mathbf{u}\|^2+\|\nabla\Phi_\theta\|^2.\tag{16.67}$$

(Flavin and Rionero, 1999a) now determine a number R_E for which $I/D\le R_E^{-1}$ with the solutions lying in a suitable space H. They also show that

$$\|\nabla\Phi_\theta\|^2\ge\pi^2\|\Phi_\theta\|^2\ge2\pi^2\int_V\Phi dV.\tag{16.68}$$

Inequality (16.68) together with the restriction on I/D then allow one to see that provided $R < R_E$,

$$\frac{dE}{dt} \leq -\Big(\frac{R_E - R}{R_E}\Big)(\|\nabla \mathbf{u}\|^2 + \|\nabla \Phi_\theta\|^2)$$
$$\leq -2\pi^2\Big(\frac{R_E - R}{R_E}\Big)\Big(\frac{1}{2}\|\mathbf{u}\|^2 + \int_V \Phi dV\Big). \qquad (16.69)$$

From (16.69) one then deduces

$$E(t) \leq E(0)\exp\big[-2\pi^2(1 - R/R_E)t\big], \quad \text{when } Pr < 1,$$
$$E(t) \leq E(0)\exp\big[-2\pi^2 Pr^{-1}(1 - R/R_E)t\big], \quad \text{when } Pr \geq 1.$$

Thus, for $R < R_E$, (Flavin and Rionero, 1999a) have obtained unconditional nonlinear stability.

(Flavin and Rionero, 1999a) continue to show how one may derive an estimate for their number R_E. They solve this explicitly in a number of important cases.

16.6 Unconditional nonlinear stability in porous temperature dependent convection

In this section we concentrate on an analysis of nonlinear stability of thermal convection in a porous layer when the viscosity depends on temperature. The section focusses on unconditional nonlinear stability results. The theory mostly used here is porous media theory which includes inertial drag terms of Forchheimer type, cf. section 3.4.3.

To the best of our knowledge the first study of the problem of nonlinear energy stability in Bénard convection in a porous layer with a viscosity linear in temperature is due to (Richardson, 1993) who employed a Brinkman term in the momentum equation for the porous medium. Richardson's analysis is conditional on the size of the initial data. (Qin et al., 1995) and (Qin and Chadam, 1996) argue that it is more appropriate to include an inertial drag term as in Forchheimer theory rather than use Brinkman theory. The analysis of (Qin and Chadam, 1996) is also conditional. The first study of which we are aware which obtains unconditional nonlinear energy stability in porous media theory with a temperature dependent viscosity is that of (Payne and Straughan, 2000b). These writers use Forchheimer terms. (Rionero, 2000) uses the maximum principle on the temperature to derive an unconditional result using Darcy theory. His Rayleigh number threshold is $R_L \exp(-\Gamma\theta_s/2)$ where R_L is the critical Rayleigh number of linear theory, Γ represents the viscosity variation, and θ_s is the maximum value of the initial temperature perturbation.

In this section we describe results of (Payne and Straughan, 2000b) who derive unconditional nonlinear stability bounds which are identical or very close to the linear instability ones. These writers work with a linear viscosity of form

$$\mu(T) = \mu_0\big[1 - \gamma(T - T_0)\big] \tag{16.70}$$

for T_0, μ_0, γ constant. The Forchheimer equations for a perturbation to the steady state in which velocity is zero and temperature is a linear function across the layer $z \in (0, d)$ heated from below, are

$$\begin{aligned} \frac{\partial \pi}{\partial x_i} &= -\frac{\mu_0}{k}\,(1 + \gamma\beta z)u_i + \frac{\mu_0\gamma}{k}\,\theta u_i \\ &\qquad + g\rho_0\alpha\theta\delta_{i3} - b'|\mathbf{u}|u_i - c'|\mathbf{u}|^2 u_i, \qquad (16.71) \\ \frac{\partial u_i}{\partial x_i} &= 0, \qquad \frac{\partial \theta}{\partial t} + u_i\,\frac{\partial \theta}{\partial x_i} = \beta w + \kappa\Delta\theta. \end{aligned}$$

In equations (16.71) the terms (u_i, θ, π) are perturbations to velocity, temperature and pressure, $w = u_3$, and $k, \beta, g, \alpha, \kappa$ are permeability, temperature gradient, gravity, thermal expansion coefficient, and thermal diffusivity. The coefficients b' and c' represent Forchheimer terms, described in e.g. (Firdaouss et al., 1997). Mathematical analysis of steady flows of equations (16.71) is performed by (Néel and Nemrouch, 2002) where references to further work justifying the use of $b' \neq 0$ and/or $c' \neq 0$ are given.

The Forchheimer equations, cf. section 3.4.3, are usually written with only the b' term and $c' = 0$. However, as (Firdaouss et al., 1997) point out, the original (Forchheimer, 1901) theory allows for a polynomial expression, and hence a c' term may also be included. (Néel, 1998) studies a differentially heated convection problem with imposed throughflow using a Forchheimer theory when $b' = 0$ but $c' \neq 0$.

(Payne and Straughan, 2000b) non-dimensionalize equations (16.71) via the scalings $T^\sharp = Ud\sqrt{\mu_0\beta/\kappa\rho_0 g\alpha k}$, $U = \kappa/d$, $R = \sqrt{d^2\beta g\alpha k/\kappa(\mu_0/\rho_0)}$, $\mathcal{T} = d^2/\kappa$, $\Gamma = \gamma\beta d$, $b = b'k/\rho_0 d$, $c = c'k\mu_0/\rho_0^3 d^2$, where $T^\sharp, U$ and $\mathcal{T}$ are typical temperature, velocity and time scales. The perturbations then satisfy

$$\begin{aligned} &\frac{\partial \pi}{\partial x_i} = -(1 + \Gamma z)u_i + \frac{\Gamma}{R}\,\theta u_i + R\theta\delta_{i3} - b|\mathbf{u}|u_i - c|\mathbf{u}|^2 u_i, \\ &\frac{\partial u_i}{\partial x_i} = 0, \\ &\frac{\partial \theta}{\partial t} + u_i\,\frac{\partial \theta}{\partial x_i} = Rw + \Delta\theta, \end{aligned} \tag{16.72}$$

where $Ra = R^2$ is the Rayleigh number. The domain for (16.72) is $\{(x, y) \in \mathbb{R}^2\} \times \{z \in (0, 1)\} \times \{t > 0\}$. The boundary conditions are

$$w = \theta = 0, \qquad z = 0, 1, \tag{16.73}$$

with (u_i, θ, π) satisfying a periodic plan-form shape which tiles the plane. The period cell which so arises is denoted by V.

16.6.1 The quadratic Forchheimer model

The standard Forchheimer equations have $c = 0$ which corresponds to the situation of (Qin and Chadam, 1996). This case is analysed first in (Payne and Straughan, 2000b).

The idea is to commence with a standard energy analysis of (16.72) to obtain

$$\begin{aligned}\frac{d}{dt}\frac{1}{2}\|\theta\|^2 =& 2R(w,\theta) + \frac{\Gamma}{R}\int_V \theta|\mathbf{u}|^2 dx \\ &- \int_V (1+\Gamma z)|\mathbf{u}|^2 dx - \|\nabla\theta\|^2 - b\int_V |\mathbf{u}|^3 dx.\end{aligned} \tag{16.74}$$

(Payne and Straughan, 2000b) show that (16.74) does not lead to a meaningful energy theory because the appropriate I/D ratio is unbounded. To see this they define I and D by

$$I = 2R(w,\theta) + \frac{\Gamma}{R}\int_V \theta|\mathbf{u}|^2 dx, \tag{16.75}$$

$$D = \int_V (1+\Gamma z)|\mathbf{u}|^2 dx + \|\nabla\theta\|^2 + b\int_V |\mathbf{u}|^3 dx. \tag{16.76}$$

Then, (16.74) formally leads to

$$\frac{d}{dt}\frac{1}{2}\|\theta\|^2 \le -D\Big(1 - \max_H \frac{I}{D}\Big), \tag{16.77}$$

where H is the space of admissible solutions in the stability problem.

To analyse the ratio I/D we note that the space H is linear and so we may replace u_i by αu_i and θ by $\beta\theta$ for arbitrary real numbers α, β. This yields

$$\frac{I}{D} = \frac{A\alpha\beta + B\beta\alpha^2}{C\alpha^2 + D\beta^2 + E\alpha^3}$$

where $A, \dots, E$ are constants, being the appropriate integrals in (16.75) and (16.76). If one now puts $\beta = \alpha^\epsilon$, $\epsilon = 1 + \omega$, for positive numbers ϵ and ω, I/D reduces to

$$\frac{I}{D} = \frac{A\alpha^{2+\omega} + B\alpha^{3+\omega}}{C\alpha^2 + D\alpha^{2+2\omega} + E\alpha^3}.$$

We may now select $0 < \omega < 1$, e.g. $\omega = 1/2$, so that

$$\frac{I}{D} = \frac{A\alpha^{1/2} + B\alpha^{3/2}}{C + (D+E)\alpha}. \tag{16.78}$$

The conclusion is that as $\alpha \to \infty$, $I/D \to \infty$, and the maximum of I/D in H cannot be achieved.

(Payne and Straughan, 2000b) conclude that any straightforward approach to derive global nonlinear stability via an $L^2(V)$ theory appears doomed. They instead derive an energy equation for the $L^3(V)$ norm of θ. Thus

$$\frac{d}{dt}\frac{1}{3}\|\theta\|_3^3 = R\int_V (\operatorname{sign}\theta)w\theta^2 dx - \frac{8}{9}\int_V |\theta|_{,i}^{3/2}|\theta|_{,i}^{3/2}\, dx,$$

which after use of Poincaré's inequality gives

$$\frac{d}{dt}\frac{1}{3}\|\theta\|_3^3 \le R\int_V (\operatorname{sign}\theta)w\theta^2 dx - \frac{8\pi^2}{9}\|\theta\|_3^3. \tag{16.79}$$

(Payne and Straughan, 2000b) next select a Lyapunov functional $E(t)$ by

$$E(t) = \frac{1}{2}\|\theta\|^2 + \frac{a}{3}\|\theta\|_3^3 \tag{16.80}$$

where $a(>0)$ is a coupling parameter.

The idea is to add (16.74) together with a multiple of (16.79) but first estimate the $w\theta^2$ and $\theta|\mathbf{u}|^2$ terms by using Young's inequality as follows

$$\frac{\Gamma}{R}\int_V \theta|\mathbf{u}|^2 dx \le \frac{2\Gamma\alpha}{3R}\int_V |\mathbf{u}|^3 dx + \frac{\Gamma}{3R\alpha^2}\int_V |\theta|^3 dx, \tag{16.81}$$

$$R\int_V (\operatorname{sign}\theta)w\theta^2 dx \le \frac{R\beta^2}{3}\int_V |\mathbf{u}|^3 dx + \frac{2R}{3\beta}\int_V |\theta|^3 dx, \tag{16.82}$$

where $\alpha, \beta(>0)$ are constants to be selected. The upshot of this calculation is the inequality

$$\begin{aligned}\frac{dE}{dt} \le\, & 2R(w,\theta) - \int_V \mu(z)|\mathbf{u}|^2 dx - \|\nabla\theta\|^2\\ & - \int_V |\mathbf{u}|^3 dx\left\{b - \frac{2\Gamma\alpha}{3R} - \frac{R\beta^2 a}{3}\right\}\\ & - \int_V |\theta|^3 dx\left\{\frac{8\pi^2 a}{9} - \frac{\Gamma}{3R\alpha^2} - \frac{2Ra}{3\beta}\right\},\end{aligned} \tag{16.83}$$

where $\mu(z) = 1 + \Gamma z$. The first three terms are special since they represent the theory associated with the linear instability problem. (Payne and Straughan, 2000b) select a, α, β, such that

$$a = \frac{8\pi^2\Gamma}{9R^3} + \frac{27}{8\pi^2}\epsilon, \qquad \beta = \frac{9R}{8\pi^2}, \qquad \alpha = \frac{9R}{8\pi^2}. \tag{16.84}$$

These values optimize in a certain sense the cubic terms in (16.83).

The resulting inequality from (16.83) is

$$\frac{dE}{dt} \le 2R(w,\theta) - \int_V \mu(z)|\mathbf{u}|^2 dx - \|\nabla\theta\|^2 - \hat{b}\int_V |\mathbf{u}|^3 dx - \epsilon\int_V |\theta|^3 dx, \tag{16.85}$$

where $\hat{b} = b - 9\Gamma/8\pi^2 - 729R^3\epsilon/512\pi^6$.

Next define $\mathcal{I}$, $\mathcal{D}$ and R_E by

$$\mathcal{I} = 2(w,\theta), \qquad \mathcal{D} = \int_V \mu(z)|\mathbf{u}|^2 dx + \|\nabla\theta\|^2, \tag{16.86}$$

$$\frac{1}{R_E} = \max_H \frac{\mathcal{I}}{\mathcal{D}}. \tag{16.87}$$

(Payne and Straughan, 2000b) show that if b is restricted so that

$$b > \frac{9\Gamma}{8\pi^2}, \tag{16.88}$$

then (16.85) leads to

$$\frac{dE}{dt} \le -\Big(\frac{R_E - R}{R_E}\Big)\mathcal{D} - \hat{b}\int_V |\mathbf{u}|^3 dx - \epsilon\int_V |\theta|^3 dx. \tag{16.89}$$

Hence, provided $R < R_E$ and b satisfies (16.88) unconditional nonlinear stability follows from (16.89).

(Payne and Straughan, 2000b) show that the energy inequality $dE/dt \le -kE$, follows from (16.89) with $k = \min\{3\epsilon/a, 2\pi^2(R_E - R)/R_E\}$. This leads to exponential decay of θ in L^3 (and L^2) measure. Decay in $\|\mathbf{u}\|$ and $\|\mathbf{u}\|_3$ is also obtained in (Payne and Straughan, 2000b).

The above result shows that the conditions $R < R_E$ and $b > 9\Gamma/8\pi^2$ yield global nonlinear stability. This is a very strong result because the bound on b is weak and the critical Rayleigh number R_E is shown in (Payne and Straughan, 2000b) to be exactly the same as the linear instability one. Thus, if the Forchheimer coefficient satisfies the b restriction, sub-critical instabilities cannot arise.

16.6.2 *Unconditional nonlinear stability for other problems*

(Payne and Straughan, 2000b) also study what improvements or changes arise when the cubic Forchheimer term, the c term, is present. They show that an L^2 theory will now work due to the presence of the c term, and the numerical results demonstrate that very small c values lead to sharp bounds. However, they also show that an even sharper energy stability theory arises when one uses a more intricate energy analysis which employs the L^3 and L^4 integrals of θ. The energy they use has form

$$E(t) = \frac{1}{2}\int_V \theta^2 dx + \frac{a}{3}\int_V |\theta|^3 dx + \frac{\omega}{4}\int_V \theta^4 dx, \tag{16.90}$$

where a and ω are positive constants selected judiciously.

(Payne and Straughan, 2000b) also show how their approach extends to penetrative convection, cf. section 17.1. They utilize a quadratic density relation appropriate to penetrative convection with water being the saturating fluid and allow a viscosity linear in temperature,

$$\mu(T) = \mu_0\big[1 - \gamma_1(T-4)\big], \tag{16.91}$$

$$\rho(T) = \rho_0\big[1 - \alpha(T-4)^2\big], \tag{16.92}$$

where α is an expansion coefficient and (16.92) incorporates the maximum density at 4°C. (Payne and Straughan, 2000b) consider the situation of a porous layer saturated with water with the lower boundary, $z = 0$, held at constant temperature 0°C while the upper, $z = d$, is held at constant temperature $T_u (\geq 4°\text{C})$. The steady solution is

$$\bar{T} = \frac{T_u}{d}\, z, \qquad \bar{u}_i = 0. \tag{16.93}$$

They use the Forchheimer system with $b' \neq 0$ but $c' = 0$. For this the equations for a perturbation (u_i, θ, π) to the stationary solution are in non-dimensional form

$$\begin{aligned} \frac{\partial \pi}{\partial x_i} &= -\big[1 + \Gamma(\xi - z)\big]u_i - 2RM\delta_{i3}\theta + \frac{\Gamma}{R}\,\theta u_i + \theta^2\delta_{i3} - b|\mathbf{u}|u_i, \\ \frac{\partial u_i}{\partial x_i} &= 0, \qquad \frac{\partial \theta}{\partial t} + u_i\frac{\partial \theta}{\partial x_i} = -Rw + \Delta\theta, \end{aligned} \tag{16.94}$$

where $\xi = 4/T_u$.

(Payne and Straughan, 2000b) develop an unconditional energy stability theory for (16.94) using the energy functional

$$E = \frac{\zeta}{2}\int_V \theta^2 dx + \frac{\lambda}{3}\int_V |\theta|^3 dx. \tag{16.95}$$

Since (Qin and Chadam, 1996) also allowed a viscosity which is quadratic in temperature, (Payne and Straughan, 2000b) showed how to obtain global nonlinear stability for this situation when the quadratic and cubic Forchheimer terms are present. The viscosity has form

$$\mu(T) = \mu_0\big[1 - \gamma_1(T - T_0) - \gamma_2(T-T_0)^2\big]. \tag{16.96}$$

With $\Gamma_1 = \gamma_1\beta d$ and $\Gamma_2 = \gamma_2\beta^2 d^2$, the non-dimensional perturbation equations are

$$\begin{aligned} \frac{\partial \pi}{\partial x_i} &= -(1 + \Gamma_1 z - \Gamma_2 z^2)u_i + \left(\frac{\Gamma_1}{R} - \frac{2\Gamma_2}{R}\,z\right)\theta u_i \\ &\quad + \frac{\Gamma_2}{R^2}\,\theta^2 u_i + R\theta\delta_{i3} - b|\mathbf{u}|u_i - c|\mathbf{u}|^2 u_i, \\ \frac{\partial u_i}{\partial x_i} &= 0, \qquad \frac{\partial \theta}{\partial t} + u_i\frac{\partial \theta}{\partial x_i} = Rw + \Delta\theta. \end{aligned} \tag{16.97}$$

(Payne and Straughan, 2000b) show that by using an energy functional of form (16.90) a sharp unconditional nonlinear stability boundary may be derived even for equations (16.97).

17
Penetrative convection

17.1 Unconditional nonlinear energy stability for a quadratic buoyancy

A pioneering piece of work on penetrative convection is the beautiful paper of (Veronis, 1963). Our description of penetrative convection relies much on his paper.

In many natural phenomena the process of thermal convection involves a penetrative motion into a stably stratified fluid. Examples of penetrative convection are to be found in several areas of geo- and astrophysical fluid dynamics. For example, the Earth's atmosphere is bounded below by the ground or ocean; this bounding surface is heated by solar radiation, and the air close to the surface then becomes warmer than the upper air, and so a gravitationally unstable system results. When convection occurs, the warm air rises and penetrates into regions that are stably stratified.

In the oceans evaporation is the main cause of gravitational instability near the sea surface. As the cool surface water is carried downward, it too enters regions that are stably stratified. In a star the surface layer is stable. (Veronis, 1963) explains that in the interior of a star the temperature gradient may rise to a value which is greater than the adiabatic gradient at locations which are well below the surface. The cause of this is adiabatic compression which causes negative hydrogen to form. The larger temperature gradient gives rise to an unstable region. The superadiabatic gradient so formed may extend deeply into the star's interior leading to a very high temperature. This high temperature may in turn cause the gas to be ionized which then results in loss of the superadiabatic gradient. The end result is an unstable layer with stable fluid both above and below.

To study penetrative convection on a laboratory scale (Veronis, 1963) proposed a simplified model. Consider a layer of water with the bottom maintained at 0°C and the top at a temperature greater than 4°C. Due to the fact that the density of water below 4°C is a decreasing function of temperature, the above situation results in a gravitationally unstable layer of fluid lying below a stably stratified one. When convection occurs in the lower layer the motions will penetrate into the upper layer.

The phenomenon of penetrative convection is lucidly described by (Veronis, 1963) who analysed in detail the linearized system and developed a weakly nonlinear finite amplitude analysis for two stress free boundaries. (Veronis, 1963) finds that penetrative convection differs markedly from the classical case of Bénard convection for which sub-critical instabilities are not possible and shows that a finite amplitude solution exists for subcritical Rayleigh numbers. He notes that a finite amplitude instability may commence for a Rayleigh number which is below that given by the linear instability threshold.

The phenomenon of penetrative convection and, in particular, applications to geophysics and to astrophysics is still a subject of active research, see e.g., (Alex et al., 2001), (Ames and Cobb, 1994), (Azouni, 1983), (Azouni and Normand, 1983a; Azouni and Normand, 1983b), (Chasnov and Tse, 2001), (Gentile and Rionero, 2000; Gentile and Rionero, 2002), (George et al., 1989), (Hansen and Yuen, 1989), (Kondo and Unno, 1982; Kondo and Unno, 1983), (Lindsay and Straughan, 1992), (McKay, 1992a; McKay, 1996), (McKay and Straughan, 1991; McKay and Straughan, 1992; McKay and Straughan, 1993), (Martin and Kauffman, 1974), (Matthews, 1988), (Merker et al., 1979), (Mlaouah et al., 1997), (Payne et al., 1988), (Niedrauer and Martin, 1979), (Oldenburg and Pruess, 1998), (Selak and Lebon, 1997), (Tse and Chasnov, 1998), (Walden and Ahlers, 1981), (Whitehead, 1971), (Whitehead and Chen, 1970), and (Zahn et al., 1982). Many other references may be found in the book by (Straughan, 1993a).

The equations of motion in the layer differ from those derived by the Boussinesq approximation in section 3.2, but only in the form of the dependence of density on temperature in the body force term. The density is still assumed constant everywhere except in the body force term. Because the density of water is almost parabolic in the range $0-8$°C Veronis (1963) chooses as equation of state,

$$\rho = \rho_0 \big[1 - \alpha(T-4)^2\big], \tag{17.1}$$

where $\rho(T)$ is the density, ρ_0 the density at 4°C, $\alpha \approx 7.68 \times 10^{-6}$ (°C^{-2}), and Veronis points out that even at 14°C (17.1) involves only a 10% error.

Suppose the fluid is contained in the layer $z \in (0, d)$ and the surface temperatures are

$$T = 0°\mathrm{C} \quad \text{at} \quad z = 0, \qquad T = T_2 \quad \text{at} \quad z = d, \tag{17.2}$$

where T_2 is a constant temperature not less than 4°C, and the phase change effect at the lower boundary is ignored. The perturbation equations to the motionless solution

$$\bar{u}_i = 0, \quad \bar{T} = \beta z, \qquad \beta = \frac{T_2}{d},$$

are in non-dimensional form

$$u_{i,t} + u_j u_{i,j} = -\pi_{,i} + \Delta u_i - 2R\theta(\xi - z)k_i + Pr\theta^2 k_i\,, \tag{17.3}$$

$$u_{i,i} = 0, \tag{17.4}$$

$$Pr(\theta_{,t} + u_i\theta_{,i}) = -Rw + \Delta\theta, \tag{17.5}$$

where R^2 is a Rayleigh number, $\xi = 4/T_2$, and $k_i = \delta_{i3}$.

The energy analysis of (Straughan, 1985) proceeds via an L^2 method using a Sobolev inequality to handle the resulting cubic nonlinear term; however, this analysis is only conditional. In section 2.4 it was shown for the one-dimensional equation with a quadratic force that boundedness could be achieved by employing a weighted energy. (Payne and Straughan, 1987), in fact, employ a weighted energy precisely to overcome the conditional restriction. They select a *generalized energy* of the form

$$E(t) = \frac{1}{2} < u_i u_i > + \frac{1}{2} Pr < (\mu - 2z)\theta^2 >, \tag{17.6}$$

in which $\mu\,(\geq 2)$ is a coupling parameter to be chosen optimally. The next step is to differentiate E and substitute for $u_{i,t}, \theta_{,t}$ from (17.3), (17.5). In the standard energy technique the $\theta u_i \theta_{,i}$ term that arises integrates to zero, but the effect of the weight in this case gives rise to a term $-Pr < \theta^2 w >$, which exactly cancels the corresponding destabilizing term in the momentum equation. Having removed the offending cubic term, the *weighted* energy equation then has the same form as usual,

$$\frac{dE}{dt} = -\mathcal{D} + IR,$$

where now,

$$\mathcal{D} = < u_{i,j}u_{i,j} > + < (\mu - 2z)\theta_{,i}\theta_{,i} >$$

and

$$I = - < (\mu + 2\xi - 4z)\theta w > .$$

The energy limit R_E is defined in the usual way, $1/R_E = \max_{\mathcal{H}} I/\mathcal{D}$, and again it may be shown that $R < R_E$ is a sufficient condition for nonlinear stability. We stress that the stability obtained is *unconditional;* i.e., for all initial amplitude disturbances, this being achieved by the use of the spatial weight in the definition of $E(t)$. The unconditional nonlinear energy stability boundary is, in fact, very close to the linear instability one; correct numerical results are given in (Payne and Straughan, 1988).

17.2 Unconditional nonlinear energy stability with an internal heat source

There is an alternative way to obtain the penetrative effect of convection in the sense that one part of the layer is stabilizing while the other wishes to overturn and create convective instability. This is achieved by having an internal heat source in the layer but keeping the density linear in the temperature field. For this situation the governing equations are

$$\begin{aligned} v_{i,t} + v_j v_{i,j} &= -\frac{1}{\rho_0} p_{,i} + \nu \Delta v_i + \alpha g k_i T, \\ v_{i,i} &= 0, \\ T_{,t} + v_i T_{,i} &= \kappa \Delta T + Q. \end{aligned} \tag{17.7}$$

The notation is as in section 3.2 except the term Q represents a heat source ($Q > 0$) or a heat sink ($Q < 0$). The steady state obtained depends on the choice of Q. We wish to derive a theory in some sense analogous to that of section 17.1 and so restrict attention to $Q = -A < 0$ where A is a constant. This will yield the upper part of the layer stabilizing while the lower part wishes to convect. If one employs another choice of $Q(z)$ then various stabilizing/destabilizing layer situations can be constructed.

Thus equations (17.7) are defined on the layer $\mathbb{R}^2 \times \{z \in (0,d)\}$ for $t > 0$. On the boundaries we impose the conditions

$$T = T_L, \; z = 0, \qquad T = T_U, \; z = d, \tag{17.8}$$

where $T_L > T_U$. The basic steady state in whose stability we are interested is

$$\bar{T}(z) = -\frac{Q}{2\kappa} z^2 + \left(\frac{Qd}{2\kappa} - \frac{\Delta T}{d} \right) z + T_L \,, \qquad \bar{v}_i \equiv 0,$$

and this yields

$$\frac{d\bar{T}}{dz} = -\frac{Q}{\kappa} z + \frac{Qd}{2\kappa} - \frac{\Delta T}{d} \,. \tag{17.9}$$

Perturbations (u_i, θ, π) to (17.9) thus satisfy the equations

$$\begin{aligned} u_{i,t} + u_j u_{i,j} &= -\frac{1}{\rho_0} \pi_{,i} + \nu \Delta u_i + \alpha g k_i \theta, \\ u_{i,i} &= 0, \\ \theta_{,t} + u_i \theta_{,i} &= -\bar{T}' w + \kappa \Delta \theta \,. \end{aligned} \tag{17.10}$$

We non-dimensionalize these equations with the scalings

$$x_i = x_i^* d, \quad t = t^* \mathcal{T}, \quad u_i = u_i^* U, \quad \mathcal{T} = \frac{d^2}{\nu}, \quad U = \frac{\nu}{d},$$
$$\pi = P\pi^*, \quad P = \frac{\rho_0 \nu U}{d}, \quad \xi = \frac{\kappa \Delta T}{2Ad^2}, \quad \Delta T = T_L - T_U,$$

and define the Rayleigh number, $Ra = R^2$, by

$$R^2 = \frac{A\alpha g d^5}{2\nu\kappa^2}. \tag{17.11}$$

The non-dimensional perturbation equations are then (stars omitted)

$$\begin{aligned} &u_{i,t} + u_j u_{i,j} = -\pi_{,i} + \Delta u_i + Rk_i\theta, \\ &u_{i,i} = 0, \\ &Pr(\theta_{,t} + u_i\theta_{,i}) = 2R(\xi - z)w + \Delta\theta. \end{aligned} \tag{17.12}$$

Equations (17.12) have been deliberately arranged to be analogous to (17.3) - (17.5) as used in the alternative model for penetrative convection in section 17.1. We have identified coefficients between the two sets of equations and to achieve this requires

$$A = \frac{2T_U\kappa(T_L - T_U)}{(8 - T_U)d^2}. \tag{17.13}$$

This allows a direct comparison with section 17.1.

Firstly we note that if we transform $\theta \to -\theta$ in (17.12) the linearized systems (17.3) - (17.5) and (17.12) are adjoint. To see this, we let

$$M = \xi - z \tag{17.14}$$

and denote by $\mathcal{L}_{T^2}$ and $\mathcal{L}_Q$ the linear operators on the right of (17.3) - (17.5) and (17.12), respectively, with the pressure term momentarily omitted. Then

$$\mathcal{L}_{T^2}\begin{pmatrix}\mathbf{u}\\ \theta\end{pmatrix} = \begin{pmatrix}\Delta & 0 & 0 & 0\\ 0 & \Delta & 0 & 0\\ 0 & 0 & \Delta & -2RM\\ 0 & 0 & -R & \Delta\end{pmatrix}\begin{pmatrix}u\\ v\\ w\\ \theta\end{pmatrix}$$

and

$$\mathcal{L}_Q\begin{pmatrix}\mathbf{u}\\ \theta\end{pmatrix} = \begin{pmatrix}\Delta & 0 & 0 & 0\\ 0 & \Delta & 0 & 0\\ 0 & 0 & \Delta & -R\\ 0 & 0 & -2RM & \Delta\end{pmatrix}\begin{pmatrix}u\\ v\\ w\\ \theta\end{pmatrix}$$

The adjointness is visible from these expressions.

One advantage (17.12) have over (17.3) - (17.5) is that $(17.12)_1$ does not involve a θ^2 term. This allows us to use an unweighted L^2 norm in θ to obtain unconditional nonlinear energy stability. However, the model of this section is a less physical model than the nonlinear density model of section 17.1. Nevertheless, use of an internal heat source/sink is a good mathematical way to produce a model of convection which has part of the layer stabilizing with the rest destabilizing.

We give brief details of an energy analysis for (17.12). The energy, E, dissipation, D, and production, I, are

$$\begin{aligned} E &= \frac{1}{2}\|\mathbf{u}\|^2 + \frac{1}{2}\lambda Pr\|\theta\|^2, \\ D &= \|\nabla \mathbf{u}\|^2 + \lambda\|\nabla\theta\|^2, \qquad I = \big([1 + 2\lambda(\xi - z)]w, \theta\big), \end{aligned} \tag{17.15}$$

where λ is a positive coupling parameter. The energy equation is

$$\frac{dE}{dt} = RI - D,$$

and by defining $R_E^{-1} = \max_H (I/D)$ one may develop an unconditional energy stability theory as elsewhere in this book. The Euler-Lagrange equations may be formed for R_E and solved numerically and then compared with the linear instability critical Rayleigh numbers R_L. We do not do this specifically here, but not dissimilar calculations in porous media have been carried out by (Straughan and Walker, 1996a) and by (Carr and de Putter, 2003), see also (Carr, 2003a).

17.3 More general equations of state

In general, nonlinear equations of state of the form (17.1) form part of the theory of so-called non-Boussinesq convection. Non-Boussinesq convection essentially employs the Boussinesq approximation in the sense that the only variations in density are through the body force, but the equation of state is no longer linear in the temperature.

In a recent series of developments, several studies have questioned whether the quadratic equation (17.1) is accurate enough for detailed comparison with field studies and experiments involving convection in a layer where the fluid has a density maximum, see e.g., (Merker et al., 1979) and the references therein. These writers suggest using a density law like

$$\rho = \rho_0[1 + AT - BT^2 + CT^3], \tag{17.16}$$

or even like

$$\rho = \rho_0[1 + AT - BT^2 + CT^3 - DT^4 + ET^5], \tag{17.17}$$

where ρ_0 now denotes the density of water at 0°C and the coefficients A, B, C, D, E are constants obtained by curve fitting to data points.

(Merker et al., 1979) suggest for water, values in (17.16) of

$$A = 6.85650 \times 10^{-5}\ (^\circ\mathrm{C}^{-1}),$$
$$B = 8.82063 \times 10^{-6}\ (^\circ\mathrm{C}^{-2}),$$
$$C = 4.16668 \times 10^{-8}\ (^\circ\mathrm{C}^{-3}),$$

while in (17.17)

$$A = 6.79939 \times 10^{-5}\ (^\circ\mathrm{C}^{-1}),$$
$$B = 9.10749 \times 10^{-6}\ (^\circ\mathrm{C}^{-2}),$$
$$C = 1.00543 \times 10^{-7}\ (^\circ\mathrm{C}^{-3}),$$
$$D = 1.12689 \times 10^{-9}\ (^\circ\mathrm{C}^{-4}),$$
$$E = 6.59285 \times 10^{-12}\ (^\circ\mathrm{C}^{-5}).$$

(Merker et al., 1979) further regard (17.17) as *exact* in the range 0°C to 40°C. It is important to note that for highly accurate predictions on the onset of convection Merker et al. (1979) suggest (*on the basis of linear theory*) that (17.16) is about 10% more accurate than (17.1) whereas (17.17) yields approximately a 3% improvement over (17.16). The numerical results from the weighted energy analysis of (McKay and Straughan, 1992) suggest that it is preferable to employ (17.16), but (17.17) for such a small gain in accuracy leads to much greater mathematical complications. We take this opportunity to point out that (17.16) has been employed by (Niedrauer and Martin, 1979) in their investigation of the convective motion of brine in channels formed in sea ice. Equation (17.16) has also been advocated by (Ruddick and Shirtcliffe, 1979).

(Merker et al., 1979) give only linear instability results. However, the analytical methods of (Veronis, 1963) clearly indicate that sub-critical instabilities will exist for (17.16) and (17.17); (Busse, 1967) points this out in the general case where either the thermal expansion coefficient, or the viscosity depend on temperature. Thus a nonlinear energy stability result is desirable in that it represents a useful *quantitative threshold* (i.e., lower bound) for the hexagon sub-critical instability curve drawn qualitatively by (Busse, 1967), p. 644. The above objective of estimating the turning point of "inverted bifurcation" has been investigated *experimentally* in the work of (Walden and Ahlers, 1981) in liquid Helium I and of (Azouni and Normand, 1983a) in water.

We now describe analysis appropriate to (17.16); further details, and analysis pertinent to (17.17) may be found in (McKay and Straughan, 1992). Suppose the boundary conditions for the temperature are

$$T = T_1 \quad \text{at} \quad z = 0, \qquad T = T_2 \quad (\geq 4^\circ\mathrm{C}) \quad \text{at} \quad z = d,$$

where T_1 and T_2 are constants with $0 \leq T_1 \leq 4$. The equations for thermally driven convective motion in a layer are (3.41), and we use them with $\rho(T)$ in the buoyancy term given by (17.16). The resulting system then possesses the motionless solution,

$$\bar{u}_i = 0, \qquad \bar{T} = T_1 + \beta z, \qquad \beta = \frac{\Delta T}{d}, \qquad \Delta T = T_2 - T_1, \qquad (17.18)$$

with the hydrostatic pressure determined from what amounts to $(3.41)_1$ but with the buoyancy density law (17.16).

Under a suitable non-dimensionalization the equations for a perturbation to this solution are

$$\begin{aligned} u_{i,t} + u_j u_{i,j} =& - p_{,i} + \Delta u_i - R f_1 \theta k_i \\ & + Pr k_i (f_2 \theta^2 - a_2 \frac{Pr}{R} \theta^3) , \qquad (17.19) \\ u_{i,i} =& 0, \qquad (17.20) \\ Pr(\theta_{,t} + u_i \theta_{,i}) =& - Rw + \Delta \theta, \qquad (17.21) \end{aligned}$$

where $Ra = R^2$ is the Rayleigh number, $\xi = T_1/\Delta T$, $f_1(z) = 1 - 2a_1(\xi + z) + 3a_2(\xi + z)^2$, $f_2(z) = a_1 - 3a_2(\xi + z)$, $a_1 = (B/A)\Delta T$, and $a_2 = (C/A)(\Delta T)^2$. The standard boundary conditions apply, i.e., $u_i = \theta = 0$ on $z = 0, 1$, and u_i, θ, p are periodic in x, y.

It is quite possible that exchange of stabilities holds for the linearized version of (17.19) - (17.21). Indeed, (Merker et al., 1979) *assume* $\sigma \in \mathbb{R}$. For two free surfaces it is easy to adapt the Spiegel method to prove exchange of stabilities, c.f. section 4.3.

We may obtain L^2 energy identities from (17.19) and (17.21), and these are

$$\begin{aligned} \frac{1}{2}\frac{d}{dt}\|\mathbf{u}\|^2 =& - R < f_1 \theta w > - < u_{i,j} u_{i,j} > \\ & + Pr < f_2 \theta^2 w > - a_2 \frac{Pr^2}{R} < w\theta^3 >, \\ \frac{1}{2} Pr \frac{d}{dt}\|\theta\|^2 =& - R < w\theta > - < \theta_{,i}\theta_{,i} > . \end{aligned}$$

I do not see how to proceed *directly* from here since I do not see a way of dominating the $< w\theta^3 >$ term (arising from the cubic term in the density) by means of the stabilizing terms $< u_{i,j} u_{i,j} >$ and $< \theta_{,i}\theta_{,i} >$. Instead, two approaches using a generalized energy are suggested.

One alternative is to employ a weighted energy, the other defines an energy that also includes a piece of the L^4 integral of θ to control the relevant destabilizing term. The L^4 part is used only to dominate the non-linear terms, while the stability boundary is determined by that part of the energy involving the L^2 integrals. (It is worth noting that (17.17) leads to worse problems and evidently a yet more complicated energy must be employed which involves either both L^4 and L^6 integrals of θ, or a more involved weight function.)

For (17.19) - (17.21) we may choose,

$$E(t) = \frac{1}{2}\|\mathbf{u}\|^2 + \frac{1}{2}\lambda Pr\|\theta\|^2 + \frac{\mu}{4} Pr\|\phi\|^2,$$

where $\phi = \theta^2$, and λ, μ are positive coupling parameters, then conditional nonlinear stability can be achieved. To obtain *unconditional* nonlinear stability one approach is to utilize the weighted energy

$$E(t) = \frac{1}{2}\|\mathbf{u}\|^2 + \frac{1}{2}Pr\mu_1\|\theta\|^2 + \frac{1}{3}Pr < \mu_2(z)\theta^3 > + \frac{\lambda}{4R^2} Pr\|\phi\|^2,$$

where $\mu_2(z) = 3(\lambda + a_2 Pr^2)z/PrR$. The coupling parameters μ_1 and λ have to be opportunely selected. The numerical results so far available are very satisfactory, at least in the range $0 - 9°C$. Complete details may be found in (McKay and Straughan, 1992).

The above analysis has been extended to patterned ground formation by (McKay and Straughan, 1991) and to porous media convection by (Gentile and Rionero, 2000).

17.4 Convection with radiation heating

In sections 17.1 and 17.2 we have investigated the nonlinear stability problem for models of penetrative convection which are driven by a quadratic density-temperature relationship or by an internal heat source, respectively. In both sections, the temperature was prescribed on both the upper boundary, $z = d$, and on the lower boundary, $z = 0$, of a horizontal layer, boundary conditions (17.2) and (17.8). The purpose of this section is to study the effect of radiation heating at the lower boundary on the stability of the solution to the equations for an incompressible heat conducting viscous fluid, allowing for penetrative convection effects. Another penetrative convection model which involves radiation heating is the clever one of (Krishnamurti, 1997) where a heat source depends on a chemical concentration. This is discussed in section 17.8.

We again apply an energy stability analysis to obtain *quantitative nonlinear* stability estimates which guarantee nonlinear stability for the problem of convection in a plane layer with a constant temperature upper surface, while the lower surface is subject to a prescribed heat flux. These boundary conditions were suggested by (Veronis, 1963), and we believe them to be the physically correct ones for many applications. It is of especial relevance to this section to observe, for example, that the Earth's atmosphere is bounded below by the ground or the ocean. The bounding surface is heated by solar radiation and the air close to the surface then becomes warmer than the upper air and so a gravitationally unstable system results. When the convective motion begins, the warm air rises and penetrates into regions which are stably stratified. Therefore, in this section we consider boundary conditions appropriate to the above situation rather than those of (17.2).

Before progressing to the problem of penetrative convection with the heat flux prescribed at the lower boundary we digress into the classical Bénard problem with this boundary condition. For the classical Bénard problem the equations are given by (3.41) where the linear density-temperature relationship (3.40) has been employed. For clarity we rewrite these equations here as

$$\begin{aligned} v_{i,t} + v_j v_{i,j} &= -\frac{p_{,i}}{\rho_0} + \nu \Delta v_i + \alpha g k_i T, \\ v_{i,i} &= 0, \\ T_{,t} + v_i T_{,i} &= \kappa \Delta T, \end{aligned} \tag{17.22}$$

where the notation is as in section 3.2 and the constant $g(1 - \alpha T_R)$ term which arises in the buoyancy force from equation (3.40) has been absorbed in the pressure term. We wish to consider the stability of the steady conduction solution to (17.22) but instead of employing the boundary conditions (3.42) which presribe the temperature, T, on both $z = 0$ and $z = d$, we employ correct boundary conditions for radiation heating at the lower boundary. These are

$$v_i = 0, \quad z = 0, d; \qquad T = T_U, \quad z = d; \qquad \frac{\partial T}{\partial z} = \gamma, \quad z = 0. \tag{17.23}$$

In (17.23) γ is a constant. To understand the meaning of this boundary condition recall that the heat flux, q_i, is given by Fourier's law as $q_i = -\kappa T_{,i}$ and the flux out of the boundary is $q_i n_i$ where n_i is the unit outward normal to the boundary. When the boundary is the plane $z = 0$, $\mathbf{n} = (0, 0, -1)$, and so the heat flux out of the boundary is given by

$$\mathbf{q} \cdot \mathbf{n} = -q_3 = \kappa \frac{\partial T}{\partial z}, \quad \text{at } z = 0. \tag{17.24}$$

Thus, from (17.23), the heat flux out of the boundary $z = 0$ is $\kappa \partial T / \partial z = \kappa \gamma$. Hence, when $\gamma > 0$ this means heat is being taken out of the fluid layer $\mathbb{R}^2 \times \{z \in (0, d)\}$ whereas $\gamma < 0$ means that heat is being input into the fluid layer through the boundary $z = 0$.

The steady solution of equations (17.22) which satisfies the boundary equations (17.23) is

$$\bar{v}_i \equiv 0, \qquad \bar{T} = T_U - \gamma(d - z). \tag{17.25}$$

The corresponding steady pressure $\bar{p}$ is then determined from the equation $(17.22)_1$ as

$$\bar{p} = \rho_0 g \alpha \int_0^z \bar{T}(s) ds.$$

We can draw a one to one correspondence between solution (17.25) and the analogous solution for the classical Bénard problem with prescribed temperatures, namely solution (3.43). If we replace the lower temperature T_0 in (3.42) by T_L and the upper temperature T_1 by T_U then the steady state temperature field in (3.43) has form

$$\bar{T} = T_L - \left(\frac{T_L - T_U}{d}\right) z. \tag{17.26}$$

The expressions for $\bar{T}$ in (17.25) and (17.26) are equivalent if $\gamma = (T_U - T_L)/d$. While, however, there is a direct correspondence between the steady solutions the two thermal convection problems are distinct. This is because of the boundary conditions. If θ denotes a perturbation to $\bar{T}$ then $\theta = 0$ on the planes $z = 0, d$ in the classical problem studied in section 3.3. For the present situation, we put $T = \bar{T}(z) + \theta$ and then note that $\bar{T} = T_U$ on $z = d$, $\partial\bar{T}/\partial z = \gamma$ on $z = 0$. This means that the temperature perturbation must satisfy the boundary conditions

$$\theta = 0 \text{ on } z = d, \qquad \frac{\partial\theta}{\partial z} = 0 \quad \text{on } z = 0. \tag{17.27}$$

As a consequence the analysis and quantitative stability results will be different from those of section 3.3. We can likewise expect a difference in the penetrative convection problem with boundary conditions (17.23) from the theory of sections 17.1 and 17.2.

To investigate the stability problem for (17.22) with the steady solution (17.25) and boundary conditions (17.23), we let (u_i, θ, π) be perturbations to $(\bar{v}_i, \bar{T}, \bar{p})$. We then non-dimensionalize with the time, velocity, pressure and temperature scales $\mathcal{T} = d^2/\nu$, $U = \nu/d$, $P = \rho_0\nu U/d$ and $T^\sharp = U\sqrt{|\gamma|\nu/\kappa\alpha g}$. We define $H = \mathrm{sgn}(\gamma)$ so that $\gamma = H|\gamma|$ and observe $H > 0$ corresponds to heat being taken out of the layer whereas $H < 0$ means heat is input into the layer. The Prandtl number and Rayleigh number are defined as $Pr = \nu/\kappa$, $Ra = R^2 = g\alpha d^4|\gamma|/\kappa\nu$. Then, the nonlinear non-dimensional perturbation equations are

$$\begin{aligned} &u_{i,t} + u_j u_{i,j} = -\pi_{,i} + \Delta u_i + Rk_i\theta, \\ &u_{i,i} = 0, \\ &Pr(\theta_{,t} + u_i\theta_{,i}) = -HRw + \Delta\theta. \end{aligned} \tag{17.28}$$

The non-dimensional boundary conditions become

$$\begin{aligned} &u_i = 0, \quad z = 0, 1; \qquad \theta = 0, \quad z = 1; \\ &\frac{\partial\theta}{\partial z} = 0, \quad z = 0; \qquad \mathbf{u}, \theta, p \quad \text{have a periodic structure in } x, y, \end{aligned} \tag{17.29}$$

where the assumed periodicity in (17.29) is meant to reflect the cellular structure observed in practical convection patterns. The fact that $u_i = 0$ on $z = 0, 1$ means we are considering two fixed surfaces.

It is instructive to understand the outcome of (17.28) and (17.29) before moving to the analogous problems in penetrative convection. When the heat flux is out of the layer and so heat is being removed, then $H > 0$. In this case we can multiply $(17.28)_1$ by u_i, $(17.28)_3$ by θ and integrate over a period cell V. This leads to the energy equation

$$\frac{d}{dt}\Big(\frac{1}{2}\|\mathbf{u}\|^2 + \frac{1}{2}Pr\|\theta\|^2\Big) = -\|\nabla\mathbf{u}\|^2 - \|\nabla\theta\|^2. \qquad (17.30)$$

Even though boundary conditions (17.29) are in force, Poincaré's inequality still holds and so we can find a positive constant k such that $\|\nabla\mathbf{u}\|^2 + \|\nabla\theta\|^2 \geq kE$, where we have set $E(t) = \|\mathbf{u}\|^2/2 + (Pr/2)\|\theta\|^2$. Thus, from (17.30) we find $dE/dt \leq -kE$ and exponential decay of $E(t)$ follows for all initial data. Thus, when $\gamma > 0$ there is always nonlinear stability.

When $\gamma < 0$, $H < 0$ and then (17.28) and (17.29) define a symmetric system as discussed in section 4.3. Instability is possible, but we have the optimal result that the nonlinear stability critical Rayleigh number, Ra_E, and the linear instability critical Rayleigh number, Ra_L, are the same. Thus, sub-critical instabilities do not occur since the nonlinear result is unconditional. We do not calculate the stability/instability boundary here since our goal is to study analogous problems in penetrative convection.

17.4.1 Models for penetrative convection with boundary heating

We now study the effect of radiation heating at the lower boundary with the models of section 17.1 or 17.2 where penetrative convection is introduced via a nonlinear density law or an internal heat source, respectively. For the former, the equation we adopt is no longer one where the density is a linear function of temperature, but is the non-Boussinesq quadratic model of (Veronis, 1963), as given in (17.1). We extend (17.1) to allow a maximum temperature T_m, rather than 4°C, so choose

$$\rho = \rho_m[1 - \alpha(T - T_m)^2], \qquad (17.31)$$

where α is a constant, ρ is density and ρ_m is the maximum value of ρ, at temperature $T = T_m$.

We also study a model with a heat source, and so employ equations (17.7), or equations (4.91). In this section Q is taken to be constant.

When the density is nonlinear as in (17.31) or a heat source is present as in (17.7), or (4.91), sub-critical instabilities may occur and so in addition to our nonlinear stability results, which establish a critical Rayleigh number below which convection cannot occur, it is prudent to calculate the linear value above which convection occurs. We include the results of such a calculation for the quadratic density model and it is found that the parameter region of possible sub-critical instability is small. The problem of radiation heating at the lower boundary with the density (17.31) is studied in (Straughan, 1991b).

The mathematical picture is then that of a layer of heat-conducting viscous fluid with the quadratic equation of state (17.31) being employed as the buoyancy term in the body force or a layer of heat conducting viscous fluid governed by the heat source model (17.7). The fluid is taken to occupy the horizontal layer $z \in (0, d)$ with the lower boundary $z = 0$ heated by radiation. We select the temperature scale such that the temperature at $z = d$ remains a constant, T_U. The partial differential equations governing the quadratic density model are as in section 17.1, namely

$$\begin{aligned} v_{i,t} + v_j v_{i,j} &= -\frac{1}{\rho_m}p_{,i} + \nu\Delta v_i - gk_i[1 - \alpha(T - T_m)^2], \\ v_{i,i} &= 0, \\ T_{,t} + v_i T_{,i} &= \kappa\Delta T, \end{aligned} \tag{17.32}$$

where $v_i, p, T, \nu, g, \alpha, \kappa$ are velocity, pressure, temperature, viscosity, gravity, a thermal expansion coefficient and thermal diffusivity, and again $\mathbf{k} = (0, 0, 1)$. The spatial region of definition of (17.32) is $\mathbb{R}^2 \times \{z \in (0, d)\}$. The partial differential equations governing the heat source model are as in section 17.2, namely

$$\begin{aligned} v_{i,t} + v_j v_{i,j} &= -\frac{1}{\rho_m}p_{,i} + \nu\Delta v_i - gk_i[1 - \alpha(T - T_m)], \\ v_{i,i} &= 0, \\ T_{,t} + v_i T_{,i} &= \kappa\Delta T + Q. \end{aligned} \tag{17.33}$$

The heat source is here Q (constant) and the spatial domain of (17.33) is $\mathbb{R}^2 \times \{z \in (0, d)\}$.

For radiation heating at the lower boundary the correct boundary conditions are (17.23). The steady (conduction) solution $(\bar{v}_i, \bar{T}, \bar{p})$, whose stability we investigate, which satisfies equations (17.32) and the boundary conditions (17.23) is:

$$\bar{v}_i \equiv 0; \qquad \bar{T} = T_U - \gamma(d - z), \tag{17.34}$$

with the pressure $\bar{p}$ being determined from equation $(17.32)_1$ as:

$$\bar{p} = -\rho_m g \int_0^z \left[1 - \alpha(\bar{T}(s) - T_m)^2\right] ds.$$

The analogous steady solution $(\bar{v}_i, \bar{T}, \bar{p})$, to equations (17.33) together with the boundary conditions (17.23) is given by

$$\bar{v}_i \equiv 0; \qquad \bar{T} = \frac{Q}{2\kappa}(d^2 - z^2) - \gamma(d - z) + T_U. \tag{17.35}$$

The steady pressure $\bar{p}$ is now found as

$$\bar{p} = -\rho_m g \int_0^z \left[1 - \alpha(\bar{T}(s) - T_m)\right] ds.$$

We point out that (17.34) still allows the possibility of penetrative convection due to the nonlinear density relation (17.31). A penetrative effect is also allowed by (17.35) because the parabolic profile allows a temperature maximum or minimum inside the layer. In fact, with (17.35), one can show penetrative convection is definitely possible if Q and γ are positive so that heat is being extracted through the boundary but input into the layer via a heat source. The same is true for Q and γ both negative wherein heat is input through the lower boundary and extracted from the layer via an internal heat sink Q. Let us note that for (17.35) the steady temperature gradient is given by

$$\bar{T}'(z) = -\frac{Qz}{\kappa} + \gamma. \tag{17.36}$$

The stability analysis of either the conduction solution (17.34) or (17.35) is facilitated by the introduction of the perturbations (u_i, θ, π) defined by $v_i = \bar{v}_i + u_i$, $T = \bar{T} + \theta$, $p = \bar{p} + \pi$. We non-dimensionalize the perturbation equations using the following scalings which are common to both the quadratic density and heat source models (where stars denote dimensionless quantities):

$$t = t^* \frac{d^2}{\nu}, \qquad U = \frac{\nu}{d}, \qquad P = \frac{\rho_m \nu U}{d},$$
$$\mathbf{x} = \mathbf{x}^* d, \qquad \theta = \theta^* T^\#, \qquad u_i = U u_i^*.$$

$Pr = \nu/\kappa$ is again the Prandtl number.

For the quadratic density model, (17.32), we choose the temperature scale $T^\sharp$, the Rayleigh number $Ra = R^2$, and numbers δ and ζ by

$$T^\# = U\sqrt{\frac{\nu}{\alpha g d \kappa}}, \quad R^2 = \frac{\gamma^2 d^5 g \alpha}{\kappa \nu}, \quad \delta = \frac{T_U - T_m}{\gamma d}, \quad \zeta = 1 - \delta. \tag{17.37}$$

For the heat source model, equations (17.33), we put $Q = H|Q|$ with $H = \mathrm{sgn}\,(Q)$ and in addition to definitions of $T^\sharp$ and $Ra = R^2$ we need a parameter ξ. These are defined by

$$T^\# = \frac{U}{\kappa}\sqrt{\frac{|Q| d \nu}{\alpha g}}, \quad R^2 = \frac{|Q| d^5 g \alpha}{\kappa \nu}, \quad \xi = \frac{\gamma \kappa}{Q d}. \tag{17.38}$$

In writing down the fully nonlinear non-dimensional perturbation equations we now omit all stars although the variables are understood to be dimensionless. We denote the components of the velocity perturbation $\mathbf{u}$ as $\mathbf{u} = (u, v, w)$. Then, for the quadratic density model one derives the perturbation equations from (17.32) as

$$\begin{aligned} u_{i,t} + u_j u_{i,j} &= -\pi_{,i} + \Delta u_i - 2R(\zeta - z)\theta k_i + Pr\theta^2 k_i; \\ u_{i,i} &= 0; \\ Pr(\theta_{,t} + u_i \theta_{,i}) &= -Rw + \Delta\theta. \end{aligned} \tag{17.39}$$

For the heat source model the perturbation equations arising from (17.33) are

$$\begin{aligned} u_{i,t} + u_j u_{i,j} &= -\pi_{,i} + \Delta u_i + R\theta k_i; \\ u_{i,i} &= 0; \\ Pr(\theta_{,t} + u_i\theta_{,i}) &= H(z-\xi)Rw + \Delta\theta. \end{aligned} \tag{17.40}$$

Both sets of equations (17.39) and (17.40) are defined on the domain $\mathbb{R}^2 \times (0,1) \times \{t > 0\}$. The non-dimensional boundary conditions for both sets of equations are those of (17.29).

17.4.2 Linear instability equations

The equations governing linear instability are derived for both models by neglecting terms involving squares and then by seeking a time dependence like $e^{\sigma t}$, i.e. we write $u_i(\mathbf{x},t) = e^{\sigma t}u_i(\mathbf{x})$, $\theta(\mathbf{x},t) = e^{\sigma t}\theta(\mathbf{x})$, and $\pi(\mathbf{x},t) = e^{\sigma t}\pi(\mathbf{x})$.

For the heat source model we find from equations (17.40) that the linear theory is governed by

$$\begin{aligned} &\sigma u_i = -\pi_{,i} + \Delta u_i + R\theta k_i, \qquad u_{i,i} = 0, \\ &\sigma Pr\theta = H(z-\xi)Rw + \Delta\theta. \end{aligned} \tag{17.41}$$

We may eliminate π from equation $(17.41)_1$ to see that w satisfies the equation

$$\sigma\Delta w = \Delta^2 w + R\Delta^*\theta. \tag{17.42}$$

On seeking a normal mode representation of form $w = W(z)f(x,y)$, $\theta = \Theta(z)f(x,y)$, where f is a plane tiling form as in (3.48), with $\Delta^* f = -a^2 f$ for a wavenumber a, we find W and Θ satisfy the system

$$\begin{aligned} (D^2-a^2)(D^2-a^2-\sigma)W &= Ra^2\Theta, \\ (D^2-a^2-\sigma Pr)\Theta &= -HR(z-\xi)W, \end{aligned} \tag{17.43}$$

where $D = d/dz$, (17.43) are defined on $(0,1)$, and the boundary conditions (17.29) convert into

$$W = DW = 0,\ z = 0,1; \quad \Theta = 0,\ z = 1; \quad D\Theta = 0,\ z = 0. \tag{17.44}$$

For the quadratic density model the linear instability equations which arise from (17.39) are

$$\begin{aligned} &\sigma u_i = -\pi_{,i} + \Delta u_i - 2R(\zeta - z)\theta k_i, \qquad u_{i,i} = 0, \\ &\sigma Pr\theta = -Rw + \Delta\theta. \end{aligned} \tag{17.45}$$

When π is eliminated from $(17.45)_1$ this yields the equation for w,

$$\sigma\Delta w = \Delta^2 w - 2R(\zeta - z)\Delta^*\theta. \tag{17.46}$$

The normal mode representation involving W and Θ in this case satisfies the equations

$$\begin{aligned}(D^2-a^2)(D^2-a^2-\sigma)W&=-2Ra^2(\zeta-z)\Theta,\\(D^2-a^2-\sigma Pr)\Theta&=RW.\end{aligned}\qquad(17.47)$$

The boundary conditions for (17.47) are again (17.44).

The critical Rayleigh number of linear instability theory, Ra_L, for either (17.43) or (17.47) together with the boundary conditions (17.44) is found by putting $\sigma=\sigma_r+i\sigma_i$, locating where $\sigma_r=0$, and then finding $Ra_L=\min_{a^2}R^2(a^2)$.

17.4.3 Energy stability for the Q model

To derive a nonlinear energy stability theory for the heat source model we return to equations (17.40) and the boundary conditions (17.29). We multiply $(17.40)_1$ by u_i and integrate over a period cell V, and then multiply $(17.40)_3$ by θ and integrate over the same cell. Upon using the boundary conditions (17.29), this leads to the separate equations

$$\begin{aligned}\frac{d}{dt}\frac{1}{2}\|\mathbf{u}\|^2&=R(\theta,w)-\|\nabla\mathbf{u}\|^2,\\\frac{d}{dt}\frac{1}{2}Pr\|\theta\|^2&=HR\big((z-\xi)\theta,w\big)-\|\nabla\theta\|^2.\end{aligned}$$

Let $\lambda>0$ be a coupling parameter and then define $E(t)$ by

$$E(t)=\frac{1}{2}\|\mathbf{u}\|^2+\frac{\lambda Pr}{2}\|\theta\|^2.\qquad(17.48)$$

If we now define I and D by

$$I=\big(\theta,w[1+\lambda H(z-\xi)]\big),\qquad D=\|\nabla\mathbf{u}\|^2+\lambda\|\nabla\theta\|^2,\qquad(17.49)$$

we may derive the energy equation

$$\frac{dE}{dt}=RI-D.\qquad(17.50)$$

Next, define R_E by

$$R_E=\max_H\frac{I}{D}\qquad(17.51)$$

where H is the space of admissible solutions. Then from (17.50) we derive

$$\frac{dE}{dt}\le-D\left(\frac{R_E-R}{R_E}\right).\qquad(17.52)$$

Suppose now $R<R_E$, define $a=(R_E-R)/R_E$, and employ Poincaré's inequality on D in the form $D\ge kE$, where k is a positive constant, to

deduce from (17.52)

$$\frac{dE}{dt} \leq -kaE. \tag{17.53}$$

Unconditional nonlinear stability follows in energy measure from (17.53). The nonlinear stability analysis, therefore, reduces to finding R_E and so we must solve the maximum problem (17.51). The Euler-Lagrange equations for R_E are

$$\begin{aligned} &\Delta u_i + R_E J\theta k_i = -\omega_{,i}, \qquad u_{i,i} = 0, \\ &\Delta\theta + R_E J w = 0, \end{aligned} \tag{17.54}$$

where $J(z) = [1 + \lambda H(z-\xi)]/2\sqrt{\lambda}$ and ω is a Lagrange multiplier. If we eliminate ω from (17.54) then we must solve the reduced system

$$\begin{aligned} &\Delta^2 w + R_E J \Delta^* \theta = 0, \\ &\Delta\theta + R_E J w = 0. \end{aligned} \tag{17.55}$$

Equations (17.55) are reduced by normal modes to the following equations for $W(z)$, $\Theta(z)$, in which $D = d/dz$ and a is the wavenumber,

$$\begin{aligned} &(D^2 - a^2)^2 W - R_E a^2 J\Theta = 0, \\ &(D^2 - a^2)\Theta + R_E J W = 0. \end{aligned} \tag{17.56}$$

Equations (17.56) are defined on $z \in (0,1)$ and W, Θ satisfy the boundary conditions (17.44). We must then solve the system (17.56) and (17.44) for the lowest eigenvalue $R_E(a^2;\lambda)$ and then find

$$Ra_E = \max_{\lambda>0} \min_{a^2} R_E^2(a^2;\lambda).$$

The number Ra_E is the critical Rayleigh number of energy theory for the Q model.

17.4.4 Energy stability for the $\rho(T^2)$ model

The $Pr\theta^2 k_i$ term in $(17.39)_1$ prevents a straightforward L^2 energy analysis from yielding global results, i.e. for all initial data. Despite the imposition of the heat flux boundary condition $(17.23)_3$ we are able to utilize a weighted analysis. Such an analysis commences by introducing the weighted energy

$$E = \frac{1}{2}\|\mathbf{u}\|^2 + \frac{1}{2}Pr < (\mu - 2z)\theta^2 >, \tag{17.57}$$

for $\mu(>2)$ a coupling parameter to be optimally selected. To see how this generalized energy leads to unconditional stability we first multiply $(17.39)_1$ by u_i, and integrate over a period cell V to derive the identity

$$\frac{d}{dt}\frac{1}{2}\|\mathbf{u}\|^2 = -\|\nabla\mathbf{u}\|^2 - 2R\big([\zeta - z]\theta, w\big) + Pr(\theta^2, w). \tag{17.58}$$

The weight $\mu - 2z$ is added in (17.57) to lead to a term to balance out the cubic term in (17.58). In fact, we may see this by multiplying $(17.39)_3$ by $(\mu - 2z)\theta$, and integrating over a period cell V. This then yields

$$\begin{aligned}\frac{Pr}{2}\frac{d}{dt}\big\langle(\mu - 2z)\theta^2\big\rangle =&\big([\mu - 2z]\theta, Pr\theta_{,t}\big)\\ =&\big([\mu - 2z]\theta, -Pru_i\theta_{,i} - Rw + \Delta\theta\big)\\ =&-Pr(w,\theta^2) - R\big([\mu - 2z]\theta, w\big)\\ &-\big\langle(\mu - 2z)\theta_{,i}\theta_{,i}\big\rangle - \big\langle(\mu - 2z)_{,z}\theta\theta_{,z}\big\rangle\end{aligned} \tag{17.59}$$

where we have used the boundary conditions $u_i = 0$, on $z = 0, d$, $\theta = 0$ on $z = 1$ and $\theta_{,z} = 0$ on $z = 0$. The last term is further reduced to

$$\begin{aligned}-\big\langle(\mu - 2z)_{,z}\theta\theta_{,z}\big\rangle =& \int_V \frac{\partial\theta^2}{\partial z}\,dx\\ =&-\int_\Gamma \theta^2 dA,\end{aligned} \tag{17.60}$$

where Γ is the part of the boundary of V which lies in the plane $z = 0$. Hence, (17.59) and (17.60) yield

$$\begin{aligned}\frac{Pr}{2}\frac{d}{dt}\big\langle(\mu - 2z)\theta^2\big\rangle =&-Pr(w,\theta^2) - R\big([\mu - 2z]\theta, w\big)\\ &-\big\langle(\mu - 2z)\theta_{,i}\theta_{,i}\big\rangle - \int_\Gamma \theta^2 dA.\end{aligned} \tag{17.61}$$

Upon addition of (17.58) and (17.61) the cubic terms disappear and we are left with an energy equation involving only quadratic terms, of form

$$\frac{dE}{dt} = RI - \mathcal{D} - \int_\Gamma \theta^2 dA. \tag{17.62}$$

The functions I and $\mathcal{D}$ in this case are given by

$$I = -\big\langle(\mu + 2\zeta - 4z)\theta w\big\rangle, \qquad \mathcal{D} = \|\nabla\mathbf{u}\|^2 + \big\langle(\mu - 2z)|\nabla\theta|^2\big\rangle.$$

As in section 17.4.3 we derive an estimate for dE/dt and from (17.62) we next conclude:

$$\frac{dE}{dt} \le -\mathcal{D}\Big(\frac{R_E - R}{R_E}\Big), \tag{17.63}$$

where the number R_E is defined by the variational problem

$$\frac{1}{R_E} = \max_{\mathcal{H}} \frac{I}{\mathcal{D}}, \tag{17.64}$$

with $\mathcal{H}$ the space of admissible functions. The analysis from this point follows as that following (17.52). We, therefore, find global nonlinear stability (i.e. unconditional stability in the energy measure) provided $R < R_E$. We point out that decay in $\|\mathbf{u}\|$ and $\|\theta\|$ is deduced, because $\mu > 2$. The nonlinear stability analysis, therefore, reduces to finding R_E and so we must solve the maximum problem (17.64). The Euler-Lagrange equations for R_E for the present problem become

$$\Delta u_i - KR_E k_i\theta = \varpi_{,i}, \qquad \nabla\big([\mu - 2z]\nabla\theta\big) - KR_E w = 0, \qquad (17.65)$$

where u_i is solenoidal, ϖ is a Lagrange multiplier, and $K = K(z)$ is given by $K = \zeta + (\mu/2) - 2z$. The resolution of R_E follows by eliminating ϖ and introducing normal modes to reduce to a system in W and Θ as in section 17.4.3. The system (17.65) reduces to:

$$(D^2-a^2)^2W = -R_E K a^2\Theta, \qquad (D^2-a^2)\Theta = LR_E W + \frac{2D\Theta}{(\mu - 2z)}, \qquad (17.66)$$

where $L(z) = K(z)/(\mu - 2z)$. Equations (17.66) are to be solved with the boundary conditions (17.44). We must then solve the system (17.66), (17.44) for the lowest eigenvalue $R_E(a^2;\mu)$ and since we require $\mu - 2z > 0$ we then perform the optimization

$$Ra_E = \max_{\mu>2}\min_{a^2} R_E^2(a^2;\lambda).$$

The number Ra_E is the critical Rayleigh number of energy theory in this section.

In figures 17.1, 17.2, 17.3 we display Ra, a^2 and the optimal value of the coupling parameter μ, against the parameter δ. The upper curve marked by circles denotes the linear instability threshold Ra_L, whereas the lower one (filled in squares) represents the nonlinear energy stability boundary Ra_E. The same convention applies to figure 17.2 with the wavenumbers. In calculating the linear instability curve we have set $\sigma = 0$ and calculated the threshold only for stationary convection. Even so, the range where subcritical instabilities may possibly occur is small (at least for $\delta \le 0.3$). The Rayleigh number graphs for $\delta > 0$ employ a modified Rayleigh number which reflects the depth of that part of the fluid layer which can be destabilizing. In fact, for $\delta > 0$ figure 17.1 uses the definition $Ra = R^2(T_m/T_U)^5$.

17.4.5 Radiant heating, internal heat generation and nonlinear density

In this subsection we obtain *quantitative nonlinear* stability estimates which guarantee nonlinear stability for the problem of penetrative convection in a plane layer with a non-uniform heat source, and a constant temperature upper surface, while the lower surface is subject to a prescribed heat flux. The buoyancy force is given by the Veronis equation of state (17.31).

We have studied in the previous subsections the problems of convection with the heat flux prescribed at the lower boundary for the situations in which penetrative convection is caused either by a heat source or by a nonlinear density when the heat flux is prescribed on the lower plane $z = 0$.

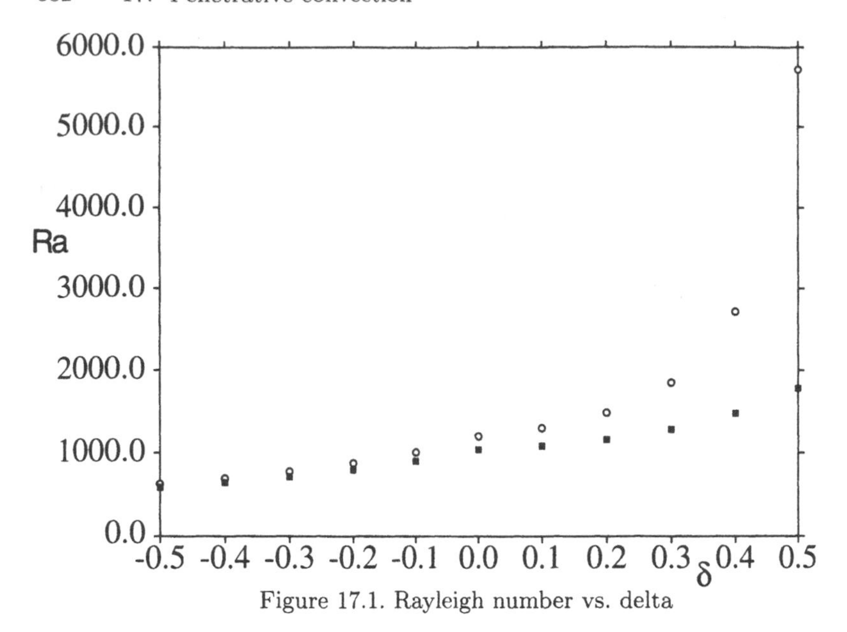

Figure 17.1. Rayleigh number vs. delta

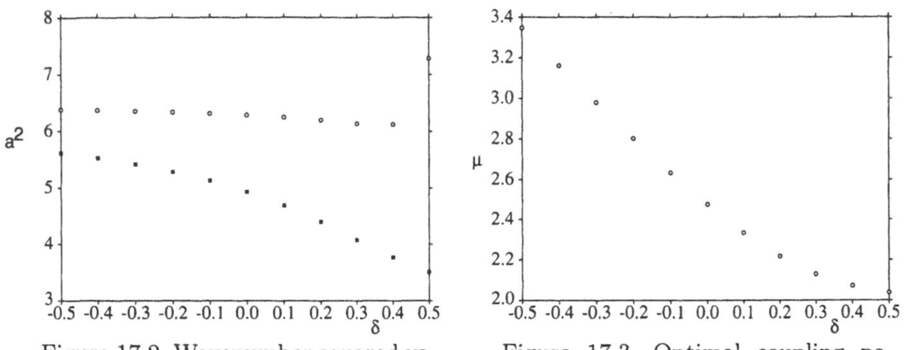

Figure 17.2. Wavenumber squared vs. delta

Figure 17.3. Optimal coupling parameter vs. delta

Since subcritical instabilities may occur in the individual convection problems, for an internal heat source, see (Joseph and Carmi, 1966), (Joseph and Shir, 1966), and when the density is quadratic in temperature, see (Veronis, 1963), we might reasonably expect the same to be true when both effects are simultaneously present. Unconditional nonlinear energy stability results can be very useful in this situation, especially if they are close to the corresponding linear instability ones. In addition, in heat source driven convection in the laboratory it is difficult to achieve a constant heat source experimentally, see (Tritton and Zarraga, 1967), (Roberts, 1967a). We allow the possibility of a non-uniform heat source and obtain the nonlinear energy stability and linear instability critical Rayleigh numbers.

Consider a layer of heat-conducting viscous fluid, with a quadratic equation of state in the body force, equation (17.31), and a heat source/sink Q.

The equations are then a combination of (17.32) and (17.33). The fluid occupies the horizontal layer $z \in (0,d)$, with the lower boundary $z=0$ heated by radiation and with the temperature scale selected so that the temperature at $z=d$ remains a constant, T_U. The equations for this situation are:

$$\begin{aligned} v_{i,t} + v_j v_{i,j} &= -\frac{1}{\rho_m} p_{,i} + \nu \Delta v_i - g k_i [1 - \alpha (T - T_m)^2], \\ v_{i,i} = 0, \qquad T_{,t} + v_i T_{,i} &= \kappa \Delta T + Q, \end{aligned} \tag{17.67}$$

where the notation is as given earlier in this section. Equations (17.67) are defined on the region $\mathbb{R}^2 \times \{z \in (0,d)\} \times \{t > 0\}$.

The appropriate boundary conditions are those of (17.23) which for clarity are restated,

$$v_i = 0, \ z = 0, d; \qquad T = T_U, \ z = d; \qquad \frac{\partial T}{\partial z} = \gamma, \ z = 0. \tag{17.68}$$

In order to investigate the spatial variation of the heat source, we consider three specific heat supply functions,

$$\begin{aligned} &\text{A.} \quad Q = Q_0 \Big(\frac{1}{2} + \frac{z}{d} \Big), \\ &\text{B.} \quad Q = Q_0 \Big(2 + \frac{3z^2}{2d^2} - 3\frac{z}{d} \Big), \\ &\text{C.} \quad Q = Q_0 \Big(1 + \sin \frac{2\pi z}{d} + \sin \frac{4\pi z}{d} \Big), \end{aligned} \tag{17.69}$$

where Q_0 is a constant. We observe that other choices for Q are studied in (Straughan, 1991a).

The steady (conduction only) solutions $(\bar{v}_i, \bar{T}, \bar{p})$ to (17.67), which are in accordance with the boundary conditions (17.68) are:

$$\text{Case A}: \quad \bar{v}_i \equiv 0; \quad \bar{T} = \frac{Q_0}{2\kappa} \Big(\frac{5d^2}{6} - \frac{1}{2} z^2 - \frac{z^3}{3d} \Big) - \gamma (d - z) + T_U,$$

$$\text{Case B}: \quad \bar{v}_i \equiv 0; \quad \bar{T} = \frac{Q_0 d^2}{\kappa} \Big(\frac{5}{8} - \frac{z^2}{d^2} - \frac{z^4}{8d^4} + \frac{z^3}{2d^3} \Big) - \gamma (d - z) + T_U,$$

$$\text{Case C}: \quad \bar{v}_i \equiv 0;$$

$$\bar{T} = \frac{Q_0 d^2}{2\pi\kappa} \Big(\pi + \frac{3}{2} - \frac{3z}{2d} - \pi \frac{z^2}{d^2} + \frac{1}{2\pi} \sin \frac{2\pi z}{d} + \frac{1}{8\pi} \sin \frac{4\pi z}{d} \Big) - \gamma (d - z) + T_U,$$

with $\bar{p}$ determined from:

$$\bar{p} = -\rho_m g \int_0^z \Big[1 - \alpha \big\{ \bar{T}(s) - T_m \big\}^2 \Big] ds,$$

where the $\bar{T}$ appropriate to the case in hand is understood to be employed.

To investigate the stability of these solutions we introduce perturbations (u_i, θ, π) by the definitions $v_i = \bar{v}_i + u_i, T = \bar{T} + \theta, p = \bar{p} + \pi$. The following scalings are now utilized, $t = t^* d^2/\nu$, $u_i = u_i^* U$, $U = \nu/d$, $\mathbf{x} = \mathbf{x}^* d$,

$\theta = \theta^* T^\#$, $T^\# = U\sqrt{\nu/\alpha g d\kappa}$, and $Pr = \nu/\kappa$, where Pr is the Prandtl number. The Rayleigh number $Ra = R^2$ is defined below together with the parameters δ and ϵ, by

$$Ra = R^2 = \frac{Q_0^2 d^7 g\alpha}{\kappa^3 \nu}, \qquad \delta = \frac{(T_m - T_U)\kappa}{Q_0 d^2}, \qquad \epsilon = \frac{\gamma\kappa}{Q_0 d}.$$

Then the *non-dimensional* perturbation equations become (*omitting all stars*):

$$\begin{aligned} &u_{i,t} + u_j u_{i,j} = -\pi_{,i} + \Delta u_i - 2Rf_1(z)\theta k_i + Pr\theta^2 k_i; \\ &u_{i,i} = 0; \qquad Pr(\theta_{,t} + u_i\theta_{,i}) = -Rf_2(z)w + \Delta\theta; \end{aligned} \tag{17.70}$$

where $\mathbf{k} = (0,0,1), w = u_3$, and the functions f_1 and f_2 are defined by $f_1(z) = \delta + \epsilon(1-z) - m(z)$, $f_2(z) = \epsilon - \ell(z)$. The functions m and ℓ are non-dimensional forms of the steady temperature field and steady temperature gradient and are given by

$$\ell(z) = \begin{cases} \frac{1}{2}(z+z^2), & \text{case A;} \\ 2z + \frac{1}{2}z^3 - \frac{3}{2}z^2, & \text{case B;} \\ \frac{1}{2\pi}\left(\frac{3}{2} + 2\pi z - \cos 2\pi z - \frac{1}{2}\cos 4\pi z\right), & \text{case C;} \end{cases} \tag{17.71}$$

$$m(z) = \begin{cases} \frac{5}{12} - \frac{1}{4}z^2 - \frac{1}{6}z^3, & \text{A;} \\ \frac{5}{8} - z^2 - \frac{1}{8}z^4 + \frac{1}{2}z^3, & \text{B;} \\ \frac{1}{2\pi}\left(\pi + \frac{3}{2} - \frac{3}{2}z - \pi z^2 + \frac{1}{2\pi}\sin 2\pi z + \frac{1}{8\pi}\sin 4\pi z\right), & \text{C.} \end{cases} \tag{17.72}$$

The layer is scaled to $(0,1)$, and so equations (17.70) hold on $\mathbb{R}^2 \times (0,1) \times \{t > 0\}$. The non-dimensionalized boundary conditions for the perturbations are (17.29).

Equation $(17.70)_1$ contains a quadratic term in θ and so to obtain unconditional nonlinear stability we follow the method of section 17.4.4 and employ a weighted energy. Hence, we multiply $(17.70)_1$ by u_i, $(17.70)_3$ by $(\mu - 2z)\theta$, add, and integrate over a period cell V, for $\mu(> 2)$ a coupling parameter to be chosen. The details of this calculation are similar to those of section 17.4.4 and yield, after use of the boundary conditions (17.29),

$$\frac{dE}{dt} = RI - \mathcal{D} - \int_\Gamma \theta^2 dA. \tag{17.73}$$

In this case, the weighted energy E, the production term I, and the dissipation $\mathcal{D}$ are given by:

$$\begin{aligned} E &= \frac{1}{2}\|\mathbf{u}\|^2 + \frac{1}{2}Pr\int_V (\mu - 2z)\theta^2\,dx, \\ I &= -\Big(\big[2f_1 + (\mu - 2z)f_2\big]\theta, w\Big), \\ \mathcal{D} &= \|\nabla\mathbf{u}\|^2 + \int_V (\mu - 2z)|\nabla\theta|^2\,dx. \end{aligned} \tag{17.74}$$

The analysis after (17.73) follows that earlier in this section (e.g. after (17.62)) and we deduce

$$\frac{dE}{dt} \leq -\mathcal{D}\left(\frac{R_E - R}{R_E}\right), \tag{17.75}$$

where R_E is defined by:

$$\frac{1}{R_E} = \max_{\mathcal{H}} \frac{I}{\mathcal{D}}, \tag{17.76}$$

with $\mathcal{H}$ being the space of admissible functions. When $R < R_E$ we may deduce exponential decay of $E(t)$ from (17.75) and this in turn leads to unconditional nonlinear stability.

The Euler-Lagrange equations which arise from (17.76) are

$$\Delta u_i - GR_E k_i \theta = \varpi_{,i}, \qquad \nabla\big([\mu - 2z]\nabla\theta\big) - GR_E w = 0, \tag{17.77}$$

where ϖ is a Lagrange multiplier and $G(z)$ is given by $G = f_1(z) + f_2(z)([\mu/2] - z)$. After eliminating ϖ and employing normal modes these are reduced to

$$(D^2-a^2)^2 W = -R_E G a^2 \Theta, \qquad (D^2-a^2)\Theta = FR_E W + \frac{2D\Theta}{(\mu - 2z)}, \tag{17.78}$$

where $F(z) = G(z)/(\mu - 2z)$. Equations (17.78) are to be solved for $0 < z < 1$ subject to the boundary conditions:

$$W = DW = 0, \ z = 0,1; \quad \Theta = 0, \ z = 1; \quad D\Theta = 0, \ z = 0. \tag{17.79}$$

We determine the critical Rayleigh number of energy theory, Ra_E, from the relation

$$Ra_E = \max_{\mu > 2} \min_{a^2} R_E^2(a^2; \mu). \tag{17.80}$$

System (17.78), (17.79) is solved numerically by the compound matrix method, and some results are now included. We also include numerical results for the linear instability boundary with $\sigma = 0$, i.e. the stationary convection boundary. In general, one should calculate the linear instability threshold since σ may be complex in this problem, cf. section 17.5.

Tables 17.1 -17.3 demonstrate that the linear instability and energy stability results are reasonably close. We are unaware of experimental results for cases A-C, or for situations closely approximating such internal heat sources for a fluid possessing a density maximum. It would, however, be highly useful to obtain such experimental data to determine whether subcritical bifurcation is observed, thereby testing the nonlinear energy boundary.

Ra_L	Ra_E	a_L^2	a_E^2	δ	ϵ	μ^*
5682.76	5375.78	6.673	5.981	0	0	2.981
9427.71	8926.36	6.904	6.378	0	0.1	3.119
18748.63	17315.06	7.544	7.214	0	0.2	3.305
4324.87	4138.58	6.754	6.235	-0.1	0	3.472
6837.62	6522.95	7.007	6.646	-0.1	0.1	3.776
12686.02	11730.03	7.685	7.474	-0.1	0.2	4.241
3490.17	3360.25	6.805	6.397	-0.2	0	3.977
5363.05	5133.28	7.068	6.803	-0.2	0.1	4.450
9584.74	8860.25	7.760	7.609	-0.2	0.2	5.195

Table 17.1. Critical Rayleigh numbers of linear theory, Ra_L, and energy theory, Ra_E, together with the respective critical wavenumbers a_L, a_E, and the best value of μ, μ^*. Case A.

Ra_L	Ra_E	a_L^2	a_E^2	δ	ϵ	μ^*
2391.80	2205.30	6.456	5.451	0	0	2.671
3247.77	3013.36	6.501	5.552	0	0.1	2.703
4663.88	4357.49	6.571	5.701	0	0.2	2.746
1939.72	1821.31	6.512	5.682	-0.1	0	2.940
2563.67	2422.07	6.567	5.807	-0.1	0.1	3.025
3554.55	3378.91	6.651	5.983	-0.1	0.2	3.144
1631.10	1549.05	6.552	5.850	-0.2	0	3.224
2117.22	2021.67	6.612	5.983	-0.2	0.1	3.364
2870.95	2754.67	6.703	6.166	-0.2	0.2	3.561

Table 17.2. Critical Rayleigh numbers of linear theory, Ra_L, and energy theory, Ra_E, together with the respective critical wavenumbers a_L, a_E, and the best value of μ, μ^*. Case B.

Ra_L	Ra_E	a_L^2	a_E^2	δ	ϵ	μ^*
1663.45	1519.88	6.423	5.342	0	0	2.600
2137.87	1962.69	6.450	5.407	0	0.1	2.615
2848.74	2629.68	6.490	5.494	0	0.2	2.634
1384.92	1288.86	6.473	5.554	-0.1	0	2.812
1744.09	1631.64	6.507	5.639	-0.1	0.1	2.860
2265.66	2131.70	6.557	5.751	-0.1	0.2	2.921
1186.10	1117.33	6.510	5.715	-0.2	0	3.036
1472.54	1394.10	6.549	5.810	-0.2	0.1	3.118
1880.34	1789.35	6.604	5.932	-0.2	0.2	3.225

Table 17.3. Critical Rayleigh numbers of linear theory, Ra_L, and energy theory, Ra_E, together with the respective critical wavenumbers a_L, a_E, and the best value of μ, μ^*. Case C.

17.5 Combined effect of a heat source and nonlinear density

In section 17.4 we have already examined a convection problem where the density is quadratic in temperature and a heat source is present. The purpose of this section is to revisit this problem but primarily report some very interesting findings of (Normand and Azouni, 1992). These writers analyse the problem of a horizontal layer of water of height h bounded by two rigid surfaces. The upper boundary, $z = h$, is held at fixed temperature 4°C, which is the temperature where the density of water is a maximum. The lower boundary, $z = 0$, is maintained at constant temperature $T_0 \neq T_m = 4°C$, and the water nearby is therefore less dense. They adopt the (Veronis, 1963) (17.1) equation for the density - temperature relationship but additionally suppose a constant heat source is acting throughout the fluid. This yields a steady solution for which $\bar{v}_i \equiv 0$ and the steady temperature has form

$$\bar{T}(z) = T_0 + \frac{\Delta T_1}{h^2}(1-\mu)z^2 + \frac{\Delta T_1}{h}\mu z. \tag{17.81}$$

In (17.81) $\Delta T_1 = T_m - T_0$ where $T_m = 4°C$ and $\mu = (2/\gamma)[1+\gamma+(1+\gamma)^{1/2}]$, γ being given by $\gamma = \Delta T_1/\Delta T_2$. The constant $\Delta T_2 = T_{ex} - T_m$ where T_{ex} is the temperature where $d\bar{T}/dz = 0$ at a depth d. In fact, this results in a three layer situation in which there is an unstable layer of depth $d_2 = h - d$ lying above a stable layer, which in turn lies above an unstable layer of depth $d_1 = 1/(\mu - 1)$ which commences at $z = 0$. Hence, there is a stable layer of fluid bounded above and below by two potentially unstable layers. This means that convection may commence in one or other of the layers and penetrate into the rest of the layer. However, for certain depths there arises the possibility that convection will switch from one layer to the other or essentially commence in both layers simultaneously like a resonance effect. (Normand and Azouni, 1992) demonstrate this beautifully. They show by a linear instability analysis that for values of μ between $\mu = 4.6$ and $\mu = 4.7$ the instability curve switches from one mode to another and there is a region in between where oscillatory convection occurs. This provides an answer to the question raised in section 4.3 concerning equations (4.97) - (4.99). For a general $H(z)$ and $N(z)$ as given there oscillatory convection may occur and exchange of stabilities does not hold.

The equations governing the (Normand and Azouni, 1992) problem are

$$\begin{aligned} v_{i,t} + v_j v_{i,j} &= -\frac{1}{\rho_m} p_{,i} + \nu \Delta v_i - g k_i \big[1 - \alpha(T - T_m)^2\big], \\ v_{i,i} = 0, \qquad T_{,t} &+ v_i T_{,i} = \kappa \Delta T + Q. \end{aligned} \tag{17.82}$$

The perturbation equations which arise from this are non-dimensionalized with length scale h, time scale h^2/ν, velocity scale ν/h, pressure scale $\nu\rho_m/h$, and temperature scale $U\sqrt{\nu/\kappa\alpha g h}$. The resulting non-dimensional perturbation equations are

$$\begin{aligned} &u_{i,t} + u_j u_{i,j} = -\pi_{,i} + \Delta u_i + 2RF(z)\theta k_i + Pr\theta^2 k_i, \\ &u_{i,i} = 0, \qquad Pr(\theta_{,t} + u_i\theta_{,i}) = -RF'(z)w + \Delta\theta, \end{aligned} \tag{17.83}$$

where

$$F(z) = (1-\mu)z^2 + \mu z - 1. \tag{17.84}$$

We may develop an unconditional nonlinear energy stability analysis for these equations by employing a weighted energy of form

$$E(t) = \frac{1}{2}\|\mathbf{u}\|^2 + < (\delta - 2z)\theta^2 >, \tag{17.85}$$

for $\delta > 2$ a coupling parameter. One may show that this leads to

$$\frac{dE}{dt} = RI - D, \tag{17.86}$$

where

$$I =< (2F - [\delta - 2z]F')w\theta >, \qquad D = \|\nabla\mathbf{u}\|^2 + < (\delta - 2z)|\nabla\theta|^2 > .$$

Unconditional nonlinear stability follows from (17.86) in the usual way with the nonlinear critical Rayleigh number being determined by $R_E^{-1} = \max_H I/D$. Numerical solution of the Euler-Lagrange equations which arise from this variational problem yields the critical nonlinear stability threshold.

17.6 Penetrative convection in porous media

The subject of penetrative convection in a porous medium is one with many geophysical applications. One of these has already been discussed in section 7.1. In fact, in many ways the subject of thermal and even solutal convection in porous media is much richer than it is in classical fluid dynamics. In porous media models one has to contend with effects of anisotropy in the porous matrix, boundary layers due to a Brinkman effect, and nonlinearity due to Forchheimer terms. Such porous media models have been the subject of much investigation, see e.g. (Mahidjiba et al., 2003), (Payne et al., 2001), (Qin and Kaloni, 1992a; Qin and Kaloni, 1993b; Qin and Kaloni, 1994a), (Qin et al., 1995), (Storesletten, 1993), (Straughan, 1993a), (Straughan and Walker, 1996a), (Tyvand and Storesletten, 1991), and the references therein. In this chapter we concentrate on penetrative convection in a layer which has transversely isotropic permeability but with the axis of isotropy allowed to be aligned obliquely to the layer. The energy and linear analyses for this were developed by (Straughan and Walker, 1996a). (Carr and de Putter, 2003), see also (Carr, 2003a) develop a sharp analysis for a similar problem, but when the isotropy axis is vertical. (Tyvand and

Storesletten, 1991) initiated a very interesting sequence of papers to study when convection will begin in an anisotropic porous medium with oblique principal axes. They chose a porous medium in which the permeability is transversely isotropic with respect to an axis which makes an angle of β to the horizontal. The angle β may have an arbitrary value between 0 and 90°. (Tyvand and Storesletten, 1991) discovered several differences compared to convection in an isotropic porous medium. One notable difference is that in the isotropic situation the preferred motion at the onset of convection is in the form of rolls with square cross sections, but in the anisotropic case the preferred motion may well be in tilted cells.

(Straughan and Walker, 1996a) used the configuration of (Tyvand and Storesletten, 1991) but assumed a non-Boussinesq density in the buoyancy force. They also investigated another model for penetrative convection in an anisotropic porous medium. This is analogous to that of section 17.2 where the penetration effect is introduced through a heat supply.

17.6.1 A model for anisotropic penetrative convection

We report on the findings of (Straughan and Walker, 1996a) who consider a fluid saturated porous layer contained between horizontal planes $z = 0, d$ with gravity in the negative z direction, $-\mathbf{k}$. The upper boundary is maintained at a constant temperature $T_1 (> 0°C)$ while the lower one is maintained at temperature $0°C$. The density of water is taken to be modelled by the relation

$$\rho(T) = \rho_m\big[1 - \gamma(T - T_m)^2\big], \tag{17.87}$$

where ρ_m is the density of water at $T_m = 4°C$, T is temperature, and $\gamma(\approx 7.68 \times 10^{-6}(°\mathrm{C}^{-2}))$ is the expansion coefficient, cf. (Veronis, 1963). The permeability tensor $\mathbf{K}^*$ is that of (Tyvand and Storesletten, 1991),

$$\mathbf{K}^* = K_{\|}\mathbf{i}'\mathbf{i}' + K_{\perp}(\mathbf{j}'\mathbf{j}' + \mathbf{k}'\mathbf{k}'). \tag{17.88}$$

Here $K_{\|}$ and $K_{\perp}$ are the longitudinal and transverse components of permeability. The vectors $\mathbf{i}'$, $\mathbf{j}'$ and $\mathbf{k}'$ are an orthogonal system with $\mathbf{j} = \mathbf{j}'$ and $\mathbf{i}$ is aligned with the projection of $\mathbf{i}'$ on the (x, y) plane. The angle β is the angle between the vectors $\mathbf{i}$ and $\mathbf{i}'$. The porous layer is contained between the horizontal planes $z = 0$ and $z = d$, z being in the $\mathbf{k}-$direction. (Tyvand and Storesletten, 1991) show that there is an inverse permeability tensor $\mathbf{M}$ such that

$$\mathbf{M}.\mathbf{K}^* = K_{\perp}\mathbf{I}, \tag{17.89}$$

with $\mathbf{I}$ the identity tensor. They further show

$$\mathbf{M} = \xi\mathbf{i}'\mathbf{i}' + \mathbf{j}'\mathbf{j}' + \mathbf{k}'\mathbf{k}' = M_{11}\mathbf{ii} + M_{13}(\mathbf{ik} + \mathbf{ki}) + M_{33}\mathbf{kk} + \mathbf{jj} \tag{17.90}$$

where the non-zero coefficients M_{ij} are given by

$$M_{11} = \xi\cos^2\beta + \sin^2\beta, \qquad M_{13} = (\xi - 1)\cos\beta\sin\beta,$$
$$M_{33} = \cos^2\beta + \xi\sin^2\beta. \tag{17.91}$$

The coefficient $\xi = K_\perp/K_\parallel$ is the anisotropy parameter.

Darcy's law is taken to govern the fluid motion in the porous layer so that

$$K^*_{ij}p_{,j} = -\mu v_i - gK^*_{ij}k_j(2\rho_m\gamma T_m T - \gamma\rho_m T^2), \tag{17.92}$$

where p is effectively the pressure, μ dynamic viscosity, and v_i velocity. The full system of partial differential equations governing the velocity, temperature, and pressure fields are then written in terms of the inverse permeability tensor $\mathbf{M}$, and are

$$p_{,i} = -\frac{\mu}{K_\perp} M_{ij}v_j - 2\rho_m\gamma T_m g k_i T + \gamma\rho_m g k_i T^2, \tag{17.93}$$
$$v_{i,i} = 0, \tag{17.94}$$
$$T_{,t} + v_i T_{,i} = \kappa\Delta T. \tag{17.95}$$

Here κ is the (constant) thermal diffusivity.

The basic solution to (17.93) - (17.95) whose stability is investigated is given by

$$\bar{v}_i = 0, \qquad \bar{T} = \frac{T_1}{d}\, z. \tag{17.96}$$

The steady pressure $\bar{p}$ is found from (17.93). Perturbations (u_i, θ, π) to $(\bar{v}_i, \bar{T}, \bar{p})$ are introduced by $v_i = \bar{v}_i + u_i$, $T = \bar{T} + \theta$, $p = \bar{p} + \pi$ and the equations for the perturbations are non-dimensionalized with the scalings $t^* = d^2/\kappa$, $U = \kappa/d$, $\mathbf{x} = \mathbf{x}^* d$, $T^\sharp = (\beta\mu\kappa/\rho_m\gamma g T_1 K_\perp)^{1/2}$, $P = \mu\kappa/K_\perp$. The Rayleigh number R^2 is defined by

$$R^2 = T_1^2\,\frac{dK_\perp\rho_m\gamma g}{\mu\kappa},$$

and a penetration parameter $\hat{\xi}$ necessarily arises, $\hat{\xi} = T_m/T_1$. The non-dimensional equations for the perturbations are

$$\pi_{,i} = -M_{ij}u_j - 2R(\hat{\xi} - z)k_i\theta + k_i\theta^2, \tag{17.97}$$
$$u_{i,i} = 0, \tag{17.98}$$
$$\theta_{,t} + u_i\theta_{,i} = -Rw + \Delta\theta, \tag{17.99}$$

in which $w = u_3$.

The boundary conditions are

$$w = \theta = 0 \qquad \text{at} \qquad z = 0, 1, \tag{17.100}$$

with (u_i, θ, π) satisfying a plane tiling shape in the x, y directions. The three - dimensional periodicity cell which results is denoted by V.

17.6.2 Linearized instability

The linearized instability equations are derived from (17.97) - (17.99) and are

$$\begin{aligned} \pi_{,i} &= -M_{ij}u_j - 2RMk_i\theta, \\ u_{i,i} &= 0, \qquad \sigma\theta = -Rw + \Delta\theta, \end{aligned} \tag{17.101}$$

where σ is the growth rate in a temporal form $e^{\sigma t}$.

(Tyvand and Storesletten, 1991) argue that $\sigma \in \mathbb{R}$ in the equivalent anisotropic problem without penetrative convection. (Straughan and Walker, 1996a) show that the (Tyvand and Storesletten, 1991) result may be extended in that $\sigma \in \mathbb{R}$ even for an anisotropic, but symmetric thermal conductivity, with the viscosity μ dependent on temperature T, and a temperature dependent thermal conductivity κ_{ij}. (Straughan and Walker, 1996a) also note that by using Spiegel's method one may show $\sigma \in \mathbb{R}$ in the *isotropic* situation but *with* penetrative convection, However, the striking result of (Straughan and Walker, 1996a) is that $\sigma \in \mathbb{C}$ in the *anisotropic* case, i.e. when $\beta \neq 90°$, $\xi \neq 1$, with penetrative convection.

They eliminate the pressure perturbation π from (17.101) to obtain the coupled system for w and θ,

$$\mathcal{L}w = -2RML_1\theta, \qquad \sigma\theta = -Rw + \Delta\theta, \tag{17.102}$$

where $\mathcal{L}$ and L_1 are the (Tyvand and Storesletten, 1991) operators defined by

$$\begin{aligned} \mathcal{L} &= \xi\frac{\partial^2}{\partial y^2} + M_{11}\frac{\partial^2}{\partial z^2} + M_{33}\frac{\partial^2}{\partial x^2} - 2M_{13}\frac{\partial^2}{\partial x\partial z}, \\ L_1 &= M_{11}\frac{\partial^2}{\partial y^2} + \frac{\partial^2}{\partial x^2}. \end{aligned}$$

(Straughan and Walker, 1996a) solve the eigenvalue problem (17.102) numerically by introducing normal modes $w = W(z)e^{i(kx+my)}$, $\theta = \Theta(z)e^{i(kx+my)}$. This means they solve the system

$$\begin{aligned} D^2W =& 2ik\frac{M_{13}}{M_{11}}DW + \left(\frac{M_{33}k^2 + \xi m^2}{M_{11}}\right)W \\ &+ 2RM(z)\left(m^2 + \frac{k^2}{M_{11}}\right)\Theta, \\ D^2\Theta =& (a^2 + \sigma)\Theta + RW, \end{aligned} \tag{17.103}$$

where $a^2 = k^2 + m^2$, and $D = d/dz$.

(Straughan and Walker, 1996a) solve the above system by two entirely different numerical methods, the compound matrix method, and the Chebyshev tau method, see e.g. chapter 19.

17.6.3 Unconditional nonlinear stability and weighted energy

(Straughan and Walker, 1996a) are interested in unconditional nonlinear stability and due to the presence of the θ^2 term in (17.97) they resort to a weighted energy technique.

They multiply (17.97) by u_i and integrate over V to obtain

$$0 = - < M_{ij}u_iu_j > -2R < M\theta w > + < \theta^2 w > . \qquad (17.104)$$

For $\mu(> 2)$ a constant to be chosen, the weight function $\hat{\mu}$ is introduced as $\hat{\mu} = \mu - 2z$, and the energy functional $E(t)$ is defined as

$$E(t) = \frac{1}{2} < \hat{\mu}\theta^2 > . \qquad (17.105)$$

By differentiation, use of the differential equation and the boundary conditions one finds

$$\frac{dE}{dt} = - < w\theta^2 > -R < \hat{\mu}\theta w > - < \hat{\mu}|\nabla\theta|^2 > . \qquad (17.106)$$

The effect of the weight is to yield the cubic term in (17.106). Upon addition of (17.104) and (17.106) the cubic terms disappear and we derive

$$\frac{dE}{dt} = RI - D, \qquad (17.107)$$

where

$$I = - < (2M + \hat{\mu})\theta w >, \qquad D =< M_{ij}u_iu_j > + < \hat{\mu}|\nabla\theta|^2 > .$$

A weighted energy Rayleigh number, R_W, is defined by

$$\frac{1}{R_W} = \max_{\mathcal{H}} \frac{I}{D}, \qquad (17.108)$$

where $\mathcal{H}$ is the space of admissible solutions. For $R < R_W$ one now shows from (17.107) that

$$\frac{dE}{dt} \le -D\Big(\frac{R_W - R}{R_W}\Big) . \qquad (17.109)$$

To make progress from (17.109) (Straughan and Walker, 1996a) show that for $0 < \xi < 1$,

$$M_{ij}u_iu_j \ge v^2 + \xi(u^2 + w^2), \qquad (17.110)$$

so that $M_{ij}u_iu_j$ is positive definite. Furthermore, for $\xi \ge 1$ they deduce a similar conclusion.

Hence, since $\mu > 2$ use of Poincaré's inequality shows there is a constant $\delta > 0$ such that $D \ge \delta E$, and from (17.109), provided $R < R_W$, it follows that $E \to 0$ at least exponentially, as $t \to \infty$. Thus, the number R_W represents a threshold for unconditional nonlinear stability.

(Straughan and Walker, 1996a) calculate the Euler Lagrange equations from (17.108) as

$$\begin{aligned} & -R_W F\theta k_i - M_{ij}u_j = \pi_{,i}\,, \\ & u_{i,i} = 0, \\ & \hat{\mu}\Delta\theta - 2\theta_{,z} - R_W F w = 0, \end{aligned} \tag{17.111}$$

where $F(z) = M(z) + \frac{1}{2}\hat{\mu}(z)$.

These writers develop a numerical routine to calculate R_W for fixed k, m and μ, and then minimize in k or m, then maximize in μ.

(Straughan and Walker, 1996a) also consider another energy technique appropriate to the boundary condition $\partial\theta/\partial z = 0$ on $z = 1$. This results in a conditional nonlinear stability threshold, although this may be more appropriate for a model for patterned ground formation.

17.6.4 Anisotropic porous penetrative convection via an internal heat source

(Straughan and Walker, 1996a) also develop a heat source model for penetrative convection in an anisotropic porous layer.

They assume that the density is linear in the temperature field, i.e.

$$\rho(T) = \rho_0\big[1 - \alpha(T - T_0)\big]\,,$$

ρ_0, T_0 constants, and let Q (constant) be a heat source or sink.

The equations are

$$\begin{aligned} & \nabla p = -\mu(\mathbf{K}^*)^{-1}\mathbf{v} + g\rho_0\alpha T\mathbf{k}, \\ & \mathrm{div}\mathbf{v} = 0, \quad T_{,t} + (\mathbf{v}.\nabla)T = \kappa\Delta T + Q. \end{aligned} \tag{17.112}$$

The basic solution whose stability is under investigation is

$$\bar{v}_i = 0, \qquad \bar{T} = -\frac{Q}{2\kappa}z^2 + \left[\frac{(T_1 - T_0)}{d} + \frac{Qd}{2\kappa}\right]z + T_0,$$

where T_1, T_0 are the temperatures of the planes $z = d$, $z = 0$, respectively. They treat explicitly the case $A = -Q(> 0)$, with T_0, T_1 prescribed in such a way as to yield a layer in which the upper half is gravitationally stable whereas the lower half will be unstable.

The perturbation equations for this situation are, see (Straughan and Walker, 1996a)

$$\begin{aligned} & \pi_{,i} = -M_{ij}u_j + Rk_i\theta, \\ & u_{i,i} = 0, \qquad \theta_{,t} + u_i\theta_{,i} = Rf(z)w + \Delta\theta, \end{aligned} \tag{17.113}$$

where $f(z) = \frac{1}{2} - \epsilon - z$, $\epsilon = \Delta T\,\kappa/d^2A$, and $\Delta T = T_1 - T_0$.

Extensive numerical results for equations (17.97) - (17.99) are given in (Straughan and Walker, 1996a). Some numerical results are provided for (17.113). In particular, (Straughan and Walker, 1996a) concentrate on the

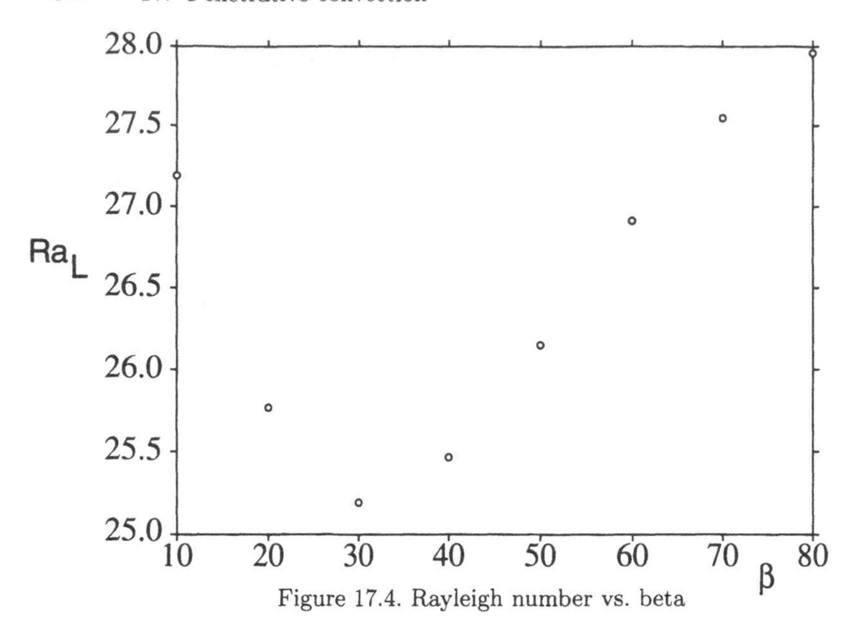

Figure 17.4. Rayleigh number vs. beta

case $0 < \xi < 1$ since this is where they find the significant changes are with non-penetrative convection. The situation $0 < \xi < 1$ means $K_{\|} > K_{\perp}$, i.e. the fluid flows easiest along the direction of the major principal axis, and this is the situation probably of most interest physically. When $0 < \xi < 1$ (Straughan and Walker, 1996a) find $\sigma_1 \neq 0$ at criticality. We refer the reader to (Straughan and Walker, 1996a) for a detailed discussion of the numerical results. We do, however, include the figure 17.4.

The interpretation of figure 17.4 is that Ra_L achieves a minimum for β between 30° and 40°, a result in agreement with (Tyvand and Storesletten, 1991). This shows that in applying this theory to insulation techniques one is best with a parallel or orthogonal orientation of the fibres, since an oblique orientation always reduces the critical Rayleigh number and hence convection will begin sooner and heat transfer will be enhanced at lower Rayleigh number.

17.7 Radiation effects

(Larson, 2000; Larson, 2001) presents a very interesting study involving linear instability techniques and energy methods for a model which is more realistic for atmospheric convection than the standard Bénard one of section 3.2. He argues that the Bénard problem is deficient in several respects as a model to capture the convective motions which occur in the Earth's atmosphere. He essentially includes two new effects, these being a flux of radiative energy, and an adiabatic lapse rate. He argues that latent heating may be simulated by choosing an adiabatic lapse rate which is that of the climatological average. Thus, the model of (Larson, 2000; Larson, 2001) consists of fluid in the infinite horizontal region between the boundaries $z = 0$ and $z = 1$. The boundary conditions are those appropriate to stress free surfaces. The momentum equation is effectively that employed in Bénard convection, the effects of adiabatic lapse rate and radiation being included in the energy balance equation.

The equations of (Larson, 2000; Larson, 2001) are

$$\begin{aligned} &\chi(v_{i,t} + v_j v_{i,j}) = -\gamma p'_{,i} + \gamma T' k_i + \Delta v_i, \\ &v_{i,i} = 0, \\ &T_{,t} + v_i T_{,i} + \Gamma w = -F_{i,i} + \kappa \Delta T, \\ &\Big(\frac{1}{\alpha} F_{j,j}\Big)_{,i} - 3\alpha F_i = \frac{3}{4} T^4_{,i}. \end{aligned} \tag{17.114}$$

The new terms in these equations are the term Γ which is an adiabatic lapse rate, and F_i is the thermal radiative flux. The last equation in (17.114) is a radiative transfer equation.

(Larson, 2000; Larson, 2001) derives a steady state $\bar{v}_i = 0$, $\bar{F}(z)$, $\bar{T}(z)$, which satisfies $(17.114)_3$ and $(17.114)_4$. The basic temperature profile is, in general, a nonlinear function of z. He shows that with a suitable approximation to $(17.114)_4$ the perturbation fields (u_i, θ, π) (the flux is eliminated) satisfy the equations

$$\begin{aligned} &u_{i,t} + u_j u_{i,j} = -\frac{1}{\rho_*}\pi_{,i} + \gamma\theta k_i + \nu\Delta u_i, \\ &u_{i,i} = 0, \\ &\theta_{,t} + u_i\theta_{,i} = -(\Gamma + \bar{T}_{,z})w + \kappa\Delta\theta - 3\alpha\theta. \end{aligned} \tag{17.115}$$

(Larson, 2000; Larson, 2001) develops a detailed linear instability analysis for cases of $\kappa = 0$ and $\kappa \neq 0$. Many interesting numerical results are presented and these are related to the physical atmospheric convection problem. However, of particular interest to the current book are the nonlinear energy stability results (Larson, 2000; Larson, 2001) derives. He works with an energy of form

$$E = \frac{\chi}{2}\|\mathbf{u}\|^2 + \frac{\eta}{2}\|\theta\|^2,$$

for a coupling parameter η, where the norm is on $L^2(V)$, V being a period cell for the disturbance.

The energy equation of (Larson, 2001) has form

$$\chi\frac{dE}{dt} = \sqrt{\gamma} I - D,$$

with

$$I =< (1 - \eta \bar{T}_{,z} - \eta \Gamma) w\theta >,$$
$$D = \|\nabla \mathbf{u}\|^2 + \kappa\eta \|\nabla\theta\|^2 + 3\eta < \alpha\theta^2 > .$$

Here $< \cdot >$ denotes integration over V. His nonlinear energy stability threshold γ_E is defined as $\gamma_E^{-1/2} = \max_H (I/D)$. He calculates the Euler-Lagrange equations for this and shows γ_E is indeed a threshold for unconditional nonlinear stability.

The results of (Larson, 2000; Larson, 2001) are very sharp. In the limit $\kappa \to 0$ the linear instability threshold coincides with the nonlinear energy stability one and so no subcritical instabilities can exist. For $\kappa \neq 0$, the two boundaries are close and the nonlinear results represent valuable information. Indeed, the papers of (Larson, 2000; Larson, 2001) represent a valuable contribution to the extension of the Bénard problem as applied to convective atmospheric physics.

17.7.1 *An optically thin model*

Another very interesting nonlinear energy stability analysis for a thermal convection model with radiation is due to (Neitzel et al., 1994). These writers treat an optically thin layer and use careful bounding arguments to derive useful unconditional nonlinear stability results.

Their model involves a fluid in the layer $\mathbb{R}^2 \times \{z \in (0, d)\}$ with lower plane hotter than the upper, both planes at constant temperatures, $T(z = 0) = T_H$, $T(z = d) = T_C$. The equations governing the model of (Neitzel et al., 1994) are

$$\begin{aligned} v_{i,t} + v_j v_{i,j} &= -\frac{1}{\rho_C} p_{,i} + \nu\Delta v_i - \frac{\rho}{\rho_C} g k_i, \\ v_{i,i} &= 0, \\ T_{,t} + v_i T_{,i} &= \kappa\Delta T - \sigma'\big[T^4 - \frac{1}{2}(T_C^2 + T_H^2)\big], \end{aligned} \tag{17.116}$$

where σ' is a constant and ρ is given by $\rho = \rho_C[1 - \beta(T - T_C)]$. These equations are believed appropriate to model convection in a layer of optically thin fluid subject to radiative heat transfer. In this case the radiation effect is felt through the σ' term in (17.116).

The steady state temperature for this problem is not linear in z. (Neitzel et al., 1994) calculate this numerically. Their non-dimensional perturbation equations are

$$\begin{aligned} Pr^{-1}(u_{i,t} + u_j u_{i,j}) &= -\pi_{,i} + \Delta u_i + R\theta k_i, \\ u_{i,i} &= 0, \\ \theta_{,t} + u_i\theta_{,i} &= \ R\bar{T}_{,z} w + \Delta\theta - N\Psi, \end{aligned} \tag{17.117}$$

where R^2 is the Rayleigh number, N is a non-dimensional constant, and Ψ is a nonlinear function of θ given by

$$\Psi = R^{-3}\theta^4 + \frac{4}{R}\Big(\bar{T} + \frac{T_C}{\Delta T}\Big)\theta^3 + \frac{6}{R}\Big(\bar{T} + \frac{T_C}{\Delta T}\Big)^2\theta^2 + 4\Big(\bar{T} + \frac{T_C}{\Delta T}\Big)^3\theta.$$

Here ΔT is the temperature difference $T_H - T_C$.

The nonlinear energy analysis of (Neitzel et al., 1994) defines $E(t) = \|\mathbf{u}\|^2/2Pr + \lambda\|\theta\|^2/2$ and derives an energy identity of form

$$\frac{dE}{dt} = -D + RI - \lambda N(\theta, \Psi). \tag{17.118}$$

The production term I and dissipation D are given by

$$I = < (1 - \lambda\bar{T}_{,z})w\theta >, \qquad D = \|\nabla\mathbf{u}\|^2 + \lambda\|\nabla\theta\|^2.$$

They employ the reformulated energy theory of (Davis and von Kerczek, 1973) but they interestingly handle the nonlinear term in (θ, Ψ) to obtain unconditional nonlinear stability. To do this they rewrite $\theta\Psi$ as

$$\theta\Psi = \frac{\theta^2}{R^3}\,[x^3 + \zeta Rx^2 + \zeta^2 Rx + \zeta^3 R^3],$$

for x and ζ suitably chosen. They show the term in square brackets is non-negative. They then find a lower bound for this term.

Many detailed numerical results are provided by (Neitzel et al., 1994) for their model, and closeness of the nonlinear energy thresholds to the linear instability ones is achieved. Thus, the nonlinear energy theory of (Neitzel et al., 1994) represents a highly useful contribution to convection with radiative heat transfer.

We close this section by mentioning two other related areas of work which do not specifically deal with energy stability but are, nevertheless, connected and very interesting. The first is the work of (Ghosal and Spiegel, 1991). This is for convection in a compressible fluid but the energy balance equation may have a heat/mass source. This model may be applied to a He^3 layer in the Sun's core. The second topic involves papers of (Zhang and Schubert, 2000; Zhang and Schubert, 2002) which derive extremely interesting numerical results for convection in rotating stratified spherical layers. They extensively investigate a situation where an inner spherical fluid layer which is convectively unstable is bounded by a corotating outer stable spherical layer. Convective motions may penetrate into the stable layer. This model is of much interest because it may apply to fluid motions seen on the surface of planets such as Jupiter.

17.8 Krishnamurti's model

(Krishnamurti, 1997) is a very interesting contribution to the theory of penetrative convection. She starts with the premise that internal heating in a fluid layer can induce convective overturning even when the steady state is stably stratified. Thus, she wants to have a situation where a fluid layer has a species dissolved in it and the density gradient due to the species is stabilizing. Furthermore, the temperature at the lower plane can even be greater than that at the upper. Thus, both salt and temperature effects are stabilizing and any instability will be caused by a penetrative effect due to internal heating. To illustrate this she performs a beautiful experiment with a layer of water heated above and cooled below. In the water is dissolved the chemical thymol blue. This colours the water yellow when the pH is low and blue for a high pH. The colouration helps to visualize the experiment. She applies a DC voltage across the layer and hydroxyl ions form near the bottom of the layer which turn the fluid blue. Internal heating is produced by illuminating from above with a sodium lamp which generates radiation in the layer. The radiation is only absorbed readily by the blue fluid which in turn warms. This results in the blue fluid rising with falling yellow fluid and a convective motion ensues.

In addition (Krishnamurti, 1997) develops a very appealing theoretical model for her experiment and this assumes the heat source is a function of the species concentration. She additionally includes a linear instability analysis and a weakly nonlinear analysis to analyse theoretically the form of convection patterns. Her theoretical work assumes two stress free surfaces and employs an approximation to avoid extensive numerical computation.

(Straughan, 2002b) further investigates the model of (Krishnamurti, 1997) but takes the boundaries to be fixed. [1] He performs a linear instability analysis and complementary nonlinear energy stability analysis. His analysis confirms that the model of (Krishnamurti, 1997) is a very effective one.

The theory of (Krishnamurti, 1997) uses the momentum, conservation of mass, and energy balance equations for a linear viscous fluid,

$$\frac{\partial v_i}{\partial t} + v_j \frac{\partial v_i}{\partial x_j} = -\frac{1}{\rho_0}\frac{\partial p}{\partial x_i} - k_i g\,\frac{\rho}{\rho_0} + \nu\Delta v_i, \tag{17.119}$$

$$\frac{\partial v_i}{\partial x_i} = 0, \tag{17.120}$$

$$\frac{\partial T}{\partial t} + v_i \frac{\partial T}{\partial x_i} = \kappa\Delta T + \frac{Q}{\rho_0 c_p}, \tag{17.121}$$

where Q is an internal heat source. In addition she needs an equation for the thymol blue concentration, of form

$$\frac{\partial C}{\partial t} + v_i \frac{\partial C}{\partial x_i} = \kappa_C \Delta C\,. \tag{17.122}$$

[1] Some of the material in this section is reprinted from Dynamics of Atmospheres and Oceans, Vol. 35, B. Straughan, Global stability for convection induced by absorption of radiation, pp. 351-361, Copyright (2002), with permission from Elsevier Science.

In these equations v_i, p, T, C, are velocity, pressure, temperature, and concentration, $\mathbf{k} = (0,0,1)$, g is gravity, ν, κ, κ_C are kinematic viscosity, thermal diffusivity, and thymol blue diffusivity, ρ_0 is a constant density, and c_p is the specific heat at constant pressure of the fluid. Constitutive equations are needed for the density in (17.119) and for the heat source in (17.121) and those chosen by (Krishnamurti, 1997) are

$$\rho = \rho_0(1 - \alpha T), \qquad Q = \rho_0 c_p \beta C. \tag{17.123}$$

Here β is a constant, and α is the coefficient of thermal expansion. This means that the density in the buoyancy force does not depend on the thymol blue concentration; however, the heat source is linear in this concentration.

A linear instability and complementary nonlinear global analysis for the Krishnamurti model with (17.123) is contained in (Straughan, 2002b).

In this section we develop the analysis for a linear instability and a nonlinear energy stability study for an alternate model for Krishnamurti's problem. We commence with the basic governing equations (17.119) - (17.122) but employ a different constitutive theory from (17.123). We still take the heat source Q to be linear in C but now allow the density in the buoyanct force to depend linearly on both temperature, T, and thymol blue concentration, C. Thus, we assume

$$\rho = \rho_0(1 - \alpha T + \zeta C), \quad \text{and} \quad Q = \rho_0 c_p \beta C. \tag{17.124}$$

The buoyancy force must to a certain extent be influenced by the thymol blue concentration, and so (17.124) will certainly be more accurate. It may be that in practice the heat source is a nonlinear function of C as opposed to $(17.124)_2$. We could develop our analysis with a polynomial function of C, but how one then develops an unconditional nonlinear energy stability analysis is not clear. With (17.124), equations (17.119) - (17.122) become

$$\begin{aligned}
&\frac{\partial v_i}{\partial t} + v_j \frac{\partial v_i}{\partial x_j} = -\frac{1}{\rho_0}\frac{\partial p}{\partial x_i} - k_i g(1 - \alpha T + \zeta C) + \nu \Delta v_i,\\
&\frac{\partial v_i}{\partial x_i} = 0,\\
&\frac{\partial T}{\partial t} + v_i \frac{\partial T}{\partial x_i} = \kappa \Delta T + \beta C,\\
&\frac{\partial C}{\partial t} + v_i \frac{\partial C}{\partial x_i} = \kappa_C \Delta C.
\end{aligned} \tag{17.125}$$

To model the (Krishnamurti, 1997) experiment we assume that the fluid occupies the layer $\mathbb{R}^2 \times \{z \in (0,d)\}$, i.e. (17.125) hold on $\mathbb{R}^2 \times (0,d) \times \{t > 0\}$. On the boundaries $z = 0, d$, the temperature is prescribed with constant values T_L and T_U, respectively. The concentration of thymol blue on the upper boundary is zero while that at the lower boundary $z = 0$ takes the constant value C_L. Since the fluid is viscous the velocity is zero at the boundaries, so $v_i = 0$ at $z = 0, d$. The steady solution $(\bar{v}_i, \bar{T}, \bar{C})$ which satisfies these conditions is,

$$\bar{v}_i \equiv 0, \qquad \bar{C} = (d-z)\,\frac{C_L}{d}\,,$$
$$\bar{T} = -\frac{\beta C_L}{d\kappa}\Big(\frac{dz^2}{2} - \frac{z^3}{6}\Big) + \Big[-\frac{\Delta T}{d} + \frac{\beta C_L d}{3\kappa}\Big]\,z + T_L. \tag{17.126}$$

The steady pressure $\bar{p}$ may be found from $(17.125)_1$. Perturbations (u_i, θ, ϕ, π) to $(\bar{v}_i, \bar{T}, \bar{C}, \bar{p})$ are sought and then equations for these are derived from (17.125) recalling (17.126). We put $\Delta T = H|T_L - T_U|$, where $H = \mathrm{sgn}\,(T_L - T_U)$ and non-dimensionalize with the time, velocity, pressure, temperature and concentration scales as $\mathcal{T} = d^2/\kappa$, $U = \kappa/d$, $P = \nu U \rho_0/d$, $T^\sharp = U\sqrt{\nu |T_L - T_U|/g\alpha d\kappa}$ and $C^\sharp = C_L$. The Rayleigh number is defined as $Ra = R^2 = g\alpha d^3|T_L - T_U|/\kappa\nu$ and parameters η and γ are introduced as $\eta = \kappa_C/\kappa$, $\gamma = \beta C_L d^2/\kappa|T_L - T_U|$.

Then the nonlinear perturbation equations may be non-dimensionalized and written as,

$$\begin{aligned}
&Pr^{-1}(u_{i,t} + u_j u_{i,j}) = -\pi_{,i} + \Delta u_i + k_i R\theta - k_i R_S^2 \phi,\\
&u_{i,i} = 0,\\
&\theta_{,t} + u_i \theta_{,i} = RF(z)w + \gamma R\phi + \Delta\theta,\\
&\phi_{,t} + u_i \phi_{,i} = w + \eta\Delta\phi.
\end{aligned} \tag{17.127}$$

These equations hold on $\{(x,y) \in \mathbb{R}^2\} \times \{z \in (0,1)\} \times \{t > 0\}$. Furthermore, $Pr = \nu/\kappa$ is the Prandtl number, $R_S^2 = g\zeta C_L d^3/\kappa\nu$ is the thymol blue Rayleigh number, and the function $F(z)$ has form $F(z) = H - \gamma/3 + \gamma(z - z^2/2)$. Note that H takes the value +1 when the layer is heated from below and -1 when heated from above.

The boundary conditions are those for two fixed surfaces, so

$$u_i \equiv 0, \quad \theta = 0, \quad \phi = 0, \qquad z = 0, 1, \tag{17.128}$$

where in addition, u_i, π, θ and ϕ satisfy a plane tiling period planform in x and y. The boundary condition on ϕ is achieved in (Krishnamurti, 1997) by maintaining a voltage on the boundaries of the fluid layer.

If we write (17.127) in the abstract form of section 4.3 then the operator A is given by $A = \mathrm{diag}\,\{Pr^{-1}, Pr^{-1}, Pr^{-1}, 1, 1\}$ with the solution being (u, v, w, θ, ϕ). The linear operator L can be split into its symmetric and skew-symmetric parts L_S and L_A and here

$$L_S = \begin{pmatrix} \Delta & 0 & 0 & 0 & 0\\ 0 & \Delta & 0 & 0 & 0\\ 0 & 0 & \Delta & R(1+F)/2 & (1-R_S^2)/2\\ 0 & 0 & R(1+F)/2 & \Delta & R\gamma/2\\ 0 & 0 & (1-R_S^2)/2 & R\gamma/2 & \eta\Delta \end{pmatrix}$$

and

$$L_A = \begin{pmatrix} 0 & 0 & 0 & 0 & 0 \\ 0 & 0 & 0 & 0 & 0 \\ 0 & 0 & 0 & R(1-F)/2 & -(1+R_S^2)/2 \\ 0 & 0 & R(F-1)/2 & 0 & R\gamma/2 \\ 0 & 0 & (1+R_S^2)/2 & -R\gamma/2 & 0 \end{pmatrix}$$

This clearly demonstrates a lack of symmetry due to the 3-4, 3-5 and 4-5 elements in the matrix of the linear operator. Thus, the theory of section 4.3 does not apply and a nonlinear energy analysis will be very useful in delineating where possible subcritical instabilities may arise. In fact, the nonlinear energy stability analysis of (Straughan, 2002b), which is different and only applies when $R_S = 0$, yields unconditional nonlinear stability thresholds which are extremely close to the boundaries of linear theory.

The linearized instability equations which arise from (17.127) assume $u_i = e^{\sigma t}u_i(\mathbf{x})$, $\theta = e^{\sigma t}\theta(\mathbf{x})$, $\phi = e^{\sigma t}\phi(\mathbf{x})$, with a similar representation for π and then one shows the relevant linearized equations are

$$\begin{aligned} &\frac{\sigma}{Pr}\Delta w = R\Delta^*\theta - R_S^2\Delta^*\phi + \Delta^2 w, \\ &\sigma\theta = RF(z)w + \gamma R\phi + \Delta\theta, \\ &\sigma\phi = w + \eta\Delta\phi, \end{aligned} \tag{17.129}$$

where $\Delta^* = \partial^2/\partial x^2 + \partial^2/\partial y^2$. Normal modes are introduced, $w = W(z)f(x,y), \theta = \Theta(z)f(x,y), \phi = \Phi(z)f(x,y)$, with f a horizontal planform satisfying $\Delta^* f = -a^2 f$, where a is the wavenumber of the disturbance. Equations (17.129) with $R_S = 0$ are solved numerically for many parameter values in (Straughan, 2002b). We stress that this is different from the case considered here where $R_S \neq 0$. We do not include details of the numerical computations of (Straughan, 2002b) here but we observe that the numerical findings are very much in agreement with the work of (Krishnamurti, 1997) who suggests that the critical Rayleigh number, Ra, is proportional to η/γ, and the critical wavenumber is $\sqrt{2/3}$ times larger than that of classical Bénard convection.

To study nonlinear energy stability for system (17.127), let V be a period cell for the disturbance, multiply $(17.127)_1$ by u_i, $(17.127)_3$ by θ, and $(17.127)_4$ by ϕ. Integrate over V and one finds

$$\frac{d}{dt}\frac{1}{2Pr}\|\mathbf{u}\|^2 = R(\theta,w) - R_S^2(\phi,w) - \|\nabla\mathbf{u}\|^2, \tag{17.130}$$

$$\frac{d}{dt}\frac{1}{2}\|\theta\|^2 = R(F\theta,w) + \gamma R(\phi,\theta) - \|\nabla\theta\|^2, \tag{17.131}$$

$$\frac{d}{dt}\frac{1}{2}\|\phi\|^2 = (w,\phi) - \eta\|\nabla\phi\|^2. \tag{17.132}$$

These separate energy identities lead to an unconditional stability analysis. Let $\lambda_1, \lambda_2 (> 0)$ be coupling parameters. Define the energy functional $E(t)$

by

$$E(t) = \frac{1}{2Pr}\,\|\mathbf{u}\|^2 + \frac{\lambda_1}{2}\,\|\theta\|^2 + \frac{\lambda_2}{2}\,\|\phi\|^2. \tag{17.133}$$

The production term I and dissipation term D are

$$I = R\big([1+\lambda_1 F]w,\theta\big) + \lambda_1\gamma R(\phi,\theta) + (\lambda_2 - R_S^2)(w,\phi), \tag{17.134}$$

$$D = \|\nabla\mathbf{u}\|^2 + \lambda_1\|\nabla\theta\|^2 + \lambda_2\eta\|\nabla\phi\|^2. \tag{17.135}$$

Equations (17.130) - (17.132) lead to the energy equation

$$\frac{dE}{dt} = I - D. \tag{17.136}$$

We define R_E by

$$\frac{1}{R_E} = \max_H \frac{I}{D} \tag{17.137}$$

with H being the space of admissible solutions, and one shows

$$\frac{dE}{dt} \le -D\Big(1 - \frac{1}{R_E}\Big). \tag{17.138}$$

By using Poincaré's inequality, $D \ge cE$, and then for $R_E > 1$ one shows

$$\frac{dE}{dt} \le -\Big(\frac{R_E - 1}{R_E}\Big)\,cE. \tag{17.139}$$

Integration of this leads to

$$E(t) \le E(0)\exp\left[-(R_E - 1)ct/R_E\right]$$

and this in turn yields unconditional nonlinear stability.

The Euler-Lagrange equations which arise from the maximum problem (17.137) are at the threshold $R_E = 1$,

$$\begin{aligned} &2\Delta u_i + R(1+\lambda_1 F)\theta k_i + (\lambda_2 - R_S^2)\phi k_i = \delta_{,i}\,, \qquad u_{i,i} = 0,\\ &2\lambda_1\Delta\theta + R(1+\lambda_1 F)w + \lambda_1\gamma R\phi = 0,\\ &2\lambda_2\eta\Delta\phi + \lambda_1\gamma R\theta + (\lambda_2 - R_S^2)w = 0, \end{aligned} \tag{17.140}$$

where δ is a Lagrange multiplier. Equations (17.140) yield an eighth order eigenvalue problem for the nonlinear critical Rayleigh number Ra_E and then one performs the optimization

$$Ra_E = \max_{\lambda_1,\lambda_2}\ \min_{a^2}\ R^2(a^2;\lambda_1,\lambda_2). \tag{17.141}$$

The analogous problem to (17.140) and (17.141) when $R_S = 0$ is solved numerically in (Straughan, 2002b). We refer to this paper for details. However, we do note that the nonlinear energy results found there are very close to the linear instability ones. Furthermore, to achieve this requires careful selection of λ_1 and λ_2. These parameters may vary by $O(10^3)$.

The conclusions drawn in (Straughan, 2002b) are essentially that the results of linear instability theory show that the (Krishnamurti, 1997) relation $Ra = k\eta/\gamma$ for the instability threshold is reasonably accurate even when fixed boundaries are considered. Also, the results of unconditional nonlinear energy stability analysis which guarantee global stability when the Rayleigh number is below this are so close to those of linear theory as to allow one to deduce that linear instability theory will be sufficiently accurate to predict the onset of convective motion.

We point out that the system studied by (Straughan, 2002b) is different from (17.140). The presence of the terms in R_S will make a significant difference. For example, (Straughan, 2002b) and (Krishnamurti, 1997) both find the growth rate σ of the linearized eigenvalue problem is real. When $R_S \neq 0$ then (17.129) are expected to lead to some parameter situations for which $\sigma \in \mathbb{C}$. We can be sure of this because in the heated below - salted below problem of section 14.1 it is known that σ may be complex. When $\gamma = 0$, (17.129) and (17.140) reduce to the equations covering that scenario. By continuity arguments we expect $\sigma \in \mathbb{C}$ for (17.129) in some cases, at least for small γ. Thus an energy analysis and the numerical results one will achieve by solving (17.140) should prove revealing, especially when compared critically with the linear instability results one will derive from solving (17.129) numerically.

18
Nonlinear stability in ocean circulation models

The purpose of this chapter is to review and examine work on nonlinear energy stability theory applied to models for circulation within the oceans. This is a highly applicable subject and one which may lead to bounds which could be of use in environmental engineering. Since it is believed thermohaline circulation in the ocean may have an effect on climate change and even global warming, cf. (Clark et al., 2002), any useful thresholds energy stability bounds can yield will be welcome.

In this chapter we restrict attention to work on energy stability theory in ocean models and, in particular, to the work of (Lebovich and Paolucci, 1980; Lebovich and Paolucci, 1981), (Crisciani and Mosetti, 1990; Crisciani and Mosetti, 1991; Crisciani and Mosetti, 1992; Crisciani and Mosetti, 1994), (Crisciani and Purini, 1995; Crisciani and Purini, 1996), (Crisciani et al., 1994; Crisciani et al., 1995), and (Crisciani, 1998). Our aim is to describe this work involving energy stability theory and so we do not dwell on the models themselves. However, this is a very important part of the problem and very good expositions of ocean circulation models in general may be found in the books by (Mellor, 1996), (Pedlosky, 1996).

18.1 Langmuir circulations

This section is based on interesting work of (Lebovich and Paolucci, 1980; Lebovich and Paolucci, 1981), who developed linear instability and nonlinear energy stability analyses of models for so-called Langmuir circulations in an ocean.

The paper of (Lebovich and Paolucci, 1981) concentrates on two-dimensional disturbances whereas that of (Lebovich and Paolucci, 1980) deals with three-dimensional ones. We here devote attention to the work of (Lebovich and Paolucci, 1981). This paper investigates the stability of wind driven convective motions in the upper layers of the ocean, known as Langmuir circulations. From the mathematical viewpoint they employ a half space $z < 0$ to be the domain of the (infinitely deep) ocean. The basic temperature profile $\bar{T}(z)$ is linearly increasing. Due to the wind, a drift, $\mathbf{u}_s$, is imposed in the $x-$direction. This has form $\mathbf{u}_s = U_s(z)\mathbf{i}$ although for specific calculations (Lebovich and Paolucci, 1981) adopt the form $U_s(z) = 2e^{2z}$. The dimensionless equations of (Lebovich and Paolucci, 1981) are then

$$\begin{aligned} v_{i,t} + v_j v_{i,j} &= U_s u_{,i} + Ri\,\theta\delta_{i3} - p_{,i} + La\Delta v_i, \\ v_{i,i} &= 0, \\ \theta_{,t} + v_i\theta_{,i} &= -v_i\delta_{i3} + \frac{La}{Pr}\,\Delta\theta. \end{aligned} \tag{18.1}$$

In these equations v_i is the velocity, after adjusting for the drift velocity $\mathbf{u}_s$, θ is the temperature perturbation to $\bar{T}$, $\mathbf{v} = (u, v, w)$, Pr is the Prandtl number, and Ri, La are non-dimensional numbers given specifically in (Lebovich and Paolucci, 1981).

To equations (18.1) must be added the boundary conditions

$$\begin{aligned} & v_i(\mathbf{x}, t) \to 0 \text{ as } z \to -\infty, \quad \forall t < \infty, \\ & \frac{\partial}{\partial z}(\mathbf{k} \times \mathbf{v}) = (0, H(t), 0), \quad \mathbf{k}.\mathbf{v} = 0, \quad \text{on } z = 0, \\ & \theta = 0 \text{ at } z = 0, \quad \theta \to 0 \text{ as } z \to -\infty, \end{aligned} \tag{18.2}$$

where $\mathbf{k} = (0, 0, 1)$ and $H(t)$ is the Heaviside function, and the initial conditions

$$v_i(\mathbf{x}, 0) = 0, \ \theta(\mathbf{x}, 0) = 0. \tag{18.3}$$

There is a basic space-time dependent solution to this problem found by (Lebovich, 1977) and this is

$$\begin{aligned} \bar{\mathbf{v}} &= (U, 0, 0) = 2\sqrt{\frac{tLa}{\pi}}\left[\exp(-\eta^2) - \eta\,\mathrm{erfc}(\eta)\sqrt{\pi}\right]\mathbf{i}, \\ \bar{\theta} &= 0, \end{aligned} \tag{18.4}$$

where the basic pressure gradient has form $\nabla\bar{p} = U_s\nabla U$ and the variable η is given by $\eta = -z/2\sqrt{tLa}$. The papers of (Lebovich and Paolucci, 1980; Lebovich and Paolucci, 1981) are occupied with an investigation of the nonlinear stability and instability of (18.4).

To study stability (Lebovich and Paolucci, 1981) introduce perturbations (u_i, ϕ, π) to (18.4) and the perturbation equations for the perturbation velocity u_i, and perturbation temperature ϕ are then

$$\begin{aligned} &u_{i,t} + u_j u_{i,j} = U_s u_{,i} - U'\delta_{i1} w + Ri\,\phi\delta_{i3} - \pi_{,i} + La\Delta u_i, \\ &u_{i,i} = 0, \\ &\phi_{,t} + u_i \phi_{,i} = -w + \frac{La}{Pr}\Delta\phi, \end{aligned} \tag{18.5}$$

where a prime denotes $\partial/\partial z$ and $\mathbf{u} = (u, v, w)$. The perturbation boundary conditions are

$$\begin{aligned} &\mathbf{k}.\mathbf{u} = (\mathbf{u} \times \mathbf{k})' = \phi \text{ on } z = 0, \\ &\mathbf{u} \to \mathbf{0},\ \phi \to 0 \text{ as } z \to -\infty, \end{aligned} \tag{18.6}$$

with u_i given at $t = 0$ and $\phi = 0$ at $t = 0$.

We stress that the paper of (Lebovich and Paolucci, 1981) develops a detailed linearised instability analysis for (18.5), (18.6). Since the basic solution U is a function of both z and t this is non-trivial. It is, however, very necessary for comparison with the nonlinear energy stability results. We concentrate here only on the energy theory. (Lebovich and Paolucci, 1981) consider two-dimensional spatial disturbances in the sense that (u, v, w, ϕ, π) are functions of y and z only. The three-dimensional case and associated computations are reported in (Lebovich and Paolucci, 1980). We do note, however, that the results of (Lebovich and Paolucci, 1981) are sharp when compared to those of linearized instability theory. (Lebovich and Paolucci, 1981) decompose u_i as $\mathbf{u} = u\mathbf{i} + \mathbf{w}$ where $\mathbf{w} = v\mathbf{j} + w\mathbf{k}$. Thus, if we denote $w_\alpha \equiv (v, w)$, they work with three energy equations derived from (18.5), (18.6). These are equations for u^2, $w_\alpha w_\alpha$, and ϕ^2. These equations have form

$$\frac{d}{dt}\frac{1}{2}\int_V w_\alpha w_\alpha dV = Ri\int_V w\phi dV - \int_V U_s' uw\,dV - La\int_V w_{\alpha,i} w_{\alpha,i} dV\,,$$

$$\frac{d}{dt}\frac{1}{2}\int_V u^2 dV = -\int_V U' uw\,dV - La\int_V u_{,i} u_{,i} dV\,,$$

$$\frac{d}{dt}\frac{1}{2}\int_V \phi^2 dV = -\int_V w\phi\,dV - \frac{La}{Pr}\int_V \phi_{,i}\phi_{,i} dV\,,$$

where V is the half-space $z < 0$, and we continue to use subscript $,i$ for spatial differentiation even though the $x-$dependence is neglected.

(Lebovich and Paolucci, 1981) introduce two coupling parameters λ_1 and λ_2 and work with the energy functional

$$E(t) = \frac{1}{2}\lambda_1\int_V u^2 dV + \frac{1}{2}\int_V w_\alpha w_\alpha dV + \lambda_2 Ri\int_V \phi^2 dV.$$

This leads then to the energy equation

$$\frac{dE}{dt} = I - LaD, \tag{18.7}$$

where the production and dissipation terms are given by

$$I = -\int_V (U_s' + \lambda_1 U')uw\,dV - Ri(\lambda_2 - 1)\int_V w\phi\,dV,$$

$$D = \lambda_1 \int_V u_{,i}u_{,i}dV + \int_V w_{\alpha,i}w_{\alpha,i}dV + \frac{\lambda_2 Ri}{Pr}\int_V \phi_{,i}\phi_{,i}dV.$$

The nonlinear energy stability analysis of (Lebovich and Paolucci, 1981) is complicated by two factors. Firstly, the maximum of I/D depends also on time, but secondly, since the domain V is a half space care must be taken to ensure the maximum exists and in the subsequent treatment of decay of $E(t)$. Thus (Lebovich and Paolucci, 1981) define

$$La_\tau(t) = \max_H \frac{I}{D}$$

where H is the space of admissible solutions and they determine

$$La_G = \max_t La_\tau(t).$$

They refer to work of Galdi and Rionero on the half space maximum problem, see e.g. chapter 5, and from (18.7) they write

$$\frac{dE}{dt} = -LaD\Big(1 - \frac{I}{LaD}\Big).$$

They then deduce that

$$E(t) + La\int_0^t \Big(1 - \frac{I}{LaD}\Big)D(s)ds = E(0).$$

They thus base their nonlinear stability criterion on the condition

$$0 \le \int_0^\infty D(s)ds < \frac{E(0)}{(La - La_G)}.$$

This shows that for $0 < La < La_G$, the function $D(t) \in L^1(0,\infty)$. This by itself does not guarantee $D \to 0$ as $t \to \infty$ (consider e.g. the series $\sum_{n=1}^\infty n^{-2} < \infty$; any function $D(t)$ lying below $\sum_{n=1}^\infty n^{-2}$ is integrable but need not decay). However, it is some measure of nonlinear stability.

(Lebovich and Paolucci, 1981) develop the Euler-Lagrange equations for their max I/D problem and solve these numerically. The numerical results found by (Lebovich and Paolucci, 1981) are displayed in their paper for various values of Ri believed to be realistic and they give very useful thresholds when compared to the linearized (time-dependent) instability results.

18.2 Stommel-Veronis (quasigeostrophic) model

In a series of papers Crisciani and his co-workers have made progress with nonlinear energy stability criteria for a variety of ocean circulation models, these being mostly of quasigeostrophic type, see (Crisciani and Mosetti, 1990; Crisciani and Mosetti, 1991; Crisciani and Mosetti, 1992; Crisciani and Mosetti, 1994), (Crisciani et al., 1994; Crisciani et al., 1995), (Crisciani and Purini, 1995; Crisciani and Purini, 1996), (Crisciani, 1998). They basically study a two-dimensional circulation model in a closed basin (i.e. a finite two-dimensional spatial domain), denoted by D. For practical calculations this may be taken as the non-dimensional region $[0, \pi]^2$.

The first quasigeostrophic model we investigate is that of Stommel-Veronis, see (Crisciani et al., 1994). If ψ denotes the streamfunction of the two-dimensional motion, $u = \psi_y$, $v = -\psi_x$, where (u, v) is the velocity, then the governing equations of the Stommel-Veronis model are

$$\frac{\partial}{\partial t}\Delta\psi + RJ(\psi, \Delta\psi) + \frac{\partial\psi}{\partial x} = (\nabla \times \boldsymbol{\tau})_z - \epsilon\Delta\psi, \tag{18.8}$$

where Δ is the two-dimensional Laplacian, R, ϵ are non-dimensional parameters, J is the Jacobian, and $\nabla \times \boldsymbol{\tau}$ is a term associated with externally supplied wind stress, see (Pedlosky, 1996), eq. (1.2.9). This equation is developed and discussed in detail in chapter 2 of (Pedlosky, 1996). (Crisciani et al., 1994) study (18.8) on a bounded two-dimensional spatial domain D with $t > 0$, subject to the boundary conditions

$$\psi = 0 \quad \text{on } \partial D, \tag{18.9}$$

∂D being the boundary of D.

The solution whose stability is investigated is a steady solution, ψ_0, to (18.8), (18.9), so ψ_0 satisfies

$$\begin{aligned} &RJ(\psi_0, \Delta\psi_0) + \frac{\partial\psi_0}{\partial x} = (\nabla \times \boldsymbol{\tau})_z - \epsilon\Delta\psi_0, \quad \mathbf{x} \in D, \\ &\psi_0 = 0, \quad \mathbf{x} \in \partial D. \end{aligned} \tag{18.10}$$

For ϕ a non-dimensional perturbation to ψ_0, it is convenient to define q and Q by $q = \Delta\phi$, $Q = \Delta\psi_0 + y/R$. Then the boundary initial value problem for ϕ is

$$\begin{aligned} &\frac{\partial\Delta\phi}{\partial t} + RJ(\psi_0, q) + RJ(\phi, Q) \\ &\qquad + RJ(\phi, q) + \epsilon\Delta\phi = 0, \quad \mathbf{x} \in D,\, t > 0, \\ &\phi = 0, \quad \mathbf{x} \in \partial D. \end{aligned} \tag{18.11}$$

The energy stability analysis of (Crisciani et al., 1994) employs the energy functional

$$E(t) = \frac{1}{2}\|\Delta\phi\|^2 + \frac{a}{2}\|\nabla\phi\|^2,$$

where $\|\cdot\|$ is the L^2 norm on D, and $a > 0$ is a coupling parameter. They derive the energy equation

$$\frac{dE}{dt} = -R\int_D qu_i(Q - a\psi_0)_{,i}dA - \epsilon\|\Delta\phi\|^2 - \epsilon a\|\mathbf{u}\|^2, \qquad (18.12)$$

where $u_i \equiv (u, v) = (\phi_y, -\phi_x)$ represents the perturbation velocity field. We note in passing that $u_i u_i = |\nabla\phi|^2$.

To proceed from (18.12) these writers use the arithmetic-geometric mean inequality to write

$$\int_D qu_i(Q - a\psi_0)_{,i}dA \leq \mu(a)\|q\|\,\|\mathbf{u}\|, \qquad (18.13)$$

where $\mu(a) = \max_D|\nabla(Q - a\psi_0)|$. (Crisciani et al., 1994) deduce stability from (18.12) by using (18.13) to derive

$$\frac{dE}{dt} \leq \mu(a)\|q\|\,\|\mathbf{u}\| - \epsilon\|q\|^2 - \epsilon a\|\mathbf{u}\|^2, \qquad (18.14)$$

and then optimizing the right hand side. This leads to the nonlinear energy stability criterion

$$R\max_D\left|\nabla\Big(\Delta\psi_0 + \frac{y}{R} - a\psi_0\Big)\right| < 2\epsilon\sqrt{a}. \qquad (18.15)$$

The presence of the coupling parameter a in (18.15) allows an optimal R against ϵ result to be achieved from this inequality.

(Crisciani et al., 1994) note that (Veronis, 1966b) derives an approximate solution to ψ_0 of form

$$\psi_0 = 2(1 - e^{-x/\epsilon})\sin y - \frac{2R}{\epsilon^3}xe^{-x/\epsilon}\sin 2y. \qquad (18.16)$$

They apply their criterion to this solution in the non-dimensional ocean basin $(0,\pi)^2$. They take realistic values of R, ϵ as $R = 10^{-6}$, $\epsilon = \pi\times 10^{-2}$ and then $R = 10^{-4}$, $\epsilon = \pi\times 10^{-2}$ and calculate the best value of a in (18.15). A parameter range of a is found to guarantee nonlinear energy stability of (18.16).

We observe that one can derive a nonlinear energy stability criterion directly using $\|\mathbf{u}\|^2 = \|\nabla\phi\|^2$. To do this multiply (18.11) by ϕ and integrate over D (arbitrary). If we use the fact that $\psi_0 = 0$ on ∂D so that $\partial\psi_0/\partial s = 0$ also on ∂D, $\partial/\partial s$ denoting the tangential derivative, then we may derive an energy equation of form

$$\frac{d}{dt}\frac{1}{2}\|\nabla\phi\|^2 = R\int_D(\nabla\psi_x^0\phi_y\nabla\phi - \nabla\psi_y^0\phi_x\nabla\phi)\,dA - \epsilon\|\nabla\phi\|^2. \qquad (18.17)$$

We quickly find that

$$\int_D(\nabla\psi_x^0\phi_y\nabla\phi - \nabla\psi_y^0\phi_x\nabla\phi)\,dA \leq \int_D\left\{\frac{1}{2}(|\psi_{xx}^0| + |\psi_{yy}^0|) + |\psi_{xy}^0|\right\}|\nabla\phi|^2\,dA.$$

Thus, from (18.17) a criterion for unconditional nonlinear stability (exponential decay of $\|\mathbf{u}\|^2$) is

$$R \max_D \left\{ \frac{1}{2}(|\psi^0_{xx}| + |\psi^0_{yy}|) + |\psi^0_{xy}| \right\} < \epsilon. \tag{18.18}$$

If we apply (18.18) to the Veronis solution (18.16) then one may show that nonlinear stability is guaranteed for $0 < \epsilon \leq \pi$ when

$$R\left[2 + \frac{4R}{\epsilon^2 e} + \frac{1}{\epsilon^2} + \frac{2R}{\epsilon^4} + \frac{R}{\epsilon^4 e} + \frac{2}{\epsilon} + \frac{4R}{\epsilon^3} + \frac{4R}{\epsilon^3 e}\right] < \epsilon,$$

or for $\epsilon > \pi$ when

$$R\left[2 + \frac{4R\pi}{\epsilon^3}e^{-\pi/\epsilon} + \frac{1}{\epsilon^2} + \frac{2R}{\epsilon^4} + \frac{R\pi}{\epsilon^5}e^{-\pi/\epsilon} + \frac{2}{\epsilon} + \frac{4R}{\epsilon^3} + \frac{4R\pi}{\epsilon^4}e^{-\pi/\epsilon}\right] < \epsilon.$$

Of course, these criteria will be crude and could easily be bettered by using (18.18) and (18.16) and maximizing over $D = (0, \pi)^2$.

We close this section by observing that to the best of our knowledge no one has developed a variational (i.e. I/D) analysis for the Stommel-Veronis model. This would involve solving a two-dimensional eigenvalue problem numerically, but the result would undoubtedly be well worth it.

18.3 Quasigeostrophic model

The term geostrophic arises because ocean models involve a momentum equation in a rotating frame. This gives rise to a geostrophic balance between the horizontal pressure gradient, the horizontal Coriolis acceleration, and a term representing turbulent (small scale) mixing in the ocean. This is explained succinctly by (Pedlosky, 1996), pp. 5-8. The model discussed in section 18.2 is already a quasigeostrophic model with an approximation. In this section we concentrate on work of (Crisciani et al., 1995) who develop a nonlinear energy stability analysis for a more general model than that of section 18.2.

If we let ψ be the streamfunction then the quasigeostrophic model studied by (Crisciani et al., 1995) has the governing non-dimensional equation

$$\begin{aligned}\left(\frac{\delta_I}{L}\right)^2 \left[\frac{\partial}{\partial t}(\Delta\psi - \underline{F\psi}) + J(\psi, \Delta\psi)\right] + \frac{\partial\psi}{\partial x}& \\ = (\nabla \times \boldsymbol{\tau})_z + \underline{\left(\frac{\delta_I}{L}\right)^2 \frac{1}{Re}\,\Delta^4\psi}.&\end{aligned} \tag{18.19}$$

Mathematically, the difference between (18.19) and the model (18.8) of section 18.2 lies in the presence of the underlined terms in (18.19). The F term will be stabilizing while the $\Delta^4\psi$ term is higher order than before, although still stabilizing. The coefficients δ_I, L, F, Re are positive constants.

(Crisciani et al., 1995) study (18.19) on the domain $D = (0, \pi)^2$ for $t > 0$. The boundary conditions they adopt are

$$\psi = 0 \text{ on } \partial D, \quad \nabla\psi = 0 \text{ for } x = 0, \pi, \quad \Delta\psi = 0 \text{ for } y = 0, \pi, \tag{18.20}$$

i.e. no slip on the lateral walls $x = 0, \pi$ and stress free boundaries at $y = 0, \pi$.

The steady state solution, ψ_0, whose stability is under investigation satisfies (18.19), (18.20) with the $\partial/\partial t$ term missing. Of course, ψ_0 is still a solution to a nonlinear partial differential equation.

A perturbation is introduced so that $\psi = \phi + \psi_0$ and then from (18.19), (18.20) ϕ is seen to satisfy the system

$$\begin{aligned}\left(\frac{\delta_I}{L}\right)^2\left[\frac{\partial}{\partial t}(\Delta\phi - F\phi) + J(\psi_0, q) + J(\phi, \Delta\psi_0) + J(\phi, q)\right]& \\ + \frac{\partial\phi}{\partial x} = \left(\frac{\delta_I}{L}\right)^2 \frac{1}{Re}\,\Delta^4\phi,& \qquad (18.21)\\ \phi = 0 \text{ on } \partial D, \quad \nabla\phi = 0 \text{ for } x = 0, \pi,& \\ \Delta\phi = 0 \text{ for } y = 0, \pi.&\end{aligned}$$

Here, as in section 18.2, $q = \Delta\phi$.

The nonlinear energy analysis of (Crisciani et al., 1995) multiplies $(18.21)_1$ by ϕ and integrates to derive the energy equation

$$\frac{d}{dt}\left(\frac{1}{2}\|\mathbf{u}\|^2 + \frac{1}{2}F\|\phi\|^2\right) = (\phi, J(\psi_0, q)) - \frac{1}{Re}\|q\|^2, \tag{18.22}$$

where $\|\mathbf{u}\| \equiv \|\nabla\phi\|$ and $(\cdot,\cdot)$ denotes the inner product on $L^2(D)$. (Crisciani et al., 1995) now bound the term (ϕ, J) by a variety of inequalities. They show that

$$(\phi, J(\psi_0, q)) = -\int_D q u_i^0 \phi_{,i} dA \equiv I,$$

where $\mathbf{u}^0$ is the base velocity corresponding to ψ_0. They then use Hölder's inequality to show

$$\begin{aligned} I \leq& \|q\|\, \|\mathbf{u}_0\|_4\, \|\nabla\phi\|_4 \\ \leq& 2^{1/4}\, \|\mathbf{u}_0\|_4\, \|q\|^2, \end{aligned} \tag{18.23}$$

where $\|\cdot\|_4$ denotes the norm on $L^4(D)$. From (18.23) and (18.22), (Crisciani et al., 1995) deduce nonlinear asymptotic energy stability provided

$$\int_D (\mathbf{u}_0 \cdot \mathbf{u}_0)^2 dA < \frac{1}{2Re^4}\,. \tag{18.24}$$

The quantity *Re* is a Reynolds number and so (18.24) is a testable criterion.

The paper of (Crisciani et al., 1995) employs an approximate solution to the steady state version of (18.19), (18.20), derived by (Munk, 1950).

They deduce that under appropriate conditions on the coefficients in the equations then nonlinear asymptotic stability of the Munk solution holds.

(Crisciani, 1998) develops the analyses of sections 18.2, 18.3 further and derives a nonlinear energy stability analysis for an ocean circulation model due to Niiler. This again leads to useful criteria delimiting a region of certain stability. We point out that we are unaware of any variational treatments of the maximum problem in the ocean circulation models discussed in sections 18.2, 18.3. This will certainly involve numerical computation of two-dimensional eigenvalue problems with a base solution which has to be determined numerically. However, the improved stability criteria to be expected are likely to be very worthwhile.

To close this chapter we mention interesting work of (Fukuta and Murakami, 1993; Fukuta and Murakami, 1995) who develop a clever nonlinear energy stability analysis of a solution, in a two-dimensional spatial domain, which they refer to as Kolmogorov flow. Their solution is again determined in terms of a streamfunction. Due to the boundary conditions the base solution is also time-dependent and a variational analysis is developed for the I/D maximum problem. The nonlinear and linear results of (Fukuta and Murakami, 1995) are developed in much detail.

19 Numerical solution of eigenvalue problems

19.1 The shooting method

The purpose of this chapter is to describe two very efficient methods for solving eigenvalue problems of the type encountered in linear and energy stability convection problems. The techniques referred to are the compound matrix method, which is simple to implement, and the Chebyshev tau technique. The chapter is intended to be a practical guide as to how to solve relevant eigenvalue problems. Several examples from fluid mechanics and porous convection are included. First we briefly describe a standard shooting method.

We begin at an elementary level and investigate the second order eigenvalue problem,

$$\begin{aligned}\frac{d^2u}{dx^2}+\lambda u=0, \qquad 0<x<1,\\ u(0)=u(1)=0.\end{aligned}\tag{19.1}$$

This problem arises in section 2.2.3, where it was seen that

$$\lambda_n=n^2\pi^2.\tag{19.2}$$

For stability studies it is often only the smallest eigenvalue that is of interest, i.e., λ_1.

To determine λ_1 numerically we retain $u(0)=0$, but replace $u(1)=0$ by a prescribed condition on $u'(0)$, a convenient choice being $u'(0)=1$, thereby converting the *boundary value problem* to an *initial value* one. Two values of λ_1 are selected, say $0 < \lambda_1^{(1)} < \lambda_1^{(2)}$, and then the initial value problem is integrated numerically to find $u_1(1)$, $u_2(1)$, where u_i denotes the solution corresponding to $\lambda_1^{(i)}$: (a high order, variable step Runge-Kutta-Verner technique is often adequate for the numerical integration). The idea is to use $u_1(1)$, $u_2(1)$ so found to ensure $u(1)$ is as close to zero as required by some pre-specified degree of accuracy. An iteration technique is then employed to find a sequence $u_k(1)$, corresponding to $\lambda_1^{(k)}$, such that

$$|u_k(1)| < \epsilon, \qquad k \text{ large enough}, \tag{19.3}$$

where ϵ is a user specified tolerance. I have found the secant method, see e.g., (Cheney and Kincaid, 1985) p. 97, is a suitable routine for this purpose. Once a $u_k(1)$ is determined to satisfy (19.3), $\lambda_1^{(k)}$ is then the required numerical estimate of the first eigenvalue to (19.1).

For practical purposes it is usually necessary to have some guide as to what values to select for $\lambda_1^{(1)}$, $\lambda_1^{(2)}$. However, in energy theory one can usually use linear stability theory as a guide for this. For linear stability theory it is often possible to use known results from related problems that can be solved analytically.

19.1.1 *A system: the Viola eigenvalue problem*

Of course, the eigenvalue problems encountered in convection are more complicated than that discussed in section 19.1, but the basic numerical method is the same. To illustrate how the shooting method works on a system we use an eigenvalue problem studied by (Viola, 1941); see also (Fichera, 1978). This is

$$\begin{aligned} &\frac{d^2}{dx^2}\Big[(1-\theta x)^3\frac{d^2u}{dx^2}\Big] - \lambda(1-\theta x)u = 0, \quad && x \in (0,1), \\ &u = \frac{d^2u}{dx^2} = 0, && x = 0,1, \end{aligned} \tag{19.4}$$

where the constant θ satisfies $0 \le \theta < 1$.

To solve (19.4) for the first (lowest) eigenvalue λ_1 does not appear possible analytically, in general, i.e., for $\theta \neq 0$. This problem is described in detail by (Fichera, 1978), pp. 41–43, who gives an exposition of the orthogonal invariants method for obtaining very accurate upper and lower bounds to λ_k, $k = 1, 2, \dots$. To solve (19.4) by a shooting method we first write it as a system of four first order differential equations in the vector $\mathbf{u} = (u, u', u'', u''')$, where $u' = du/dx$, $u'' = d^2u/dx^2$, etc. The boundary conditions at $x = 1$ on u, u'' are replaced in turn by $u' = 1$, $u''' = 0$, and then $u' = 0$, $u''' = 1$, at $x = 0$. The two initial value problems thereby obtained are then integrated numerically. Let the solution so found be written as a linear combination of the two solutions so obtained, say $\mathbf{u} = \alpha\mathbf{v} + \beta\mathbf{w}$. Then, the correct boundary condition $u = u'' = 0$ at $x = 1$ is imposed, and this requires

$$\det \begin{pmatrix} v & v'' \\ w & w'' \end{pmatrix} = 0 \tag{19.5}$$

to hold at $x = 1$.

While the shooting method is easy to understand and implement, it suffers from a serious drawback. This is that one has numerically to locate the zero of a determinant; e.g., in (19.5), one has to locate

$$v(1)w''(1) - w(1)v''(1) = 0. \tag{19.6}$$

The two quantities $v(1)w''(1)$ and $w(1)v''(1)$ must, therefore, be very close although neither need be close to zero (and generally will not). One is thus faced with subtracting two nearly identical quantities, and this can lead to very large round off errors and significant error build up during the solution of a convection eigenvalue problem. There are many ways to overcome this: we describe only one. This is the compound matrix technique, which has distinct advantages for energy eigenvalue problems. Basically the idea is to remove the troublesome location of the zero of a determinant by converting to a system of ordinary differential equations in the determinants themselves.

19.2 The compound matrix method

We begin this section with an accurate numerical calculation of eigenvalues to (19.4). For $0 \le \theta < 0.9$ accurate results are evidently easily found by the standard shooting method described in section 19.1. However, for the case where $\theta \to 1^-$ for which (19.4) becomes singular, other methods must be employed. Some useful information may be gleaned in this case with the compound matrix technique.

To solve (19.4) by the compound matrix method we let $\mathbf{U} = (u, u', u'', u''')^T$, and then suppose $\mathbf{U}_1$ and $\mathbf{U}_2$ are solutions to (19.4) with values at $x = 0$ of $(0, 1, 0, 0)^T$ and $(0, 0, 0, 1)^T$, respectively. A new six vector

$$\mathbf{Y} = (y_1, y_2, y_3, y_4, y_5, y_6)^T$$

is defined as the 2×2 minors of the 4×2 solution matrix whose first column is $\mathbf{U}_1$ and second $\mathbf{U}_2$. So,

$$\begin{aligned} y_1 &= u_1 u_2' - u_1' u_2, \\ y_2 &= u_1 u_2'' - u_1'' u_2, \\ y_3 &= u_1 u_2''' - u_1''' u_2, \\ y_4 &= u_1' u_2'' - u_1'' u_2', \\ y_5 &= u_1' u_2''' - u_1''' u_2', \\ y_6 &= u_1'' u_2''' - u_1''' u_2''. \end{aligned} \tag{19.7}$$

The 2×2 minors are the determinants we refer to at the end of section 19.1.1. The variable y_2 corresponds to the quantity in (19.6). With the compound matrix method we replace (19.6) by $y_2(1) = 0$ and thus avoid the problem of round off error due to subtraction.

By direct calculation from (19.4) the initial value problem for the y_i is found to be

$$\begin{aligned} y_1' &= y_2, \\ y_2' &= y_3 + y_4, \\ y_3' &= y_5 + 6\frac{\theta}{M}y_3 - 6\left(\frac{\theta}{M}\right)^2 y_2, \\ y_4' &= y_5, \\ y_5' &= y_6 + 6\frac{\theta}{M}y_5 - 6\left(\frac{\theta}{M}\right)^2 y_4 - \frac{\lambda}{M^2}y_1, \\ y_6' &= 6\frac{\theta}{M}y_6 - \frac{\lambda}{M^2}y_2, \end{aligned} \tag{19.8}$$

where $M = 1 - \theta x$. From the initial conditions on $\mathbf{U}_1$ and $\mathbf{U}_2$ we see that system (19.8) is to be integrated numerically subject to the initial condition

$$y_5(0) = 1 \tag{19.9}$$

and the final condition

$$y_2(1) = 0. \tag{19.10}$$

Again the zero in (19.10) is located to a pre-assigned degree of accuracy.

The numerical computations we report were all done on the University of Wyoming's CDC Cyber 760. The numerical results are very accurate in comparison with the available bounds of (Fichera, 1978), p. 43. For $\theta = 0.5$, Fichera finds

$$50.71623063 \leq \lambda_1 \leq 50.71623066, \qquad 838.2089 \leq \lambda_2 \leq 838.2091;$$

we find

$$\lambda_1 = 50.71623064799, \qquad \lambda_2 = 838.2090471111.$$

For $\theta = 0$, the exact values are $\lambda_1 = \pi^4$, $\lambda_2 = 16\pi^4$; we find

$$\lambda_1 = 97.40909103339, \qquad \lambda_2 = 1558.545456538.$$

Table 19.1 gives what we believe are accurate values of λ_1 for θ between 0 and 0.9. We have also calculated λ_2 for other values of θ, e.g.,

$$\begin{aligned} \theta &= 0.05, \qquad & \lambda_2 &= 1481.214398565, \\ \theta &= 0.1, \qquad & \lambda_2 &= 1405.075275206. \end{aligned}$$

What is clearly an evident fact is that the asymptotic behaviour of λ as $\theta \to 1^-$ appears to be a very interesting problem. The orthogonal invariants method of (Fichera, 1978) clearly needs a careful analysis for $\theta \sim 1^-$, since

λ_1	θ	λ_1	θ
97.40909	0	59.74500	0.4
92.55916	0.05	50.71623	0.5
87.75027	0.1	41.81693	0.6
82.98231	0.15	33.00917	0.7
78.25510	0.2	24.20406	0.8
73.56833	0.25	19.73541	0.85
68.92154	0.3	15.12813	0.9

Table 19.1. Numerical values of λ_1 for various θ.

terms like $\log(1-\theta)$ are involved. The numerical results *indicate* that λ_1 may approach 0 as $\theta \to 1^-$. The behaviour of λ_1 as $\theta \to 1^-$ is an open and interesting problem of analysis.

We have already noted that the compound matrix method is designed to avoid round off error, and this technique works well if the system of differential equations is stiff.

To describe the compound matrix method further we consider the general linear system

$$\begin{aligned} w'' &= \alpha_1 w' + \alpha_2 w + \alpha_3 \theta' + \alpha_4 \theta, \\ \theta'' &= \beta_1 w' + \beta_2 w + \beta_3 \theta' + \beta_4 \theta, \end{aligned} \tag{19.11}$$

where a prime denotes differentiation with respect to z, $\alpha_1, \dots, \alpha_4$, $\beta_1, \dots, \beta_4$, are known coefficients which may depend on z, they may be complex, and $z \in (0,1)$. One or more of the coefficients contains an eigenvalue, σ say. The boundary conditions we consider are

$$w = \theta = 0 \qquad \text{at} \qquad z = 0, 1. \tag{19.12}$$

Other boundary conditions are easily incorporated. System 19.11 is typical of the eigenvalue problems which occur in porous convection stability problems.

We introduce the variables $y_1, \dots, y_6$, being the 2×2 minors arising from w and θ, i.e.

$$\begin{aligned} y_1 &= w_1 w_2' - w_2 w_1', & y_4 &= w_1' \theta_2 - w_2' \theta_1, \\ y_2 &= w_1 \theta_2 - w_2 \theta_1, & y_5 &= w_1' \theta_2' - w_2' \theta_1', \\ y_3 &= w_1 \theta_2' - w_2 \theta_1', & y_6 &= \theta_1 \theta_2' - \theta_2 \theta_1'. \end{aligned} \tag{19.13}$$

The y_i variables satisfy the matrix equation

$$\mathbf{y}' = A\mathbf{y}, \tag{19.14}$$

where A is the 6×6 matrix

$$A = \begin{pmatrix} \alpha_1 & \alpha_4 & \alpha_3 & 0 & 0 & 0 \\ 0 & 0 & 1 & 1 & 0 & 0 \\ \beta_1 & \beta_4 & \beta_3 & 0 & 1 & 0 \\ 0 & \alpha_2 & 0 & \alpha_1 & 1 & -\alpha_3 \\ -\beta_2 & 0 & \alpha_2 & \beta_4 & \alpha_1 + \beta_3 & \alpha_4 \\ 0 & -\beta_2 & 0 & -\beta_1 & 0 & \beta_3 \end{pmatrix}$$

Due to the conditions on w, θ at $z = 0$, we take $w_1'(0) = 1$, $\theta_2'(0) = 1$, and then the eigenvalues σ are found by integrating (19.14) from 0 to 1 employing the initial condition

$$y_5(0) = 1. \tag{19.15}$$

To satisfy the final conditions $w(1) = \theta(1) = 0$, we must iterate on the condition on y_2,

$$y_2(1) = 0. \tag{19.16}$$

By having to satisfy condition (19.16) on the single variable y_2 we avoid the problem inherent in the standard shooting method where one has to subtract nearly equal quantities.

The determination of σ_i is relatively straightforward. The procedure whereby one calculates the corresponding eigenfunctions (w_i, θ_i) is described in (Straughan and Walker, 1996b). Rigorous mathematical results involving the compound matrix method for application to eigenvalue problems may be found in the interesting papers of (Greenberg and Marletta, 2000; Greenberg and Marletta, 2001), and (Allen and Bridges, 2002).

19.3 The Chebyshev tau method

We describe the Chebyshev tau method in the context of system (19.11). Thus, we define the operators L_1 and L_2 by

$$\begin{aligned} L_1(u,v) &= u'' - \alpha_2 u - \alpha_4 v - \alpha_1 u' - \alpha_3 v', \\ L_2(u,v) &= v'' - \beta_2 u - \beta_4 v - \beta_1 u' - \beta_3 v'. \end{aligned} \tag{19.17}$$

The Chebyshev tau method is very general, and the coefficients α_i, β_i may depend on z and may also be complex. The eigenvalue σ appears in one or more coefficients. System (19.11) is equivalent to

$$L_1(u,v) = 0, \qquad L_2(u,v) = 0, \tag{19.18}$$

on the domain (-1,1) together with the boundary conditions

$$u = v = 0 \qquad \text{at} \quad z = \pm 1. \tag{19.19}$$

The system (19.11) has been transformed from (0,1) to (-1,1) as this is the natural domain in which to use Chebyshev polynomials. The above choice of boundary conditions is not necessary and other boundary conditions may be handled, cf. (Straughan and Walker, 1996b).

We now describe the procedure for finding eigenvalues and eigenfunctions to (19.18), (19.19). It is important to realise that other boundary conditions may be handled, and, in particular, higher order systems of differential equations are naturally dealt with by the same technique.

The key idea is to write u, v as a finite series of Chebyshev polynomials

$$u = \sum_{k=0}^{N+2} a_k T_k(z), \qquad v = \sum_{k=0}^{N+2} b_k T_k(z). \tag{19.20}$$

The exact solution to the differential equation is an infinite series, i.e. let $N \to \infty$ in (19.20). Due to the truncation, to solve (19.18) with the approximate form (19.20), we solve

$$L_1(u,v) = \tau_1 T_{N+1} + \tau_2 T_{N+2}, \qquad L_2(u,v) = \hat{\tau}_1 T_{N+1} + \hat{\tau}_2 T_{N+2}. \tag{19.21}$$

In (19.21) the parameters $\tau_1, \tau_2, \hat{\tau}_1, \hat{\tau}_2$ are effectively error indicators for the truncation in (19.20).

To determine the unknown coefficients a_i and b_i we take the inner product with T_i of (19.21) in the weighted $L^2(-1,1)$ space with inner product

$$(f,g) = \int_{-1}^{1} \frac{fg}{\sqrt{1-z^2}}\, dz\,.$$

Let us denote the associated norm by $\|\cdot\|$. The Chebyshev polynomials are orthogonal in this space, and thus (19.21) leads to the $2(N+1)$ equations

$$(L_1(u,v), T_i) = 0 \quad (L_2(u,v), T_i) = 0 \qquad i = 0, 1, \dots, N. \tag{19.22}$$

There are four more conditions which arise by taking inner products, and these are

$$(L_1(u,v), T_{N+j}) = \tau_j \|T_{N+j}\|^2, \;\; (L_2(u,v), T_{N+j}) = \hat{\tau}_j \|T_{N+j}\|^2, \;\; j = 1, 2.$$

These four equations yield the tau coefficients $\tau_1, \tau_2, \hat{\tau}_1, \hat{\tau}_2$, which in turn are measures of the error involved in the truncation (19.20). To derive four more equations for a_i and b_i to add to (19.22) we employ the boundary conditions. The Chebyshev polynomials $T_n(z)$ satisfy $T_n(\pm 1) = (\pm 1)^n$, and this together with (19.19) and (19.20) yield

$$\begin{aligned} &\sum_{n=0}^{N+2} (-1)^n a_n = 0, \quad \sum_{n=0}^{N+2} a_n = 0, \\ &\sum_{n=0}^{N+2} (-1)^n b_n = 0, \quad \sum_{n=0}^{N+2} b_n = 0. \end{aligned} \tag{19.23}$$

Equations (19.22) and (19.23) yield a system of $2(N+3)$ equations for the $2(N+3)$ unknowns a_i, b_i, $i = 0, \dots, N+2$. We now suppose α_i, β_i

are constant. If they are functions of z then they must be expanded in a series of Chebyshev polynomials, cf. (Orszag, 1971), p. 702. One then uses the relation $2T_mT_n = (T_{m+n} + T_{|m-n|})$ to write expressions as a linear combination of the T_i. For many convection problems the coefficients are linear, quadratic or third order polynomials and these are easily handled.

To calculate the coefficients in (19.22) we observe that the derivative of a Chebyshev polynomial is a linear combination of lower order Chebyshev polynomials and it may be shown that

$$T_n' = \begin{cases} 2n(T_{n-1} + \ldots + T_1), & n \text{ even}, \\ 2n(T_{n-1} + \ldots + T_2) + nT_0, & n \text{ odd}. \end{cases} \tag{19.24}$$

By recalling (19.17) and using (19.20) and (19.24), (19.22) are reduced to the $2(N+1)$ algebraic equations

$$\begin{aligned} a_i^{(2)} - \alpha_2 a_i - \alpha_4 b_i - \alpha_1 a_i^{(1)} - \alpha_3 b_i^{(1)} &= 0, \quad i = 0, \ldots, N, \\ b_i^{(2)} - \beta_2 a_i - \beta_4 b_i - \beta_1 a_i^{(1)} - \beta_3 b_i^{(1)} &= 0, \quad i = 0, \ldots, N. \end{aligned} \tag{19.25}$$

The coefficients $a_i^{(1)}$ and $a_i^{(2)}$ are given by

$$a_i^{(1)} = \frac{2}{c_i} \sum_{\substack{p=i+1 \\ p+i \text{ odd}}}^{p=N+2} p a_p, \quad a_i^{(2)} = \frac{2}{c_i} \sum_{\substack{p=i+2 \\ p+i \text{ even}}}^{p=N+2} p(p^2 - i^2) a_p. \tag{19.26}$$

A similar representation holds for $b_i^{(1)}, b_i^{(2)}$. The coefficients c_i have form $c_0 = 2, c_i = 1, i = 1, 2, \ldots$. The $2(N+1)$ equations (19.25) together with the four equations (19.23) form a system of simultaneous linear equations for the $2(N+3)$ unknowns (a_i, b_i). This may be written in matrix form as

$$A\mathbf{x} = \sigma B\mathbf{x}, \tag{19.27}$$

where $\mathbf{x} = (a_0, \ldots, a_{N+2}, b_0, \ldots, b_{N+2})^T$.

The matrices involved in the definition of $a_i^{(1)}$ and $a_i^{(2)}$ may alternatively be derived as follows. We know that

$$\begin{aligned} u' &= \sum_{s=0}^{N+2} a_s T_s'(z) \\ &= \sum_{s=0}^{N+2} a_s \Big(\sum_{r=0}^{N+2} D_{rs} T_r \Big) \\ &= \sum_{r=0}^{N+2} \Big(\sum_{s=0}^{N+2} D_{rs} a_s \Big) T_r \end{aligned}$$

and so

$$a_r^{(1)} = \sum_{s=0}^{N+2} D_{rs} a_s.$$

In addition,

$$u'' = \sum_{r=0}^{N+2} \left(\sum_{s=0}^{N+2} D_{rs} a_s^{(1)} \right) T_r .$$

Therefore,

$$\begin{aligned} a_r^{(2)} &= \sum_{s=0}^{N+2} D_{rs} a_s^{(1)} \\ &= \sum_{s=0}^{N+2} D_{rs} \sum_{k=0}^{N+2} D_{sk} a_k \\ &= \sum_{s=0}^{N+2} \sum_{k=0}^{N+2} D_{rs} D_{sk} a_k . \end{aligned}$$

The differentiation matrix D, and second differentiation matrix D^2 thus arise naturally. These matrices and their coefficients take the form

$$\begin{aligned} D_{0,2j-1} &= 2j-1, && j \geq 1, \\ D_{i,i+2j-1} &= 2(i+2j-1), && i \geq 1, j \geq 1, \\ D^2_{0,2j} &= \frac{1}{2}(2j)^3, && j \geq 1, \\ D^2_{i,i+2j} &= (i+2j)4j(i+j), && i \geq 1, j \geq 1, \end{aligned} \tag{19.28}$$

or in matrix form

$$D = \begin{pmatrix} 0 & 1 & 0 & 3 & 0 & 5 & 0 & 7 & 0 & 9 & \dots \\ 0 & 0 & 4 & 0 & 8 & 0 & 12 & 0 & 16 & 0 & \dots \\ 0 & 0 & 0 & 6 & 0 & 10 & 0 & 14 & 0 & 18 & \dots \\ 0 & 0 & 0 & 0 & 8 & 0 & 12 & 0 & 16 & 0 & \dots \\ 0 & 0 & 0 & 0 & 0 & 10 & 0 & 14 & 0 & 18 & \dots \\ \dots & \dots & \dots & \dots & \dots & \dots & \dots & \dots & \dots & \dots & \dots \end{pmatrix}$$

$$D^2 = \begin{pmatrix} 0 & 0 & 4 & 0 & 32 & 0 & 108 & \dots \\ 0 & 0 & 0 & 24 & 0 & 120 & 0 & \dots \\ 0 & 0 & 0 & 0 & 48 & 0 & 192 & \dots \\ \dots & \dots & \dots & \dots & \dots & \dots & \dots & \dots \end{pmatrix}$$

Note that in the matrix sense $D^2 = D \cdot D$. The B matrix in (19.27) is singular due to the way the boundary condition rows are added to A. When it is possible, it is usually best to remove the singular behaviour since this can result in the formation of spurious eigenvalues (i.e. numbers which appear in the eigenvalue list, but which are not eigenvalues).

For the boundary conditions (19.19) we may easily eliminate a_{N+1}, a_{N+2}, b_{N+1}, b_{N+2}. Suppose N is odd, then

$$a_{N+1} = -(a_0 + a_2 + \ldots + a_{N-1}), \qquad a_{N+2} = -(a_1 + a_3 + \ldots + a_N). \tag{19.29}$$

Similar forms hold for the b's. This allows us to remove the $N+1$ and $N+2$ rows of D^2 and eliminate the $N+1$, $N+2$ columns. This yields $(N+1) \times (N+1)$ matrices D^2, and the matrix problem resulting from (19.27) does not suffer from B being singular because of zero boundary condition rows.

The equation which results has again form (19.27) but now A and B are $(N+1) \times (N+1)$ matrices and $\mathbf{x} = (a_0, \ldots, a_N, b_0, \ldots, b_N)$. Explicit details of A, B are given in sections 19.5 and 19.7. The eigenvalues of the generalised eigenvalue problem (19.27) are found efficiently using the QZ algorithm. This algorithm is available in many standard libraries, e.g. in the routines F02BJF, F02GJF of the NAG library. Since u and v have the forms (19.20) the calculation of the eigenfunctions using the Chebyshev tau method is really efficient. As soon as we know the coefficients a_k and b_k, u and v follow immediately from (19.20).

If one works with a system of first order differential equations and consequently employs a $D-$Chebyshev tau method rather than a D^2 one, details of how to remove boundary condition rows in D are given in (Payne and Straughan, 2000a).

19.4 Compound matrix calculation for a thermal convection problem in a fluid

The equations for a perturbation to the conduction solution of Bénard convection are given by (3.46) and are

$$\begin{aligned} &u_{i,t} + u_j u_{i,j} = -p_{,i} + \Delta u_i + R\theta\delta_{i3}, \qquad u_{i,i} = 0, \\ &Pr(\theta_{,t} + u_i\theta_{,i}) = Rw + \Delta\theta, \end{aligned} \tag{19.30}$$

where the notation is as throughout the book. Equations (19.30) hold for positive time on the spatial domain $\mathbb{R}^2 \times (0,1)$.

When (19.30) are linearized, a normal mode representation is employed, and the time behaviour is assumed like $e^{\sigma t}$ there results,

$$\begin{aligned} (D^2 - a^2)^2 W - Ra^2\Theta &= \sigma(D^2 - a^2)W, \\ (D^2 - a^2)\Theta + RW &= Pr\sigma\Theta, \end{aligned} \tag{19.31}$$

where $D = d/dz$, a is the wavenumber, and $W(z)$, $\Theta(z)$, represent the z-dependent parts of w and θ. If the surfaces $z = 0, 1$ are free from tangential stress then to (19.31) we append the boundary conditions

$$W = D^2W = \Theta = 0, \qquad z = 0, 1. \tag{19.32}$$

System (19.31), (19.32) is an eigenvalue problem for σ, given R, a, and Pr. We now discuss a very accurate scheme for its solution (and many similar free-free boundary condition eigenvalue problems).

Recall that in section 3.3 we derived the linearised equations for instability in the Bénard problem, equations (3.49), and we showed that the growth rate σ is real. Then, the instability boundary is found from the eigenvalue problem for the linear critical Rayleigh number, R^2, governed by (3.49) with $\sigma = 0$, namely

$$\begin{aligned} 0 &= -p_{,i} + \Delta u_i + k_i R\theta, \\ u_{i,i} &= 0, \\ 0 &= Rw + \Delta\theta. \end{aligned} \tag{19.33}$$

Employing normal modes as in (19.31), (19.33) becomes

$$\begin{aligned} (D^2 - a^2)^2 W - Ra^2\Theta = 0, \\ (D^2 - a^2)\Theta + RW = 0. \end{aligned} \tag{19.34}$$

We describe the compound matrix method applied to (19.34), (19.32), rather than keep σ in. A complex problem where σ is retained is discussed in section 19.8. To employ the compound matrix method it is convenient to rewrite (19.34) as

$$\begin{aligned} D^4 W &= 2a^2 D^2 W - a^4 W + Ra^2\Theta, \\ D^2\Theta &= a^2\Theta - RW. \end{aligned} \tag{19.35}$$

The compound matrix method for (19.35) works with the 3×3 minors of the 6×3 solution matrix formed from $\mathbf{W}_1 = (W_1, W_1', W_1'', W_1''', \Theta_1, \Theta_1')$, $\mathbf{W}_2 = (W_2, W_2', W_2'', W_2''', \Theta_2, \Theta_2')$, $\mathbf{W}_3 = (W_3, W_3', W_3'', W_3''', \Theta_3, \Theta_3')$. The solutions $\mathbf{W}_i$ are independent solutions to (19.35) for different initial values. If the bounding surfaces $z = 0, 1$ are stress free then $W = W'' = \Theta = 0$ at $z = 0$. Thus, we would choose $W_1'(0) = 1$, $W_2'''(0) = 1$, and $\Theta_3'(0) = 1$, i.e. $\mathbf{W}_1, \mathbf{W}_2, \mathbf{W}_3$ would correspond to solutions for starting values $(0,1,0,0,0,0)^T$, $(0,0,0,1,0,0)^T$, and $(0,0,0,0,0,1)^T$, respectively. We define twenty new variables $y_1 - y_{20}$ as the 3×3 minors. For example,

$$\begin{aligned} y_1 = &\begin{vmatrix} W_1 & W_2 & W_3 \\ W_1' & W_2' & W_3' \\ W_1'' & W_2'' & W_3'' \end{vmatrix} \\ = &W_1 W_2' W_3'' + W_2 W_3' W_1'' + W_3 W_1' W_2'' \\ &- W_1 W_3' W_2'' - W_2 W_1' W_3'' - W_3 W_2' W_1''. \end{aligned} \tag{19.36}$$

The idea is to define $y_2 - y_{20}$ similarly and then obtain differential equations for the y_i by differentiation. With a little practice there is no need to write out the whole determinant each time. The first term suffices. Thus, we write

$$\begin{aligned}
y_1 &= W_1 W_2' W_3'' + \dots & y_{11} &= W_1' W_2'' W_3''' + \dots \\
y_2 &= W_1 W_2' W_3''' + \dots & y_{12} &= W_1' W_2'' \Theta_3 + \dots \\
y_3 &= W_1 W_2' \Theta_3 + \dots & y_{13} &= W_1' W_2'' \Theta_3' + \dots \\
y_4 &= W_1 W_2' \Theta_3' + \dots & y_{14} &= W_1' W_2''' \Theta_3 + \dots \\
y_5 &= W_1 W_2'' W_3''' + \dots & y_{15} &= W_1' W_2''' \Theta_3' + \dots \\
y_6 &= W_1 W_2'' \Theta_3 + \dots & y_{16} &= W_1' \Theta_2 \Theta_3' + \dots \\
y_7 &= W_1 W_2'' \Theta_3' + \dots & y_{17} &= W_1'' W_2''' \Theta_3 + \dots \\
y_8 &= W_1 W_2''' \Theta_3 + \dots & y_{18} &= W_1'' W_2''' \Theta_3' + \dots \\
y_9 &= W_1 W_2''' \Theta_3' + \dots & y_{19} &= W_1'' \Theta_2 \Theta_3' + \dots \\
y_{10} &= W_1 \Theta_2 \Theta_3' + \dots & y_{20} &= W_1''' \Theta_2 \Theta_3' + \dots
\end{aligned} \tag{19.37}$$

By differentiating each y_i in turn and substituting from equations (19.35) we arrive at the following differential equations for the y_i:

$$\begin{aligned}
y_1' &= y_2 & y_2' &= 2a^2 y_1 + Ra^2 y_3 + y_5 \\
y_3' &= y_4 + y_6 & y_4' &= a^2 y_3 + y_7 \\
y_5' &= Ra^2 y_6 + y_{11} & y_6' &= y_7 + y_8 + y_{12} \\
y_7' &= a^2 y_6 + y_9 + y_{13} & y_8' &= 2a^2 y_6 + y_9 + y_{14} \\
y_9' &= 2a^2 y_7 + a^2 y_8 & y_{10}' &= y_{16} \\
&\quad + Ra^2 y_{10} + y_{15} & & \\
y_{11}' &= -a^4 y_1 + Ra^2 y_{12} & y_{12}' &= y_{13} + y_{14} \\
y_{13}' &= -Ry_1 + a^2 y_{12} + y_{15} & y_{14}' &= a^4 y_3 + 2a^2 y_{12} \\
& & &\quad + y_{15} + y_{17} \\
y_{15}' &= -Ry_2 + a^4 y_4 & y_{16}' &= -Ry_3 + y_{19} \\
&\quad + 2a^2 y_{13} + a^2 y_{14} & & \\
&\quad + Ra^2 y_{16} + y_{18} & & \\
y_{17}' &= a^4 y_6 + y_{18} & y_{18}' &= -Ry_5 + a^4 y_7 \\
& & &\quad + a^2 y_{17} + Ra^2 y_{19} \\
y_{19}' &= -Ry_6 + y_{20} & y_{20}' &= -Ry_8 - a^4 y_{10} + 2a^2 y_{19}
\end{aligned} \tag{19.38}$$

These equations are integrated numerically from 0 to 1.

For two free surfaces the boundary conditions on $\mathbf{W}$ are (19.32). We keep the boundary conditions at $z = 0$ and replace the ones at $z = 1$ by

$$W_1'(0) = 1, \quad W_2'''(0) = 1, \quad \Theta_3'(0) = 1, \tag{19.39}$$

which using (19.37) yields the initial condition for (19.38) as

$$y_{15}(0) = 1. \tag{19.40}$$

The final condition which satisfies (19.32) is seen using (19.37) to be

$$y_6(1) = 0. \tag{19.41}$$

If the boundary conditions are for fixed surfaces then (19.32) are replaced by

$$W = W' = \Theta = 0 \qquad \text{at } z = 0, 1. \tag{19.42}$$

This yields the initial condition

$$y_{18}(0) = 1, \tag{19.43}$$

and the final condition

$$y_3(1) = 0. \tag{19.44}$$

The eigenvalue R is varied until (19.41) or (19.44) (as appropriate) is satisfied to some pre-assigned tolerance; a technique such as the secant method works well to determine R. In solving (19.38) we keep a^2 fixed and then find numerically

$$Ra_L = \min_{a^2} R^2(a^2).$$

A very reliable method for determining the minimum is that of golden section search, see e.g., (Cheney and Kincaid, 1985), p. 462.

19.5 Chebyshev tau calculation for a thermal convection problem in a fluid

The key to the numerical resolution of (19.31), (19.32), using the Chebyshev tau method is to write (19.31) as three second order equations and then represent the functions as a series of Chebyshev polynomials. To do this define the operators L_1, L_2, L_3 by

$$\begin{aligned} L_1 u =& (D^2 - a^2)W - A, \\ L_2 u =& (D^2 - a^2)A - Ra^2\Theta - \sigma A, \\ L_3 u =& (D^2 - a^2)\Theta + RW - Pr\sigma\Theta, \end{aligned} \tag{19.45}$$

where u denotes the vector $(W, A, \Theta)^T$. Then, we need to solve

$$L_\alpha u = 0, \qquad \alpha = 1, 2, 3, \tag{19.46}$$

together with the boundary conditions

$$W = A = \Theta = 0, \qquad z = 0, 1. \tag{19.47}$$

(Dongarra et al., 1996) refer to the above form as a D^2–Chebyshev tau method.

Even though (19.45) - (19.47) refer to the three-dimensional situation, the way a occurs means it is analogous to a two-dimensional one. In the equivalent two - dimensional case if one introduces a stream function ψ by $u = \psi_{,z}, w = -\psi_{,x}$, then from (19.30) we obtain

$$(D^2 - a^2)\Psi = \Omega,$$
$$(D^2 - a^2)\Omega - Ria\Theta = \sigma\Omega,$$
$$(D^2 - a^2)\Theta - Ria\Psi = Pr\sigma\Theta,$$

Ψ and Ω being the z−parts of the stream function and vorticity $\omega = u_{,z} - w_{,x}$. The transformation $\Theta \to -ia\tilde{\Theta}$ then transforms this to a system equivalent to (19.46). Thus, we are effectively using a stream function - vorticity Chebyshev tau method, see (McFadden et al., 1990). However, as is pointed out in detail in (Dongarra et al., 1996), the lowering of order method as in (19.45), (19.46) applies to three-dimensional problems such as normal velocity - normal vorticity interactions, (Butler and Farrell, 1992), anisotropic porous penetrative convection, see section 17.6, or Hadley flow, cf. section 7.9 or section 19.8. This lowering of order is referred to in (Dongarra et al., 1996) as a D^2 method and its strength is that the matrices which arise from (19.46) have only $O(M^3)$ growth, where M is the number of Chebyshev polynomials, rather than $O(M^7)$ which is the case for a Chebyshev tau method which discretizes the fourth order differentiation operator. For convection problems typically a handful of Chebyshev polynomials are needed, but for calculating spectra of Orr - Sommerfeld like problems such as those studied in (Dongarra et al., 1996) at high Reynolds number, several hundred polynomials are required and the problems due to round off error because of growth of matrix terms are serious. One way of avoiding round off error due to coefficient growth is to use a lowering of order method like (19.45)-(19.47).

To solve (19.46), (19.47) by a Chebyshev tau method we first transform to the interval $(-1, 1)$. One has to be careful with the Chebyshev tau method because spurious eigenvalues may arise. Indeed, the articles of (Dawkins et al., 1998), (Gardner et al., 1989), (McFadden et al., 1990), (Straughan and Walker, 1996b), and (Zebib, 1978) address this point. We do not find any trouble with spurious eigenvalues employing the method outlined below, which also hinges on the QZ algorithm of (Moler and Stewart, 1973), because we believe the removal of boundary condition rows in the matrix A of (19.55) and consequently the removal of extraneous rows of zeros in the B matrix of (19.55) stabilizes the numerical eigenvalue problem; although the removal of boundary condition rows is straightforward we have not seen this pointed out explicitly in the literature in connection with other free surface boundary condition problems. Before detailing the analysis we should point out that use of Chebyshev polynomials in hydrodynamic stability problems has a long history, at least since the work of (Clenshaw and Elliott, 1960) and the fundamental paper of (Orszag, 1971) on the Orr-Sommerfeld equation.

We now write W, A, Θ as a finite series of Chebyshev polynomials

$$W = \sum_{k=0}^{N+2} W_k T_k(z), \quad A = \sum_{k=0}^{N+2} A_k T_k(z), \quad \Theta = \sum_{k=0}^{N+2} \Theta_k T_k(z), \tag{19.48}$$

where the underlying idea is that (19.48) are truncations of infinite series. Due to the truncation, we solve

$$\begin{aligned} L_1 u &= \tau_1 T_{N+1} + \tau_2 T_{N+2}, \\ L_2 u &= \hat{\tau}_1 T_{N+1} + \hat{\tau}_2 T_{N+2}, \\ L_3 u &= \tilde{\tau}_1 T_{N+1} + \tilde{\tau}_2 T_{N+2}, \end{aligned} \tag{19.49}$$

where the tau coefficients τ_1, τ_2, $\hat{\tau}_1, \hat{\tau}_2$, $\tilde{\tau}_1, \tilde{\tau}_2$ may be determined and used as error indicators.

The infinite dimensional system (19.46), (19.47) is reduced by taking the inner product of each of (19.46) with T_i in the weighted $L^2(-1,1)$ space with inner product defined by

$$(f,g) = \int_{-1}^{1} \frac{fg}{\sqrt{1-z^2}}\, dz\,.$$

The Chebyshev polynomials are orthogonal in this space and this leads to $3(N+1)$ algebraic equations

$$(L_\alpha u, T_i) = 0, \qquad \alpha = 1,2,3, \quad i = 0,1,\dots,N, \tag{19.50}$$

together with six further equations

$$\begin{aligned} (L_1 u, T_{N+j}) &= \|T_{N+j}\|^2 \tau_j, \qquad & j &= 1,2, \\ (L_2 u, T_{N+j}) &= \|T_{N+j}\|^2 \hat{\tau}_j, \qquad & j &= 1,2, \\ (L_3 u, T_{N+j}) &= \|T_{N+j}\|^2 \tilde{\tau}_j, \qquad & j &= 1,2. \end{aligned} \tag{19.51}$$

Equations (19.51) may be used to determine the tau coefficients and used in an error analysis. To determine the coefficients in (19.48) we need six further equations and these come from the boundary conditions (19.47) which are (since $T_n(\pm 1) = (\pm 1)^n$),

$$\begin{aligned} \sum_{n=0}^{N+2} (-1)^n W_n &= 0, \quad & \sum_{n=0}^{N+2} W_n &= 0, \\ \sum_{n=0}^{N+2} (-1)^n A_n &= 0, \quad & \sum_{n=0}^{N+2} A_n &= 0, \\ \sum_{n=0}^{N+2} (-1)^n \Theta_n &= 0, \quad & \sum_{n=0}^{N+2} \Theta_n &= 0. \end{aligned} \tag{19.52}$$

In this way, equations (19.50) and (19.52) yield a system of $3(N+3)$ algebraic equations for the $3(N+3)$ unknowns W_i, A_i, Θ_i, $i = 0, \dots, N+2$. In this manner, the high frequency behaviour of the solution is determined not by the dynamical equations but rather by the boundary conditions, as is concisely observed by (Orszag, 1971).

The individual terms in (19.50) are easily calculated using the fact that

$$D^2W = \sum_{k=0}^{N+2} W_k^{(2)} T_k(z), \tag{19.53}$$

where

$$W_i^{(2)} = \frac{2}{c_i} \sum_{\substack{p=i+2 \\ p+i \text{ even}}}^{p=N+2} p(p^2 - i^2) W_p, \tag{19.54}$$

with $c_0 = 2, c_i = 1, i = 1, 2, \dots$, and where analogous expressions hold for $D^2A, D^2\Theta$. In fact, if we denote the coefficients of the second differentiation matrix (the matrix which arises in the Chebyshev representation of D^2) by $D^2_{i,j}$, then from (19.53), (19.54) we may show $D^2_{i,j}$ are given by (19.28).

The conclusion is that (19.50) and (19.52) represent a matrix equation of form

$$A\mathbf{x} = \sigma B\mathbf{x}. \tag{19.55}$$

Before describing $A, B, \mathbf{x}$, however, we write D^2 as an $(N+1) \times (N+3)$ matrix and remove the $N+2$, $N+3$ columns by using the boundary conditions (19.52), cf. (Straughan and Walker, 1996b) for further details, and (Haidvogel and Zang, 1979). It remains to solve (19.55) where now A and B are $(N+1) \times (N+1)$ matrices of form

$$A = \begin{pmatrix} D^2 - a^2I & -I & 0 \\ 0 & D^2 - a^2I & -Ra^2I \\ RI & 0 & D^2 - a^2I \end{pmatrix} \qquad B = \begin{pmatrix} 0 & 0 & 0 \\ 0 & I & 0 \\ 0 & 0 & PrI \end{pmatrix}$$

where D^2 refers to (19.53) form *but* with the $N+2, N+3$ columns removed due to the boundary condition rows. The vector $\mathbf{x}$ is $\mathbf{x} = (W_0, \dots, W_N, A_0, \dots, A_N, \Theta_0, \dots, \Theta_N)^T$.

We have found the solution to (19.55) is achieved very accurately by employing the QZ algorithm of (Moler and Stewart, 1973). This algorithm relies on the fact that there are unitary matrices Q and Z such that QAZ and QBZ are both upper triangular. The algorithm then yields sets of values α_i, β_i which are the diagonal elements of QAZ and QBZ. The eigenvalues σ_i of (19.55) are then obtained from the relation $\sigma_i = \alpha_i/\beta_i$, provided $\beta_i \neq 0$. This is very important, since the way we have constructed B means it contains a singular band and we find one third of the $\beta_i = 0$; these β_i must be filtered out. Indeed, with the technique advocated here one ought always to consider the α_i and β_i, since as (Moler and Stewart, 1973) point out, the α_i and β_i contain more information than the eigenvalues themselves.

Accurate formulations of the QZ algorithm are available in standard libraries, for example in the routines ZGGHRD, ZHGEQZ and ZTGEVC of the LAPACK library, (Anderson et al., 1995), or the routines F02BJF and F02GJF of the NAG library, and so the method outlined here is readily accessible. We do not describe results for the Bénard problem except we point out very high accuracy is achieved with only a handful of polynomials. The essential details of the method outlined above apply to the convection eigenvalue problems described in this book.

19.6 Convection in thawing subsea permafrost

This topic is covered in section 7.4. For the present chapter the model of convection in thawing subsea permafrost is useful as it illustrates boundary conditions which are not just homogeneous in the functions themselves. The equations for linear instability theory, which in chapter 7 are shown to be identical to those of nonlinear energy stability theory in a certain sense, are

$$Rs\delta_{i3} - p_{,i} = u_i, \quad u_{i,i} = 0, \quad Rw + \Delta s = 0, \tag{19.56}$$

in the layer $\mathbb{R}^2 \times (0,1)$, together with the boundary conditions

$$\begin{aligned} u_i n_i &= 0, && \text{on} \quad z = 0,1, \\ s &= 0, && \text{on} \quad z = 0, \\ \frac{\partial s}{\partial z} + \alpha s &= 0, && \text{on} \quad z = 1, \end{aligned} \tag{19.57}$$

where R^2 is the Rayleigh number, s, u_i, p are perturbations in salt field, velocity, pressure, respectively, $w = u_3$ and $\alpha(\geq 0)$ is a parameter related to the melting of the permafrost interface. Upon employing normal modes the boundary value problem (19.56), (19.57) reduces to the eigenvalue problem

$$(D^2 - a^2)W + Ra^2 S = 0, \qquad (D^2 - a^2)S + RW = 0, \tag{19.58}$$

$z \in (0,1)$, $D = d/dz$, $a^2 = k^2 + m^2$ being the square of the wavenumber, and the boundary conditions are

$$W = 0, \quad z = 0,1; \qquad S = 0, \quad z = 0; \qquad DS + \alpha S = 0, \quad z = 1. \tag{19.59}$$

The constants k and m are, respectively, the wavenumbers in the x and y directions.

To solve (19.58), (19.59) using the compound matrix method we solve the system (19.14) with the matrix A given by

$$A = \begin{pmatrix} 0 & -Ra^2 & 0 & 0 & 0 & 0 \\ 0 & 0 & 1 & 1 & 0 & 0 \\ 0 & a^2 & 0 & 0 & 1 & 0 \\ 0 & a^2 & 0 & 0 & 1 & 0 \\ R & 0 & a^2 & a^2 & 0 & -Ra^2 \\ 0 & R & 0 & 0 & 0 & 0 \end{pmatrix}$$

The initial condition is

$$y_5(0) = 1, \tag{19.60}$$

while the appropriate final condition which transpires from (19.59) is

$$y_3(1) + \alpha y_2(1) = 0. \tag{19.61}$$

The critical Rayleigh number is found by iterating on (19.61) and minimizing over a^2. Critical values of the Rayleigh number and the wavenumber are given in table 1 of (Galdi et al., 1987); no eigenfunctions are given there and we include them here.

The equivalent analysis of (19.58), (19.59) by the Chebyshev tau method reduces to solving the matrix equation

$$A\mathbf{x} = RB\mathbf{x}, \tag{19.62}$$

where $(W_0, \ldots, W_N, S_0, \ldots, S_N)^T$,

$$A = \begin{pmatrix} D^2 - a^2 I & 0 \\ 0 & \tilde{D}^2 - a^2 I \end{pmatrix} \qquad B = \begin{pmatrix} 0 & -a^2 I \\ -I & 0 \end{pmatrix}$$

where D^2 is given by (19.28) with the last two columns overwritten from the boundary conditions $W = 0, \quad z = 0, 1$, i.e.

$$\begin{aligned} W_{N+1} &= -(W_0 + W_2 + \ldots + W_{N-1}), \\ W_{N+2} &= -(W_1 + W_3 + \ldots + W_N). \end{aligned} \tag{19.63}$$

The matrix $\tilde{D}^2$ derives also from (19.28) but the boundary conditions are (since $T_n'(\pm 1) = (\pm 1)^{n-1} n^2$)

$$S_0 - S_1 \ldots - S_N + S_{N+1} = S_{N+2}, \qquad \sum_{n=0}^{N+2} n^2 S_n + \alpha \sum_{n=0}^{N+2} S_n = 0. \tag{19.64}$$

Ra	a^2	α
27.098	5.411	0
30.269	6.751	1
34.130	8.233	4
36.538	9.040	10
39.097	9.773	100

Table 19.2. Critical parameters of linear instability theory, derived from (19.58), (19.59).

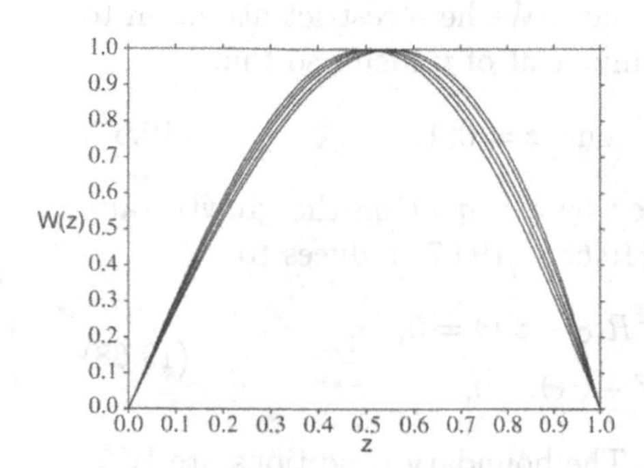

Figure 19.1. Plots of $W(z)$. The right most curve is for $\alpha = 0$. The curves for $\alpha = 1, 4, 10, 100$ then follow in succession as one moves left.

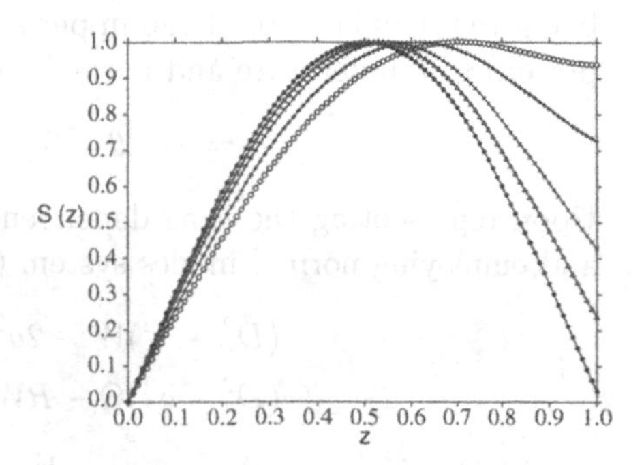

Figure 19.2. Plots of $S(z)$. Open circle (∘) $\alpha = 0$; Plus (+) $\alpha = 1$; Cross (×) $\alpha = 4$; Triangle (△) $\alpha = 10$; Filled circle (•) $\alpha = 100$.

To usefully employ (19.64) we might put $u_{m+1} = S_m$, $M = N + 1$, and then from (19.60) we may derive the conditions (useful for programming)

$$\begin{aligned} u_{M+1} &= -\frac{1}{(2M^2 + 2M + 2\alpha + 1)} \sum_{i=1}^{M} \Big[(i-1)^2 \\ &\qquad + \alpha + (-1)^{i+1} \big\{ (M+1)^2 + \alpha \big\} \Big] u_i , \qquad (19.65) \\ u_{M+2} &= \sum_{i=1}^{M} (-1)^{i+1} u_i - u_{M+1} , \end{aligned}$$

where it is to be understood that in $(19.65)_2$ expression $(19.65)_1$ is utilized for u_{M+1}. Equations (19.65) give u_{M+1}, u_{M+2} as a linear combination of $u_1, \dots, u_M$.

For $\alpha = 0, 1, 4, 10, 100$ the critical values of Ra, a^2 are given in table 19.2, cf. (Galdi et al., 1987), and the W and S eigenfunctions are shown in figures 19.1, 19.2 below; the eigenfunctions are normalised to have largest value 1. The eigenfunctions are easily found since from (19.62) we determine the coefficients $W_0, \dots, W_N, S_0, \dots, S_N$.

19.7 Penetrative convection in a porous medium

This problem is studied in section 17.6. The equations for linearised instability for penetrative convection in an isotropic porous medium are

$$\begin{aligned} p_{,i} &= -u_i - 2R\theta(\xi - z)\delta_{i3}, \qquad u_{i,i} = 0, \\ \theta_{,t} &= -Rw + \Delta\theta, \end{aligned} \tag{19.66}$$

for $\mathbf{x} \in \mathbb{R}^2 \times (0,1)$, where p, u_i, θ are perturbations of pressure, velocity, temperature, $w = u_3$, R^2 is the Rayleigh number, $\xi = 4/T_u$, with T_u being the temperature of the upper surface. We here restrict attention to prescribed temperature and normal component of velocity so that

$$w = 0, \quad \theta = 0 \qquad \text{on} \quad z = 0, 1. \tag{19.67}$$

Upon representing the time dependency by $e^{\sigma t}$, σ being the growth rate, and employing normal modes system (19.66), (19.67) reduces to

$$\begin{aligned} &(D^2 - a^2)W - 2a^2R(\xi - z)\Theta = 0, \\ &(D^2 - a^2)\Theta - RW - \sigma\Theta = 0, \end{aligned} \tag{19.68}$$

$z \in (0,1)$, where a is the wavenumber. The boundary conditions are $W = \Theta = 0$, $z = 0, 1$.

For the purpose of this chapter this is an example of a system where the coefficients are functions of the spatial variable z. Also, as T_u increases the coefficient involving $(\xi - z)$ has a strong effect leading to a stiff system and the eigenfunctions vary strongly.

One can show exchange of stabilities holds for (19.66), (19.67), and then to find the instability boundary it is sufficient to take $\sigma = 0$ in (19.68). The compound matrix equations and boundary conditions are then (19.14) - (19.16) with A here given by

$$A = \begin{pmatrix} 0 & 2Ra^2M & 0 & 0 & 0 & 0 \\ 0 & 0 & 1 & 1 & 0 & 0 \\ 0 & a^2 & 0 & 0 & 1 & 0 \\ 0 & a^2 & 0 & 0 & 1 & 0 \\ -R & 0 & a^2 & a^2 & 0 & 2a^2RM \\ 0 & -R & 0 & 0 & 0 & 0 \end{pmatrix}$$

The Chebyshev tau method requires solution of (19.62) where now $\mathbf{x} = (W_0, \ldots, W_N, \Theta_0, \ldots, \Theta_N)$, and the matrices A, B are given by

$$A = \begin{pmatrix} D^2 - a^2I & 0 \\ 0 & D^2 - a^2I \end{pmatrix} \qquad B = \begin{pmatrix} 0 & 2a^2(\xi I - M) \\ I & 0 \end{pmatrix}.$$

For coding purposes we work with $N \times N$ matrices and with $i = 1, \ldots, N$, and then in the above M is the $N \times N$ matrix arising from the Chebyshev

Ra	Ra_V	a	T_u(°C)
99.01505	29.33779	3.50914	6
235.69233	29.46154	4.67519	8
460.96361	29.50167	5.87808	10
796.51749	29.50065	7.05235	12
796.51749	29.50065	7.05235	12
1264.84132	29.50067	8.22779	14

Table 19.3. Critical Rayleigh and wave numbers for porous penetrative convection.

representation of z, i.e.

$$\begin{aligned} &M_{i,i+1} = \frac{1}{2}, \qquad i = 1, \ldots, N-1, \\ &M_{21} = 1; \quad M_{i+1,i} = \frac{1}{2}, \quad i = 2, \ldots, N-1; \qquad \text{rest } 0. \end{aligned} \tag{19.69}$$

The matrix equation (19.62) is conveniently solved by the QZ algorithm of (Moler and Stewart, 1973). This routine yields all eigenvalues and eigenfunctions with no trouble.

We present critical Rayleigh and wave numbers for (19.66), (19.67) here. The values of a^2, Ra represent those at which instability commences. The Rayleigh number Ra_V is adapted from (Veronis, 1963) study of penetrative convection in a fluid; this Rayleigh number reflects the depth of fluid which is actually destabilizing and is given by $Ra_V = \xi^3 Ra$.

The functions $W(z), \Theta(z)$ for the same range of upper temperatures as those given in table 19.3 are presented in figures 19.3 and 19.4. It is noticeable how the penetration effect, W changing sign, is evident in the velocity eigenfunction and how the convection is strongest in the $0-4$°C domain. While the strong effect of convection in the $0-4$°C region is also apparent for the temperature eigenfunction, this field does not have such a strong variation in the region which is stable before convection commences.

19.8 Convection in a porous medium with inclined temperature gradient

(Nield, 1990) initiated an interesting study of convection in a layer of porous material when there is a temperature gradient in the vertical direction but the same gradient varies as the layer is traversed in one of the horizontal directions, This problem is discussed in section 7.9. For the present chapter this is an interesting example because the equations involve complex coefficients which depend on the z variable. The compound matrix and Chebyshev tau methods are applied to this problem in some detail in (Straughan and Walker, 1996b).

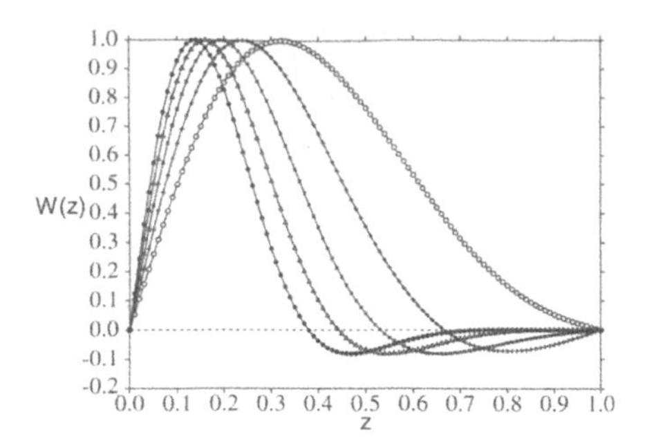

Figure 19.3. Plots of $W(z)$. Open circle ($\circ$) $T_u = 6°C$; Plus ($+$) $T_u = 8°C$; Cross ($\times$) $T_u = 10°C$; Triangle ($\triangle$) $T_u = 12°C$; Filled circle ($\bullet$) $T_u = 14°C$.

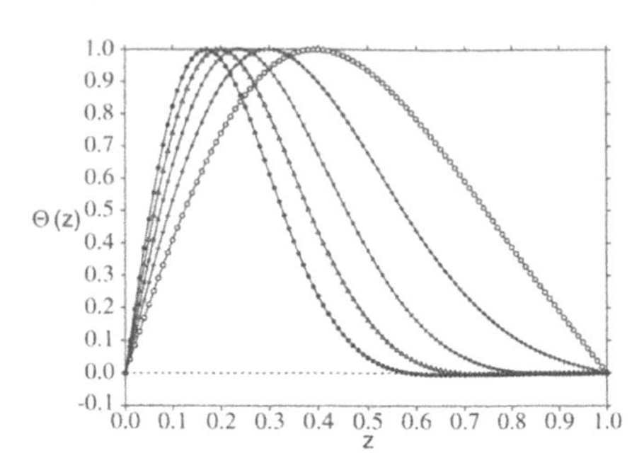

Figure 19.4. Plots of $\Theta(z)$. Open circle ($\circ$) $T_u = 6°C$; Plus ($+$) $T_u = 8°C$; Cross ($\times$) $T_u = 10°C$; Triangle ($\triangle$) $T_u = 12°C$; Filled circle ($\bullet$) $T_u = 14°C$.

The boundary conditions on the temperature field in non-dimensional form are

$$T = \mp\frac{1}{2}R_V - R_H x, \qquad z = \pm\frac{1}{2},$$

where R_V and R_H are the vertical and horizontal Rayleigh numbers. The steady solution is

$$\bar{U} = R_H z, \qquad \bar{T} = -R_V z + \frac{1}{24}R_H^2(z - 4z^3) - R_H x, \tag{19.70}$$

with $z \in (-\frac{1}{2}, \frac{1}{2})$. Of course, this is an interesting steady solution because the horizontal velocity field is not zero. (Nield, 1994a) refers to solution (19.70) as Hadley flow.

The non-dimensionalised perturbation equations yield the following eigenvalue problem, where k, m are the x and y wavenumbers, and $a^2 = k^2 + m^2$,

$$\begin{aligned} &(D^2 - a^2)W + a^2\Theta = 0,\\ &\big[D^2 - a^2 - i\sigma - ik\bar{U}(z)\big]\Theta + \frac{ik}{a^2}R_H DW - (D\bar{T})W = 0, \end{aligned} \tag{19.71}$$

$z \in (-\frac{1}{2}, \frac{1}{2})$. The boundary conditions become

$$W = \Theta = 0, \qquad z = \pm\frac{1}{2}. \tag{19.72}$$

19.8.1 The compound matrix method for Hadley flow

To derive the compound matrix equations we introduce the variables

$$\begin{aligned} u_1 &= W_1W_2' - W_2W_1', & u_2 &= W_1\Theta_2 - W_2\Theta_1, \\ u_3 &= W_1\Theta_2' - W_2\Theta_1', & u_4 &= W_1'\Theta_2 - W_2'\Theta_1, \\ u_5 &= W_1'\Theta_2' - W_2'\Theta_1', & u_6 &= \Theta_1\Theta_2' - \Theta_2\Theta_1'. \end{aligned} \tag{19.73}$$

Since the coefficients in (19.71) are complex we assume at the outset that u_i are complex. If we set $\sigma = \sigma_r + i\sigma_i$, $\alpha_r = a^2 - \sigma_i$, $\alpha_i = \sigma_r + k\bar{U}(z)$, $\beta = kR_H/a^2$, then the differential equations u_i satisfy may be shown to be

$$\begin{aligned} u_1' &= -a^2u_2, \\ u_2' &= u_3 + u_4, \\ u_3' &= -i\beta u_1 + (\alpha_r + i\alpha_i)u_2 + u_5, \\ u_4' &= a^2u_2 + u_5, \\ u_5' &= D\bar{T}\, u_1 + a^2u_3 + (\alpha_r + i\alpha_i)u_4 - a^2u_6, \\ u_6' &= D\bar{T}\, u_2 + i\beta u_4. \end{aligned} \tag{19.74}$$

We transform the interval $(-1/2, 1/2)$ to $(0,1)$ and integrate (19.74) over $(0,1)$. The coefficients in (19.74) then involve the functions

$$\bar{U}(z) = R_H(z - \frac{1}{2}), \quad D\bar{T} = -R_V - \frac{R_H^2}{12}(1 - 6z + 6z^2).$$

(Straughan and Walker, 1996b) give details of this procedure but they put $u_1 = y_1 + iy_7$, $u_2 = y_2 + iy_8$, $u_3 = y_3 + iy_9$, $u_4 = y_4 + iy_{10}$, $u_5 = y_5 + iy_{11}$, $u_6 = y_6 + iy_{12}$ and work with twelve compound matrix equations for the real and imaginary parts y_i.

The boundary conditions dictate that the initial condition is

$$u_5(0) = 1 \qquad (\text{or } y_5(0) + iy_{11}(0) = 1 + i0)$$

whereas the final condition is

$$u_2(1) = 0 \qquad (\text{or } y_2(1) + iy_8(1) = 0 + i0).$$

19.8.2 The Chebyshev tau method for Hadley flow

The Chebyshev tau method applied to (19.71), (19.72) reduces to solving

$$A\mathbf{x} = \sigma B\mathbf{x} \tag{19.75}$$

in which $\mathbf{x} = (W_0, \dots, W_N, \Theta_0, \dots, \Theta_N)$ and the matrices A and B are given by

$$A = \begin{pmatrix} D^2 - a^2I & a^2I \\ \frac{ik}{a^2}R_HD + \left(R_V - \frac{R_H^2}{24}\right)I + \frac{R_H^2}{8}P & D^2 - a^2I - \frac{1}{2}ikR_HM \end{pmatrix}$$

$$B = \begin{pmatrix} 0 & 0 \\ 0 & I \end{pmatrix}$$

In the expression for A, P is the matrix which arises from the Chebyshev representation of z^2. If the code involves $N \times N$ matrices, then the non-zero entries of P are

$$P_{11} = P_{ii} = \frac{1}{2}, \quad i = 3, \dots, N; \qquad P_{22} = \frac{3}{4}; \qquad P_{31} = \frac{1}{2};$$

$$P_{i,i+2} = \frac{1}{4}, \quad i = 1, \dots, N-2; \quad P_{i+2,i} = \frac{1}{4}, \quad i = 2, \dots, N-2.$$

D^2 is the second differentiation matrix allowing for the boundary conditions (19.72).

Results of computations using the compound matrix and Chebyshev tau methods on the Hadley flow problem are discussed in (Straughan and Walker, 1996b).

Appendix A

Useful inequalities in energy stability theory

The purpose of this appendix is to collect some of the inequalities that have been found to be very useful in energy stability theory. The inequalities presented are not necessarily the most general forms available, but are given in the form in which I have seen them used.

A.1 The Poincaré inequality

Let V be a "cell" in three dimensions. Suppose for simplicity V is the cell $0 \le x < 2a_1$, $0 \le y < 2a_2$, $0 < z < 1$, and suppose u is a function periodic in x, y, of period $2a_1, 2a_2$, respectively, and $u = 0$ on $z = 0, 1$. Then the Poincaré inequality may be written

$$< u^2 > \le \frac{1}{\pi^2} < u_{i,j} u_{i,j} >, \tag{A.1}$$

where $< \cdot >$ denotes integration over V. In general, the constant, $1/\pi^2$ in (A.1), depends on the geometry and size of the domain V.

A.2 The Wirtinger inequality

Suppose the boundary conditions on u above are replaced by

$$\frac{\partial u}{\partial z} = 0, \qquad \text{on} \qquad z = 0, 1. \tag{A.2}$$

For functions such that

$$< u > = 0, \tag{A.3}$$

u is periodic in x, y with periods $2a_1$, $2a_2$; the Wirtinger inequality is

$$< u^2 > \leq k < u_{i,j} u_{i,j} > . \tag{A.4}$$

(Kaiser and Xu, 1998) show that the constant k is given by $k = \max\{\pi^{-2}, a_1^{-2}, a_2^{-2}\}$. They show that the function which gives rise to this value of k is $f = \sqrt{a_1 a_2 / 2\pi^2} \cos\{n\pi(z + 1/2)\} \exp[i(a_1 \kappa_1 x + a_2 \kappa_2 y)]$. This function is a solution to the variational problem

$$\lambda_1 = \sup \frac{\|f\|^2}{\|\nabla f\|^2}.$$

(When no boundary conditions are specified, but the zero mean condition $< u > = 0$ is still required, (Hardy et al., 1934), p. 184, show that $\lambda_1 = 4/\pi^2$, for functions of one variable.)

A.3 The Sobolev inequality

The general Sobolev embedding inequality may be found in, e.g., (Gilbarg and Trudinger, 1977), pp. 148–157. The one of frequent use in energy stability theory is the following. Let Ω be a bounded domain in $\mathbf{R}^3$ with boundary $\partial\Omega$. Then for functions u with $u = 0$ on $\partial\Omega$,

$$\left(\int_\Omega u^6 \, dV \right)^{1/3} \leq C \int_\Omega |\nabla u|^2 \, dV , \tag{A.5}$$

where the constant C is independent of the domain; in fact, $C = 2^{2/3}/(3^{1/2}\pi^{2/3})$.

A.4 An inequality for the supremum of a function

In three space dimensions we have the inequality

$$\sup_V |u(\mathbf{x})| \leq C \|\Delta u\|. \tag{A.6}$$

This inequality is dealt with under various conditions by (Doering and Titi, 1995), (Mulone and Rionero, 2003), (Xie, 1991). (Mulone and Rionero, 2003) show that when $u = 0$ on $z = 0, 1$ with u periodic in x and y then

$$C = \frac{\sqrt{3}}{\left[\pi^5 h^3 \sqrt{2}(2^{1/2} - 1)\right]^{1/2}} + \frac{2^{5/2} 5 (1 + \pi^2)^{1/2} h^{3/5}}{3\pi} , \tag{A.7}$$

where $h = \min\{a_1, a_2, 1\}$. When the boundary condition on u is replaced by $\partial u/\partial z = 0$, $z = 0, 1$, then

$$C = 1.277\frac{k}{h_1^{3/2}} + 9.428h_1^{3/5}\sqrt{1+k}, \tag{A.8}$$

where $h_1 = \min\{\pi/a_1, \pi/a_2, 1\}$.

A.5 A Sobolev inequality for u^4

For a cell V we employ Hölder's inequality to deduce for functions u with x periodicity a_1, y periodicity a_2,

$$\int_V u^4\, dV \le (a_1a_2)^{1/3}\Big(\int_V u^6\, dV\Big)^{2/3}.$$

If we combine this with the Sobolev inequality $\|w\|_{L^6} \le 2^{5/2}\|w\|_{H^1}$, (Adams, 1975), p. 104, we find

$$< u^4 >^{1/2} \le 32(a_1a_2)^{1/6}(< u^2 > + < u_{,i}u_{,i} >). \tag{A.9}$$

This is a general inequality that does *not* require boundary conditions to be specified on u at $z = 0, 1$.

We now present a version of (A.9) for a general cell shape.

A.6 A Sobolev inequality for u^4 over a more general cell

We first present a general inequality, cf. (Galdi et al., 1987). For simplicity we assume:

(i) Λ is a cross section of a cell in the layer $\{z \in (0,1)\}$ and is star shaped with respect to an origin in Λ;

(ii) lines parallel to the x and y axes intersect $\partial\Lambda$ in at most two points.

A typical configuration is shown in Figure A.1.

Let u be any function that is zero on $z = 0$. Then

$$u^2(x,y,z) = u^2(x_1,y,z) - 2\int_x^{x_1} u(s,y,z)\frac{\partial u}{\partial s}(s,y,z)ds,$$

$$u^2(x,y,z) = u^2(x_0,y,z) + 2\int_{x_0}^{x} u(s,y,z)\frac{\partial u}{\partial s}(s,y,z)ds,$$

from which we find

$$u^2(x,y,z) \le \frac{1}{2}\Big[u^2(x_0,y,z) + u^2(x_1,y,z) + 2\int_{x_0}^{x_1} |u|\,|u_s|ds\Big].$$

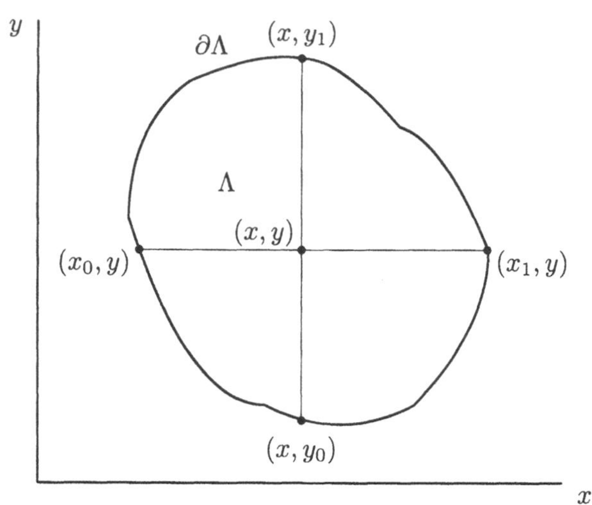

Figure A.1. Sketch of the cross section of a cell

In a similar manner we may show

$$u^2(x,y,z) \le \frac{1}{2}\Big[u^2(x,y_0,z) + u^2(x,y_1,z) + 2\int_{y_0}^{y_1} |u|\,|u_t|dt\Big].$$

These inequalities are multiplied, and we integrate to obtain

$$\begin{aligned}\int_\Lambda u^4\,dA \le &\Big[\frac{1}{2}\int_{\partial\Lambda} u^2|n_1|ds + \int_\Lambda |u|\,|u_x|\,dA\Big]\\ &\times\Big[\frac{1}{2}\int_{\partial\Lambda} u^2|n_2|ds + \int_\Lambda |u|\,|u_y|\,dA\Big],\end{aligned} \tag{A.10}$$

where n_i are the components of the unit outward normal to $\partial\Lambda$ and ds denotes integration along $\partial\Lambda$. We next employ the arithmetic-geometric mean inequality in (A.10), and then observe that

$$\int_{\partial\Lambda} u^2|n_1|ds \int_{\partial\Lambda} u^2|n_2|ds \le \frac{1}{2}\Big(\int_{\partial\Lambda} u^2 ds\Big)^2,$$

to derive from (A.10)

$$\int_\Lambda u^4\,dA \le \frac{1}{4}\Big(\int_{\partial\Lambda} u^2 ds\Big)^2 + \int_\Lambda u^2\,dA \int_\Lambda u_{,\alpha}u_{,\alpha}\,dA, \tag{A.11}$$

where α sums over 1 and 2. Furthermore,

$$\int_{\partial\Lambda} u^2 ds \le \frac{1}{\sigma}\int_{\partial\Lambda} x_\alpha n_\alpha u^2 ds = \frac{2}{\sigma}\int_\Lambda u^2\,dA + \frac{2}{\sigma}\int_\Lambda x_\alpha u u_{,\alpha}\,dA, \tag{A.12}$$

in which

$$\sigma = \min_{\partial\Lambda} x_\alpha n_\alpha .$$

Let

$$M = \max_{\Lambda} |x_\alpha|,$$

and then from (A.12) we may derive

$$\int_{\partial\Lambda} u^2 ds \leq \frac{2}{\sigma}\int_\Lambda u^2\, dA + \frac{2M}{\sigma}\Big[\int_\Lambda u^2\, dA \int_\Lambda u_{,\alpha}u_{,\alpha}\, dA\Big]^{1/2}. \tag{A.13}$$

We put (A.13) in (A.11) and then integrate with respect to z to obtain

$$\begin{aligned} < u^4 > \leq \max_z\Big(\int_\Lambda u^2\, dA\Big)\Big[\frac{1}{\sigma^2}\|u\|^2 \\ + \Big\{1 + \Big(\frac{M}{\sigma}\Big)^2\Big\}\|\nabla^* u\|^2 + \frac{2M}{\sigma^2}\|u\|\,\|\nabla^* u\|\Big], \end{aligned} \tag{A.14}$$

where $\nabla^* \equiv (\partial/\partial x, \partial/\partial y)$ and $\|\cdot\|$ dentoes the L^2 norm on the cell V. To estimate the Λ term in (A.14) we note

$$\max_z\Big(\int_\Lambda u^2\, dA\Big) \leq 2\|u\|\,\|u_z\|. \tag{A.15}$$

Also, for the boundary conditions on u, the Wirtinger inequality is

$$\|u\|^2 \leq \frac{4}{\pi^2}\|u_z\|^2. \tag{A.16}$$

Then, use of (A.15), (A.16) in (A.14) provides

$$< u^4 > \leq \gamma^4 \|\nabla u\|^2, \tag{A.17}$$

where

$$\gamma^4 = \frac{16}{\sigma^2\pi^3}\Big[1 + M\pi + \frac{1}{4}\pi^2(\sigma^2 + M^2)\Big]. \tag{A.18}$$

It should be noted that the above derivation does not assume periodicity in x and y. Furthermore, when $u = 0$ on both $z = 0, 1$, inequality (A.17) holds but with the constant in (A.18) sharpened to

$$\gamma^4 = \frac{1 + 2M\pi + \pi^2(\sigma^2 + M^2)}{\sigma^2\pi^3}.$$

When the periodicity conditions are taken into account we may further sharpen (A.14) and (A.17). For a hexagon with side length $4\pi/3a$ and solution wavelength a, (A.14), (A.17) continue to hold but the constants may be replaced by

$$\gamma^4 = \frac{36a^2 + 48a\pi^2 + 28\pi^4}{3\pi^5} \quad \text{or} \quad = \frac{9a^2 + 24a\pi^2 + 28\pi^4}{12\pi^5}, \tag{A.19}$$

respectively.

A.7 A two-dimensional surface inequality

The following two inequalities and the counterexample were established in (Galdi et al., 1987).

We establish a value for c in the *two-dimensional* inequality, namely,

$$\int_\Gamma u^3 dA \le c\|u\| D(u), \tag{A.20}$$

for $u = 0$ at $z = 0$, and where Γ is the region $x \in (0,k)$ and V the two-dimensional cell $\Gamma \times (0,1)$. $\|\cdot\|$ denotes the L^2 norm on V and $D(\cdot)$ the Dirichlet integral there.

The procedure begins by noting

$$\int_\Gamma u^3 dA = 3\int_V u^2 \frac{\partial u}{\partial z}\, dV.$$

The Cauchy-Schwarz inequality is applied to obtain

$$\int_\Gamma u^3 dA \le 3D^{1/2}(u)\Big(\int_V u^4\, dV\Big)^{1/2}. \tag{A.21}$$

The u^4 term is handled by the method of (Joseph, 1976b). Since $u(x,0) = 0$ and u is $x-$periodic of period k, it follows that for $x \in (\hat{x}, \hat{x}+k)$,with $\hat{x}$ arbitrary but fixed,

$$u^2(x,z) = 2\int_0^z u(x,t)\frac{\partial u}{\partial t}(x,t)dt,$$

$$u^2(x,z) \le \int_{\hat{x}}^{\hat{x}+k} |u(s,z)|\Big|\frac{\partial u}{\partial s}\Big| ds + u^2(\hat{x},z).$$

We multiply these expressions and integrate over $(\hat{x}, \hat{x}+k)\times(0,1)$ to find

$$\begin{aligned}\int_0^1\int_{\hat{x}}^{\hat{x}+k} u^4(x,z)\, dx\, dz \le & 2\int_0^1\int_{\hat{x}}^{\hat{x}+k} |u|\,|u_z|\, dx\, dz \int_0^1\int_{\hat{x}}^{\hat{x}+k} |u|\,|u_x|\, dx\, dz \\ & + 2\int_0^1 u^2(\hat{x},z)\, dz \int_0^1\int_{\hat{x}}^{\hat{x}+k} |u|\,|u_x|\, dx\, dz.\end{aligned}$$

Only the single integral in the expression above depends on $\hat{x}$, since u is a periodic function in x, of period k. Therefore, we integrate this inequality over $(\hat{x}, \hat{x}+k)$ and set $\hat{x} = 0$, and then use the Cauchy-Schwarz inequality to find

$$\int_0^1\int_0^k u^4(x,z)\, dx\, dz \le 2\|u\|^2\, \|u_x\|\, \|u_z\| + \frac{2}{k}\|u\|^3\, \|u_x\|.$$

Finally, we use Poincaré's inequality (for a function that vanishes only at $z = 0$) to deduce

$$\int_0^1\int_0^k u^4(x,z)\, dx\, dz \le 2\|u\|^2\, \|\nabla u\|^2 + \frac{4}{\pi k}\|u\|^2\, \|\nabla u\|^2. \tag{A.22}$$

Inequality (A.20) follows by using (A.22) in (A.21), with a value of c given by

$$c = 3\sqrt{2 + \frac{4}{\pi k}}. \tag{A.23}$$

A.8 Inequality (A.20) is false in three-dimensions

To provide a counterexample to (A.20) in three dimensions consider the region

$$V = (0,2) \times (0,2) \times (0,1).$$

Define a C^∞ cut-off function by

$$h = \begin{cases} 1, & r \equiv |\mathbf{x} - P| \le \frac{1}{2}, \\ \exp\left[\frac{4}{3} + (r^2 - 1)^{-1}\right], & \frac{1}{2} \le r \le 1, \\ 0, & r \ge 1, \end{cases}$$

where $P = (1,1,1)$, on the cell V. For $\alpha > 1$, a constant to be specified, define f by $f(\mathbf{x}) = h \exp(-\alpha r)$, and define the quantity F by

$$F = \frac{\int_\Gamma |f|^3 dA}{\|f\| \, \|\nabla f\|^2},$$

where Γ is the $z = 1$ boundary of V. Since $|h|, |h'|$ are bounded and $h \equiv 1$ for $r \in (0, \frac{1}{2})$, we put $\rho^2 = (x-1)^2 + (y-1)^2$ to deduce

$$F \ge B \, \frac{\int_0^{\frac{1}{2}} \exp(-3\alpha\rho)\, \rho \, d\rho}{\alpha^2} \left(\int_0^1 \exp(-2\alpha r)\, r^2 \, dr \right)^{3/2}.$$

In this expression B is a positive constant, independent of α. The change of variables $u = \alpha\rho$, $v = \alpha r$, leads to

$$F \ge \text{constant} \times \sqrt{\alpha}, \qquad \alpha \to \infty.$$

This establishes (A.20) is not true in three dimensions.

A.9 A boundary estimate for u^2

Let now V be a three-dimensional periodicity cell contained in $z \in (0,1)$. Let also Γ be that part of the boundary of V that intersects $z = 1$.

We establish the following *isoperimetric* inequality for functions u that vanish on $z = 0$,

$$\int_\Gamma u^2 \, dA \le 2\|u\| \, \|\nabla u\|. \tag{A.24}$$

This follows since

$$u^2(x,y,1) = 2\int_0^1 u\,\frac{\partial u}{\partial z}\,dz;$$

integration over V and use of the Cauchy-Schwarz inequality completes the derivation. To see inequality (A.24) is the best possible, choose $u = \sinh \alpha z$, then

$$\frac{\|u\|\,\|\nabla u\|}{\int_\Gamma u^2\,dA} \to \frac{1}{2},$$

for $\alpha \to \infty$.

A.10 A surface inequality for u^4

Suppose $u(x,y)$ is periodic in x, y of $x-$period k and of $y-$period m. Define Γ to be the rectangle $(\hat{x}, \hat{x}+k) \times (\hat{y}, \hat{y}+m)$, $\hat{x}$, $\hat{y}$ fixed. We now show that provided

$$\int_\Gamma u\,dA = 0,$$

$$\int_\Gamma u^4\,dA \le \left[1 + \frac{4}{\pi^2} + \frac{2(k+m)}{\pi\sqrt{km}}\right] \int_\Gamma u^2\,dA \int_\Gamma u_{;\alpha} u_{;\alpha}\,dA. \tag{A.25}$$

The proof employs the method of (Joseph, 1976b), p. 249. Since u has $x-$period k, $y-$period m, we write

$$\begin{aligned} 2\int_{\hat{x}}^{x} u(s,y)u_s(s,y)\,ds + u^2(\hat{x},y) &= u^2(x,y) \\ &= -2\int_x^{\hat{x}+k} u(s,y)u_s(s,y)\,ds + u^2(\hat{x},y) \end{aligned}$$

and

$$\begin{aligned} 2\int_{\hat{y}}^{y} u(x,t)u_t(x,t)\,dt + u^2(x,\hat{y}) &= u^2(x,y) \\ &= -2\int_y^{\hat{y}+m} u(x,t)u_t(x,t)\,dt + u^2(x,\hat{y}). \end{aligned}$$

From these expressions it is easily seen that

$$u^2(x,y) \le \int_{\hat{x}}^{\hat{x}+k} |u(s,y)|\,|u_s(s,y)|\,ds + u^2(\hat{x},y), \tag{A.26}$$

$$u^2(x,y) \le \int_{\hat{y}}^{\hat{y}+m} |u(x,t)|\,|u_t(x,t)|\,dt + u^2(x,\hat{y}). \tag{A.27}$$

We now multiply (A.26) and (A.27) together and integrate over Γ twice to find, with the help of the Cauchy-Schwarz inequality,

$$\begin{aligned} km \int_\Gamma u^4 \, dA \leq & km \int_\Gamma u^2 \, dA \int_\Gamma u_{;\alpha} u_{;\alpha} \, dA + \left(\int_\Gamma u^2 \, dA \right)^2 \\ & + (k+m) \left(\int_\Gamma u^2 \, dA \right)^{3/2} \left(\int_\Gamma u_{;\alpha} u_{;\alpha} \, dA \right)^{1/2} . \end{aligned} \tag{A.28}$$

The Wirtinger inequality, in two dimensions, shows

$$\int_\Gamma u^2 \, dA \leq \frac{4km}{\pi^2} \int_\Gamma u_{;\alpha} u_{;\alpha} \, dA.$$

We put this into (A.28) and divide by km to obtain (A.25).

References

Abo-Eldahab, E. and El Gendy, M. (2000). Radiation effect on convective heat transfer in an electrically conducting fluid at a stretching surface with variable viscosity and uniform free stream. *Physica Scripta*, 62:321–325.

Abo-Eldahab, E. and Salem, A. (2001). Radiation effect on mhd free convection flow of a gas past a semi-infinite vertical plate with variable viscosity. *Int. J. Computational Fluid Dynamics*, 14:243–252.

Abraham, A. (2002). Rayleigh-Bénard convection in a micropolar ferromagnetic fluid. *Int. J. Engng. Sci.*, 40:449–460.

Adams, R. (1975). *Sobolev spaces.* Academic Press, New York.

Ahmadi, G. (1976). Stability of a micropolar fluid layer heated from below. *Int. J. Engng. Sci.*, 14:81–89.

Alex, S. and Patil, P. (2002a). Effect of a variable gravity field on convection in an anisotropic porous medium with internal heat source and inclined temperature gradient. *J. Heat Transfer*, 124:144–150.

Alex, S. and Patil, P. (2002b). Effect of variable gravity field on thermal instability in a porous medium with inclined temperature gradient and vertical throughflow. *J. Porous Media*, 5:137–147.

Alex, S., Patil, P., and Venkatakrishnan, K. (2001). Variable gravity effects on thermal instability in a porous medium with internal heat source and inclined temperature gradient. *Fluid Dyn. Res.*, 29:1–6.

Allen, L. and Bridges, T. (2002). Numerical exterior algebra and the compound matrix method. *Numer. Math.*, 92:197–232.

Allen, M. (1984). *Collocation Techniques for Modeling Compositional Flows in Oil Reservoirs.* Springer, Heidelberg.

Allen, M. (1986). Mechanics of multiphase fluid flows in variably saturated porous media. *Int. J. Engng. Sci.*, 24:339–351.

Allen, M., Behie, A., and Trangenstein, J. (1988). *Multiphase Flow in Porous Media: Mechanics, Mathematics, and Numerics.* Springer, New York.

Allen, M., Ewing, R., and Lu, P. (1992). Well conditioned iterative schemes for mixed finite-element models of porous-media flows. *SIAM Jour. Sci. Stat. Comp.*, 13:794–814.

Alvarez-Ramirez, J., Puebla, H., and Ochoa-Tapia, J. (2001). Linear boundary control for a class of nonlinear pde processes. *Systems and Control Letters*, 44:395–403.

Ambrosi, D. (2002). Infiltration through deformable porous media. *ZAMM*, 82:115–124.

Ames, K. and Cobb, S. (1994). Penetrative convection in a porous medium with internal heat sources. *Int. J. Engng. Sci.*, 32:95–105.

Ames, K., Payne, L., and Song, J. (2001). Spatial decay in pipe flow of a viscous fluid interfacing a porous medium. *Math. Models Meth. Appl. Sci.*, 11:1547–1562.

Anderdeck, C., Colovas, P., Degen, M., and Renardy, Y. (1998). Instabilities in two layer Rayleigh-Bénard convection: overview and outlook. *Int. J. Engng. Sci.*, 36:1451–1470.

Andersland, O. and Anderson, D. (1978). *Geotechnical engineering for cold regions.* McGraw-Hill, New York.

Anderson, E., Bai, Z., Bischof, C., Demmel, J., Dongarra, J., Du Croz, J., Greenbaum, A., Hammarling, S., McKenney, A., Ostrouchov, S., and Sorensen, D. (1995). *LAPACK Users' guide.* SIAM, second edition.

Antar, B., Collins, F., and Fichtl, G. (1980). Influence of solidification on surface tension driven convection. *Int. J. Heat Mass Transfer*, 23:191–201.

Antontsev, S., Diaz, J., and Shmarev, S. (2001). *Energy methods for free boundary problems.* Birkhauser, Boston.

Arnold, V. (1965a). Conditions for nonlinear stability of stationary plane curvilinear flows of an ideal fluid. *Dokl. Akad. Nauk SSSR*, 162:975–978.

Arnold, V. (1965b). Variational principle for three-dimensional steady state flows of an ideal fluid. *Prikl. Mat. Mekh.*, 29:846–851.

Arnold, V. (1966a). An a priori estimate in the theory of hydrodynamic stability. *Izv. VUZ Mat.*, 54:3–5.

Arnold, V. (1966b). Sur un principe variationnel pour les écoulements des liquides parfaits et ses applications aux problèmes de stabilité nonlineaires. *J. Mécanique*, 5:29–43.

Azouni, M. (1983). Hysteresis loop in water between 0°C and 4°C. *Geophys. Astrophys. Fluid Dyn.*, 24:137–142.

Azouni, M. and Normand, C. (1983a). Thermoconvective instabilities in a vertical cylinder of water with maximum density effects. I. Experiments. *Geophys. Astrophys. Fluid Dyn.*, 23:209–222.

Azouni, M. and Normand, C. (1983b). Thermoconvective instabilities in a vertical cylinder of water with maximum density effects. II. Theory. *Geophys. Astrophys. Fluid Dyn.*, 23:223–245.

Bailey, P., Chen, P., and Straughan, B. (1984). Stabilization criteria for thermally explosive materials. *Acta Mechanica*, 53:73–79.

Baines, P. and Gill, A. (1969). On thermohaline convection with linear gradients. *J. Fluid Mech.*, 37:289–306.

Bardan, G., Bergeon, A., Knobloch, E., and Mojtabi, A. (2000). Nonlinear doubly diffusive convection in vertical enclosures. *Physica D*, 138:91–113.

Bardan, G., Knobloch, E., Mojtabi, A., and Khallouf, H. (2001). Nonlinear doubly diffusive convection with vibration. *Fluid Dyn. Res.*, 28:159–187.

Bardan, G. and Mojtabi, A. (1998). Theoretical stability study of double-diffusive convection in a square cavity. *Comptes Rendues de l'Academie des Sciences*, 326:851–857.

Bartuccelli, M. (2002). On the asymptotic positivity of solutions for the extended Fisher-Kolmogorov equation with nonlinear diffusion. *Math. Methods Appl. Sci.*, 25:701–708.

Bartuccelli, M., Doering, C., Gibbon, J., and Malham, S. (1993). Length scales in solutions of the Navier-Stokes equations. *Nonlinearity*, 6:549–568.

Bartuccelli, M. and Woolcock, C. (2001). On the positivity of solutions for a generalized diffusion model. *Nuovo Cimento B*, 116:1365–1373.

Batchelor, G. (1967). *An introduction to fluid dynamics.* Cambridge University Press, Cambridge.

Bear, J. and Gilman, A. (1995). Migration of salts in the unsaturated zone caused by heating. *Transport in Porous Media*, 19:139–156.

Beavers, G. and Joseph, D. (1967). Boundary conditions at a naturally impermeable wall. *J. Fluid Mech.*, 30:197–207.

Bees, M., Andrésen, P., Mosekilde, E., and Givskov, M. (2000). The interaction of thin-film flow, bacterial swarming and cell differentiation. *Journal of Mathematical Biology*, 40:27–63.

Bees, M. and Hill, N. (1999). Non-linear bioconvection in a deep suspension of gyrotactic swimming micro-organisms. *Journal of Mathematical Biology*, 38:135–168.

Bees, M., Pons, A., Sorensen, P., and Sagués, F. (2001). Chemoconvection: a chemically driven hydrodynamic instability. *Journal of Chemical Physics*, 114:1932–1943.

Beirao da Veiga, H. (1983). Diffusion on viscous fluids. existence and asymptotic properties of solutions. *Ann. Scuola Norm. Sup. Pisa*, 10:341–351.

Beirao da Veiga, H., Serapioni, R., and Valli, A. (1982). On the motion of non-homogeneous fluids in the presence of diffusion. *J. Math. Anal. Appl.*, 85:179–191.

Bénard, H. (1900). Les tourbillons cellulaires dans une nappe liquide. *Revue Gén. Sci. Pure Appl.*, 11:1261–1271.

Berezin, Y. and Hutter, K. (1997). On large scale vortical structures in (incompressible) fluids with thermal expansion. *Math. Models Meth. Appl. Sci.*, 7:113–123.

Berezin, Y., Hutter, K., and Spodareva, L. (1998). Stability analysis of gravity driven shear flows with free surface for power-law fluids. *Arch. Appl. Mech.*, 68:169–178.

Berg, J. and Acrivos, A. (1965). The effect of surface active agents on convection cells induced by surface tension. *Chem. Engng. Sci.*, 20:737–745.

Blennerhassett, P., Lin, F., and Stiles, P. (1991). Heat transfer through strongly magnetized ferrofluids. *Proc. Roy. Soc. London A*, 433:165–177.

Borchers, W. and Miyakawa, T. (1995). On stability of exterior stationary Navier-Stokes flow. *Acta Math.*, 174:311–382.

Bormann, A. (2001). The onset of convection in the Rayleigh - Bénard problem for compressible fluids. *Continuum Mech. Thermodyn.*, 13:9–23.

Bradley, R. (1978). Overstable electroconvective instabilities. *Q. J. Mech. Appl. Math.*, 31:381–390.

Bresch, D., Essoufi, E., and Sy, M. (2002). Some new Kazhikhov-Smagulov type systems: pollutant spread and low mach number combustion models. *Comptes Rendus Math.*, 335:973–978.

Briley, P., Deemer, A., and Slattery, J. (1976). Blunt knife-edge and disk surface viscometers. *J. Colloid Interface Sci.*, 56:1–18.

Brinkman, H. (1957). A calculation of viscous force exerted by a flowing fluid on a dense swarm of particles. *Appl. Sci. Res.*, 1:27–34.

Budu, P. (2002). *Conditional and unconditional nonlinear stability in fluid dynamics.* PhD thesis, University of Durham.

Buonomo, B. and Rionero, S. (2001). Some nonlinear stability results for bioconvection of upswimming cells. In Ciancio, V., Donato, A., Oliveri, F., and Rionero, S., editors, *Proc. WASCOM99 10th conference on Waves and Stability in Continuous Media*, pages 66–76.

Busse, F. (1967). The stability of finite amplitude cellular convection and its relation to an extremum principle. *J. Fluid Mech.*, 30:625–649.

Butler, K. and Farrell, B. (1992). Three-dimensional optimal perturbations in viscous shear flow. *Phys. Fluids A*, 4:1637–1650.

Calmelet-Eluhu, C. (1996). Energy stability of a stationary conducting thermomicropolar fluid layer under the influence of a magnetic field. *Math. Models Meth. Appl. Sci.*, 6:385–403.

Caltagirone, J. (1975). Thermoconvective instabilities in a horizontal porous layer. *J. Fluid Mech.*, 72:269–287.

Caltagirone, J. (1980). Stability of a saturated porous layer subject to a sudden rise in surface temperature: comparison between the linear and energy methods. *Q. J. Mech. Appl. Math.*, 33:47–58.

Capone, F. (2001). On the onset of convection in porous media: temperature dependent viscosity. *Boll. Unione Matem. Ital.*, 4:143–156.

Capone, F. and De Angelis, M. (1993). On the energy stability of fluid motions in the exterior of a sphere under free boundary like conditions. *Rend. Accad. Sci. Fis. Matem. Napoli*, 60:7–26.

Capone, F. and Gentile, M. (1994). Nonlinear stability analysis of convection for fluids with exponentially temperature-dependent viscosity. *Acta Mechanica*, 107:53–64.

Capone, F. and Gentile, M. (1995). Nonlinear stability analysis of the Bénard problem for fluids with a convex nonincreasing temperature depending viscosity. *Continuum Mech. Thermodyn.*, 7:297–309.

Capone, F. and Rionero, S. (1999). Temperature depedent viscosity and its influence on the onset of convection in a porous medium. *Rend. Accad. Sci. Fis. Matem. Napoli*, 66:159–172.

Capone, F. and Rionero, S. (2000). Thermal convection with horizontally periodic temperature gradient. *Rend. Accad. Sci. Fis. Matem. Napoli*, 67:119–128.

Capone, F. and Rionero, S. (2001). Temperature depedent viscosity and its influence on the onset of convection in a porous medium. In Ciancio, V., Donato, A., Oliveri, F., and Rionero, S., editors, *Proceedings of Wascom 99*, pages 77–85. World Scientific.

Carmi, S. (1974). Energy stability of modulated flows. *Phys. Fluids*, 17:1951–1955.

Carr, M. (2003a). *Convection in porous media flows.* PhD thesis, University of Durham.

Carr, M. (2003b). A model for convection in the evolution of under-ice melt ponds. *Continuum Mech. Thermodyn.*, 15:45–54.

Carr, M. (2003c). Unconditional nonlinear stability for temperature dependent density flow in a porous medium. *Math. Models Meth. Appl. Sci.*, 13:207–220.

Carr, M. and de Putter, S. (2003). Penetrative convection in a horizontally isotropic porous layer. *Continuum Mech. Thermodyn.*, 15:33–43.

Carr, M. and Straughan, B. (2003). Penetrative convection in fluid overlying a porous layer. *Advances in Water Resources*, 26:263–276.

Carrigan, C. and Cygan, R. (1986). Implications of magma chamber dynamics for Soret-related fractionation. *J. Geophys. Res.*, 91:1451–1461.

Castellanos, A., Atten, P., and Velarde, M. (1984a). Electrothermal convection: Felici's hydraulic model and the Landau picture of non-equilibrium phase transistions. *J. Non-Equilibrium Thermodyn.*, 9:235–244.

Castellanos, A., Atten, P., and Velarde, M. (1984b). Oscillatory and steady convection in dielectric layers subjected to unipolar injection and temperature gradient. *Phys. Fluids*, 27:1607–1615.

Castillo, J. and Velarde, M. (1983). Buoyancy-thermocapillary instability: the role of interfacial deformation in one and two - component fluid layers heated from below or above. *J. Fluid Mech.*, 125:463–474.

Cathles, L. (1990). Scales and effects of fluid flow in the upper crust. *Science*, 248:323–329.

Caviglia, G., Morro, A., and Straughan, B. (1992). Reflection and refraction at a variable porosity interface. *J. Acoustical Soc. Amer.*, 1992:1113–1119.

Celia, M., Kindred, J., and Herrera, I. (1989). Contaminant transport and biodegradation. i. A numerical model for reactive transport in porous media. *Water Resources Research*, 25:1141–1148.

Chandra, K. (1938). Instability of fluids heated from below. *Proc. Roy. Soc. London A*, 164:231–242.

Chandrasekhar, S. (1953). The instability of a layer of fluid heated from below and subject to Coriolis forces. *Proc. Roy. Soc. London A*, 217:306–327.

Chandrasekhar, S. (1981). *Hydrodynamic and Hydromagnetic Stability.* Dover, New York.

Chandrasekhar, S. and Elbert, D. (1955). The instability of a layer of fluid heated from below and subject to Coriolis forces. II. *Proc. Roy. Soc. London A*, 231:198–210.

Charki, Z. (1995). Existence and uniqueness of solutions for the steady deep Bénard convection problem. *Zeit. Angew. Math. Mech.*, 75:909–915.

Charki, Z. (1996). The initial value problem for the deep Bénard convection equations with data in L(q). *Math. Models Meth. Appl. Sci.*, 6:269–277.

Charki, Z. and Zeytounian, R. (1994). The Bénard problem for deep convection - Lorenz deep system. *Int. J. Engng. Sci.*, 32:1561–1566.

Charki, Z. and Zeytounian, R. (1995). The Bénard problem for deep convection - derivation of the Landau - Ginzburg equation. *Int. J. Engng. Sci.*, 33:1839–1847.

Charrier-Mojtabi, M., Karimi-Fard, M., Azaiez, M., and Mojtabi, A. (1998). Onset of a double-diffusive convective regime in a rectangular porous cavity. *J. Porous Media*, 1:107–121.

Chasnov, J. and Tse, K. (2001). Turbulent penetrative convection with an internal heat source. *Fluid Dyn. Res.*, 28:397–421.

Chen, B., Cunningham, A., Ewing, R., Peralta, R., and Visser, E. (1994). Two-dimensional modelling of microscale transport and biotransformation in porous media. *Numer. Meth. Part. Diff. Equations*, 10:65–83.

Chen, F. and Chen, C. (1988). Onset of finger convection in a horizontal porous layer underlying a fluid layer. *J. Heat Transfer*, 3:403–409.

Cheney, W. and Kincaid, D. (1985). *Numerical mathematics and computing.* Brooks-Cole, Monterey.

Cheng, P. (1978). Heat transfer in geothermal systems. *Advances in Heat Transfer*, 14:1–105.

Cheng, P., Chen, C., and Lai, H. (2001). Nonlinear stability analysis of the thin micropolar liquid film flowing down on a vertical cylinder. *J. Fluids Engng. ASME*, 123:411–421.

Chhuon, B. and Caltagirone, J. (1979). Stability of a horizontal porous layer with timewise periodic boundary conditions. *J. Heat Transfer*, 101:244–248.

Childress, S. (1981). *Mechanics of swimming and flying.* Cambridge University Press, Cambridge.

Childress, S., Levandowsky, M., and Spiegel, E. (1975). Pattern formation in a suspension of swimming micro-organisms: equations and stability theory. *J. Fluid Mech.*, 63:591–613.

Chirita, S. (2001). Uniqueness and continuous dependence of solutions to the incompressible micropolar flows forward and backward in time. *Int. J. Engng. Sci.*, 39:1787–1802.

Christopherson, D. (1940). Note on the vibration of membranes. *Quart. J. Math.*, 11:63–65.

Ciarletta, M. (1995). On the theory of heat-conducting micropolar fluids. *Int. J. Engng. Sci.*, 33:1403–1417.

Ciarletta, M. (2001). Spatial decay estimates for heat-conducting micropolar fluids. *Int. J. Engng. Sci.*, 39:655–668.

Ciesjko, M. and Kubik, J. (1999). Derivation of matching conditions at the contact surface between fluid-saturated porous solid and bulk fluid. *Transport in Porous Media*, 34:319–336.

Clark, P., Pisias, N., Stocker, T., and Weaver, A. (2002). The role of thermohaline circulation in abrupt climate change. *Nature*, 415:863–869.

Clenshaw, C. and Elliott, D. (1960). A numerical treatment of the Orr - Sommerfeld equation in the case of a laminar jet. *Q. Jl. Mech. Appl. Math.*, 13:300–313.

Cloot, A. and Lebon, G. (1986). Marangoni convection induced by a nonlinear temperature-dependent surface tension. *J. Physique*, 47:23–29.

Coddington, E. and Levinson, N. (1955). *Theory of ordinary differential equations.* McGraw-Hill, New York.

Coriell, S., McFadden, G., Voorhees, P., and Sekerka, R. (1987). Stability of a planar interface during solidification of a multicomponent system. *J. Crystal Growth*, 82:295–302.

Corte, A. (1966). Particle sorting by repeated freezing and thawing. *Biuletyn Peryglacjalny*, 15:175–240.

Coscia, V. and Padula, M. (1990). Nonlinear stability in a compressible atmosphere. *Geophys. Astrophys. Fluid Dyn.*, 54:49–83.

Cowley, M. and Rosensweig, R. (1967). The interfacial stability of a ferromagnetic fluid. *J. Fluid Mech.*, 30:671–688.

Crisciani, F. (1998). Stability of Niiler's solution of the general circulation problem. *J. Physical Oceanography*, 28:218–226.

Crisciani, F., Cavallini, F., and Mosetti, R. (1994). Stability of analytical solutions of Stommel-Veronis ocean circulation models. *J. Physical Oceanography*, 24:155–158.

Crisciani, F., Cavallini, F., and Mosetti, R. (1995). Nonlinear asymptotic stability of Munk's solutions of the general circulation problem. *J. Physical Oceanography*, 25:1723–1729.

Crisciani, F. and Mosetti, R. (1990). On the stability of wind-driven flows in oceanic basins. *J. Physical Oceanography*, 20:1787–1790.

Crisciani, F. and Mosetti, R. (1991). Stability criteria independent from the perturbation wavenumber for forced zonal flows. *J. Physical Oceanography*, 21:1075–1079.

Crisciani, F. and Mosetti, R. (1992). Non-linear stability of forced flows in a closed basin ocean by means of the Lyapunov method. *Stab. Appl. Anal. Continous Media*, 2:403–415.

Crisciani, F. and Mosetti, R. (1994). An extension of the Serrin non-linear universal stability criterion to the quasi-geostrophic barotropic fluid dynamics. *Il Nuovo Cimento*, 17:523–528.

Crisciani, F. and Purini, R. (1995). Non-linear stability of thermally driven flows in a closed-basin ocean. *Eur. J. Mech. B/Fluids*, 14:263–274.

Crisciani, F. and Purini, R. (1996). The effect of cross-stream topography on the inertial evolution of a coastal current. *Il Nuovo Cimento*, 19:389–398.

Curran, M. and Allen, M. (1990). Parallel computing for solute transport models via alternating direction collocation. *Advances in Water Resources*, 13:70–75.

Curtis, R. (1971). Flows and wave propagation in ferrofluids. *Phys. Fluids*, 14:2096–2102.

Darcy, H. (1856). *Les fontaines publiques de la ville de Dijon.* Dalmont, Paris.

Das, D., Nassehi, V., and Wakeman, R. (2002). A finite volume model for the hydrodynamics of combined free and porous flow in sub-surface regions. *Advances in Environmental Research*, 7:35–58.

Datta, A. and Sastry, V. (1976). Thermal instability of a horizontal layer of micropolar fluid heated from below. *Int. J. Engng. Sci.*, 14:631–637.

Dauby, P., Nelis, M., and Lebon, G. (2002). Generalized Fourier equations and thermoconvective instabilities. *Revista Mexicana de Fisica*, 48:57–62.

Davis, S. (1969a). Buoyancy-surface tension instability by the method of energy. *J. Fluid Mech.*, 39:347–359.

Davis, S. (1969b). On the principle of exchange of stabilities. *Proc. Roy. Soc. London A*, 310:341–358.

Davis, S. (1971). On the possibility of subcritical instabilities. In Leipholz, H., editor, *Instability of continuous systems*, IUTAM Symp. Herrenalb. Springer.

Davis, S. (1987). Thermocapillary instabilites. *Ann. Rev. Fluid Mech.*, 19:403–435.

Davis, S. and Homsy, G. (1980). Energy stability for free - surface problems: buoyancy - thermocapillary layers. *J. Fluid Mech.*, 98:527–553.

Davis, S. and von Kerczek, C. (1973). A reformulation of energy stability theory. *Arch. Rational Mech. Anal.*, 52:112–117.

Dawkins, P., Dunbar, S., and Douglass, R. (1998). The origin and nature of spurious eigenvalues in the spectral tau method. *J. Computational Physics*, 147:441–462.

De Angelis, M. (1990). On universal energy stability of fluid motions in unbounded domains. *Rend. Accad. Sci. Fis. Matem. Napoli*, 57:5–24.

Dennis, J. and Schnabel, R. (1983). *Numerical methods for unconstrained optimization and nonlinear equations.* Prentice-Hall, Englewood Cliffs.

Deo, B. and Richardson, A. (1983). Generalized energy methods in electrohydrodynamic stability theory. *J. Fluid Mech.*, 137:131–151.

Diaz, J. and Galiano, G. (1997). On the Boussinesq system with nonlinear thermal diffusion. *Nonlinear Analysis*, 30:3255–3263.

Diaz, J. and Galiano, G. (1998). Existence and uniqueness of solutions to the Boussinesq system with nonlinear thermal diffusion. *Topological Methods in Nonlinear Analysis*, 11:59–82.

Diaz, J. and Nagai, T. (1995). Symmetrization in a parabolic-elliptic system related to chemotaxis. *Advances in Math. Sciences Applicns.*, 5:659–680.

Diaz, J., Nagai, T., and Rakotoson, J. (1989). Symmetrization techniques on unbounded domains: application to a chemotaxis system on $\mathbb{R}^n$. *J. Differential Equations*, 145:156–183.

Diaz, J. and Quintanilla, R. (2002). Spatial and continuous dependence estimates in linear viscoelasticity. *J. Math. Anal. Appl.*, 273:1–16.

Diaz, J. and Straughan, B. (2002). Global stability for convection when the viscosity has a maximum. Manuscript.

Discacciati, M., Miglio, E., and Quarteroni, A. (2002). Mathematical and numerical models for coupling surface and groundwater flows. *Appl. Numer. Math.*, 43:57–74.

Doering, C. and Constantin, P. (1996). Variational bounds on energy dissipation in incompressible flows: III. Convection. *Phys. Rev. E*, 53:5957–5981.

Doering, C. and Constantin, P. (1998). Bounds for heat transport in a porous layer. *J. Fluid Mech.*, 376:263–296.

Doering, C. and Foias, C. (2002). Energy dissipation in body-forced turbulence. *J. Fluid Mech.*, 467:289–306.

Doering, C. and Gibbon, J. (1995). *Applied Analysis of the Navier-Stokes Equations.* Cambridge University Press, Cambridge.

Doering, C. and Gibbon, J. (2002). Bounds on moments of the energy spectrum for weak solutions of the three-dimensional Navier-Stokes equations. *Physica D*, 165:163–175.

Doering, C. and Hyman, J. (1997). Energy stability bounds on convective heat transport: Numerical study. *Phys. Rev. E*, 55:7775–7778.

Doering, C., Spiegel, E., and Worthing, R. (2000). Energy dissipation in a shear layer with suction. *Phys. Fluids*, 12:1955–1968.

Doering, C. and Titi, E. (1995). Exponential decay rate of the power spectrum for solutions of the Navier-Stokes equations. *Phys. Fluids*, 7:1384–1390.

Dongarra, J., Straughan, B., and Walker, D. (1996). Chebyshev tau - QZ algorithm methods for calculating spectra of hydrodynamic stability problems. *Appl. Numer. Math.*, 22:399–435.

Drazin, P. and Reid, W. (1981). *Hydrodynamic stability.* Cambridge University Press.

Dudis, J. and Davis, S. (1971). Energy stability of the buoyancy boundary layer. *J. Fluid Mech.*, 47:381–403.

Duong, D. and Weiland, R. (1981). Enzyme deactivation in fixed bed reactors with Michaelis - Menten kinetics. *Biotechnology and Bioengineering*, 23:691–705.

Dupuit, J. (1863). *Etudes thèoriques et pratiques sur le mouvement des eaux.* Dunod, Paris.

Embleton, C. and King, C. (1975). *Glacial and periglacial geomorpholgy.* Wiley, New York.

Eringen, A. (1964). Simple microfluids. *Int. J. Engng. Sci.*, 2:205–217.

Eringen, A. (1969). Micropolar fluids with stretch. *Int. J. Engng. Sci.*, 7:115–127.

Eringen, A. (1972). Theory of thermomicrofluids. *J. Math. Anal. Appl.*, 38:480–496.

Eringen, A. (1980). Theory of anisotropic micropolar fluids. *Int. J. Engng. Sci.*, 18:5–17.

Errafiy, M. and Zeytounian, R. K. (1991). The Bénard problem for deep convection: linear theory. *Int. J. Engng. Sci.*, 29:625–635.

Ewing, R. (1996). Multidisciplinary interactions in energy and environmental engineering. *J. Comp. Appl. Math.*, 74:193–215.

Ewing, R. (1997). Mathematical modeling and simulation for applications of fluid flow in porous media. In Alber, M., Hu, B., and Rosenthal, J., editors, *Current and Future Directions in Applied Mathematics*, pages 161–182. Birkhauser.

Ewing, R., Wang, H., Sharpley, R., and Celia, M. (1997). A three-dimensional finite element simulation of nuclear waste contamination transport in porous media. In Siriwardane, H. and Zaman, M., editors, *Computer Methods and Advances in Geomechanics*, volume 4, pages 2673–2679. A.A. Balkema.

Ewing, R. and Weekes, S. (1998). Numerical methods for contaminant transport in porous media. *Computational Mathematics*, 202:75–95.

Ezzat, M. and Othman, M. (2000). Thermal instability in a rotating micropolar fluid layer subject to an electric field. *Int. J. Engng. Sci.*, 38:1851–1867.

Fichera, G. (1978). *Numerical and quantitative analysis.* Pitman, London.

Finlayson, B. (1970). Convective instability of ferromagnetic fluids. *J. Fluid Mech.*, 40:753–767.

Finucane, R.G. & Kelly, R. (1976). Onset of instability in a fluid layer heated sinusoidally from below. *Int. J. Heat Mass Transfer*, 19:71–85.

Firdaouss, M., Guermond, J., and Le Quére, P. (1997). Nonlinear corrections to Darcy's law at low Reynolds numbers. *J. Fluid Mech.*, 343:331–350.

Flavin, J. and Rionero, S. (1995). *Qualitative Estimates for Partial Differential Equations.* CRC Press, Boca Raton.

Flavin, J. and Rionero, S. (1997). On the temperature distribution in cold ice. *Rend. Matem. Acad. Lincei*, 99:299–312.

Flavin, J. and Rionero, S. (1998). Asymptotic and other properties of a nonlinear diffusion model. *J. Math. Anal. Appl.*, 228:119–140.

Flavin, J. and Rionero, S. (1999a). The Bénard problem for nonlinear heat conduction: unconditional stability. *Q. J. Mech. Appl. Math.*, 52:441–452.

Flavin, J. and Rionero, S. (1999b). Nonlinear stability for a thermofluid in a vertical porous slab. *Continuum Mech. Thermodyn.*, 11:173–179.

Flavin, J. and Rionero, S. (2001). Some Lyapunov functionals for nonlinear diffusion and nonlinear stability. In Ciancio, V., Donato, A., Oliveri, F., and Rionero, S., editors, *Proceedings of Wascom 99*, pages 178–187. World Scientific.

Forchheimer, P. (1901). Wasserbewegung durch boden. *Z. Vereines Deutscher Ingnieure*, 50:1781–1788.

Fournier, R. (1990). Double-diffusive convection in geothermal systems: The Salton sea, California, geothermal system as a likely candidate. *Geothermics*, 19:481–496.

Franchi, F. and Straughan, B. (1988). Convection, stability and uniqueness for a fluid of third grade. *Int. J. Nonlinear Mech.*, 23:377–384.

Franchi, F. and Straughan, B. (1992a). A nonlinear energy stability analysis of a model for deep convection. *Int. J. Engng. Sci.*, 30:739–745.

Franchi, F. and Straughan, B. (1992b). Nonlinear stability for thermal convection in a micropolar fluid with temperature-dependent viscosity. *Int. J. Engng. Sci.*, 30:1349–1360.

Franchi, F. and Straughan, B. (1993). Stability and nonexistence results in the generalized theory of a fluid of second grade. *J. Math. Anal. Appl.*, 180:122–137.

Franchi, F. and Straughan, B. (1994). Thermal convection at low temperature. *J. Non-Equilibrium Thermodyn.*, 19:368–374.

Franchi, F. and Straughan, B. (2001). A comparison of the Graffi and Kazhikhov-Smagulov models for top heavy pollution instability. *Advances in Water Resources*, 24:585–594.

Frölich, J., Laure, P., and Peyret, R. (1992). Large departures from the Boussinesq approximation in the Rayleigh-Bénard problem. *Phys. Fluids A*, 4:1355–1372.

Fukuta, H. and Murakami, Y. (1993). Nonlinear stability of Kolmogorov flow with bottom-friction using the energy method. In *Proc. of analysis of nonlinear phenomena and its application*. World Scientific.

Fukuta, H. and Murakami, Y. (1995). Nonlinear stability of Kolmogorov flow with bottom-friction using the energy method. *J. Phys. Soc. Japan*, 64:3725–3739.

Gailitis, A. (1977). Formation of the hexagonal pattern on the surface of a ferrromagnetic fluid in an applied magnetic field. *J. Fluid Mech.*, 82:401–413.

Gajewski, H. and Zacharias, K. (1998). Global behaviour of a reacting-diffusion system modelling chemotaxis. *Math. Nachr.*, 195:77–114.

Galdi, G., Payne, L., Proctor, M., and Straughan, B. (1987). Convection in thawing subsea permafrost. *Proc. Roy. Soc. London A*, 414:83–102.

Galdi, G. and Rionero, S. (1985). *Weighted energy methods in fluid dynamics and elasticity*, volume 1134 of *Lect. Notes Math.* Springer.

Galdi, G. and Straughan, B. (1985a). Exchange of stabilities, symmetry and nonlinear stability. *Arch. Rational Mech. Anal.*, 89:211–228.

Galdi, G. and Straughan, B. (1985b). A nonlinear analysis of the stabilizing effect of rotation in the Bénard problem. *Proc. Roy. Soc. London A*, 402:257–283.

Galdi, G. and Straughan, B. (1987). A modified model problem of Drazin and Reid exhibiting sharp conditional stability. *Ann. Univ. Ferrara*, 32:39–43.

Galiano, G. (2000). Spatial and time localization of solutions of the Boussinesq system with nonlinear thermal diffusion. *Nonlinear Analysis*, 42:423–438.

Gardner, D., Trogdon, S., and Douglas, R. (1989). A modified tau spectral method that eliminates spurious eigenvalues. *J. Computational Physics*, 80:137–167.

Gentile, M. and Rionero, S. (2000). A note on the global nonlinear stability for penetrative convection in porous media for fluids with cubic density. *Rend. Accad. Sci. Fis. Matem. Napoli*, 67:129–142.

Gentile, M. and Rionero, S. (2002). Stability results for penetrative convection in porous media for fluids with cubic density. In Monaco, R., Bianchi, M. P., and Rionero, S., editors, *Proceedings WASCOM2001*

11th conference on waves and stability in continuous media, pages 214–219.

George, J., Gunn, R., and Straughan, B. (1989). Patterned ground formation and penetrative convection in porous media. *Geophys. Astrophys. Fluid Dyn.*, 46:135–158.

Ghorai, S. and Hill, N. (2000a). Periodic arrays of gyrotactic plumes in bioconvection. *Phys. Fluids*, 12:5–22.

Ghorai, S. and Hill, N. (2000b). Wavelengths of gyrotactic plumes in bioconvection. *Bulletin of Mathematical Biology*, 62:429–450.

Ghosal, S. and Spiegel, E. (1991). On thermonuclear convection: I. Shellular instability. *Geophys. Astrophys. Fluid Dyn.*, 61:161–178.

Gibbon, J. (1995). Length scales and ladder theorems for 2*d* and 3*d* convection. *Nonlinearity*, 8:81–92.

Gilbarg, D. and Trudinger, N. (1977). *Elliptic partial differential equations of second order*. Springer, Berlin - Heidelberg - New York.

Gill, A. (1966). The boundary layer regime for convection in a rectangular cavity. *J. Fluid Mech.*, 26:515–536.

Gill, A. (1969). A proof that convection in a porous vertical slab is stable. *J. Fluid Mech.*, 35:545–547.

Gilman, A. and Bear, J. (1996). The influence of free convection on soil salinization in arid regions. *Trans. Porous Media*, 23:275–301.

Giorgi, F. (1989). Two-dimensional simulations of possible mesoscale effects of nuclear war fires. I. Model description. *J. Geophys. Res. D*, 94:1127–1144.

Giorgi, T. (1997). Derivation of the Forchheimer law via matched asymptotic expansions. *Trans. Porous Media*, 29:191–206.

Givler, R. and Altobelli, S. (1994). A determination of effective viscosity for the Brinkman-Forchheimer flow model. *J. Fluid Mech.*, 258:355–370.

Gleason, K. (1984). Nonlinear Boussinesq convection in porous media: application to patterned ground formation. Master's thesis, University of Colorado.

Gleason, K., Krantz, W., Caine, N., George, J., and Gunn, R. (1986). Geometrical aspects of sorted patterned ground in recurrently frozen soil. *Science*, 232:216–220.

Goldthwaite, R. (1976). Frost sorted patterned ground: a review. *Quarternary Research*, 6:27–35.

Gourley, S. and Bartuccelli, M. (1995). Length scales in solutions of a scalar reaction-diffusion equation with delay. *Physics Letters A*, 202:79–87.

Graffi, D. (1953). Il teorema di unicità nella dinamica dei fluidi compressibili. *J. Rational Mech. Anal.*, 2:99–106.

Graffi, D. (1955). Il teorema di unicità per i fluidi incompressibili, perfetti, eterogenei. *Rev. Unione Mat. Argentina*, 17:73–77.

Graffi, D. (1959). Sur un théorème d'unicité pour le mouvement d'un fluide visqueux dans un domaine illimité. *Comptes Rend. Acad. Sci. Paris*, 249:1741–1743.

Graffi, D. (1960). Sul teorema di unicità per le equazioni del moto dei fluidi compressibili in un dominio illimitato. *Rend. Accad. Scienze Ist. Bologna*, 7:59–63.

Graffi, D. (1999). Opere scelte. In Fabrizio, M., Grioli, G., and Renno, P., editors, *Opere Scelte*. Consiglio Nazionale di Ricerca.

Graham, M., Müller, U., and Steen, P. (1992). Time-periodic thermal convection in Hele-Shaw slots: The diagonal oscillation. *Physics Fluids A*, 4:2382–2393.

Greenberg, L. and Marletta, M. (2000). Numerical methods for higher order Sturm-Liouville problems. *J. Computational and Applied Math.*, 125:367–383.

Greenberg, L. and Marletta, M. (2001). Numerical solution of non self-adjoint Sturm-Liouville problems and related systems. *SIAM J. Numer. Anal.*, 38:1800–1845.

Gresho, P. and Sani, R. (1970). The effects of gravity modulation on the stability of a heated fluid layer. *J. Fluid Mech.*, 40:783–806.

Griffiths, R. (1981). Layered double-diffusive convection in porous media. *J. Fluid Mech.*, 102:221–248.

Gripp, K. (1926). Uber frost und strukturboden auf Spitzbergen. *Gesell. Erdkunde Berlin Zeitschr.*, 10:351–354.

Gross, M. (1967). Electrohydrodynamic instability. In *Mantles of the Earth and terrestial planets.* Wiley.

Gumerman, R. and Homsy, G. (1975). The stability of uniformly accelerated flows with application to convection driven by surface tension. *J. Fluid Mech.*, 68:191–207.

Guo, J. and Kaloni, P. (1995a). Doubly diffusive convection in a porous medium, nonlinear stability, and the Brinkman effect. *Stud. Appl. Math.*, 94:341–358.

Guo, J. and Kaloni, P. (1995b). Nonlinear stability of convection induced by inclined thermal and solutal gradients. *ZAMP*, 46:645–654.

Guo, J. and Kaloni, P. (1995c). Nonlinear stability problem of a rotating doubly diffusive porous layer. *J. Math. Anal. Appl.*, 190:373–390.

Guo, J., Qin, Y., and Kaloni, P. (1994). Nonlinear stability problem of a rotating doubly diffusive fluid layer. *Int. J. Engng. Sci.*, 32:1207–1219.

Gurtin, M. (1971). On the thermodynamics of chemically reacting mixtures. *Arch. Rational Mech. Anal.*, 43:198–212.

Ha, V. and Lai, C. (2001). The onset of stationary Marangoni instability of an evaporating droplet. *Proc. Roy. Soc. London A*, 457:885–909.

Haidvogel, D. and Zang, T. (1979). The accurate solution of Poisson's equation by expansion in Chebyshev polynomials. *J. Computational Physics*, 30:167–180.

Hansen, U. and Yuen, D. (1989). Subcritical double-diffusive convection at infinite Prandtl number. *Geophys. Astrophys. Fluid Dyn.*, 47:199–224.

Hardy, G., Littlewood, J., and Polya, G. (1934). *Inequalities.* Cambridge Univ. Press.

Harrison, W. (1982). Formulation of a model for pore water convection in thawing subsea permafrost. *Mitteilungen der Versuchsanstalt für Wasserbau, Hydrologie und Glaziologie*, 57:1–40.

Harrison, W. and Osterkamp, T. (1982). Measurements of the electrical conductivity of interstitial water in subsea permafrost. In *Proc. Fourth Canadian Permafrost Conference*, pages 229–237. Canadian National Research Council.

Hashim, I. and Wilson, S. (1999). The effect of a uniform vertical magnetic field on the onset of oscillatory Marangoni convection in a horizontal layer of conducting fluid. *Acta Mechanica*, 132:129–146.

Hayat, T., Wang, Y., Siddiqui, A., Hutter, K., and Asghar, S. (2002). Peristaltic transport of a third-order fluid in a circular cylindrical tube. *Math. Models Meth. Appl. Sci.*, 12:1691–1706.

Helgeson, H. (1968). Geologic and thermodynamic characteristics of the Salton sea geothermal system. *Amer. J. Science*, 266:129–166.

Herron, I. (2000). On the principle of exchange of stabilities in Rayleigh-Bénard convection. *SIAM J. Appl. Math.*, 61:1362–1368.

Herron, I. (2001). Onset of convection in a porous medium with internal heat source and variable gravity. *Int. J. Engng. Sci.*, 39:201–208.

Hetnarski, R. and Ignaczak, J. (1999). Generalised thermoelasticity. *J. Thermal Stresses*, 22:451–476.

Hillesdon, A. and Pedley, T. (1996). Bioconvection in suspensions of oxytactic bacteria: linear theory. *J. Fluid Mech.*, 324:223–259.

Hillesdon, A., Pedley, T., and Kessler, J. (1995). The development of concentration gradients in a suspension of chemotactic bacteria. *Bull. Math. Biol.*, 57:299–344.

Hills, R. and Roberts, P. (1991). On the motion of a fluid that is incompressible in a generalized sense and its relationship to the Boussinesq approximation. *Stab. Appl. Anal. Cont. Media*, 1:205–212.

Holm, D., Marsden, J., Ratiu, T., and Weinstein, A. (1985). Nonlinear stability of fluid and plasma equilibria. *Physics Reports*, 123:1–116.

Homsy, G. (1973). Global stability of time-dependent flows: impulsively heated or cooled fluid layers. *J. Fluid Mech.*, 60:129–139.

Homsy, G. (1974). Global stability of time-dependent flows. Part 2. modulated fluid layers. *J. Fluid Mech.*, 62:387–403.

Hung, C., Tsai, J., and Chen, C. (1996). Nonlinear stability of the thin micropolar liquid film flowing down on a vertical plate. *J. Fluids Engng. Trans ASME*, 118:498–505.

Huppert, H. and Sparks, R. (1984). Double-diffusive convection due to crystallization in magmas. *Ann. Rev. Earth Planet. Sci.*, 12:11–37.

Hurle, D., Jakeman, E., and Pike, E. (1967). On the solution of the Bénard problem with boundaries of finite conductivity. *Proc. Roy. Soc. London A*, 296:469–475.

Hurle, D., Jakeman, E., and Wheeler, A. (1982). Effect of solutal convection on the morphological stability of a binary alloy. *J. Crystal Growth*, 58:163–179.

Hutter, K. and Straughan, B. (1997). Penetrative convection in thawing subsea permafrost. *Continuum Mech. Thermodyn.*, 9:259–272.

Hutter, K. and Straughan, B. (1999). Models for convection in thawing porous media in support of the subsea permafrost equations. *J. Geophys. Res.*, 104:29249–29260.

Ierley, G. and Worthing, R. (2001). Bound to improve: a variational approach to convective heat transport. *J. Fluid Mech.*, 441:223–253.

Iesan, D. (2002). On the theory of heat conduction in micromorphic continua. *Int. J. Engng. Sci.*, 40:1859–1878.

Jäger, W. and Mikelic, A. (1998). On the interface boundary condition by Beavers, Joseph and Saffman. Interdisziplinäres Zentrum für Wissenschaftliches Rechnen der Universität Heidelberg, Preprint 98-12.

Jäger, W., Mikelic, A., and Neuβ (1999). Asymptotic analysis of the laminar viscous flow over a porous bed. Interdisziplinäres Zentrum für Wissenschaftliches Rechnen der Universität Heidelberg, Preprint 99-33.

Jhaveri, B. and Homsy, G. (1982). The onset of convection in fluid layers heated rapidly in a time-dependent manner. *J. Fluid Mech.*, 114:251–260.

Jones, I. (1973). Low Reynolds number flow past a porous spherical shell. *Proc. Camb. Phil. Soc.*, 73:231–238.

Jordan, P. and Puri, P. (1999). Exact solutions for the flow of a dipolar fluid on a suddenly accelerated flat plate. *Acta Mechanica*, 137:183–194.

Jordan, P. and Puri, P. (2002). Exact solutions for the unsteady plane couette flow of a dipolar fluid. *Proc. Roy. Soc. London A*, 458:1245–1272.

Jordan, P. and Puri, P. (2003). Stokes' first problem for a Rivlin-Ericksen fluid of second grade in a porous half-space. *Int. J. Nonlinear Mech.*, 38:1019–1025.

Joseph, D. (1965). On the stability of the Boussinesq equations. *Arch. Rational Mech. Anal.*, 20:59–71.

Joseph, D. (1966). Nonlinear stability of the Boussinesq equations by the method of energy. *Arch. Rational Mech. Anal.*, 22:163–184.

Joseph, D. (1970). Global stability of the conduction-diffusion solution. *Arch. Rational Mech. Anal.*, 36:285–292.

Joseph, D. (1976a). *Stability of fluid motions*, volume 2. Springer.

Joseph, D. (1976b). *Stability of fluid motions*, volume 1. Springer.

Joseph, D. and Carmi, S. (1966). Subcritical convective instability. Part 2. Spherical shells. *J. Fluid Mech.*, 26:769–777.

Joseph, D. and Hung, W. (1971). Contributions to the nonlinear theory of stability of viscous flow in pipes and between rotating cylinders. *Arch. Rational Mech. Anal.*, 44:1–22.

Joseph, D. and Shir, C. (1966). Subcritical convective instability. Part 1. Fluid layers. *J. Fluid Mech.*, 26:753–768.

Jou, D., Casas-Vazquez, J., and Lebon, G. (1999). Extended irreversible thermodynamics revisited (1988-98). *Rep. Progress Phys.*, 62:1035–1142.

Kafoussias, N. and Williams, E. (1995). Thermal-diffusion and diffusion-thermo effects on mixed free-forced convective and mass transfer boundary layer flow with temperature-dependent viscosity. *Int. J. Engng. Sci.*, 33:1369–1384.

Kaiser, R. and Schmitt, B. (2001). Bounds on the energy stability limit of plane parallel shear flows. *ZAMP*, 52:573–596.

Kaiser, R. and Tilgner, A. (2002). On the generalized energy method for channel flows. In Monaco, R., Bianchi, M. P., and Rionero, S., editors, *Proceedings WASCOM2001 11th conference on waves and stability in continuous media*, pages 259–270.

Kaiser, R. and Xu, L. (1998). Nonlinear stability of the rotating Bénard problem, the case Pr=1. *Nonlin. Differ. Equ. Appl.*, 5:283–307.

Kaloni, P. (1992). Some remarks on the boundary conditions for magnetic fluids. *Int. J. Engng. Sci.*, 30:1451–1457.

Kaloni, P. and Lou, J. (2002a). On the stability of thermally driven shear flow of an Oldroyd-B fluid heated from below. *J. Non-Newtonian Fluid Mech.*, 107:97–110.

Kaloni, P. and Lou, J. (2002b). Stability of Hadley circulations in a Maxwell fluid. *J. Non-Newtonian Fluid Mech.*, 103:167–186.

Kaloni, P. and Qiao, Z. (1996). On the nonlinear stability of thermally driven shear flow heated from below. *Phys. Fluids*, 8:639–641.

Kaloni, P. and Qiao, Z. (1997a). Nonlinear convection with inclined temperature gradient and horizontal mass flow. *Int. J. Engng. Sci.*, 35:299–309.

Kaloni, P. and Qiao, Z. (1997b). Nonlinear stability of convection in a porous medium with inclined temperature gradients. *Int. J. Heat Mass Transfer*, 40:1611–1615.

Kaloni, P. and Qiao, Z. (2000). Nonlinear convection induced by inclined thermal and solutal gradients with mass flow. *Continuum Mech. Thermodyn.*, 12:185–194.

Kaloni, P. and Qiao, Z. (2001). Nonlinear convection in a porous medium with inclined temperature gradient and variable gravity effects. *Int. J. Heat Mass Transfer*, 44:1585–1591.

Karimi-Fard, M., Charrier-Mojtabi, M., and Mojtabi, A. (1999). Onset of stationary and oscillatory convection in a tilted porous cavity saturated with a binary fluid: linear stability analysis. *Phys. Fluids*, 11:1346–1358.

Kassoy, D. (1980). A guide to mathematical models of convection processes in geothermal systems. In *Fluid mechanics in energy conversion*. SIAM, Philadelphia.

Kato, T. (1976). *Perturbation theory for linear operators*. Springer, Berlin.

Kato, T. (1984). Strong L^p solutions of the Navier-Stokes equation in $\mathbb{R}^n$ with applications to weak solutions. *Math. Zeit.*, 187:471–480.

Kaye, J. and Rood, R. (1989). Chemistry and transport in a three-dimensional stratospheric model: chlorine species during a simulated stratospheric warming. *J. Geophys. Res. D*, 94:1057–1083.

Kazhikhov, A. and Smagulov, S. (1977). The correctness of boundary value problems in a diffusion model of an inhomogeneous fluid. *Sov. Phys. Dokl.*, 22:249–250.

Kelvin, L. (1887). On the stability of steady and of periodic fluid motion. *Phil. Mag.*, 23:459–464,529–539.

Kerr, O. (1989). Heating a salinity gradient from a vertical sidewall: linear theory. *J. Fluid Mech.*, 207:323–352.

Kerr, O. (1990). Heating a salinity gradient from a vertical sidewall: nonlinear theory. *J. Fluid Mech.*, 217:529–546.

Kerswell, R. (2000). Lowering dissipation bounds for turbulent shear flows using a smoothness constraint. *Physics Letters A*, 272:230–235.

Kerswell, R. (2001). New results in the variational approach to turbulent Boussinesq convection. *Phys. Fluids*, 13:192–209.

King-Hele, D. (1977). Upper atmosphere studies by ranging to satellites. *Phil. Trans. Roy. Soc. London A*, 284:555–563.

Kladias, N. and Prasad, V. (1991). Experimental verification of Darcy - Brinkman - Forchheimer flow model for natural convection in porous media. *J. Thermophys.*, 5:560–576.

Kloeden, P. and Wells, R. (1983). An explicit example of Hopf bifurcation in fluid mechanics. *Proc. Roy. Soc. London A*, 390:293–320.

Kohout, F. (1965). A hypothesis concerning cyclic flow of salt water related to geothermal heating in the Floridian aquifer. *Trans. New York Acad. Sci.*, 28:249–271.

Kondo, M. and Unno, W. (1982). Convection and gravity modes in two layer models. I. Overstable modes driven in conducting boundary layers. *Geophys. Astrophys. Fluid Dyn.*, 22:305–324.

Kondo, M. and Unno, W. (1983). Convection and gravity modes in two layer models. II. Overstable convection of large horizontal scales. *Geophys. Astrophys. Fluid Dyn.*, 27:229–252.

Kozono, H. and Ogawa, T. (1994). On stability of Navier-Stokes flow in exterior domains. *Arch. Rational Mech. Anal.*, 128:1–31.

Krantz, W., Gleason, K., and Caine, N. (1988). Patterned ground. *Scientific American*, 256:68–76.

Krishnamurti, R. (1997). Convection induced by selective absortion of radiation: A laboratory model of conditional instability. *Dynamics of Atmospheres and Oceans*, 27:367–382.

Kuhlmann, H. (1999). *Thermocapillary convection in models of crystal growth*, volume 152 of *Tracts in Modern Physics*. Springer.

Kwok, L. and Chen, C. (1987). Stability of thermal convection in a vertical porous layer. *J. Heat Transfer*, 109:889–893.

Ladyzhenskaya, O. (1967). New equations for the description of motions of viscous incompressible fluids and global solvability of their boundary value problems. *Trudy Mat. Inst. Steklov*, 102:85–104.

Ladyzhenskaya, O. (1968). On some nonlinear problems in the theory of continuous media. *Amer. Math. Soc. Translations*, 70:73–89.

Ladyzhenskaya, O. (1969). *The Mathematical Theory of Viscous Incompressible Flow.* Gordon and Breach, New York.

Lalas, D. and Carmi, S. (1971). Thermoconvective stability of ferrofluids. *Phys. Fluids*, 14:436–437.

Landau, L. and Lifshitz, E. (1959). *Fluid Mechanics.* Pergamon, London.

Landau, L., Lifshitz, E., and Pitaevskii, L. (1984). *Electrodynamics of continuous media.* Pergamon, London.

Langlois, W. (1985). Buoyancy-driven flows in crystal growth melts. *Ann. Rev. Fluid Mech.*, 17:191–215.

Larson, V. (2000). Stability properties of and scaling laws for a dry radiative-convective atmosphere. *Q. J. Royal Meteorological Soc.*, 126:145–171.

Larson, V. (2001). The effects of thermal radiation on dry convective instability. *Dynamics of Atmospheres and Oceans*, 34:45–71.

Layton, W., Schieweck, F., and Yotov, I. (2003). Coupling fluid flow with porous media flow. *SIAM J. Numer. Anal.*, 40:2195–2218.

Lebon, G. and Cloot, A. (1984). Bénard - Marangoni instability in a Maxwell - Cattaneo fluid. *Physics Letters A*, 105:361–364.

Lebon, G., Dauby, P., and Regnier, V. (2001). Role of interface deformations in Bénard-Marangoni instability. *Acta Astronautica*, 48:617–627.

Lebon, G. and Perez-Garcia, C. (1981). Convective instability of a micropolar fluid layer by the method of energy. *Int. J. Engng. Sci.*, 19:1321–1329.

Lebovich, S. (1977). On the evolution of the system of wind drift currents and Langmuir circulations in the ocean. Part 1. Theory and averaged current. *J. Fluid Mech.*, 79:715–743.

Lebovich, S. and Paolucci, S. (1980). Energy stability of the Eulerian-mean motion in the upper ocean to three-dimensional perturbations. *Phys. Fluids*, 23:1286–1290.

Lebovich, S. and Paolucci, S. (1981). The instability of the ocean to Langmuir circulations. *J. Fluid Mech.*, 102:141–167.

Levandowsky, M., Childress, S., Spiegel, E., and Hutner, S. (1975). A mathematical model of pattern formation by swimming micro-organisms. *J. Protozoology*, 22:296–306.

Levine, H. (1973). Some nonexistence and instability theorems for solutions of formally parabolic equations of the form $Pu_t = -Au + F(u)$. *Arch. Rational Mech. Anal.*, 51:371–386.

Levine, H., Payne, L., Sacks, P., and Straughan, B. (1989). Analysis of a convective reaction-diffusion equation. *SIAM J. Math. Anal.*, 20:133–147.

Lide, D., editor (1991). *Handbook of Chemistry and Physics*. CRC Press, Boca Raton.

Lietuaud, P. and Néel, M. (2001). Instabilities of an electrically driven shear flow. *Comptes Rend. Acad. Sci. IIB - Mécanque*, 329:881–887.

Lindsay, K. and Straughan, B. (1979). A thermodynamic interface theory and associated stability problems. *Arch. Rational Mech. Anal.*, 71:307–326.

Lindsay, K. and Straughan, B. (1991). Nonlinear temperature-dependent surface tension driven convection. *Stab. Appl. Anal. Cont. Media*, 1:283–305.

Lindsay, K. and Straughan, B. (1992). Penetrative convection in a micropolar fluid. *Int. J. Engng. Sci.*, 30:1683–1702.

Lombardo, S. and Mulone, G. (2002a). Double-diffusive convection in porous media: The Darcy and Brinkman models. In Monaco, R., Pandolfi-Bianchi, M., and Rionero, S., editors, *Proc. Wascom 2001*.

Lombardo, S. and Mulone, G. (2002b). Necessary and sufficient conditions for global nonlinear stability for rotating double-diffusive convection in a porous medium. *Continuum Mech. Thermodyn.*, 14:527–540.

Lombardo, S., Mulone, G., and Rionero, S. (2000). Global stability in the Bénard problem for a mixture with superimposed plane parallel shear flows. *Math. Methods Appl. Sci.*, 23:1447–1465.

Lombardo, S., Mulone, G., and Rionero, S. (2001a). Global nonlinear exponential stability of the conduction-diffusion solution for Schmidt numbers greater than Prandtl numbers. *J. Math. Anal. Appl.*, 262:191–207.

Lombardo, S., Mulone, G., and Straughan, B. (2001b). Stability in the Bénard problem for a double-diffusive mixture in a porous medium. *Math. Meth. Appl. Sci.*, 24:1229–1246.

Loper, D. and Roberts, P. (1978). On the motion of an Iron-Alloy core containing a slurry. II. General theory. *Geophys. Astrophys. Fluid Dyn.*, 9:289–321.

Loper, D. and Roberts, P. (1980). On the motion of an Iron-Alloy core containing a slurry. II. A simple model. *Geophys. Astrophys. Fluid Dyn.*, 16:83–127.

Lopez, A., Romero, L., and Pearlstein, A. (1990). Effect of rigid boundaries on the onset of convection instability in a triply diffusive fluid layer. *Phys. Fluids A*, 2:897–902.

Low, A. (1925). Instability of viscous fluid motion. *Nature*, 229:299–300.

Ludvigsen, A., Palm, E., and McKibbin, R. (1990). Convective momentum and mass transport in porous sloping layers. *J. Geophys. Res. B*, 97:12315–12325.

Lukaszewicz, G. (2001). Long time behavior of 2D micropolar fluid flows. *Math. Computer Modelling*, 34:487–509.

Lukaszewicz, G. (2003). Asymptotic behavior of micropolar fluid flows. *Int. J. Engng. Sci.*, 41:259–269.

Ly, H. and Titi, E. (1999). Global Gevrey regularity for Bénard convection in a porous medium with zero Darcy-Prandtl number. *J. Nonlinear Sci.*, 9:333–362.

Mahidjiba, A., Robillard, L., and Vasseur, P. (2003). Linear stability of cold water saturating an anisotropic porous medium - effect of confinement. *Int. J. Heat Mass Transfer*, 46:323–332.

Maiellaro, M. and Labianca, A. (2002). On the nonlinear stability in anisotropic MHD with application to Couette - Poiseuille flows. *Int. J. Engng. Sci.*, 40:1053–1068.

Malikkides, C. and Weiland, R. (1982). On the mechanism of immobilized glucose oxidase deactivation by hydrogen peroxide. *Biotechnology and Bioengineering*, 24:2491–2439.

Man, C. and Sun, Q. (1987). On the significance of normal stress effects in the flow of glaciers. *J. Glaciology*, 33:268–273.

Manole, D., Lage, J., and Nield, D. (1994). Convection induced by inclined thermal and solutal gradients with horizontal mas flow, in a shallow horizontal layer of a porous medium. *Int. J. Heat Mass Transfer*, 37:2047–2057.

Marchuk, G. and Sarkisyan, A. (1988). *Mathematical modelling of ocean circulation.* Springer, Berlin - Heidelberg - New York.

Maremonti, P. (1985). Stabilitá asintotica in media per moti fluidi viscosi in domini esterni. *Ann. Matem. Pura Appl.*, 142:57–75.

Maremonti, P. (1988). On the asymptotic behaviour of the L^2 norm of suitable weak solutions to the Navier-Stokes equations in three-dimensional exterior domains. *Commun. Math. Phys.*, 118:385–400.

Martin, P. and Richardson, A. (1984). Conductivity models of electrothermal convection in a plane layer of dielectric liquid. *J. Heat Transfer*, 106:131–136.

Martin, S. and Kauffman, P. (1974). The evolution of under-ice melt ponds, or double diffusion at the freezing point. *J. Fluid Mech.*, 64:507–527.

Massoudi, M. and Phuoc, T. (2001). Fully developed flow of a modified second grade fluid with temperature dependent viscosity. *Acta Mechanica*, 150:23–37.

Matthews, P. (1988). A model for the onset of penetrative convection. *J. Fluid Mech.*, 188:571–583.

May, A. and Bassom, A. (2000). Nonlinear convection in the boundary layer above a sinusoidally heated flat plate. *Q. J. Mech. Appl. Math.*, 53:475–495.

McFadden, G., Murray, B., and Boisvert, R. (1990). Elimination of spurious eigenvalues in the Chebyshev tau spectral method. *J. Computational Physics*, 91:228–239.

McFadden, G., Rehm, R., Coriell, S., Chuck, W., and Morrish, K. (1984). Thermosolutal convection during directional solidification. *Metall. Trans.*, 15:2125–2137.

McKay, G. (1992a). *Nonlinear stability analyses of problems in patterned ground formation and penetrative convection.* PhD thesis, Glasgow University.

McKay, G. (1992b). Patterned ground formation and solar radiation heating. *Proc. Roy. Soc. London A*, 438:249–263.

McKay, G. (1996). Patterned ground formation and convection in porous media with a phase change. *Continuum Mech. Thermodyn.*, 8:189–199.

McKay, G. (1998a). Onset of buoyancy-driven convection in superposed reacting fluid and porous layers. *J. Engng. Math.*, 33:31–46.

McKay, G. (1998b). Onset of double-diffusive convection in a saturated porous layer with time-periodic heating. *Continuum Mech. Thermodyn.*, 10:241–251.

McKay, G. (2000). Double-diffusive convective motions for a saturated porous layer subject to modulated surface heating. *Continuum Mech. Thermodyn.*, 12:69–78.

McKay, G. (2001). The Beavers and Joseph condition for velocity slip at the surface of a porous medium. In Straughan, B., Greve, R., Ehrentraut, H., and Wang, Y., editors, *Continuum Mechanics and Applications in Geophysics and the Environment.* Springer.

McKay, G. and Straughan, B. (1991). The influence of a cubic density law on patterned ground formation. *Math. Models Meths. in Appl. Sciences*, 1:27–39.

McKay, G. and Straughan, B. (1992). A nonlinear analysis of convection near the density maximum. *Acta Mechanica*, 95:9–28.

McKay, G. and Straughan, B. (1993). Patterned ground formation under water. *Continuum Mech. Thermodyn.*, 5:145–162.

McTaggart, C. (1983a). Convection driven by concentration and temperature dependent surface tension. *J. Fluid Mech.*, 134:301–310.

McTaggart, C. (1983b). *Energy and linear stability analyses of surface and second sound effects in convection problems.* PhD thesis, Glasgow University.

McTaggart, C. (1984). On the stabilizing effect of surface films in Bénard convection. *Physico Chemical Hydrodyn.*, 5:321–331.

Mellor, G. (1996). *Introduction to physical oceanography.* Springer, New York.

Mercier, J., Weisman, C., Firdaouss, M., and Le Quere, P. (2002). Heat transfer associated to natural convection flow in a partly porous cavity. *J. Heat Transfer*, 124:130–143.

Merker, G., Waas, P., and Grigull, U. (1979). Onset of convection in a horizontal water layer with maximum density effects. *Int. J. Heat Mass Transfer*, 22:505–515.

Metzler, R. and Compte, A. (1999). Stochastic foundation of normal and anomalous Cattaneo - type transport. *Physica A*, 268:454–468.

Mielke, A. (1997). Mathematical analysis of sideband instabilities with application to Rayleigh - Bénard convection. *J. Nonlinear Science*, 7:57–99.

Miglio, E., Quarteroni, A., and Saleri, F. (2003). Coupling of free surface and groundwater flows. *Computers and Fluids*, 32:73–83.

Mihaljan, J. (1962). A rigorous exposition of the Boussinesq approximation applicable to a thin layer of fluid. *Astrophys. J.*, 136:1126–1133.

Mlaouah, H., Tsuji, T., and Nagano, Y. (1997). A study of non-Boussinesq effect on transistion of thermally induced flow in a square cavity. *Int. J. Heat and Fluid Flow*, 18:100–106.

Moler, C. and Stewart, G. (1973). An algorithm for generalized matrix eigenproblems. *SIAM. J. Numerical Anal.*, 10:241–256.

Morro, A. and Straughan, B. (1990). Convective instabilities for reacting viscous flows far from equilibrium. *J. Non-Equilib. Thermodyn.*, 15:139–150.

Müller-Beck, H. (1966). Paleohunters in America: origins and diffusion. *Science*, 152:1191–1210.

Mulone, G. (1988). On the stability of a rotating fluid with Prandtl numbers less than one. *Rend. Accad. Scienze, Fisiche e Matematiche, Napoli*, 55:123–138.

Mulone, G. (1990). On the stability of plane parallel convective mixture through the Lyapunov second method. *Atti Accad. Peloritana dei Pericolanti*, 68:491–516.

Mulone, G. (1991a). On the Lyapunov stability of a plane parallel convective flow of a binary mixture. *Le Matematiche*, 46:283–294.

Mulone, G. (1991b). On the stability of plane parallel convective flow. *Acta Mechanica*, 87:153–162.

Mulone, G. (1994). On the nonlinear stability of a fluid layer of a mixture heated and salted from below. *Continuum Mech. Thermodyn.*, 6:161–184.

Mulone, G. (1998). On the nonlinear stability of the Bénard problem for a mixture: conditional and unconditional stability. *Rend. Circolo Matem. Palermo*, 57:347–356.

Mulone, G. and Rionero, S. (1989). On the nonlinear stability of the rotating Bénard problem via the Lyapunov direct method. *J. Math. Anal. Appl.*, 144:109–127.

Mulone, G. and Rionero, S. (1993). On the nonlinear stability of the magnetic Bénard problem with rotation. *Zeit. angew. Math. Mech.*, 73:35–45.

Mulone, G. and Rionero, S. (1994). On the stability of the rotating Bénard problem. *Bull. Tech. Univ. Istanbul*, 47:181–202.

Mulone, G. and Rionero, S. (1997). The rotating Bénard problem: new stability results for any Prandtl and Taylor numbers. *Continuum Mech. Thermodyn.*, 9:347–363.

Mulone, G. and Rionero, S. (1998). Unconditional nonlinear exponential stability in the Bénard problem for a mixture: necessary and sufficient conditions. *Atti Accad. Naz. Lincei*, 9:221–236.

Mulone, G. and Rionero, S. (2003). Necessary and sufficient conditions for nonlinear stability in the magnetic Bénard problem. *Arch. Rational Mech. Anal.*, 166:197–218.

Mulone, G., Rionero, S., and Straughan, B. (1996). Unconditional nonlinear stability in a polarized dielectric liquid. *Atti Accad. Naz. Lincei*, 7:241–252.

Munk, W. (1950). On the wind-driven circulation. *J. Meteor.*, 7:79–93.

Munson, B. and Joseph, D. (1971). Viscous incompressible flow between concentric rotating cylinders. Part 2. Hydrodynamic stability. *J. Fluid Mech.*, 49:305–318.

Murdoch, A. and Soliman, A. (1999). On the slip-boundary condition for liquid flow over planar porous boundaries. *Proc. Roy. Soc. London A*, 455:1315–1340.

Murty, Y. (1999). Effect of throughflow and Coriolis force on Bénard convection in micropolar fluids. *Int. J. Numer. Meth. Heat Fluid Flow*, 9:677–691.

Murty, Y. (2003). Analysis of non-uniform temperature profiles on Bénard convection in micropolar fluids. *Applied Mathematics and Computation*, 134:473–486.

Nagai, T., Senba, T., and Suzuki, T. (2000). Chemotactic collapse in a parabolic system of mathematical biology. *Hiroshima Math. J.*, 30:463–497.

Nagai, T., Senba, T., and Yoshida, K. (1997). Application of the Trudinger-Moser inequality to a parabolic system of chemotaxis. *Funkcialaj Ekvacioj*, 40:411–433.

Néel, M. (1998). Convection forcée en milieu poreux: écarts à la loi de Darcy. *C.R. Acad. Sci. Paris, série IIb*, 326:615–620.

Néel, M. and Nemrouch, F. (2001). Instabilities in an open top horizontal porous layer subjected to pulsating thermal boundary conditions. *Continuum Mech. Thermodyn.*, 13:41–58.

Néel, M. and Nemrouch, F. (2002). Sharp approximation to a filtration problem by the maximum principle. *Continuum Mech. Thermodyn.*, 14:541–548.

Neitzel, G. (1982). Onset of convection in impulsively heated or cooled fluid layers. *Phys. Fluids*, 25:210–211.

Neitzel, G., Chang, K., Jankowski, D., and Mittelmann, H. (1991a). Linear stability theory of thermocapillary convection in a model of the float-zone crystal-growth process. *Phys. Fluids A*, 5:108–110.

Neitzel, G., Law, C., Jankowski, D., and Mittelmann, H. (1991b). Energy stability of thermocapillary convection in a model of the float-zone crystal-growth process. II: Nonaxisymmetric disturbances. *Phys. Fluids A*, 3:2841–2846.

Neitzel, G., Smith, M., and Bolander, M. (1994). Thermal instability with radiation by the method of energy. *Int. J. Heat Mass Transfer*, 37:2909–2915.

Nepomnyashchy, A. and Velarde, M. (1994). A three-dimensional description of solitary waves and their interaction in Marangoni-Bénard layers. *Phys. Fluids*, 6:187–198.

Nicodemus, R., Grossmann, S., and Holthaus, M. (1998). The background flow method. Part 1. Constructive approach to bounds on energy dissipation. *J. Fluid Mech.*, 363:281–300.

Nicodemus, R., Grossmann, S., and Holthaus, M. (1999). Towards lowering dissipation bounds for turbulent flows. *European Physical J. B*, 10:385–396.

Niedrauer, T. and Martin, S. (1979). An experimental study of brine drainage and convection in young sea ice. *J. Geophys. Res. C*, 84:1176–1186.

Nield, D. (1968). Onset of thermohaline convection in a porous medium. *Water Resources Research*, 4:553–560.

Nield, D. (1977). Onset of convection in a fluid layer overlying a layer of a porous medium. *J. Fluid Mech.*, 81:513–522.

Nield, D. (1987). Throughflow effects in the Rayleigh-Bénard convective instability problem. *J. Fluid Mech.*, 185:353–360.

Nield, D. (1990). Convection in a porous medium with inclined temperature gradient and horizontal mass flow. In *Proc. Ninth. Intl. Heat Transfer Conf.*, volume 5, pages 153–158, Washington D.C. Hemisphere.

Nield, D. (1991). Convection in a porous medium with inclined temperature gradient. *Int. J. Heat Mass Transfer*, 34:87–92.

Nield, D. (1994a). Convection in a porous medium with inclined temperature gradient: additional results. *Int. J. Heat Mass Transfer*, 37:3021–3025.

Nield, D. (1994b). Convection induced by an inclined temperature gradient in a shallow horizontal layer. *Int. J. Heat Fluid Flow*, 15:157–162.

Nield, D. (1998a). Convection in a porous medium with inclined temperature gradient and vertical throughflow. *Int. J. Heat Mass Transfer*, 41:241–243.

Nield, D. (1998b). Instability and turbulence in convective flows in porous media. In Debnath, L. and Riahi, D., editors, *Nonlinear instability, chaos and turbulence*, pages 225–276. WIT Press, Boston.

Nield, D. (1998c). Modelling the effect of surface tension on the onset of natural convection in a saturated porous medium. *Transport in Porous Media*, 31:365–368.

Nield, D. (1999). Modelling the effects of a magnetic field or rotation on flow in a porous medium: momentum equation and anisotropic permeability analogy. *Int. J. Heat Mass Transfer*, 42:3715–3718.

Nield, D. (2000). Resolution of a paradox involving viscous dissipation and nonlinear drag in a porous medium. *Transport in Porous Media*, 41:349–357.

Nield, D. (2001). Some pitfalls in the modelling of convective flows in porous media. *Transport in Porous Media*, 43:597–601.

Nield, D. and Bejan, A. (1999). *Convection in Porous Media*. Springer, New York.

Nield, D., Manole, D., and Lage, J. (1993). Convection induced by inclined thermal and solutal gradients in a shallow horizontal layer of a porous medium. *J. Fluid Mech.*, 257:559–574.

Nordenskjold, O. (1909). *Die Polarwelt*. B.G. Teubner, Berlin.

Normand, C. and Azouni, A. (1992). Penetrative convection in an internally heated layer of water near the maximum density point. *Phys. Fluids A*, 4:243–253.

Noulty, R. and Leaist, D. (1989). Quaternary diffusion in aqueous KCl-KH_2PO_4-H_3PO_4 mixtures. *J. Phys. Chem.*, 202:443–465.

Okamoto, H. (1997). Exact solutions of the Navier-Stokes equations via Leray's scheme. *Japan J. Industrial Appl. Math.*, 14:169–197.

Oldenburg, C. and Pruess, K. (1998). Layered thermohaline convection in hypersaline geothermal systems. *Transport in Porous Media*, 33:29–63.

Oliver, M. and Titi, E. (2000). Gevrey regularity for the attractor of a partially dissipative model of Bénard convection in a porous medium. *J. Diff. Equations*, 163:292–311.

Orszag, S. (1971). Accurate solution of the Orr-Sommerfeld stability equation. *J. Fluid Mech.*, 50:689–703.

Osterkamp, T., Baker, G., Harrison, W., and Matawa, T. (1989). Characteristics of the active layers and shallow subsea permafrost. *J. Geophysical Res.*, 94:16227–16236.

Osterkamp, T. and Harrison, W. (1982). Temperature measurements in subsea permafrost off the coast of Alaska. In *Proc. Fourth Canadian Permafrost Conference*, pages 238–248. Canadian National Research Council.

O'Sullivan, M., Pruess, K., and Lippmann, M. (2001). State of the art of geothermal reservoir simulation. *Geothermics*, 30:395–429.

Padula, M. (1986). Nonlinear energy stability for the compressible Bénard problem. *Boll. Unione Matem. Italiana*, 5:581–602.

Palm, E., Ellingsen, T., and Gjevik, B. (1967). On the occurrence of cellular motion in Bénard convection. *J. Fluid Mech.*, 30:651–661.

Pandiz, S.N. & Seinfeld, J. (1989). Sensitivity analysis of a chemical mechanism for aqueous-phase atmospheric chemistry. *J. Geophys. Res. D*, 94:1105–1126.

Payne, L., Rodrigues, J., and Straughan, B. (2001). Effect of anisotropic permeability on Darcy's law. *Math. Meth. Appl. Sci.*, 24:427–438.

Payne, L. and Song, J. (1997). Spatial decay estimates for the Brinkman and Darcy flows in a semi-infinite cylinder. *Continuum Mech. Thermodyn.*, 9:175–190.

Payne, L. and Song, J. (2000). Spatial decay for a model of double diffusive convection in Darcy and Brinkman flows. *ZAMP*, 51:867–889.

Payne, L. and Song, J. (2002). Spatial decay bounds for the Forchheimer equations. *Int. J. Engng. Sci.*, 40:943–956.

Payne, L., Song, J., and Straughan, B. (1988). Double diffusive porous penetrative convection; thawing subsea permafrost. *Int. J. Engng. Sci.*, 26:797–809.

Payne, L., Song, J., and Straughan, B. (1999). Continuous dependence and convergence results for Brinkman and Forchheimer models with variable viscosity. *Proc. Roy. Soc. London A*, 455:2173–2190.

Payne, L. and Straughan, B. (1987). Unconditional nonlinear stability in penetrative convection. *Geophys. Astrophys. Fluid Dyn.*, 39:57–63.

Payne, L. and Straughan, B. (1988). Unconditional nonlinear stability in penetrative convection: corrected and extended numerical results. *Geophys. Astrophys. Fluid Dyn.*, 43:307–309.

Payne, L. and Straughan, B. (1989). Critical Rayleigh numbers for oscillatory and nonlinear convection in an isotropic thermomicropolar fluid. *Int. J. Engng. Sci.*, 27:827–836.

Payne, L. and Straughan, B. (1998). Analysis of the boundary condition at the interface between a viscous fluid and a porous medium and related modelling questions. *J. Math. Pures et Appl.*, 77:317–354.

Payne, L. and Straughan, B. (2000a). A naturally efficient numerical technique for porous convection stability with non-trivial boundary conditions. *Int. J. Num. Anal. Meth. Geomech.*, 24:815–836.

Payne, L. and Straughan, B. (2000b). Unconditional nonlinear stability in temperature - dependent viscosity flow in a porous medium. *Stud. Appl. Math.*, 105:59–81.

Pearlstein, A., Harris, R., and Terrones, G. (1989). The onset of convective instability in a triply diffusive fluid layer. *J. Fluid Mech.*, 202:443–465.

Pedlosky, J. (1996). *Ocean circulation theory.* Springer, Berlin, Heidelberg.

Perez Cordon, R. and Velarde, M. (1976). On the (non linear) foundations of the Boussinesq approximation applicable to a thin layer of fluid. *Journal de Physique*, 36:591–601.

Philippin, G. and Vernier-Piro, S. (2001). Decay estimates for a class of parabolic problems arising in filtration through porous media. *Boll. Unione Matem. Ital.*, 4:473–481.

Pons, A., Sorensen, P., Bees, M., and Sagués, F. (2000). Pattern formation in the methylene-blue glucose system. *Journal of Physical Chemistry*, 104:2251–2259.

Powers, D., O'Neill, K., and Colbeck, S. (1985). Theory of natural convection in snow. *J. Geophys. Res.*, 90:10641–10649.

Pradhan, G. and Samal, P. (1987). Thermal stability of a fluid layer under variable body forces. *J. Math. Anal. Appl.*, 122:487–495.

Preziosi, L. and Rionero, S. (1989). Energy stability of steady shear flows of a viscoelastic liquid. *Int. J. Engng. Sci.*, 27:1167–1181.

Proctor, M. (1981). Steady subcritical thermohaline convection. *J. Fluid Mech.*, 105:507–521.

Prodi, G. (1962). Teoremi di tipo locale per il sistema di Navier-Stokes e stabilitá delle soluzioni stazionarie. *Rend. Sem. Mat. Univ. Padova*, 32:374–397.

Prouse, G. and Zaretti, A. (1987). On the inequalities associated to a model of Graffi for the motion of a mixture of two viscous incompressible fluids. *Rendiconti Accademia Nazionale delle Scienze detta dei XL*, 11:253–275.

Puri, P. and Jordan, P. (1999a). Stokes's first problem for a dipolar fluid with nonclassical heat conduction. *J. Engng. Math.*, 36:219–240.

Puri, P. and Jordan, P. (1999b). Wave structure in Stokes' second problem for a dipolar fluid with nonclassical heat conduction. *Acta Mechanica*, 133:145–160.

Qiao, Z. and Kaloni, P. (1997). Convection in a porous medium induced by an inclined temperature gradient with mass flow. *J. Heat Trans.*, 119:366–370.

Qiao, Z. and Kaloni, P. (1998). Nonlinear convection in a porous medium with inclined temperature gradient and vertical throughflow. *Int. J. Heat Mass Transfer*, 41:2549–2552.

Qin, Y. and Chadam, J. (1995). A nonlinear stability problem for ferromagnetic fluids in a porous medium. *Appl. Math. Letters*, 8:25–29.

Qin, Y. and Chadam, J. (1996). Nonlinear convective stability in a porous medium with temperature - dependent viscosity and inertial drag. *Stud. Appl. Math.*, 96:273–288.

Qin, Y., Guo, J., and Kaloni, P. (1995). Double diffusive penetrative convection in porous media. *Int. J. Engng. Sci.*, 33:303–312.

Qin, Y. and Kaloni, P. (1992a). Steady convection in a porous medium based upon the Brinkman model. *IMA J. Appl. Math.*, 35:85–95.

Qin, Y. and Kaloni, P. (1992b). A thermal instability problem in a rotating micropolar fluid. *Int. J. Engng. Sci.*, 30:1117–1126.

Qin, Y. and Kaloni, P. (1993a). Creeping flow past a porous spherical shell. *Z. Angew. Math. Mech.*, 73:77–84.

Qin, Y. and Kaloni, P. (1993b). A nonlinear stability problem of convection in a porous vertical slab. *Phys. Fluids A*, 5:2067–2069.

Qin, Y. and Kaloni, P. (1994a). Convective instabilities in anisotropic porous media. *Stud. Appl. Math.*, 91:189–204.

Qin, Y. and Kaloni, P. (1994b). Nonlinear stability problem of a ferromagnetic fluid with surface tension effect. *Eur. J. Mech. B/Fluids*, 13:305–321.

Qin, Y. and Kaloni, P. (1995). Nonlinear stability problem of a rotating porous layer. *Quart. Appl. Math.*, 53:129–142.

Quintanilla, R. (2001). Instability and non-existence in the nonlinear theory of thermoelasticity without energy dissipation. *Continuum Mech. Thermodyn.*, 13:121–129.

Quintanilla, R. (2002a). Exponential stability for one-dimensional problem of swelling porous elastic soils with fluid saturation. *J. Comput. Appl. Math.*, 145:525–533.

Quintanilla, R. (2002b). On the linear problem of swelling porous elastic soils. *J. Math. Anal. Appl.*, 269:50–72.

Quintanilla, R. (2002c). On the linear problem of swelling porous elastic soils with incompressible fluid. *Int. J. Engng. Sci.*, 40:1485–1494.

Quintanilla, R. and Straughan, B. (2001). Growth and uniqueness in thermoelasticity. *Proc. Roy. Soc. London A*, 456:1419–1429.

Ramdath, G. (1997). Bénard-Marangoni instability in a layer of micropolar fluid. *J. Non-Equilibrium Thermodyn.*, 22:299–310.

Ray, R., Krantz, W., Caine, T., and Gunn, R. (1983). A model for sorted patterned-ground regularity. *J. Glaciology*, 29:317–337.

Reddy, B. and Voyé, H. (1988). Finite element analysis of the stability of fluid motions. *J. Computational Phys.*, 79:92–112.

Rednikov, A., Colinet, P., Velarde, M., and Legros, J. (2000). Rayleigh-Marangoni oscillatory instability in a horizontal liquid layer heated from above : coupling and mode mixing of internal and surface dilational waves. *J. Fluid Mech.*, 405:55–77.

Rees, D. and Bassom, A. (2000). The onset of Darcy-Bénard convection in an inclined layer heated from below. *Acta Mech.*, 144:103–118.

Rees, D. and Lage, J. (1997). The effect of thermal stratification on natural convection in a vertical porous insulation layer. *Int. J. Heat Mass Transfer*, 40:111–121.

Rees, D. and Postelnicu, A. (2001). The onset of convection in an inclined anisotropic porous layer. *Int. J. Heat Mass Transfer*, 44:4127–4138.

Richardson, L. (1993). *Nonlinear stability analyses for variable viscosity and compressible convection problems.* PhD thesis, Glasgow University.

Richardson, L. and Straughan, B. (1993). A nonlinear energy stability analysis of convection with temperature dependent viscosity. *Acta Mechanica*, 97:41–49.

Rionero, S. (1967a). Sulla stabilità asintotica in media in magnetoidrodinamica. *Ann. Matem. Pura Appl.*, 76:75–92.

Rionero, S. (1967b). Sulla stabilità asintotica in media in magnetoidrodinamica non isoterma. *Ricerche Matem.*, 16:250–263.

Rionero, S. (1968a). Metodi variazionali per la stabilità asintotica in media in magnetoidrodinamica. *Ann. Matem. Pura Appl.*, 78:339–364.

Rionero, S. (1968b). Sulla stabilità magnetoidrodinamica in media con vari tipi di condizioni al contorno. *Ricerche Matem.*, 17:64–78.

Rionero, S. (1970). Sulla stabilità nonlineare asintotica in media in magnetofluidodinamica. *Ricerche Matem.*, 19:269–285.

Rionero, S. (1971). Sulla stabilità magnetofluidodinamica nonlineare asintotica in media in presenza di effetto hall. *Ricerche Matem.*, 20:285–296.

Rionero, S. (2000). A new approach to the nonlinear stability of a porous medium saturated by a fluid with temperature depending viscosity. In O'Donoghue, P. and Flavin, J., editors, *Symp. Trends in the Application of Mathematics to Mechanics*, pages 10–16. Elsevier.

Rionero, S. (2001). Asymptotic and other properties of some nonlinear diffusion models. In Straughan, B., Greve, R., Ehrentraut, H., and Wang, Y., editors, *Continuum Mechanics and Applications in Geophysics and the Environment*, pages 56–78. Springer.

Rionero, S. and Maiellaro, M. (1995). On the stability of Couette-Poiseuille flows in anisotropic MHD via the Lyapunov direct method. *Rend. Acc. Sc. Fis. Mat. Napoli*, 62:315–333.

Rionero, S. and Mulone, G. (1987). On the non-linear stability of a thermodiffusive fluid mixture in a mixed problem. *J. Math. Anal. Appl.*, 124:165–168.

Rionero, S. and Mulone, G. (1988a). Non-linear stability analysis of the magnetic Bénard problem through the Lyapunov direct method. *Arch. Rational Mech. Anal.*, 103:347–368.

Rionero, S. and Mulone, G. (1988b). On a maximum problem governing the nonlinear stability of the rotating Bénard problem. *Ricerche Matem.*, 37:177–185.

Rionero, S. and Mulone, G. (1989). On the stability of a mixture in a rotating layer via the Lyapunov second method. *Zeitschrift für angewandte Mathematik und Mechanik*, 69:441–446.

Rionero, S. and Mulone, G. (1991). On the nonlinear stability of parallel shear flows. *Continuum Mech. Thermodyn.*, 3:161–184.

Rionero, S. and Straughan, B. (1990). Convection in a porous medium with internal heat source and variable gravity effects. *Int. J. Engng. Sci.*, 28:497–503.

Roberts, P. (1967a). Convection in horizontal layers with internal heat generation. theory. *J. Fluid Mech.*, 30:33–49.

Roberts, P. (1967b). *An introduction to magnetohydrodynamics*. Longman, London.

Roberts, P. (1969). Electrohydrodynamic convection. *Q. J. Mech. Appl. Math.*, 22:211–220.

Roberts, P. and Stewartson, K. (1974). On finite amplitude convection in a rotating magnetic system. *Phil. Trans. Roy. Soc. London A*, 277:287–315.

Rodrigues, J. (1986). A steady state Boussinesq-Stefan problem with continuous extraction. *Annali Matem. Pura Appl.*, 144:203–218.

Rodrigues, J. (1994). Variational methods in the Stefan problem. In *Lecture Notes in Math.*, volume 1584, pages 147–212. Springer.

Rodrigues, J. and Urbano, J. (1999). On a Darcy-Stefan problem arising in freezing and thawing of saturated porous media. *Continuum Mech. Thermodyn.*, 11:181–191.

Rodriguez-Luis, A., Castellanos, A., and Richardson, A. (1986). Stationary instabilities in a dielectric liquid layer subjected to an arbitrary unipolar injection and adverse thermal gradient. *J. Physics D*, 19:2115–2122.

Rood, R. (1987). Numerical advection algorithms and their role in atmospheric transport and chemistry models. *Reviews Geophys.*, 25:71–100.

Rosenblat, S., Davis, S., and Homsy, G. (1982a). Nonlinear Marangoni convection in bounded layers. I. Circular cylindrical containers. *J. Fluid Mech.*, 120:91–122.

Rosenblat, S., Homsy, G., and Davis, S. (1982b). Nonlinear Marangoni convection in bounded layers. II. Rectangular cylindrical containers. *J. Fluid Mech.*, 120:123–138.

Rosenblat, S. and Tanaka, G. (1971). Modulation of thermal convection instability. *Phys. Fluids*, 14:1319–1322.

Rosensweig, R. (1985). *Ferrohydrodynamics.* Cambridge Univ. Press.

Rosensweig, R., Zahn, M., and Shumovich, R. (1983). Labyrinthine instability in magnetic and dielectric fluids. *J. Magnetism and Magnetic Materials*, 39:127–132.

Rossby, H. (1969). A study of Bénard convection with and without rotation. *J. Fluid Mech.*, 36:309–335.

Rothmeyer, M. (1980). The Soret effect and salt-gradient solar ponds. *Solar Energy*, 25:567–568.

Rubin, H. (1973). Effect of nonlinear stabilizing salinity profiles on thermal convection in a porous medium layer. *Water Resources Res.*, 9:211–221.

Ruddick, B. and Shirtcliffe, T. (1979). Data for double diffusers: physical properties of aqueous salt-sugar solutions. *Deep Sea Res.*, 26:775–787.

Rudraiah, N., Shivakumara, I., and Nanjundappa, C. (2002). Effect of basic temperature gradients on Marangoni convection in ferromagnetic fluids. *Indian J. Pure Appl. Phys.*, 40:95–106.

Ruggeri, T. (2001). The binary mixtures of Euler fluids: a unified theory of second sound phenomena. In Straughan, B., Greve, R., Ehrentraut, H., and Wang, Y., editors, *Continuum Mechanics and Applications in Geophysics and the Environment.* Springer.

Saffman, P. (1971). On the boundary conditions at the surface of a porous medium. *Stud. Appl. Math.*, 50:93–101.

Saravanan, S. and Kandaswamy, P. (2002). Convection currents in a porous layer with a gravity gradient. Heat and Mass Transfer, online version. DOI/10.1007/s00231-002-0398-4.

Sattinger, D. (1970). The mathematical problem of hydrodynamic stability. *J. Math. Mech.*, 19:797–817.

Schonbek, M. (1985). L^2 decay of weak solutions of the Navier-Stokes equations. *Arch. Rational Mech. Anal.*, 88:209–222.

Schonbek, M. and Schonbek, T. (2000). On the boundedness and decay of moments of solutions to the Navier-Stokes equations. *Advances in Differential Equations*, 5:861–898.

Scriven, L. (1960). Dynamics of a fluid interface. Equation of motion for Newtonian surface fluids. *Chem. Engng. Sci.*, 12:98–108.

Secchi, P. (1982). On the initial value problem for the equations of motion of viscous, incompressible fluids in the presence of diffusion. *Boll. U.M.I.*, 1B:1117–1130.

Segel, L. and Jackson, J. (1972). Dissipative structure: an explanation and an ecological example. *J. Theoretical Biol.*, 37:545–549.

Sekar, R. and Vaidyanathan, G. (1993). Convective instability of a magnetized ferrofluid in a rotating porous medium. *Int. J. Engng. Sci.*, 31:1139–1150.

Sekar, R., Vaidyanathan, G., and Ramanathan, A. (1993). The ferroconvection in fluids saturating a rotating densely packed porous medium. *Int. J. Engng. Sci.*, 31:241–250.

Selak, R. and Lebon, G. (1997). Bénard-Marangoni thermoconvective instability with non-Boussinesq corrections. *Int. J. Heat Mass Trans.*, 40:785–798.

Senba, T. and Suzuki, T. (2000). Some structures of the solution set for a stationary system of chemotaxis. *Advances Math. Sci. Applic.*, 10:191–224.

Senba, T. and Suzuki, T. (2001). Chemotactic collapse in a parabolic-elliptic system of mathematical biology. *Advances Diff. Equations*, 6:21–50.

Serrin, J. (1959a). Mathematical principles of classical fluid mechanics. In Flügge, S., editor, *Handbuch der Physik*, volume VIII/1. Springer, Berlin - Göttingen - Heidelberg.

Serrin, J. (1959b). A note on the existence of periodic solutions of the Navier-Stokes equations. *Arch. Rational Mech. Anal.*, 3:120–122.

Serrin, J. (1959c). On the stability of viscous fluid motions. *Arch. Rational Mech. Anal.*, 3:1–13.

Serrin, J. (1959d). On the uniqueness of compressible fluid motions. *Arch. Rational Mech. Anal.*, 3:271–288.

Sharma, R. and Kumar, P. (1994). Effect of rotation on thermal convection in micropolar fluids. *Int. J. Engng. Sci.*, 32:545–551.

Shir, C. and Joseph, D. (1968). Convective instability in a temperature and concentration field. *Arch. Rational Mech. Anal.*, 30:38–80.

Shivakumara, I., Rudraiah, N., and Nanjundappa, C. (2002). Effect of non-uniform basic temperature gradient on Rayleigh - Bénard - Marangoni convection in ferrofluids. *J. Magnetism and Magnetic Materials*, 248:379–395.

Shliomis, M. (1974). Magnetic fluids. *Sov. Phys. Usp.*, 17:153–169.

Siddheshwar, P. and Pranesh, S. (1998a). Effect of a non-uniform basic temperature gradient on Rayleigh-Bénard convection in a micropolar fluid. *Int. J. Engng. Sci.*, 36:1183–1196.

Siddheshwar, P. and Pranesh, S. (1998b). Magnetoconvection in a micropolar fluid. *Int. J. Engng. Sci.*, 36:1173–1181.

Siddheshwar, P. and Pranesh, S. (1999). Effect of temperature/gravity modulation on the onset of magneto-convection in weak electrically conducting fluids with internal angular momentum. *J. Magnetism and Magnetic Materials*, 192:159–176.

Siddheshwar, P. and Pranesh, S. (2001). Effects of non-uniform temperature gradient and magnetic field on the onset of convection in fluids with suspended particles under microgravity conditions. *Indian J. Engineering and Materials Sciences*, 8:77–83.

Simo, J., Lewis, D., and Marsden, J. (1991a). Stability of relative equilibria. Part I: The reduced energy momentum method. *Arch. Rational Mech. Anal.*, 115:15–59.

Simo, J., Posbergh, T., and Marsden, J. (1990). Stability of coupled rigid body and geometrically exact rods: block diagonalization and the energy momentum method. *Physics Reports*, 193:279–360.

Simo, J., Posbergh, T., and Marsden, J. (1991b). Stability of relative equilibria. Part II: Application to nonlinear elasticity. *Arch. Rational Mech. Anal.*, 115:61–100.

Simpson, H. and Spector, S. (1989). Necessary conditions at the boundary for minimizers in finite elasticity. *Arch. Rational Mech. Anal.*, 107:105–125.

Slemrod, M. (1978). An energy stability method for simple fluids. *Arch. Rational Mech. Anal.*, 68:1–18.

Song, J. (2002). Spatial decay estimates in time-dependent double diffusive Darcy plane flow. *J. Math. Anal. Appl.*, 267:76–88.

Spiegel, E. (1965). Convective instability in a compressible atmosphere. *Astrophys. J.*, 141:1068–1090.

Spiegel, E. and Veronis, G. (1960). On the Boussinesq approximation for a compressible fluid. *Astrophys. J.*, 131:442–447.

Stern, M. (1960). The 'salt-fountain' and thermohaline convection. *Tellus*, 12:172–175.

Stiles, P. (1991). Electro-thermal convection in dielectric liquids. *Chemical Physics Letters*, 179:311–315.

Stiles, P. (1993). Convective heat transfer through polarized dielectric liquids. *Physics Fluids A*, 5:3273–3279.

Stiles, P. and Blennerhassett, P. (1993). Stability of cylindrical Couette flow of a radially magnetized ferrofluid in a radial temperature gradient. *J. Magnetism and Magnetic Materials*, 122:207–209.

Stiles, P. and Kagan, M. (1993). Stability of cylindrical Couette flow of a radially polarised dielectric liquid in a radial temperature gradient. *Physica A*, 197:583–592.

Stiles, P., Lin, F., and Blennerhassett, P. (1992). Heat transfer through weakly magnetized ferrofluids. *J. Colloid Interface Science*, 151:95–101.

Stiles, P., Lin, F., and Blennerhassett, P. (1993). Heat transfer through ferrofluids as a function of the magnetic field strength. *J. Colloid Interface Science*, 155:256–258.

Storesletten, L. (1993). Natural convection in a horizontal porous layer with anisotropic thermal diffusivity. *Trans. Porous Media*, 12:19–29.

Straughan, B. (1983). Energy stability in the Bénard problem for a fluid of second grade. *ZAMP*, 34:502–509.

Straughan, B. (1985). Finite amplitude instability thresholds in penetrative convection. *Geophys. Astrophys. Fluid Dyn.*, 34:227–242.

Straughan, B. (1986). Stability criteria for convection with large viscosity variations. *Acta Mech.*, 61:59–72.

Straughan, B. (1987). Stability of a layer of dipolar fluid heated from below. *Math. Meth. Appl. Sci.*, 9:35–45.

Straughan, B. (1988). A nonlinear analysis of convection in a porous vertical slab. *Geophys. Astrophys. Fluid Dyn.*, 42:269–275.

Straughan, B. (1989). Convection in a variable gravity field. *J. Math. Anal. Appl.*, 140:467–475.

Straughan, B. (1990). Continuous dependence on the heat source and nonlinear stability for convection with internal heat generation. *Math. Meth. Appl. Sci.*, 13:373–383.

Straughan, B. (1991a). Continuous dependence on the heat source and nonlinear stability in penetrative convection. *Int. J. Nonlin. Mech.*, 26:221–231.

Straughan, B. (1991b). Convection caused by radiation through the layer. *IMA J. Appl. Math.*, 46:211–216.

Straughan, B. (1992). Stability problems in electrohydrodynamics, ferrohydrodynamics, and thermoelectric magnetohydrodynamics. In Rodrigues, J. and Sequeira, A., editors, *Mathematical topics in fluid mechanics*. Longman.

Straughan, B. (1993a). *Mathematical aspects of penetrative convection.* Longman, Harlow.

Straughan, B. (1993b). Nonlinear stability for convection in a polarized dielectric liquid. In *Recent Advances in Mechanics of Structured Continua*. ASME.

Straughan, B. (1998). *Explosive instabilities in mechanics.* Springer, Heidelberg.

Straughan, B. (2001a). A sharp nonlinear stability threshold in rotating porous convection. *Proc. Roy. Soc. London A*, 457:87–93.

Straughan, B. (2001b). Surface tension driven convection in a fluid overlying a porous layer. *J. Computational Phys.*, 170:320–337.

Straughan, B. (2002a). Effect of property variation and modelling on convection in a fluid overlying a porous layer. *Int. J. Num. Anal. Meth. Geomech.*, 26:75–97.

Straughan, B. (2002b). Global stability for convection induced by absorption of radiation. *Dynamics of Atmospheres and Oceans*, 35:351–361.

Straughan, B. (2002c). Sharp global nonlinear stability for temperature dependent viscosity convection. *Proc. Roy. Soc. London A*, 458:1773–1782.

Straughan, B., Bampi, F., and Morro, A. (1984). Chemical convective instability and quasi-equilibrium thermodynamics. *Meccanica*, 19:291–293.

Straughan, B. and Franchi, F. (1984). Bénard convection and the Cattaneo law of heat conduction. *Proc. Roy. Soc. Edinburgh A*, 96:175–178.

Straughan, B. and Hutter, K. (1999). A priori bounds and structural stability for double diffusive convection incorporating the Soret effect. *Proc. Roy. Soc. London A*, 455:767–777.

Straughan, B. and Tracey, J. (1999). Multi-component convection-diffusion with internal heating or cooling. *Acta Mechanica*, 133:219–238.

Straughan, B. and Walker, D. (1996a). Anisotropic porous penetrative convection. *Proc. Roy. Soc. London A*, 452:97–115.

Straughan, B. and Walker, D. (1996b). Two very accurate and efficient methods for computing eigenvalues and eigenfunctions in porous convection problems. *J. Computational Phys.*, 127:128–141.

Straughan, B. and Walker, D. (1997). Multi component diffusion and penetrative convection. *Fluid Dyn. Res.*, 19:77–89.

Suchomel, B., Chen, B., and Allen, M. (1998). Network model of flow, transport and biofilm effects in porous media. *Transport in Porous Media*, 30:1–23.

Swift, D. and Harrison, W. (1984). Convective transport of brine and thaw of subsea permafrost: results of numerical simulations. *J. Geophysical Res.*, 89:2080–2086.

Swift, D., Harrison, W., and Osterkamp, T. (1983). Heat and salt transport processes in subsea permafrost at Prudhoe Bay, Alaska. In *Permafrost Fourth International Conference, Proceedings*, pages 1221–1226, Washington, D.C. National Academy Press.

Tabor, H. (1980). Non-convecting solar ponds. *Phil. Trans. Roy. Soc. London A*, 295:423–433.

Takashima, M. (1976). The effect of rotation on electrohydrodynamic instability. *Canadian J. Physics*, 54:342–347.

Takashima, M. (1980). Electrohydrodynamic instability in a dielectric fluid between two coaxial cylinders. *Q. J. Mech. Appl. Math.*, 33:93–103.

Takashima, M. and Aldridge, K. (1976). The stability of a horizontal layer of dielectric fluid under the simultaneous action of a vertical DC electric field and a vertical temeprature field. *Q. J. Mech. Appl. Math.*, 29:71–87.

Takashima, M. and Ghosh, A. (1979). Electrohydrodynamic instability in a viscoelastic liquid layer. *J. Phys. Soc. Japan*, 47:1717–1722.

Takashima, M. and Hamabata, H. (1984). The stability of natural convection in a vertical layer of dielectric fluid in the presence of a horizontal ac electric field. *J. Phys. Soc. Japan*, 53:1728–1736.

Taylor, G. (1971). A model for the boundary conditions of a porous material. *J. Fluid Mech.*, 49:319–326.

Temam, R. (1978). *The Navier-Stokes equations.* North-Holland, Amsterdam.

Thomson, J. (1882). On a changing tesselated structure in certain liquids. *Proc. Phil Soc. Glasgow*, 13:464–468.

Tigoiu, V. (2000). Weakly perturbed flows in third grade fluids. *Z. Angew. Math. Mech.*, 80:423–428.

Tippelskirch, H. (1956). über Konvektionszellen, insbesondere im flüssigen Schwefel. *Beiträge zur Physik der Atmosphäre*, 29:37–54.

Torrance, K. and Turcotte, D. (1971). Thermal convection with large viscosity variations. *J. Fluid Mech.*, 47:113–125.

Tracey, J. (1997). *Stability Analyses of Multi-Component Convection-Diffusion Problems.* PhD thesis, Glasgow University.

Tritton, D. and Zarraga, M. (1967). Convection in horizontal layers with internal heat generation. Experiments. *J. Fluid Mech.*, 30:21–31.

Truesdell, C. and Toupin, R. (1960). The classical field theories. In Flügge, S., editor, *Handbuch der Physik*, volume III/1, pages 226–793. Springer.

Tsai, T. (1998). On Leray's self-similar solutions of the Navier-Stokes equations satisfying local energy estimates. *Arch. Rational Mech. Anal.*, 143:29–51.

Tse, K. and Chasnov, J. (1998). A Fourier-Hermite pseudospectral method for penetrative convection. *J. Computational Physics*, 142:489–505.

Turnbull, R. (1968a). Electroconvective instability with a stabilizing temperature gradient. I. Theory. *Phys. Fluids*, 11:2588–2596.

Turnbull, R. (1968b). Electroconvective instability with a stabilizing temperature gradient. II. Experimental results. *Phys. Fluids*, 11:2597–2603.

Tyvand, P. and Storesletten, L. (1991). Onset of convection in an anisotropic porous medium with oblique principal axes. *J. Fluid Mech.*, 226:371–382.

Vadasz, P. (1995). Coriolis effect on free convection in a long rotating porous box subject to uniform heat generation. *Int. J. Heat Mass Transfer*, 38:2011–2018.

Vadasz, P. (1996). Convection and stability in a rotating porous layer with alternating direction of the centrifugal body force. *Int. J. Heat Mass Transfer*, 39:1639–1647.

Vadasz, P. (1997). Flow in rotating porous media. In Plessis, P. D., editor, *Fluid Transport in Porous Media*. Computational Mechanics Publications.

Vadasz, P. (1998a). Coriolis effect on gravity-driven convection in a rotating porous layer heated from below. *J. Fluid Mech.*, 376:351–375.

Vadasz, P. (1998b). Free convection in rotating porous media. In Ingham, D. and Pop, I., editors, *Transport Phenomena in Porous Media*, pages 285–312. Elsevier.

Vaidyanathan, G., Sekar, R., and Balasubramanian, R. (1991). Ferroconvective instability of fluids saturating a porous medium. *Int. J. Engng. Sci.*, 29:1259–1267.

van Duijn, C., Galiano, G., and Peletier, M. (2001). A diffusion - convection problem with drainage arising in the ecology of mangroves. *Interfaces and Free Boundaries*, 3:15–44.

Velarde, M. and Perez Cordon, R. (1976). On the (non linear) foundations of the Boussinesq approximation applicable to a thin layer of fluid. ii. Viscous dissipation and large cell gap effects. *Journal de Physique*, 37:174–182.

Venkatasubramanian, S. and Kaloni, P. (1994). Effects of rotation on the thermoconvective instability of a horizontal layer of ferrofluids. *Int. J. Engng. Sci.*, 32:237–256.

Venkatasubramanian, S. and Kaloni, P. (2002). Stability and uniqueness of magnetic fluid motions. *Proc. Roy. Soc. London A*, 458:1189–1204.

Veronis, G. (1959). Cellular convection with finite amplitude in a rotating fluid. *J. Fluid Mech.*, 5:401–435.

Veronis, G. (1963). Penetrative convection. *Astrophys. J.*, 137:641–663.

Veronis, G. (1965). On finite amplitude instability in thermohaline convection. *J. Marine Res.*, 23:1–17.

Veronis, G. (1966a). Motions at subcritical values of the Rayleigh number in a rotating fluid. *J. Fluid Mech.*, 24:545–554.

Veronis, G. (1966b). Wind driven ocean circulation - Part 1. Linear theory and perturbation analysis. *Deep Sea Res.*, 13:17–29.

Veronis, G. (1968a). Effect of a stabilizing gradient of solute on thermal convection. *J. Fluid Mech.*, 34:315–336.

Veronis, G. (1968b). Large amplitude Bénard convection in a rotating fluid. *J. Fluid Mech.*, 31:113–139.

Vincent, R. and Hill, N. (1996). Bioconvection in a suspension of phototactic algae. *J. Fluid Mech.*, 327:343–371.

Viola, T. (1941). Calcolo approssimato di autovalori. *Rend. Matematica*, 2:71–106.

Vitanov, N. and Busse, F. (2001). Bounds on the convective heat transport in a rotating layer. *Phys. Rev. E*, 63:1–8.

Walden, R. and Ahlers, G. (1981). Non-Boussinesq and penetrative convection in a cylindrical cell. *J. Fluid Mech.*, 109:89–114.

Washburn, A. (1973). *Periglacial processes and environments.* St. Martin's Press, New York.

Washburn, A. (1980). *Geocryology.* John Wiley and Sons, New York.

Whitaker, S. (1996). The Forchheimer equation: A theoretical development. *Transport in Porous Media*, 25:27–62.

Whitehead, J. (1971). Upon boundary conditions imposed by a stratified fluid. *Geophys. Fluid Dyn.*, 2:289–298.

Whitehead, J. and Chen, M. (1970). Thermal instability and convection of a thin layer bounded by a stably stratified region. *J. Fluid Mech.*, 40:549–576.

Williams, A. and McKibben, M. (1989). A brine interface in the Salton sea geothermal system California: Fluid geochemical and isotropic characteristics. *Geochimica et Cosmochimica Acta*, 53:1905–1920.

Wollkind, D. and Bdzil, J. (1971). Comments on chemical instabilities. *Phys. Fluids*, 14:1813–1814.

Wollkind, D. and Frisch, H. (1971). Chemical instabilities. I. A heated horizontal layer of dissociating fluids. *Phys. Fluids*, 14:13–18.

Worraker, W. and Richardson, A. (1979). The effect of temperature induced variations in charge carrier mobility on a stationary electrohydrodynamic instability. *J. Fluid Mech.*, 93:29–45.

Xie, W. (1991). A sharp pointwise bound for functions with L^2-Laplacians and zero boundary values on arbitrary three-dimensional domains. *Indiana Univ. Math. J.*, 40:1185–1192.

Xu, L. (2000). Unconditional nonlinear exponential stability of the motionless conduction-diffusion solution. *Acta Mechanica Sinica*, 16:113–120.

Ybarra, P. and Velarde, M. (1979). The influence of the Soret and Dufour effects on the stability of a binary gas layer heated from below or above. *Geophys. Astrophys. Fluid Dyn.*, 13:83–94.

Ydstie, B. (2002). Passivity based control via the second law. *Computers and Chemical Engineering*, 26:1037–1048.

Yih, C. and Li, C. (1972). Instability of unsteady flows or configurations. Part 2. Convective instability. *J. Fluid Mech.*, 54:143–152.

Younker, L., Kasameyer, P., and Tewhey, J. (1982). Geological, geophysical, and thermal characteristics of the Salton sea geothermal field, California. *J. Vulcanol. Geotherm. Res.*, 12:221–258.

Zahn, J., Toomre, J., and Latour, J. (1982). Nonlinear model analysis of penetrative convection. *Geophys. Astrophys. Fluid Dyn.*, 22:159–193.

Zahn, M. and Greer, D. (1995). Ferrohydrodynamic pumping in spatially uniform sinusoidally time-varying magnetic fields. *J. Magnetism and Magnetic Materials*, 149:165–173.

Zahn, M. and Pioch, L. (1999). Ferrofluid flows in AC and traveling wave magnetic fields with effective positive, zero or negative dynamic viscosity. *J. Magnetism and Magnetic Materials*, 201:144–148.

Zangrando, F. (1991). On the hydrodynamics of salt-gradient solar ponds. *Solar Energy*, 46:323–341.

Zebib, A. (1978). Removal of spurious modes encountered in solving stability problems by spectral methods. *J. Computational Physics*, 70:521–525.

Zeytounian, R. K. (1989). The Bénard problem for deep convection: rigorous derivation of approximate equations. *Int. J. Engng. Sci.*, 27:1361–1366.

Zeytounian, R. K. (1998). The Bénard-Marangoni thermocapillary-instability problem. *Uspekhi Fizicheskikh Nauk*, 168:259–286.

Zhang, K. and Schubert, G. (2000). Teleconvection: remotely driven thermal convection in rotating stratified spherical layers. *Science*, 290:1944–1947.

Zhang, K. and Schubert, G. (2002). From penetrative convection to teleconvection. *Astrophysical J.*, 572:461–476.

Zhao, A., Moates, F., and Narayanan, R. (1995). Rayleigh convection in a closed cylinder - experiments and a 3 - dimensional model with temperature - dependent viscosity effects. *Phys. Fluids*, 7:1576–1582.

Zikanov, O. and Thess, A. (1998). Direct numerical simulation of forced MHD turbulence at low magnetic Reynolds number. *J. Fluid Mech.*, 358:299–333.

Index

Applied Mathematical Sciences

(continued from page ii)

60. *Ghil/Childress:* Topics in Geophysical Dynamics: Atmospheric Dynamics, Dynamo Theory and Climate Dynamics.
61. *Sattinger/Weaver:* Lie Groups and Algebras with Applications to Physics, Geometry, and Mechanics.
62. *LaSalle:* The Stability and Control of Discrete Processes.
63. *Grasman:* Asymptotic Methods of Relaxation Oscillations and Applications.
64. *Hsu:* Cell-to-Cell Mapping: A Method of Global Analysis for Nonlinear Systems.
65. *Rand/Armbruster:* Perturbation Methods, Bifurcation Theory and Computer Algebra.
66. *Hlavácek/Haslinger/Necasl/Lovisek:* Solution of Variational Inequalities in Mechanics.
67. *Cercignani:* The Boltzmann Equation and Its Applications.
68. *Temam:* Infinite-Dimensional Dynamical Systems in Mechanics and Physics, 2nd ed.
69. *Golubitsky/Stewart/Schaeffer:* Singularities and Groups in Bifurcation Theory, Vol. II.
70. *Constantin/Foias/Nicolaenko/Temam:* Integral Manifolds and Inertial Manifolds for Dissipative Partial Differential Equations.
71. *Catlin:* Estimation, Control, and the Discrete Kalman Filter.
72. *Lochak/Meunier:* Multiphase Averaging for Classical Systems.
73. *Wiggins:* Global Bifurcations and Chaos.
74. *Mawhin/Willem:* Critical Point Theory and Hamiltonian Systems.
75. *Abraham/Marsden/Ratiu:* Manifolds, Tensor Analysis, and Applications, 2nd ed.
76. *Lagerstrom:* Matched Asymptotic Expansions: Ideas and Techniques.
77. *Aldous:* Probability Approximations via the Poisson Clumping Heuristic.
78. *Dacorogna:* Direct Methods in the Calculus of Variations.
79. *Hernández-Lerma:* Adaptive Markov Processes.
80. *Lawden:* Elliptic Functions and Applications.
81. *Bluman/Kumei:* Symmetries and Differential Equations.
82. *Kress:* Linear Integral Equations, 2nd ed.
83. *Bebernes/Eberly:* Mathematical Problems from Combustion Theory.
84. *Joseph:* Fluid Dynamics of Viscoelastic Fluids.
85. *Yang:* Wave Packets and Their Bifurcations in Geophysical Fluid Dynamics.
86. *Dendrinos/Sonis:* Chaos and Socio-Spatial Dynamics.
87. *Weder:* Spectral and Scattering Theory for Wave Propagation in Perturbed Stratified Media.
88. *Bogaevski/Povzner:* Algebraic Methods in Nonlinear Perturbation Theory.
89. *O'Malley:* Singular Perturbation Methods for Ordinary Differential Equations.
90. *Meyer/Hall:* Introduction to Hamiltonian Dynamical Systems and the N-body Problem.
91. *Straughan:* The Energy Method, Stability, and Nonlinear Convection, 2nd ed.
92. *Naber:* The Geometry of Minkowski Spacetime.
93. *Colton/Kress:* Inverse Acoustic and Electromagnetic Scattering Theory, 2nd ed.
94. *Hoppensteadt:* Analysis and Simulation of Chaotic Systems, 2nd ed.
95. *Hackbusch:* Iterative Solution of Large Sparse Systems of Equations.
96. *Marchioro/Pulvirenti:* Mathematical Theory of Incompressible Nonviscous Fluids.
97. *Lasota/Mackey:* Chaos, Fractals, and Noise: Stochastic Aspects of Dynamics, 2nd ed.
98. *de Boor/Höllig/Riemenschneider:* Box Splines.
99. *Hale/Lunel:* Introduction to Functional Differential Equations.
100. *Sirovich (ed):* Trends and Perspectives in Applied Mathematics.
101. *Nusse/Yorke:* Dynamics: Numerical Explorations, 2nd ed.
102. *Chossat/Iooss:* The Couette-Taylor Problem.
103. *Chorin:* Vorticity and Turbulence.
104. *Farkas:* Periodic Motions.
105. *Wiggins:* Normally Hyperbolic Invariant Manifolds in Dynamical Systems.
106. *Cercignani/Illner/Pulvirenti:* The Mathematical Theory of Dilute Gases.
107. *Antman:* Nonlinear Problems of Elasticity.
108. *Zeidler:* Applied Functional Analysis: Applications to Mathematical Physics.
109. *Zeidler:* Applied Functional Analysis: Main Principles and Their Applications.
110. *Diekmann/van Gils/Verduyn Lunel/Walther:* Delay Equations: Functional-, Complex-, and Nonlinear Analysis.
111. *Visintin:* Differential Models of Hysteresis.
112. *Kuznetsov:* Elements of Applied Bifurcation Theory, 2nd ed.
113. *Hislop/Sigal:* Introduction to Spectral Theory: With Applications to Schrödinger Operators.
114. *Kevorkian/Cole:* Multiple Scale and Singular Perturbation Methods.
115. *Taylor:* Partial Differential Equations I, Basic Theory.
116. *Taylor:* Partial Differential Equations II, Qualitative Studies of Linear Equations.

(continued on next page)

Applied Mathematical Sciences

(continued from previous page)

117. *Taylor:* Partial Differential Equations III, Nonlinear Equations.
118. *Godlewski/Raviart:* Numerical Approximation of Hyperbolic Systems of Conservation Laws.
119. *Wu:* Theory and Applications of Partial Functional Differential Equations.
120. *Kirsch:* An Introduction to the Mathematical Theory of Inverse Problems.
121. *Brokate/Sprekels:* Hysteresis and Phase Transitions.
122. *Gliklikh:* Global Analysis in Mathematical Physics: Geometric and Stochastic Methods.
123. *Le/Schmitt:* Global Bifurcation in Variational Inequalities: Applications to Obstacle and Unilateral Problems.
124. *Polak:* Optimization: Algorithms and Consistent Approximations.
125. *Arnold/Khesin:* Topological Methods in Hydrodynamics.
126. *Hoppensteadt/Izhikevich:* Weakly Connected Neural Networks.
127. *Isakov:* Inverse Problems for Partial Differential Equations.
128. *Li/Wiggins:* Invariant Manifolds and Fibrations for Perturbed Nonlinear Schrödinger Equations.
129. *Müller:* Analysis of Spherical Symmetries in Euclidean Spaces.
130. *Feintuch:* Robust Control Theory in Hilbert Space.
131. *Ericksen:* Introduction to the Thermodynamics of Solids, Revised ed.
132. *Ihlenburg:* Finite Element Analysis of Acoustic Scattering.
133. *Vorovich:* Nonlinear Theory of Shallow Shells.
134. *Vein/Dale:* Determinants and Their Applications in Mathematical Physics.
135. *Drew/Passman:* Theory of Multicomponent Fluids.
136. *Cioranescu/Saint Jean Paulin:* Homogenization of Reticulated Structures.
137. *Gurtin:* Configurational Forces as Basic Concepts of Continuum Physics.
138. *Haller:* Chaos Near Resonance.
139. *Sulem/Sulem:* The Nonlinear Schrödinger Equation: Self-Focusing and Wave Collapse.
140. *Cherkaev:* Variational Methods for Structural Optimization.
141. *Naber:* Topology, Geometry, and Gauge Fields: Interactions.
142. *Schmid/Henningson:* Stability and Transition in Shear Flows.
143. *Sell/You:* Dynamics of Evolutionary Equations.
144. *Nédélec:* Acoustic and Electromagnetic Equations: Integral Representations for Harmonic Problems.
145. *Newton:* The *N*-Vortex Problem: Analytical Techniques.
146. *Allaire*: Shape Optimization by the Homogenization Method.
147. *Aubert/Kornprobst:* Mathematical Problems in Image Processing: Partial Differential Equations and the Calculus of Variations.
148. *Peyret:* Spectral Methods for Incompressible Viscous Flow.
149. *Ikeda/Murota:* Imperfect Bifurcation in Structures and Materials: Engineering Use of Group-Theoretic Bifucation Theory.
150. *Skorokhod/Hoppensteadt/Salehi:* Random Perturbation Methods with Applications in Science and Engineering.
151. *Bensoussan/Frehse:* Regularity Results for Nonlinear Elliptic Systems and Applications.
152. *Holden/Risebro:* Front Tracking for Hyperbolic Conservation Laws.
153. *Osher/Fedkiw:* Level Set Methods and Dynamic Implicit Surfaces.
154. *Bluman/Anco:* Symmetry and Integration Methods for Differential Equations.
155. *Chalmond:* Modeling and Inverse Problems in Image Analysis.
156. *Kielhöfer:* Bifurcation Theory: An Introduction with Applications to PDEs.
157. *Kaczynski/Mischaikow/Mrozek:* Computational Homology.